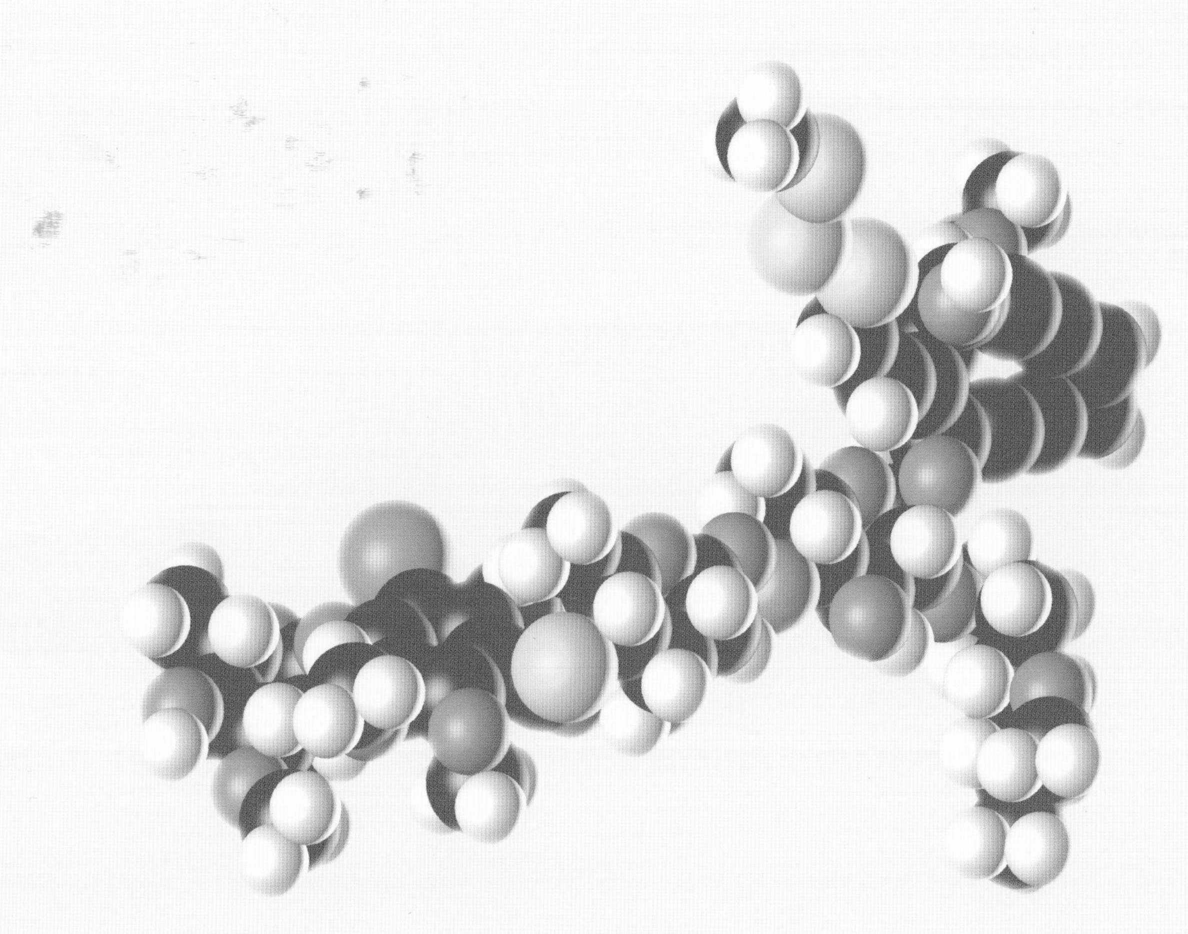

MOLECULES THAT CHANGED THE WORLD

K. C. NICOLAOU • T. MONTAGNON

Further Reading from Wiley-VCH

Nicolaou, K. C. / Sorensen, E. J.
Classics in Total Synthesis – Targets, Strategies, Methods
1996, ISBN 978-3-527-29284-4 (Hardcover)
1996, ISBN 978-3-527-29231-8 (Softcover)

Nicolaou, K. C. / Snyder, S. A.
Classics in Total Synthesis II – More Targets, Strategies, Methods
2003, ISBN 978-3-527-30685-5 (Hardcover)
2003, ISBN 978-3-527-30684-8 (Softcover)

Corey, E. J. / Cheng, X.-M.
The Logic of Chemical Synthesis
1989, ISBN 978-0-471-50979-0 (Hardcover)
1995, ISBN 978-0-471-11594-6 (Softcover)

Corey, E. J. / Czakó, B. / Kürti, L.
Molecules and Medicine
2007, ISBN 978-0-470-22749-7

Quadbeck-Seeger, H.-J.
World of the Elements – Elements of the World
2007, ISBN 978-3-527-32065-3

MOLECULES THAT CHANGED THE WORLD

A BRIEF HISTORY OF THE ART AND SCIENCE OF SYNTHESIS AND ITS IMPACT ON SOCIETY

K. C. NICOLAOU • T. MONTAGNON

With Forewords by Nobel Laureates
E. J. Corey
R. Noyori

WILEY-VCH GmbH & Co. KGaA

The Authors

Prof. Dr. K. C. Nicolaou
Department of Chemistry
The Scripps Research Institute
10550 North Torrey Pines Road
La Jolla, CA 92037
USA

and

Department of Chemistry and Biochemistry
University of California, San Diego
9500 Gilman Drive
La Jolla, CA 92093
USA

Dr. Tamsyn Montagnon
Department of Chemistry
University of Crete
Vasilika Vouton
71003 Heraklion, Crete
Greece

Research and Text Contributors
David Edmonds, Paul Bulger

Art and Design Director
K. C. Nicolaou

Copy Editor
Janise Petrey

Graphic Designers
Mary Ellen Mulshine, Robbyn Echon

Photo Editor
Janise Petrey

Printing
Betz-druck GmbH, Darmstadt

Binding
Litges & Dopf Buchbinderei GmbH, Heppenheim

Library of Congress Card No.:
applied for

British Library Cataloguing-in-Publication Data
A catalogue record for this book is available from the British Library.

Bibliographic information published by the Deutsche Nationalbibliothek
The Deutsche Nationalbibliothek lists this publication in the Deutsche Nationalbibliografie; detailed bibliographic data are available in the Internet at http://dnb.d-nb.de

Printed in the Federal Republic of Germany.
Printed on acid-free paper.

ISBN 978-3-527-30983-2

I.
Stramonia.
II.
Halimus.
III.
Botris Draconti & maior.

Dedication

This book is dedicated to Professor E. J. Corey for his enormous contributions to chemical synthesis and his outstanding mentorship of young students that so profoundly influenced and shaped the field as we know it today.

Foreword

E. J. Corey
Nobel Laureate 1990

This book is an enthusiastic celebration of many organic molecules, especially those which are of natural origin, intricate structure and biological relevance. It is also a unique tribute to the many scientists who were involved in their study and chemical synthesis, most of whom are pictured on its pages. Still another element is provided by many interesting historical details and an abundance of colorful illustrations. On top of that, there are innumerable historical vignettes that interweave chemistry and biology in a very appealing way.

Although the emphasis of this work is on chemical synthesis, it contains much that will be of interest to those outside this field and to students of chemistry - indeed to anyone with a fascination with the world of molecules. The authors have selected well over 40 prominent molecules as the key subjects of their essays. Although these represent only a small sample of the world of biologically-related molecules, they amply illustrate the importance of this field of science to humankind and the way in which the field has evolved.

I think that the authors can be confident that there will be many grateful readers who will have gained a broader perspective of the disciplines of chemical synthesis and natural products as a result of their efforts.

E. J. Corey
Harvard University
Cambridge, Massachusetts
13 August, 2007

Foreword

Ryoji Noyori
Nobel Laureate 2001

Close involvement with society is the destiny of science. Organic Synthesis has been bringing benefits to mankind since the first half of the nineteenth century, longer than almost any other scientific disciplines. *Molecules That Changed the World* has been written to tell the story of this field to future generations. Every chapter abounds with the enthusiasm of the authors, my respected friend Professor K. C. Nicolaou and Dr. T. Montagnon. They convey the intellectual fascination of chemical synthesis, its profundity, its breadth, and its importance as technology, and tell the story in an intriguing and visually appealing way. I believe this book will become a cornerstone of chemistry education.

Chemistry is beyond the science of mere observation and understanding of nature; with Organic Synthesis, we can create high-value substances from abundant natural resources such as oil, coal and biomass. Any molecule is made up of a finite number of atoms connected in accordance with a fixed set of rules. It has a single configuration and may have several conformations. Organic molecules with these precise three-dimensional structures can have amazing properties. Most importantly, we can design any molecule, and if it is sufficiently stable, we can synthesize it by using our accumulated knowledge. So, in principle, we can create molecules that have all kinds of properties. Organic compounds are particularly important in medicine. Many times during the course of human history, small molecules have cured tens of millions of people of serious diseases and improved quality of life. Organic Synthesis will always be the heart of chemistry.

Science is objective. But it is human intelligence and endeavor that discover and create interesting new substances. To date we have discovered or created more than thirty million structurally well-defined compounds. This book selects about forty molecules that have had revolutionary effects and become indispensable for human society. Blending chemistry, biology and medicine, it explains, in simple language, the scientific concepts that are necessary for their chemical synthesis (synthetic methods and path designs) and portrays the pioneering chemists who made the discoveries. The story begins with Friedrich Wöhler's synthesis of urea in 1828, continues through the strategic synthesis of structurally complex bioactive compounds, and concludes with small-molecule drugs and biologics. I am sure that many readers will be impressed and inspired by the colorful history, the beauty of the scientific theories, and the characters of the great scientists featured in this book.

Quo vadis, Organic Synthesis? Where are you going? In the 21st century, chemical synthesis is not just an intellectual challenge but also a vital field for the survival of the human race. Chemists will have to join forces with scientists in other fields, and engineers. One clear direction is for synthetic organic chemistry to merge with the biosciences to create an integrative Science of Life. Living systems all seem very complicated, but they are just collections of large and small organic molecules that function according to the laws of nature. James Watson, a 1962 Nobel Laureate in Physiology or Medicine, put it well: "Life is simply a matter of chemistry." When Watson and Crick discovered the double-helix structure of DNA in 1953, it opened the door to molecular biology; the decoding of the human genome, fifty years later, has led to another new world of science. Thanks to advanced technologies and the endeavor of scientists in many fields, we are now able to work out the precise atomic-level structures of large biomolecules such as nucleic acids, proteins, and polysaccharides. This field of Structural Biology has become a part of chemistry—in recent years, several Nobel Chemistry Prizes have been awarded for research in this area as demonstrated in this book.

The focus of chemical research has also been moving from structure to function. Dynamic interactions between large biopolymers and small organic molecules often cause and control processes in living organisms. So Chemical Biology is sure to become even more important. I am confident that in the future scientists will elucidate the chemical mechanisms of cell functions, and even thought and memory.

Organic Synthesis will remain important. But for it to continue being as useful as possible, and maintain its position as a central part of science, chemists will have to understand other fields and collaborate with other scientists. For this we need broader, suitable education. In addition, the significant mission of Organic Synthesis is to produce large quantities of important naturally occurring and artificial compounds, in a straightforward and practical way, and to provide them to society. But synthesis technology is not yet efficient enough for this. We need to develop a range of new catalysis technologies. In particular, Green Chemistry that is beneficial for both the economy and the environment needs to be promoted and supported by the scientific community as well as by governments, industry, and all other sectors of society. And Green Chemistry must not be just a slogan. It must be a "responsible science" that is put into practice to ensure the future of civilized society.

We are proud of our profession. The wonderful achievements of chemical synthesis have largely been made in Europe, America, and Japan, and mostly by men with some notable exceptions, mainly because women were not afforded the same opportunities. I hope that in the 21st century a bigger role will be played by young scientists from the rapidly progressing countries of Asia, and I believe that scientists from different cultural backgrounds and with different values, both men and women, will be able to create molecules that are even more impressive, and make life better for all of mankind. The authors of *Molecules That Changed the World* demonstrate clearly the impact we can have on our own destiny through chemistry. Their book is destined to play a major role in exciting, motivating and educating the next generation of chemists and life scientists from all over the word who are bound to make this dream a reality.

Ryoji Noyori
RIKEN and Nagoya University
Japan
12 September, 2007

Foreword

Anatolia M. Evarkiou-Kaku

As varied as the interests of landmark entrepreneur and scientist Alfred Nobel, *Molecules That Changed the World* explores the story of chemical synthesis by incorporating topics spanning from history and literature to biochemistry and pharmacy. The authors, Professor Nicolaou and Dr. Montagnon, clearly establish dialogue between the cultural background that inspires new discovery and the scientific journey of elucidating the structure and the total synthesis of molecules. For example, the chapter discussing brevetoxin B appealed most to me due to the cultural interconnectedness between the hard science and society; most specifically, I found the relation between biblical phenomena, red tides, and the destructive nature of brevetoxin B captivating. With its clear descriptions and explanations, people of all educational levels can comprehend this book.

For the nonscientist, *Molecules That Changed the World* presents the reader with a background of biochemistry's progress thus far, in addition to an understanding of social, economical, environmental, medical, and research issues the science community faces today. While exposing readers to such sobering issues, the authors convey the material clearly through both written text and colorful illustrations. In these pages, the nonscientist will find interdisciplinary and cross-cultural explanations commingled with science.

For the high school and presumably undergraduate student interested in science, this book ignites a flame of curiosity and excitement. The explanations of molecular structures and process of total synthesis unveil opportunities of research and development for young scientists. The benefits reaped from new discovery and original thinking inspire readers to explore the possibilities. Stressing the importance of forward thinking, this book motivates students and young scientists to plunge into ever-expanding fields of science to overcome new hurdles.

As this captivating book takes readers through history from ancient Greece, Rome, and Asia to today's advancements, we must remember that science will continue to accomplish the unimaginable so long as people continue to submerge their minds in the artistic world of molecules.

Anatolia M. Evarkiou-Kaku
Francis W. Parker School
Class of 2009
San Diego, California
26 July 2007

Preface

Whether you are a high school or college student, postdoctoral fellow, professor, scientist or lay person, you will find a body of enjoyable and useful information within the covers of this book. The chapters within expound upon our learned knowledge of nature's molecules and the ability of man to discover, synthesize, modify and use them to his benefit in ways not formerly envisioned. Through these pages one will also discover just how profound the influence of chemistry is in our daily lives. This book also explores some of the most exciting frontiers of modern science and medicine, and the opportunities they present to young students for future careers. Written and illustrated in a visually distinct and entertaining style, this book aims to provide insight to its readers about the role of chemistry in society in general, and how chemical synthesis, the art and science of constructing natural and designed molecules in particular, shaped and continues to shape our world. Indeed, this volume contains a wealth of information and fascinating tales about molecules and their presence in many goods such as perfumes, dyes, high tech materials, textiles, vitamins, nutritional products, pesticides, insecticides, and, above all, medicines.The history of total synthesis, the flagship of chemical synthesis, as unraveled within serves to admirably explore how chemical synthesis has enabled and facilitated world-shaping innovations since its naissance in 1828.

The inspiration for this book arose from the desire to enlighten and instill a greater appreciation in society at large about a difficult subject – chemistry, and to inspire young students to explore its fascinating, and almost infinite, applications. Many people remember chemistry as one of their most challenging subjects in college, or the class in which they struggled. For others, the mere mention of the word chemistry conjures up images of explosions, poisons, and pollution (toxic waste and dangerous fumes). The reality of chemistry, however, is far more exciting and rewarding, once these unfortunate and distressing images are dispelled. Indeed, beyond this curtain, there lies a beautiful world of molecules with a glorious history and myriad wondrous applications recognized through Nobel Prizes and other awards, prizes that acknowledge brilliant discoveries and magnificent accomplishments whose uses have alleviated untold pain and saved millions of lives. More importantly, the reader will come to greatly appreciate the skills, knowledge, and tools acquired during the campaigns to discover, synthesize, and investigate such wondrous molecules. These emerging technologies are the gift of such endeavors and of their protagonists to humanity, a gift that carries with it the awesome power to shape the pharmaceutical and biotechnology enterprises, petroleum and energy industries, nanotechnology, materials science and technology, agriculture, cosmetics, and even fashion.

Through this treatise we also aspire to capture the imagination of the next generation of students, tantalizing them with the virtues of chemistry, biology and medicine, and the intrigue of the art and science of chemical synthesis. This central discipline is both challenging and rewarding. It demands the best of human character, including a sharp intellect, originality and imagination, dexterity, and stamina. The stories of some of the personalities featured in these pages are also discussed, for they are telling of their genius and dedication. Besides the admiration and respect they deserve, such characters are wonderful role models for young students.

Few other disciplines, if any, impact their adjacent sciences so broadly and decisively as synthesis, for everything we can sense around us is an assemblage of chemicals. The manipulation of these molecules to form new variations is the subject of this science, a discipline that is also a fine art by virtue of its creative nature. However, the power gained through chemistry to shape our surroundings and create new opportunities for humankind comes with responsibilities –continued education, research, and wise application. We hope that this volume will contribute to upholding these duties, help the reader to appreciate the importance of chemistry to our everyday lives, and attract new talent to the ranks of this almost unlimited science. There is so much yet to be discovered and invented by each new generation of synthetic artisans –architects and sculptors– who choose to aim their minds and chisels towards the fashioning of the almost infinite number of yet undiscovered molecules of natural or design origins.

La Jolla
September 2007

K. C. Nicolaou
Tamsyn Montagnon

Acknowledgements

The enormity of the task to assemble this volume necessitated the assistance of many individuals whom we wish to thank. First and foremost we would like to thank Janise Petrey for her roles as copy editor, photo editor, and administrative editor that contributed considerably to its quality and timely completion. We also wish to thank Vicky Nielsen Armstrong for her encouragement and overseeing of this project that ensured its fruition in a smooth and orderly manner.

We are indebted to Drs. David Edmonds and Paul Bulger for their contributions to this project that included many hours of researching, editing, and structure drawing. We thank Drs. David Edmonds and Antonia Stepan for proofreading all the chapters of the book and making useful comments at various stages. We are grateful to Professor Albert Eschenmoser for proofreading and making substantial improvements to the vitamin B_{12} chapter and for his constant encouragement and support during this project.

Our thanks are extended to Mr. Panayiotis Gerolymatos, who kindly allowed his graphic designer, Mrs. Katerina Bakali, to design the penicillin chapter and thus set the style of the book. The design and layout of the volume was completed by Mary Ellen Mulshine and Robbyn M. Echon, to both of whom we are deeply grateful. We also thank Hao Xu, Mike Pique, and Alex Perryman for the computer-generated molecular models that appear in the book. William Brenzovich is acknowledged for his molecular modeling assistance and useful proofing during the final stages of the book.

It is also with much appreciation that we acknowledge the assistance and encouragement throughout the course of this project of the Wiley-VCH staff, especially Drs. Peter Gölitz, Elke Maase, Gudrun Walter, and Alexander Eberhard.

Finally, we wish to express the deep gratitude we owe to our immediate families (Georgette, Colette, Alex, Chris, and PJ for KCN, and George and Emmanouil for TM) for their unconditional love and support over the several years that this mission took to accomplish.

Much of the inspiration for this volume came from all those magnificent teachers-scholars of Ancient Greece whose brilliant thoughts and ideas set the foundations of the Western Civilization and which continue to shape the world today. Should this volume prove instrumental in conveying some of their spirit to young students, we will feel satisfied. It is our hope, however, that all the readers of this book will derive much knowledge and enjoyment from their reading.

Acknowledgements for Financial Support

We are grateful to the following individuals and institutions for their generous financial support of this project.

Dr. Alfred Balm	EMERGOMED
Philip Yeo	A*STAR of Singapore
Dr. Magid Abou-Gharbia	Wyeth
Dr. Narendra Mallakunta	GVK Biosciences Inc.
Dr. Eugene and Gala Vaisberg	ChemBridge Corporation
Arthur C. Cope Fund Grant	American Chemical Society
Dr. Rainer Metternich and Dr. Eckhard Ottow	Schering AG
Dr. Hiroaki Ueno	Mitsubishi Pharma Corporation
Dr. Alfred Bader	
Dr. Donald and Darlene Shiley	

The Skaggs Institute for Chemical Biology

The Scripps Research Institute

Wyeth

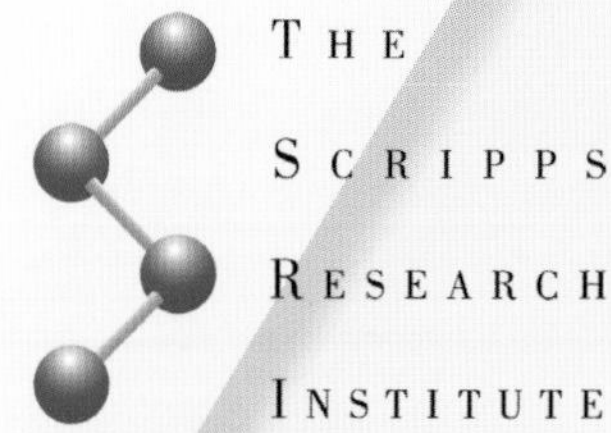

DATURA metel

About the author

K. C. Nicolaou holds joint appointments at The Scripps Research Institute, where he is the Chairman of the Department of Chemistry and holds the Darlene Shiley Chair in Chemistry and the Aline W. and L. S. Skaggs Professorship of Chemical Biology, and the University of California, San Diego, where he is Distinguished Professor of Chemistry. He is also the Program Director for the Chemical Synthesis Laboratory@ Biopolis, A*Star, Singapore.

His impact on chemistry, biology and medicine flows from his research works in chemical synthesis and chemical biology described in his numerous publications and patents. For his contributions to science and education, he was elected Fellow of the New York Academy of Sciences, Fellow of the American Association of Arts and Sciences, Member of the National Academy of Sciences, USA, and Corresponding Member of the Academy of Athens, Greece. He is the recipient of numerous prizes, awards and honors, and co-author of several other books including *Classics in Total Synthesis* (with his student Erik J. Sorensen) and *Classics in Total Synthesis II* (with his student Scott A. Snyder).

About the co-author

Tamsyn Montagnon received her B.Sc. in Chemistry with Medicinal Chemistry from the University of Leeds which was followed by a move to the University of Sussex, where she obtained a D.Phil in 2000 for research conducted under the supervision of Professor P. J. Parsons. She was awarded a GlaxoWellcome postdoctoral fellowship and joined Professor K. C. Nicolaou's group in January 2001 for three years. She is presently at the University of Crete doing research under the auspices of an EU Marie Curie Fellowship. Her research interests include natural product synthesis, medicinal chemistry, and reaction methods and mechanisms.

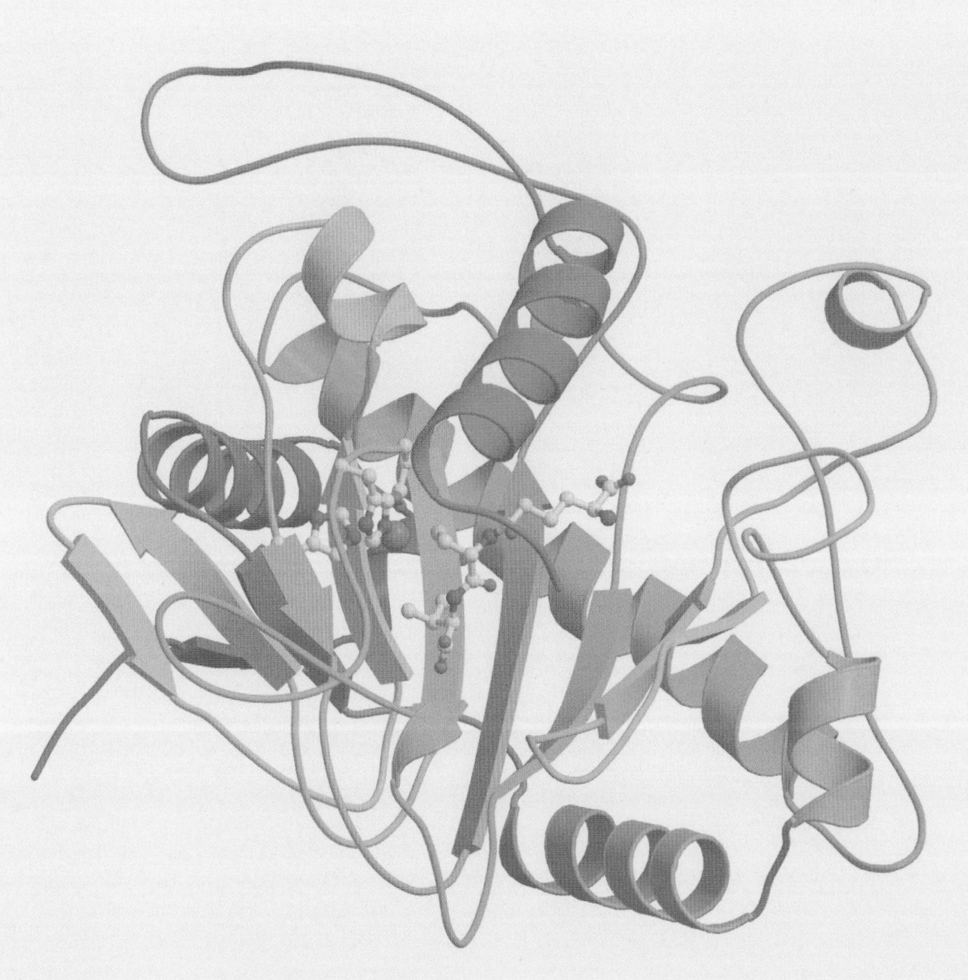

Contents

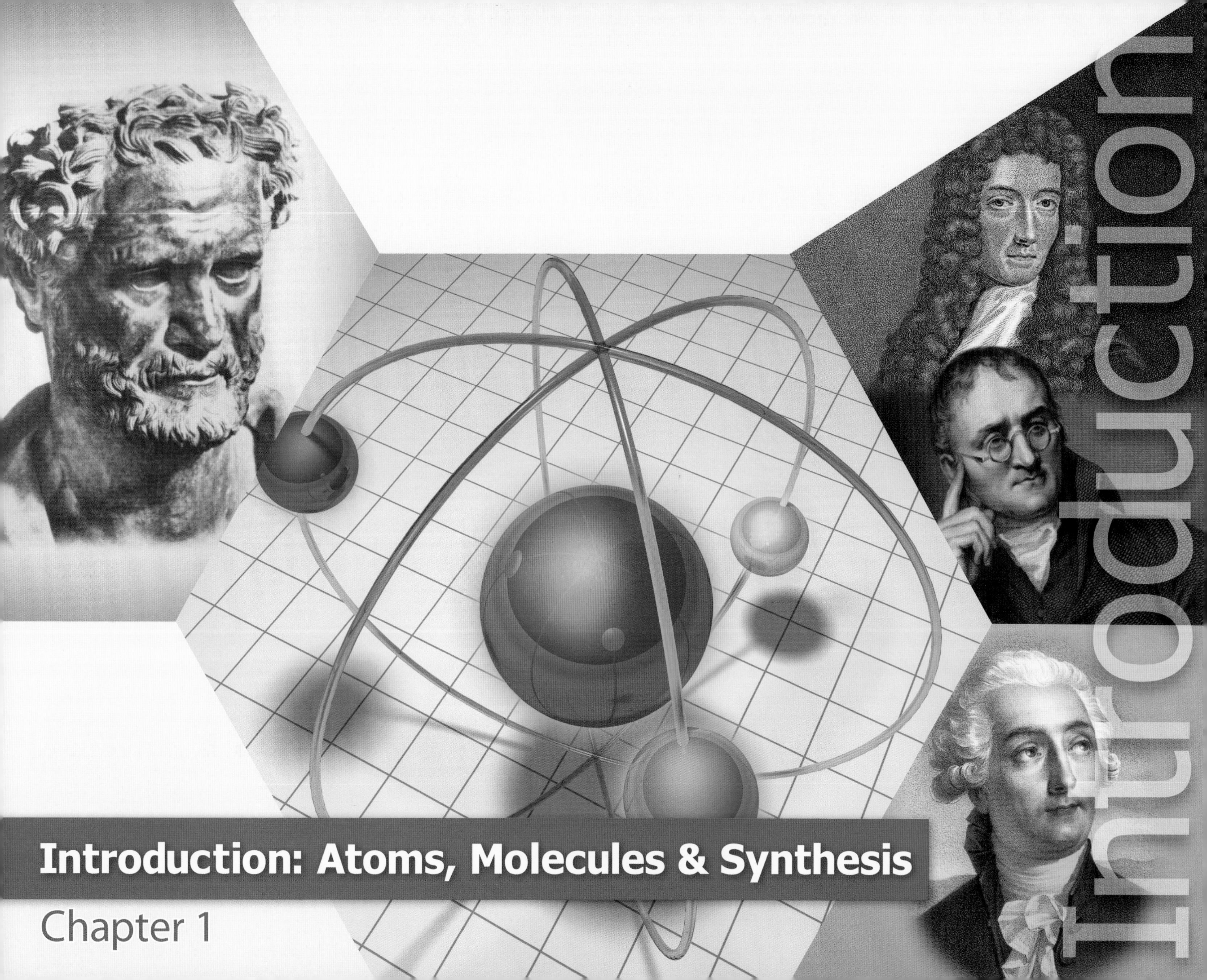
Introduction
Introduction: Atoms, Molecules & Synthesis
Chapter 1

Box 1 The Big Bang

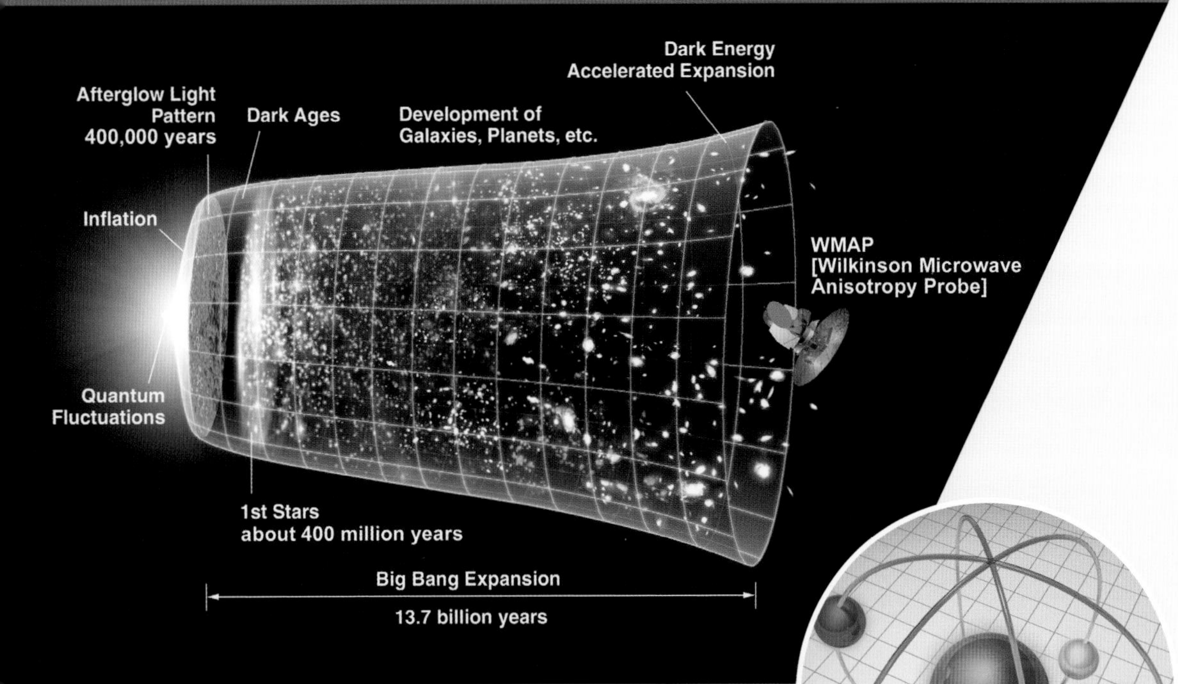

Dark Energy
Accelerated Expansion
Afterglow Light
Pattern
400,000 years
Dark Ages
Development of
Galaxies, Planets, etc.
Inflation
WMAP
[Wilkinson Microwave
Anisotropy Probe]
Quantum
Fluctuations
1st Stars
about 400 million years
Big Bang Expansion
13.7 billion years

Box 2 Planet Earth and the Universe

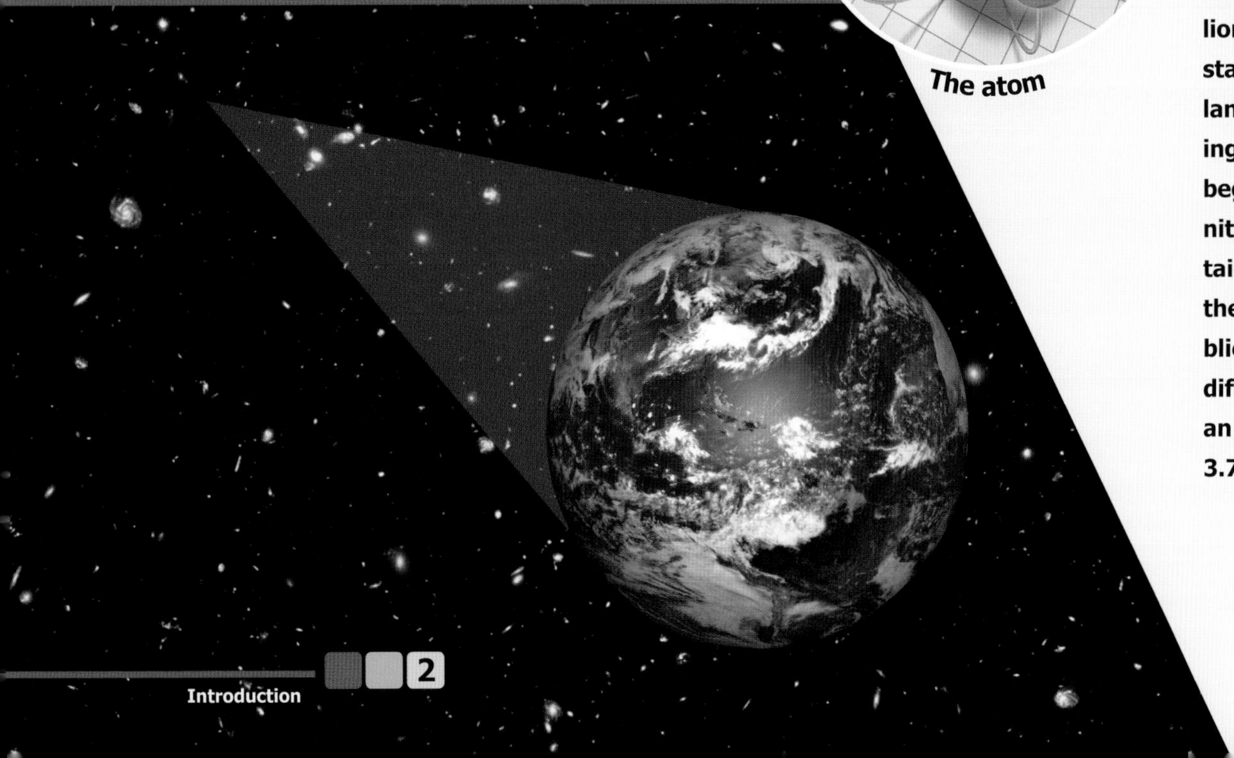

The atom

From the spectacular singularity of the Big Bang (Box 1) 13.7 billion years ago, that we consider to be the beginning of our universe, and from the infinite energy that it created, atoms and molecules were born. These tiny particles then came together to form huge galaxies and planetary systems, one of which is ours. Our planet (Box 2) emerged about 4.6 billion years ago after a struggle for existence during which it weathered collisions with competing emerging planets, meteorites and other space debris. About 4 billion years ago the Earth reached a state in which its atmosphere and landscape contained all the basic ingredients necessary for life to begin: carbon, oxygen, hydrogen, nitrogen, sulfur, phosphorous, certain metals and other elements, and their compounds, molecular assemblies formed by combining atoms of different elements. It happened as an amazing creation approximately 3.7 billion years ago as these simple chemicals came together harmoniously to form more complex chemical assemblies that were capable of surviving, replicating, and undergoing evolutionary selection to become even more sophisticated molecular assemblies. Life was switched on and continues to evolve to this day with us humans at the top of the pyramid. Both the Universe and Life have occupied the minds of countless thinkers, beginning with the ancient Greeks who pondered deeply about the nature of matter and its continuous changes.

Robert Boyle

Antoine Laurent de Lavoisier

Synthesis, a beautiful word that in its original Greek form conveys the idea of composition, is equally well suited to describing works of poetry and music, other fine arts, or the physical sciences such as chemistry. Chemical synthesis is amongst the most creative and rewarding of human endeavors, tantalizing and challenging its disciples in their continual quest for new ways to assemble multifarious and complex molecular architectures. As a discipline, it came of age only during the nineteenth and twentieth centuries, despite the fact that its theoretical foundations had been laid in the classical period of ancient Greece. (More about the birth of organic synthesis will be found in Chapter 2).

The first atomic theory of matter is a remarkable demonstration of the advanced nature of classical Greek thinking in which the power of rational thought was exercised by the philosopher-scientists of the time (Box 3). Thus, by inductive logic and without recourse to any experimentation, Demokritos and his lesser-known mentor, Leukippos, formulated this hypothesis in the fifth century B.C. to explain the behavior of various forms of matter that they observed around them. The conjecture is all the more remarkable for its accuracy, as its basic tenet still prevails today, following the resurrection of atomistic notions by a series of experimental scientists starting with Robert Boyle in the seventeenth century and proceeding through Antoine Laurent de Lavoisier to be finally cemented by John Dalton, who lucidly defined a formal set of principles in his treatise *A New System of Chemical Philosophy*, published in 1808 (Box 4).

It would seem that in the two thousand years spanning the time between Demokritos and Dalton, humanity traveled only a meager distance in its understanding of chemistry, but with these profound

Demokritos

John Dalton

Box 3 Demokritos' words

"The world is made of two parts, the full (pleres, stereon) and the empty, the vacuum (cenon, manon). The fullness is divided into small parts particles called atoms (atomon, that cannot be cut, indivisible). The atoms are infinite in number, eternal, absolutely simple; they are all alike in quality but differ in shape, order, and position. Every substance, every single object, is made up of those atoms, the possible combinations of which are infinite in an infinity of ways. The objects exist as long as the atoms constituting them remain together; they cease to exist when their atoms move away from one another. The endless changes of reality are due to the continual aggregation and disaggregation of atoms."

Demokritos, Fifth Century B.C.

Box 4 Dalton's words

"[all bodies] are constituted of a vast number of extremely small particles or atoms of matter bound together by a force of attraction...Therefore we may conclude that the ultimate particles of all homogeneous bodies are perfectly alike in weight, figure etc...Chemical analysis and synthesis go no farther than to the separation of particles one from another, and their reunion...All the changes we can produce, consist in separating particles that are in a state of cohesion or combination, and joining those that were previously at a distance."

John Dalton, 1808

Box 5 The Periodic Table of the Elements

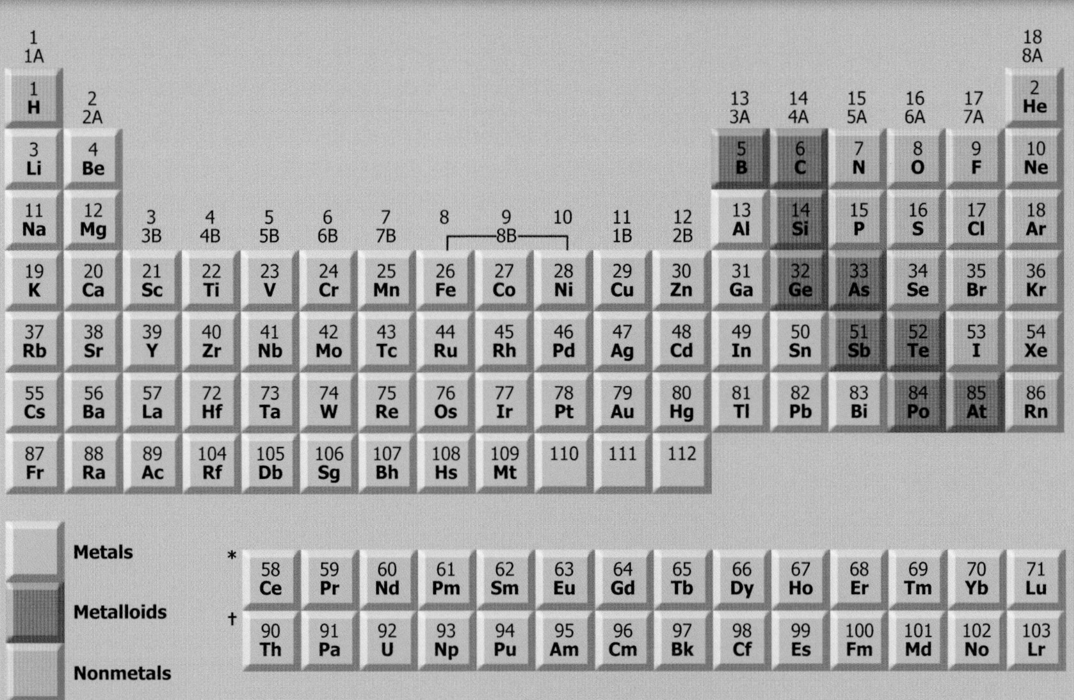

truths at last established by scientific methods, progress has since picked up pace and our ability to effect synthesis by *"joining those [particles] that were previously at a distance"* has been honed to encompass many splendid accomplishments.

1 1A
2 2A
13 3A
14 4A
15 5A
16 6A
17 7A
18 8A
3 3B
4 4B
5 5B
6 6B
7 7B
8
9 8B
10
11 1B
12 2B
1 H
2 He
3 Li
4 Be
5 B
6 C
7 N
8 O
9 F
10 Ne
11 Na
12 Mg
13 Al
14 Si
15 P
16 S
17 Cl
18 Ar
19 K
20 Ca
21 Sc
22 Ti
23 V
24 Cr
25 Mn
26 Fe
27 Co
28 Ni
29 Cu
30 Zn
31 Ga
32 Ge
33 As
34 Se
35 Br
36 Kr
37 Rb
38 Sr
39 Y
40 Zr
41 Nb
42 Mo
43 Tc
44 Ru
45 Rh
46 Pd
47 Ag
48 Cd
49 In
50 Sn
51 Sb
52 Te
53 I
54 Xe
55 Cs
56 Ba
57 La
72 Hf
73 Ta
74 W
75 Re
76 Os
77 Ir
78 Pt
79 Au
80 Hg
81 Tl
82 Pb
83 Bi
84 Po
85 At
86 Rn
87 Fr
88 Ra
89 Ac
104 Rf
105 Db
106 Sg
107 Bh
108 Hs
109 Mt
110
111
112
Metals
Metalloids
Nonmetals
*
58 Ce
59 Pr
60 Nd
61 Pm
62 Sm
63 Eu
64 Gd
65 Tb
66 Dy
67 Ho
68 Er
69 Tm
70 Yb
71 Lu
†
90 Th
91 Pa
92 U
93 Np
94 Pu
95 Am
96 Cm
97 Bk
98 Cf
99 Es
100 Fm
101 Md
102 No
103 Lr
* Lanthanides
† Actinides

Dimitri Mendeleev

The periodic table of the elements (Box 5) provides key information about each element, and is used routinely by scientists in many fields. It was first laid out by Dimitri Mendeleev in the nineteenth century and has since been expanded to include newly discovered elements (some of them manmade) and now boasts around 116 entries. Mendeleev arranged the elements according to their atomic number (i.e. the number of protons in the nucleus of the element) in columns (groups) and rows (periods). The trends he observed allowed him to spot gaps in the knowledge of his time, and he successfully predicted the discovery, and even some of the properties, of several of the then unknown elements.

As more of the world's mysteries have been unveiled through the parallel maturing of the chemical and biological sciences, a magnificent and ever-expanding library of molecular structures (Boxes 6, 7, and 8), masterpieces of nature's ingenuity, have drawn our attention and continually tested our synthetic acumen. These structures range from the self-assembled double helices of the nucleic acids (DNA and RNA, Box 7), through proteins (Box 7) and polysaccharides, to secondary metabolites (Box 8), all fiendishly complex compositions of the elements when compared to the essentials of life such as oxygen and water (Box 6). Within these remarkable molecules we see an astonishing wealth of diversity, complexity and infor-

Box 6 Elements and small molecules of life: Computer-generated and ball and stick model depictions

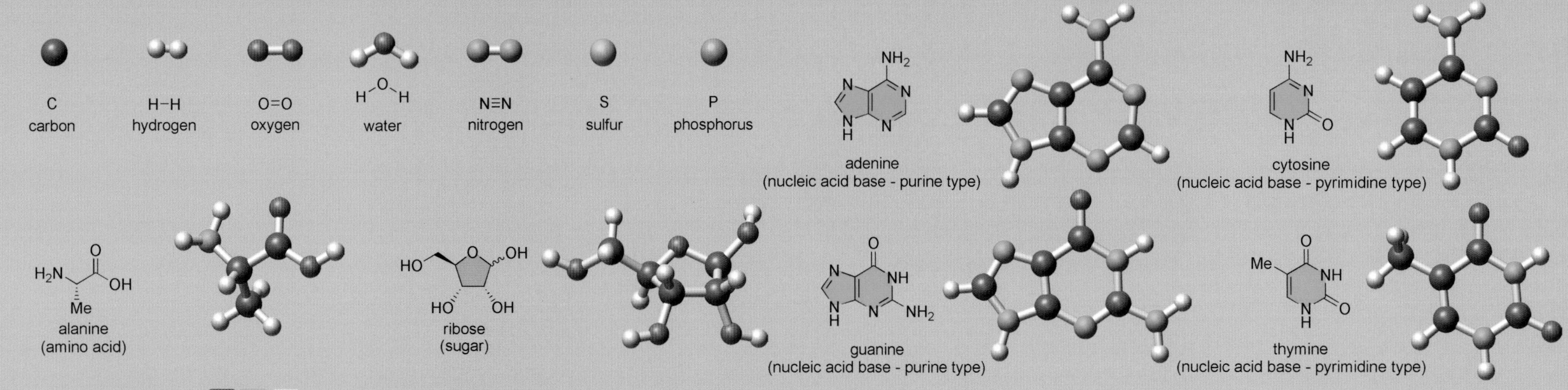

Box 7 Biopolymers of Life: Computer-generated models depicting sections of DNA, *t*-RNA and a protein

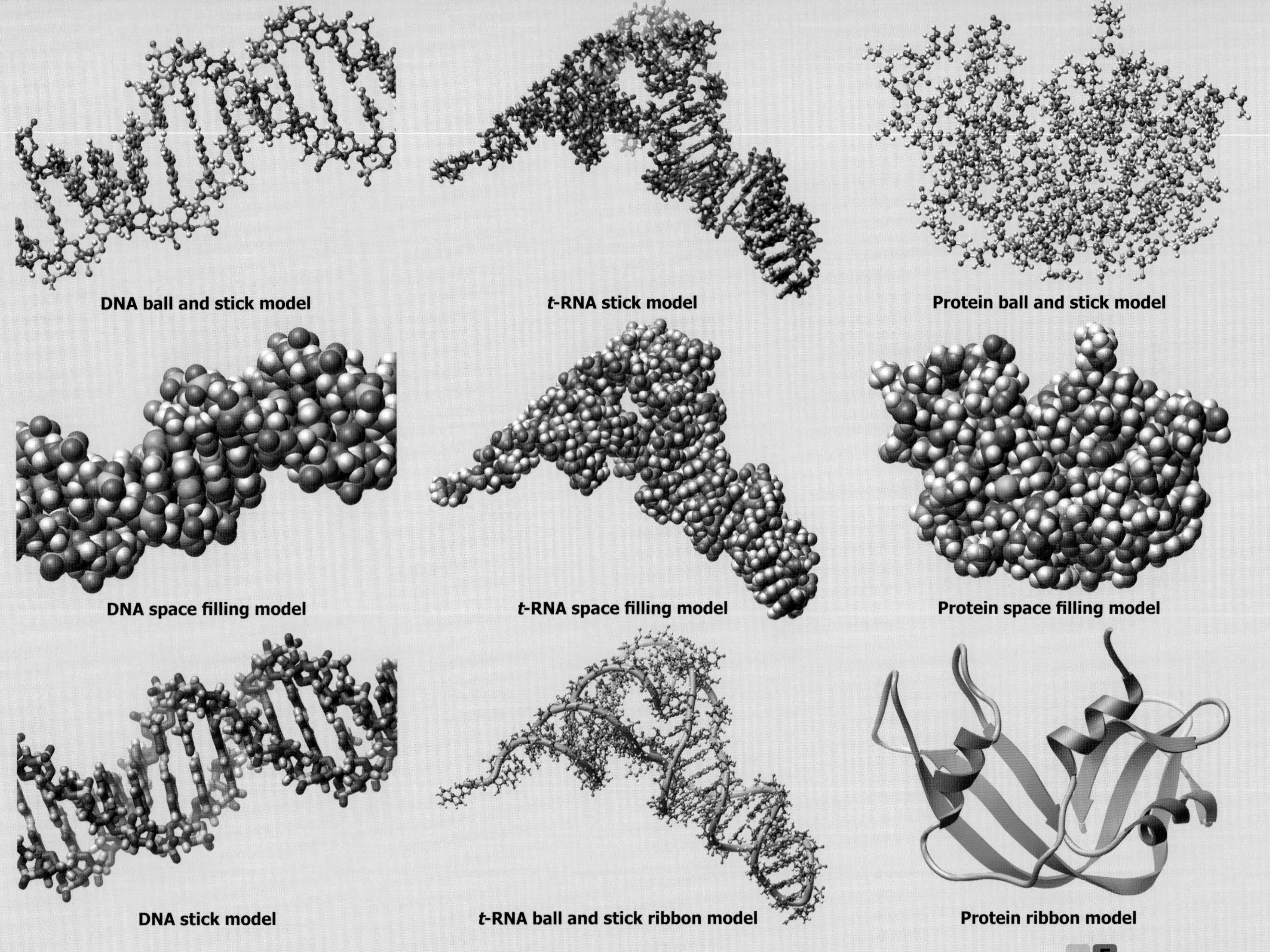

DNA ball and stick model

t-RNA stick model

Protein ball and stick model

DNA space filling model

t-RNA space filling model

Protein space filling model

DNA stick model

t-RNA ball and stick ribbon model

Protein ribbon model

Box 8 Secondary metabolites, also known as natural products

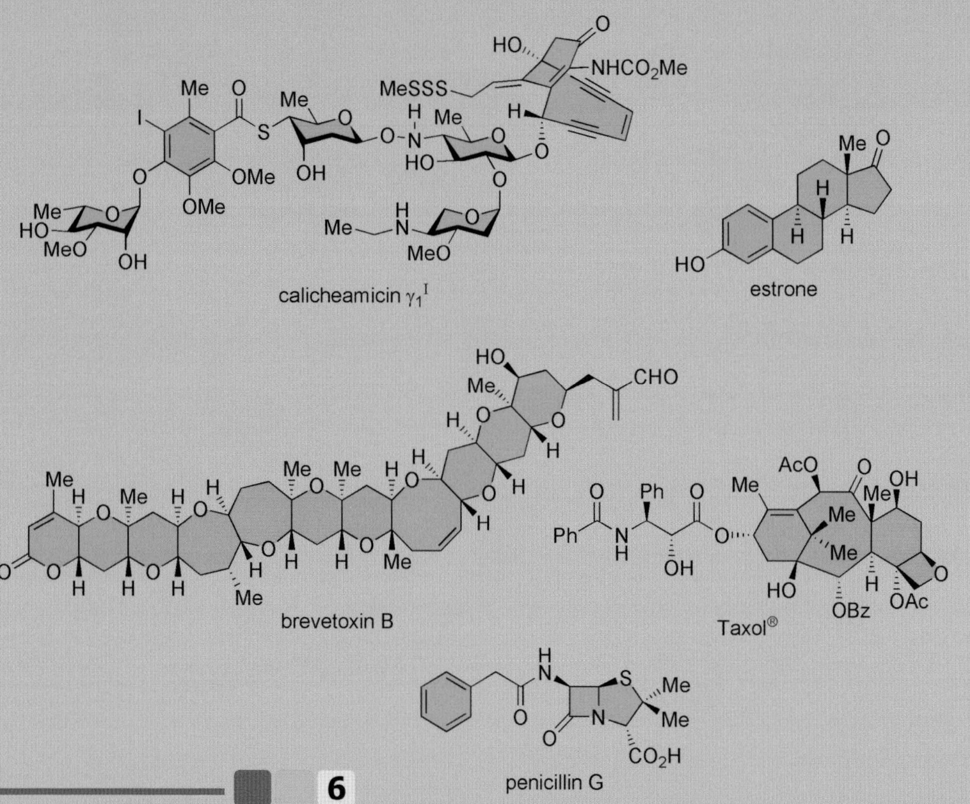

calicheamicin γ_1^I

estrone

brevetoxin B

Taxol®

penicillin G

mation that nature has so admirably engineered in order to facilitate the functions of living systems. The primary synthetic tools employed by nature are enzymes, molecular machines that deftly organize reaction sites and impart stereochemical information to the embryonic intermediates as they grow into their destined molecular structures. This seemingly effortless exercise issues a formidable gauntlet to chemists, who must search long and hard to acquire the skills necessary to compete with such exquisite laborers and artisans. Today, our ability to chemically synthesize oligonucleotides and peptides efficiently through fairly routine automated procedures shows that we may have risen to the challenge; however, secondary metabolites, more commonly known as natural products, continue to test our proficiency in these matters. The kinds of atoms, types of bonds, and connectivities seen in their structures can, and do, exhibit almost infinite variation and as such these molecules often present puzzles that dictate the design of unique strategies in order to solve the problem of their laboratory construction. However, it is these stipulations and the diversity of the challenges that ensures the vibrancy and intellectual rigor of the field of organic synthesis.

The construction of nature's molecules by chemical synthesis from simple starting materials is known as total synthesis. The same term can also be applied to the synthesis of a designed molecule of more or less complexity. Total synthesis is particularly demanding, as the practitioner is often required to address challenges set by nature for which solutions are not available from the repertoire of known synthetic tools. This sub-discipline of organic synthesis is, therefore, considered by many to be the flagship of, and the engine that drives, the more general field forward. Today, it proudly symbolizes the advanced state of the art of chemical synthesis and continues to command the center stage, as it has done ever since Friedrich Wöhler's synthesis of urea in 1828, the milestone that marks the birth of organic synthesis (Chapter 2).

Socrates

In its early years, total synthesis was motivated by the sheer fascination of making nature's molecules – something that was previously considered to be the exclusive domain of living organisms. Later, and in the era when immature analytical techniques left lingering ambiguity, total synthesis evolved beyond the "demystification" of nature and was practiced as a vehicle by which proof of structure for natural products could be obtained. Today, while the aspect of structural proof is still pertinent in certain cases, especially for absolute stereochemical assignments, total synthesis endeavors are considered to be ideal platforms for the discovery and invention of new synthetic technologies and strategies, rendering scarce substances readily available for biological investigations, and still as a source of exhilaration for the practitioner.

The allure of the creativity required and the rewards associated with the practice of chemical synthesis have attracted

Box 9 Computer-generated model depictions of secondary metabolites

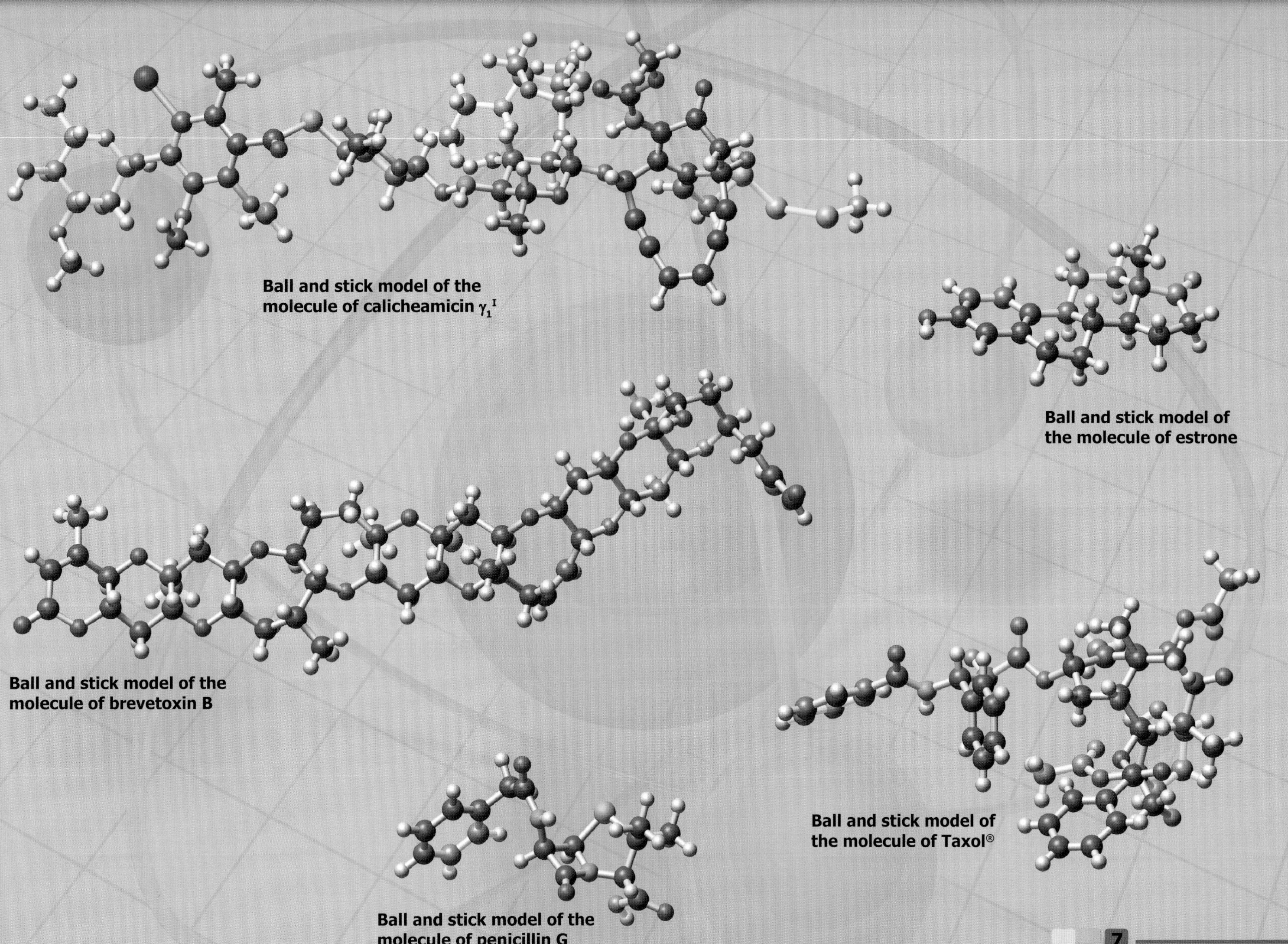

Ball and stick model of the molecule of calicheamicin γ_1^I

Ball and stick model of the molecule of estrone

Ball and stick model of the molecule of brevetoxin B

Ball and stick model of the molecule of Taxol®

Ball and stick model of the molecule of penicillin G

Further Reading

G. Sarton, *Ancient Science Through the Golden Age of Greece*, Dover Publications, New York, **1980**.

W. H. Brock, *A Norton History of Chemistry*, W W Norton, New York, **1992**.

A. Greenberg, *A Chemical History Tour*, Wiley, New York, **2000**.

K. C. Nicolaou, E. J Sorensen, *Classics in Total Synthesis*, Wiley-VCH, Weinheim, **1996**.

K. C. Nicolaou, S. A. Snyder, *Classics in Total Synthesis II*, Wiley-VCH, Weinheim, **2003**.

E. J. Corey, X.-M. Cheng, *The Logic of Chemical Synthesis*, Wiley Interscience, New York, **1989**.

K. C. Nicolaou, E. J. Sorensen, N. Winssinger, The Art and Science of Organic and Natural Products Synthesis, *J. Chem. Ed.* **1998**, *75*, 1225–1258.

K. C. Nicolaou, S. A. Snyder, The Essence of Total Synthesis, *Proc. Natl. Acad. Sci. USA* **2004**, *101*, 11929–11936.

G. R. Desiraju, Chemistry: The Middle Kingdom, *Current Science*, **2005**, *88*, 374–380.

some of the most charismatic and brilliant minds of the last two centuries. From its early nineteenth century infancy, these pioneers have advanced it to its currently impressive state, a progression made possible by rational thought and design, exquisite experimentation and, admittedly, a number of serendipitous discoveries.

We have reached a point where the impact of chemical synthesis extends far beyond chemistry, touching upon physics, material sciences, nanotechnology, and biology, an impact that, in turn, is translated to influences over many industries that can be felt everyday by people around the world in the form of dyes, fibers, plastics, foods, vitamins, high-tech materials for cars, aircraft, computers, fuels and, most importantly, medicines. It would be fair to say that the coming of age of chemical synthesis has been one of the most influential movements of the last two centuries and one where the applications have decisively shaped the world as we know it today, and will continue to do so far into the future. It took more than two thousand years for the ideas and thoughts of Demokritos, and other ancient Greek thinkers such as Socrates, Plato, and Aristotle, to bear fruit. The passage of time between the inception of an idea and practical applications resulting from it may not be so long today but, nevertheless, it is a necessary step in our quest for progress. For it is only once we have understood the fundamentals of science though basic research that we are able to enjoy the benefits of the technology that inevitably emerges from it.

With the pivotal role fulfilled by total synthesis within our world having been discussed, we wish to draw this chapter to a close with the suggestion that perhaps you may desire to hold another quote from Demokritos in the back of your mind as you imbibe the struggles that the protagonists of these trials have experienced in reaching their targets:

"*Courage is the beginning of action but chance is the master of the end.*"

For this maxim is so true not only in everyday life, but also in the enterprise of total synthesis, particularly for those cases where the outcome is far from assured or known, due to the demonic nature of the molecular architectures involved.

We hope the historical journey contained herein and the descriptions of the molecules and their syntheses will be both as enjoyable and inspiring to you as they were for us and for generations past.

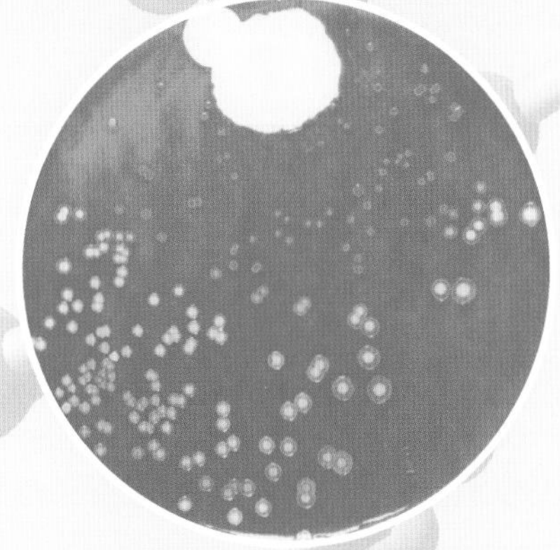

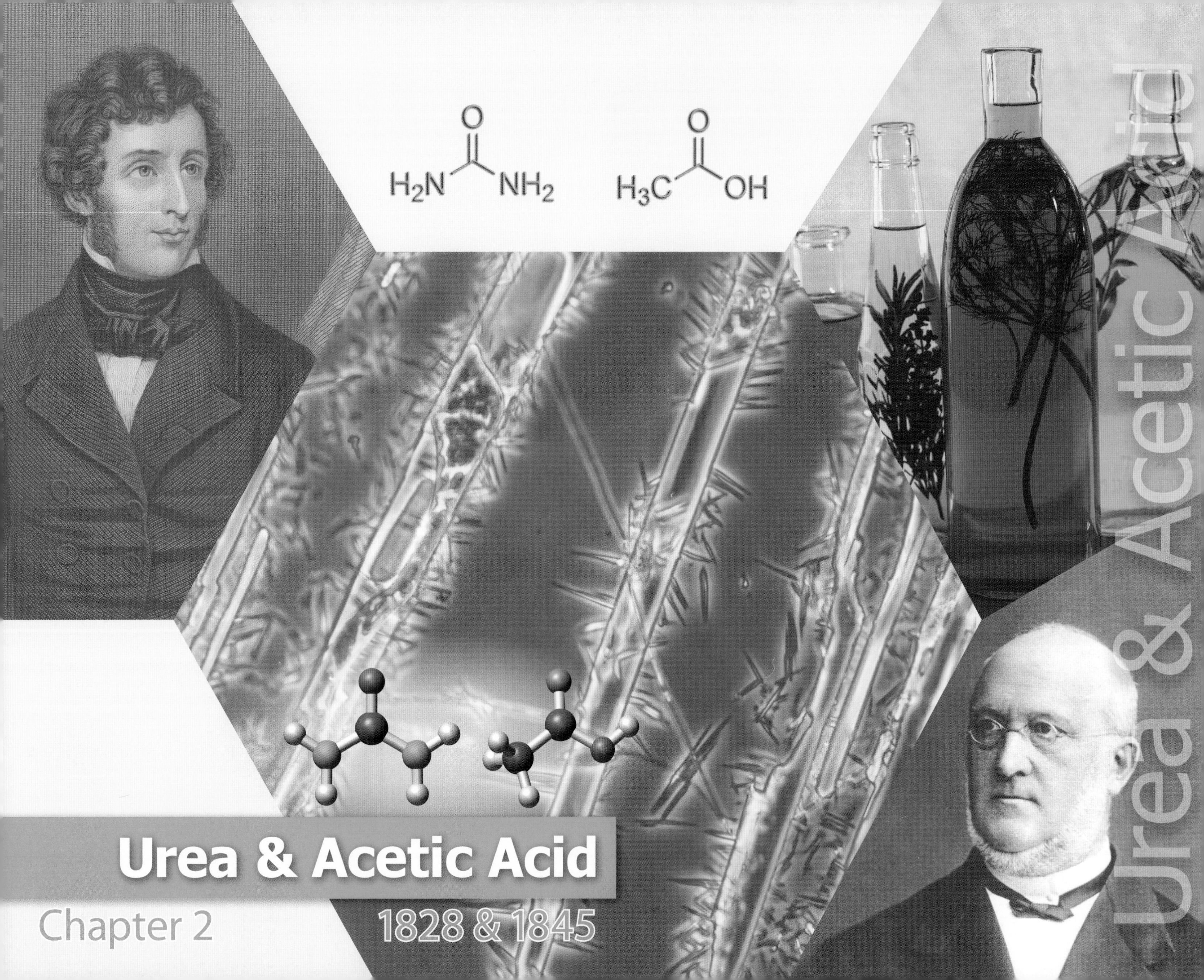

Urea & Acetic Acid

Chapter 2

1828 & 1845

Urine in test tubes

Friedrich Wöhler

Box 1 **Early comments reflecting the impact of Wöhler's synthesis of urea**

"Of special importance for organic chemistry was his [Wöhler's] discovery (1828) of the artificial production of urea. This was the first example that an organic substance could be produced through chemical means alone from inorganic materials; this discovery destroyed the till-then accepted distinction between the organic and inorganic bodies, that, namely, the first originate under the influence of vegetable or animal vitality, whereas only the latter could be produced through artiface."

Hermann Kopp, from *Volume 1* of his 1843 four part series - *Geschichte der Chemie*

"The extraordinary and to some extent inexplicable production of urea without the assistance of vital functions, for which we are indebted to Wöhler, must be considered one of the discoveries with which a new era in science has commenced"

Justus von Liebig - chemical luminary - 1837

Justus von Liebig

Urea crystals

"I cannot, so to say, hold my chemical water, and must tell you that I can make urea, without thereby needing to have kidneys, or anyhow, an animal, be it human or dog," wrote an excited Friedrich Wöhler to his mentor, the Swedish chemist Jöns Jakob Berzelius. Urea had been discovered in and isolated from human urine in 1773 by Hilaire M. Rouelle and had captivated scientists from that day forth due to the fascination surrounding what was, at that time, the mystery of all bodily functions. The first synthesis of urea was a profound moment in the history of science, marking both the death of the all-pervasive theory of vitalism, and the birth of synthetic organic chemistry as its own distinct, creative, and valuable discipline. It should also be recognized, however, that the demise of vitalism, a theory that had been championed for generations and by many luminaries, was destined to be a painfully slow process. It would take many further decades to refute these mystical ideas completely; indeed, the great biochemist Louis Pasteur was still a fan of this archaic belief system fifty years later. Irrespective of the stubborn endurance of vitalism, Wöhler's synthesis of urea must be credited as not only the earliest contribution to organic synthesis, but as the single most important blow to this vestigial theory (Box 1).

- Inorganic
- Vital Force
- Vitalism
- Organic

Despite using the terms daily, chemists often forget that the classifications of inorganic and organic compounds originally arose from the theory of vitalism, which divided matter into two classes based upon the response of the material to the application of heat. Organic compounds were transformed upon heating and could not be recovered by removing the source of warmth, the proposed explanation for this being that the "vital force of life" was evacuated by burning. Stemming from this classification was the belief that all organic compounds had to be made within living things, so as to con-

Jöns Jakob Berzelius

tain some of the maker's own vital force. However, in 1828, Wöhler's conversion of an inorganic substance, silver isocyanate, into an organic compound, urea, demonstrated that so-called 'organic compounds' could be made without the involvement of any living system (Box 2). It should be noted that even though Wöhler used silver isocyanate (AgNCO) in his synthesis of urea he incorrectly described it as silver cyanate (AgOCN), a compound with the alternate regiochemistry that even today is unknown. Significantly, during the course of these studies Wöhler also demonstrated that silver isocyanate (AgNCO) had the same chemical composition as the silver fulminate (AgCNO), studied by Justus von Liebig, but it exhibited very different properties, thus introducing the concept of isomerism, which rapidly became a new interest for all these great men of chemistry.

Friedrich Wöhler

Urea contains only one carbon atom and no carbon–carbon or carbon–hydrogen bond, and, as such, it suggested the next hurdle that the early pioneers of synthetic organic chemistry had to overcome. The feat of uniting two carbon atoms to form a carbon–carbon bond was accomplished in 1845 by the German chemist Hermann Kolbe, whose synthesis of acetic acid is remarkable from a number of perspectives. Acetic acid is familiar to everyone as a component of vinegar, the oxidation product of wines and other spirits, and its use from its natural sources is an everyday feature of our lives. Kolbe was able to synthesize acetic acid in the laboratory from elemental carbon and hydrogen, using water as the source of oxygen, and with a little assistance from iron disulfide and chlorine. This synthesis was not only an impressive

Hermann Kolbe

Box 2 Wöhler's synthesis of urea

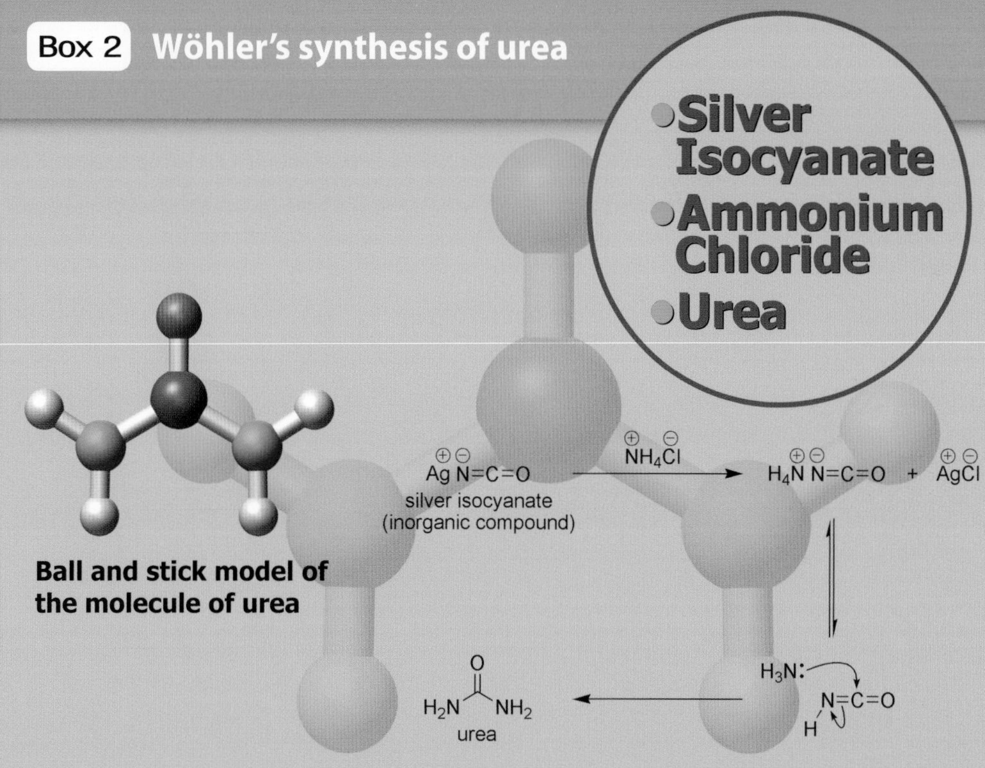

Ball and stick model of the molecule of urea

Box 3 Kolbe's synthesis of acetic acid

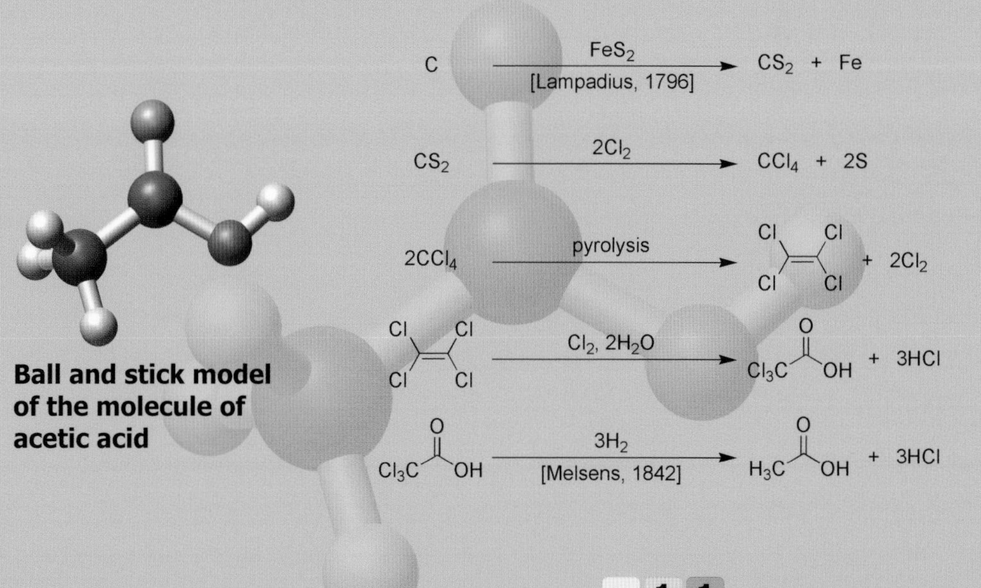

Ball and stick model of the molecule of acetic acid

Box 4 The language of organic chemistry

In the mid-nineteenth century, organic chemistry was in its infancy. With the groundbreaking work of Wöhler and Kolbe this blossoming discipline was taking its first steps. Chemists now have a keen appreciation of chemical structures, bonding, and shape, allowing the construction of some exceedingly complex molecules, as we shall be discussing in later chapters. However, when Hermann Kolbe was carrying out his landmark synthesis of acetic acid, he had no concept of the chemical bond, or even of the true structure of simple substances such as water. Today we know that the carbon atom is capable of forming four chemical bonds, which can be arranged in a number of ways. These basic arrangements are shown below: methane has a tetrahedral arrangement of four hydrogen atoms, each forming a single carbon–hydrogen bond, around the central carbon atom; ethane has a similar arrangement around each of its two carbon atoms; ethene, which has a carbon–carbon double bond, has a trigonal planar arrangement around each carbon atom; and ethyne, which has a carbon–carbon triple bond, is linear. The properties and reactivity of chemical compounds are closely linked to their atomic composition, the nature of the bonds linking their atoms, and their three-dimensional shape.

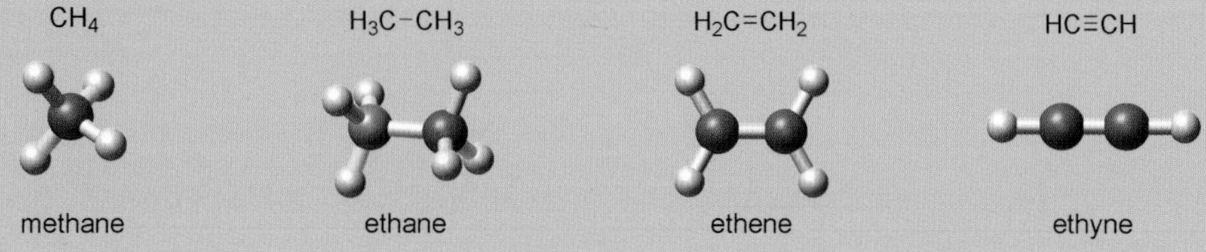

The increasingly complex nature of the compounds being investigated by organic chemists necessitated a convenient method for drawing molecular structures. Thus, the individual atoms and bonds have come to be replaced by a shorthand notation wherein the carbon framework is depicted by simple line drawings. The non-hydrogen substituents are shown, with the hydrogens themselves omitted for clarity. This convention also allows the representation of stereochemistry, using "wedged" and "hashed" bonds. A number of common groups are also given abbreviations, such as Me for methyl (CH_3) and Et for ethyl (CH_3CH_2). Drawings of ethanol, benzene, and glucose are shown below to demonstrate these conventions.

ethanol benzene glucose

Expanded and abbreviated structural formulae of organic compounds

John Dalton

Robert Boyle

Carl Wilhelm Scheele

Antoine Laurent de Lavoisier

achievement for its time, but also rare because in it one finds the construction of an organic molecule from the simplest of nature's building blocks, the elements themselves (Box 3).

The syntheses of urea and acetic acid were not isolated events, but rather occurred in the midst of a revolution of sorts that was taking place in chemistry at that time. Thus, following the remarkable thoughts of the ancient Greeks, such as Demokritos and Aristotle, on the nature of matter, followed by the rather dark age of the alchemists, great men now appeared on the scene whose work was destined to give rise to modern chemistry. Among these forefathers of chemistry were Robert Boyle (an Irishman working in Oxford), the Englishman John Dalton, and the Frenchman Antoine Laurent de Lavoisier, all of whom reintroduced theories that emanated from the ancient Greeks, but, unlike their predecessors, they now employed experimentation to discover and confirm new principles. The Swedish chemist Carl W. Scheele began to explore nature by isolating and studying several naturally occurring substances. He was succeeded by Berzelius, who advanced chemistry through his discoveries, teaching, and strong personality. The early part of the nineteenth century was dominated by men like Liebig, a German, who studied in Paris before returning to Germany and bringing with him new techniques and a new approach to chemical investigation and teaching. Liebig was a close friend of Wöhler and together they published important and influential work that was to change the face of chemistry forever. While their work set the foundations for chemical synthesis, it was their students who would proliferate the science and give it new momentum. Indeed, between them, Liebig and Wöhler educated many thousands of chemistry students. Amongst Liebig's most influential students was August Wilhelm von Hofmann, who was invited by Queen Victoria to travel to London to take charge of the newly created Royal College of Chemistry. There, Hofmann set up his laboratories and injected a new vigor into this science in Britain with his investigations into the chemistry of aromatic compounds that would lead to the founding of the dye industry (see Chapter 8). Another of Liebig's students was August Kekulé, who, along with Joseph Loschmidt and others, proposed the correct structure for benzene. The age of the molecule was thus ushered in, and by the end of the nineteenth century organic chemistry was poised for the next phase of its great march forward. The twentieth century saw a shift in the center of gravity to the United States, which began with the emigration of Joseph Priestley (1733–1804) from England to the USA in 1794, where chemistry would flourish, reaching unprecedented new heights.

Wöhler and Kolbe both made a number of other notable contributions to general chemistry during their respective long and distinguished careers. The syntheses of urea and acetic acid were, however, the most pertinent of their works from the perspective of the history of total synthesis, since these accomplishments mark the point of departure in the quest to synthesize a multitude of beautiful and complex natural products. The newly discovered discipline of fashioning substances by synthesis was sharpened and developed throughout the nineteenth and twentieth centuries, spawning a number of essential industries and evolving into a precise science and a fine art. It is fair to say that its many applications have played a pivotal role in shaping the world as we know it today. Highlights of the progression of this art are chronicled in the pages that follow; we hope you enjoy the stories.

August Wilhelm von Hofmann

Friedrich August Kekulé

Joseph Priestley

August Wilhelm von Hofmann and his students at the Royal College of Chemistry, London

Antoine Laurent de Lavoisier in his laboratory in Paris

Justus von Liebig's laboratory in Giessen, Germany

Queen Victoria

Further Reading

F. Wöhler, Über künstliche Bildung des Harnstoffs, *Poggendorf's Ann.* **1828**, *12*, 253–256.

A. J. Rocke, *The Quiet Revolution: Hermann Kolbe and the Science of Organic Chemistry*, University of California Press, Berkeley, **1993**.

P. S. Cohen, S. M. Cohen, Wöhler's Synthesis of Urea: How Do the Textbooks Report It?, *J. Chem. Ed.* **1996**, *73*, 883–886.

A. Greenberg, *A Chemical History Tour*, John Wiley & Sons, New York, **2000**.

E. J. Corey, X.-M. Cheng, *The Logic of Chemical Synthesis*, Wiley Interscience, New York, **1989**.

K. C. Nicolaou, E. J. Sorensen, *Classics in Total Synthesis*, Wiley-VCH, Weinheim, **1996**.

K. C. Nicolaou, S. A. Snyder, *Classics in Total Synthesis II*, Wiley-VCH, Weinheim, **2003**.

J. Buckingham, *Chasing the Molecule*, Sutton Publishing, Stroud, **2004**.

A World Shaping Art & Science

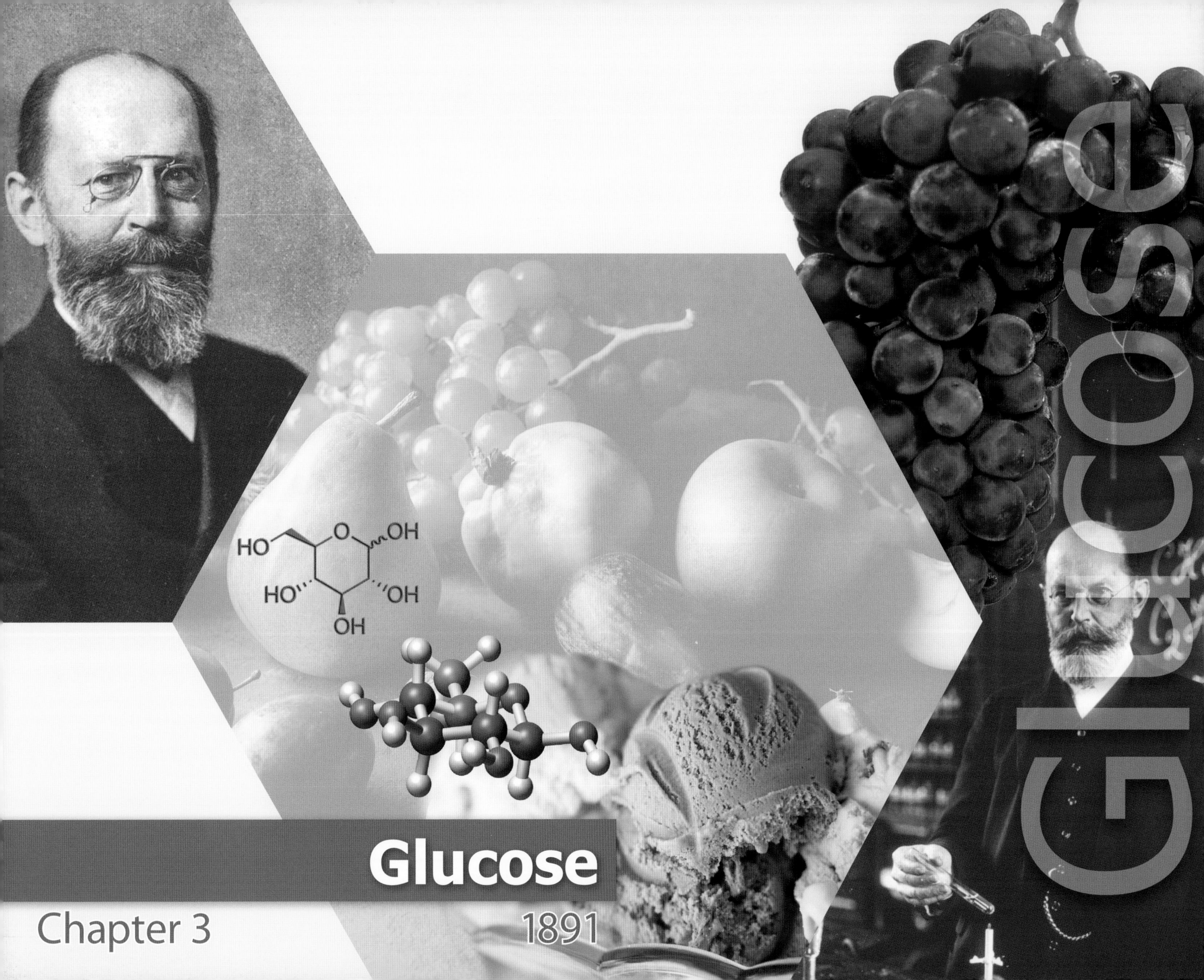
HO
O
OH
HO
OH
OH
Glucose
Glucose
Chapter 3
1891

Box 1 **Biosynthesis of D-glucose from carbon dioxide in green plants**

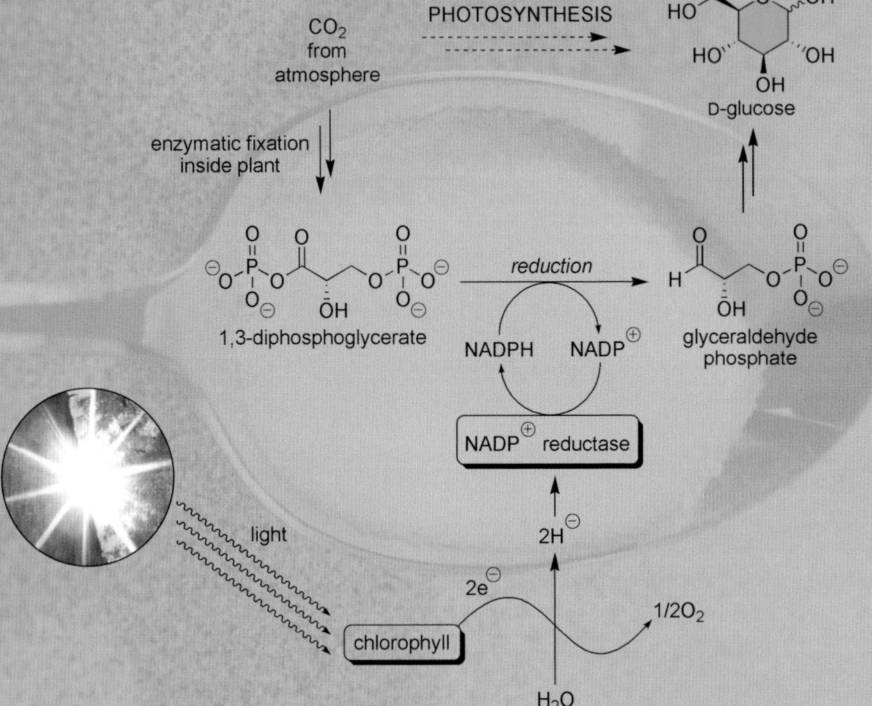

The German chemist Emil Fischer was a true pioneering scientist who defined many principles still important to present day chemistry. These include the significance of asymmetry, the intrinsic importance of organic chemistry in understanding biological mechanisms, and the art of extracting, identifying, and synthesizing naturally occurring compounds and their analogues for medicinal purposes. "I shall attempt to explain to you," he said on the occasion of the conferment of his Nobel Prize in Chemistry in 1902, "what organic chemistry is capable of as a loyal ally of physiology with refined methods of analysis and synthesis." He was well qualified to give such a speech, since he had led the way in synthesizing and examining many biologically important molecules from short peptides and purines (components of DNA and RNA) to carbohydrates, perhaps his most famous research topic of all.

Emil Fischer

Excitement about carbohydrates (also called saccharides or sugars) had first captured the interest of chemists more than a century before Fischer's work. Glucose was first isolated from raisins in 1747, by the German chemist Andreas Sigismund Marggraf, and named (from the Greek *glycos*, meaning sweet) by Jean-Baptiste Dumas in 1838. Antoine Laurent de Lavoisier, one of the founding fathers of chemistry, determined the elementary composition of sugars. As an understanding grew of their pivotal role in the supply of energy to living systems, and of their amazing biosynthetic origin from the simple molecule of carbon dioxide taken from the air (Box 1), so did the desire to synthesize these naturally occurring substances in the laboratory. Over the decades following Lavoisier's studies, only scant successes were recorded in this quest towards the synthesis of the monosaccharides (sugars containing a single saccharide unit), with the lone exception in this barren period being the significant contributions of the Russian chemist Aleksandr Mikhailovich Butlerov. In the 1860s, thirty years before Fischer's work, Butlerov treated formaldehyde (in the form of its trimer, trioxymethylene) with hot limewater and obtained a sweet syrup. However, the constitution of this medley of compounds was not definitively deconvoluted until the 1970s, when it was shown by modern chemical analysis (gas–liquid chromatography and mass spectrometry) to include various sugars with both branched and straight carbon chains. In the period since Butlerov and Fischer opened the door to this fascinating subject, carbohydrate research has continued to be at the forefront of both chemistry and biology. Indeed, glycobiology, as the field is known today (glyco indicating the connection with carbohydrates), has evolved into a rich and vibrant field of investigation of enormous importance to the life sciences. Some of the diverse structures which fall under the category of carbohydrates are shown in Box 2.

Andreas Sigismund Marggraf

The simplest sugars are mono- and disaccharides, such as glucose, fructose, sucrose, and lactose. These act as quick-release sources of energy to fuel all living cells. These simple sugars are complemented by polysaccharides, such as starch and glycogen, which provide organisms with longer-term energy storage. In addition, certain polysaccharides make up some of the strong scaffolds used within cellular architectures; examples include cellulose in plants and chitin in insect and crustacean exoskeletons. Some hybrid polymers are also involved in cellular structure; for example, polysaccharide ribbons cross-linked with short peptide chains afford the peptidoglycans (also known as glycopeptides), which make up the rigid cell walls of bacteria. The nucleic acids, DNA and RNA, are another class of hybrid polymers in which carbohydrates play a key role. The backbones of these biopolymers consist of ribose (RNA) or deoxyribose (DNA) sugar units alternating with phosphate groups. Purine and pyrimidine bases are appended to the carbohydrate rings, and it is their sequence that comprise the genetic code. The orientation of the sugar groups determines the directionality of the polymer, which is vital to decoding the genetic information. Non-polymeric sugars also serve a host of important functions which are a far cry from the simple supply of energy. For example, they can be found on extracellular surfaces where they act as antennae to assist with cellular recognition and adhesion. They have also been found within the structures of many natural products, such as the erythromycins (Chapter 17) and vancomycin (Chapter 31), where they confer specific physical and conformational attributes to the host molecules that are frequently essential for biological activity. This brief summary of the uses of sugars in nature is by no means

Aleksandr Mikhailovich Butlerov

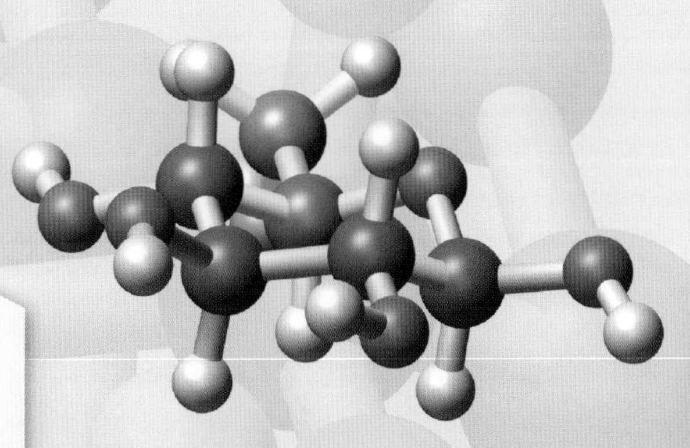

Ball and stick model of the molecule of β-D-glucose

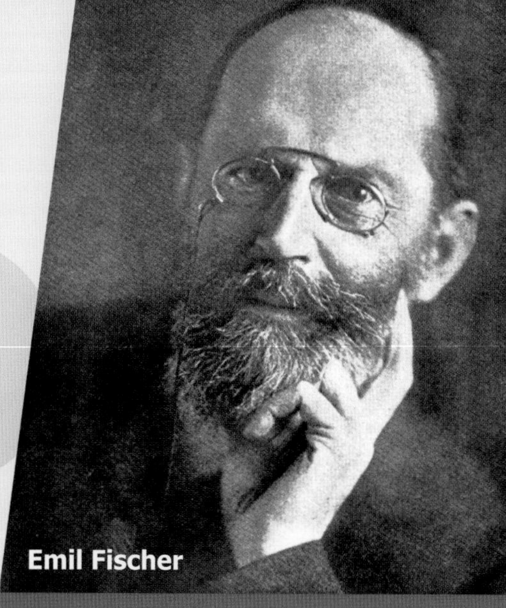

Emil Fischer

Box 2 Selected carbohydrates and carbohydrate-based polymers

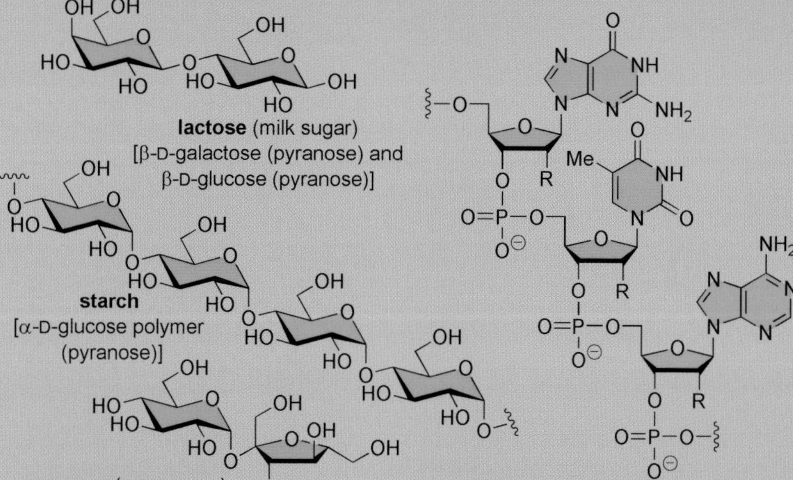

lactose (milk sugar) [β-D-galactose (pyranose) and β-D-glucose (pyranose)]

starch [α-D-glucose polymer (pyranose)]

sucrose (cane sugar) [α-D-glucose (pyranose) and β-D-fructose (furanose)]

2-deoxyribonucleic acid (DNA) (R = H) - genetic material that encodes for protein synthesis.
ribonucleic acid (RNA) (R = OH) - genetic coding, similar to DNA, but serving broader functions.

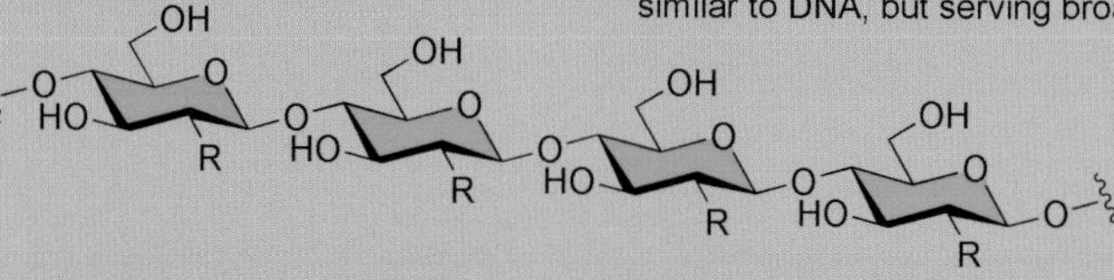

cellulose (R = OH) - structural polymer found in plants [β-D-glucose polymer (pyranose)].
chitin (R = $NHCOCH_3$) - structural polymer found in insect and crustacean exoskeletons [*N*-acetyl-β-D-glucosamine polymer (pyranose)].

Box 3 Emil Fischer's synthesis of D-glucose and related sugars

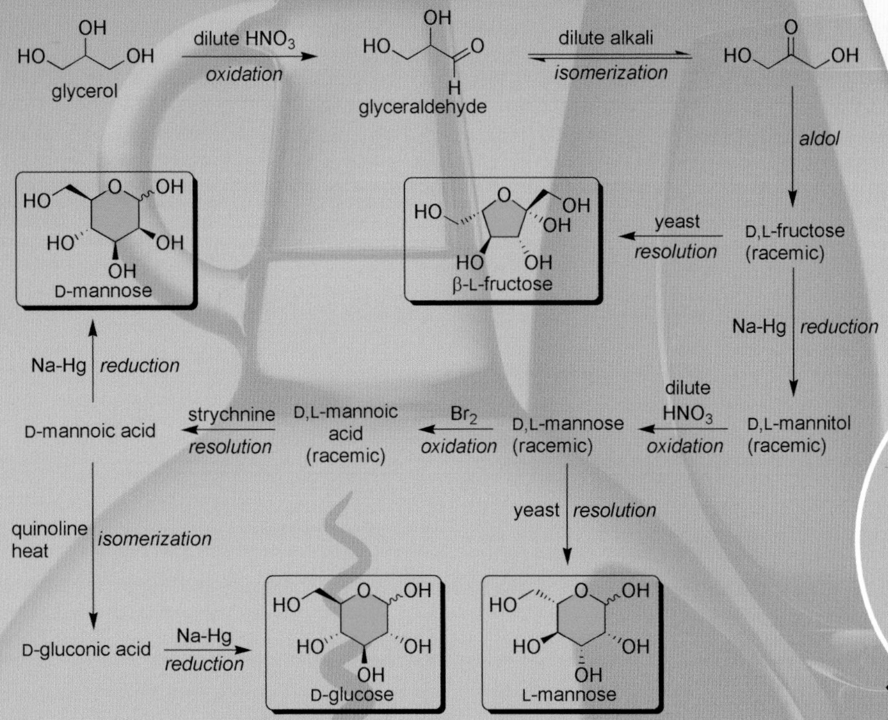

comprehensive for we have not even touched upon the many other saccharides and glycoconjugates (for example glycolipids) that contribute to a multitude of other important biological functions.

The total synthesis of glucose, Fischer's groundbreaking contribution to the field, began with glycerol, which he declared, "must be akin to grape sugar" (glucose) because of the apparent similarity in taste between the two substances. He was, however, also conscious of the fact that glycerol contained only half the requisite number of carbon atoms and did not have the necessary aldehyde functionality. Fischer hoped that through "gentle oxidation" he might be able to effect the desired conversion of glycerol into grape sugar. Indeed, his experiments bore fruit, as glycerol could be oxidized with dilute nitric acid to afford glyceraldehyde, which could be dimerized by means of an aldol reaction (see Chapter 31) upon treatment with dilute alkali to yield racemic fructose (Box 3). Fischer went on to prepare a range of other hexoses via a rather complex series of synthetic transformations, summarized in Box 3.

Fischer's accomplishments in sugar chemistry significantly advanced the understanding of stereochemistry and asymmetry around carbon atoms, taking this subject far beyond the observations of Louis Pasteur, Joseph-Achille Le Bel, and Jacobus H. van't Hoff. Pasteur had separated the two enantiomers of tartaric

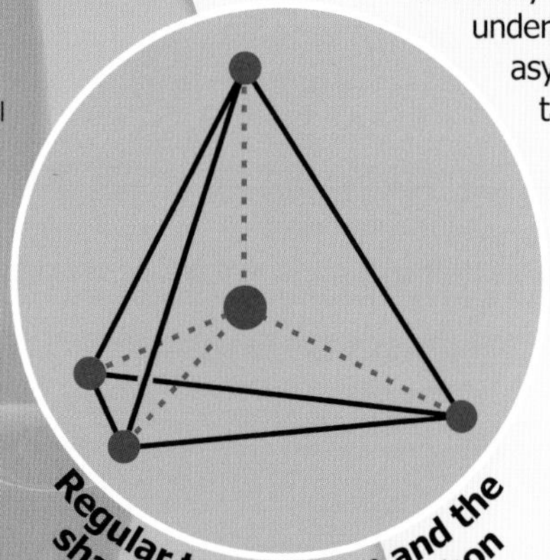

Regular tetrahedron and the shape of tetrahedral carbon

Louis Pasteur

Joseph-Achille Le Bel

Jacobus H. van't Hoff

acid in 1848 and shown that they rotated plane polarized light in opposite directions. He speculated that this phenomenon was due to the molecule having an asymmetrical structure. Le Bel and van't Hoff identified the exact source of the asymmetry in their work, which revealed the tetrahedral geometry of saturated carbon atoms. When a carbon atom in a molecule has four different substituents, that molecule can exist in two different configurations, called enantiomers. Like our hands, these two forms are mirror images that cannot be superimposed upon one another. Therein lies the source of asymmetry that dominates molecules in nature, which are generally found as single enantiomers, rather than racemic mixtures (1:1 mixtures of enantiomers). Fischer advanced the field further, beyond these initial concepts, by applying one of his earliest discoveries, the reaction of phenylhydrazine with α-hydroxy aldehydes. He used this reaction, in combination with various degradation, reconstitution, and resolution (chemical or enzymatic) reactions, to establish the configurations of the hexoses and their relationships. Through his extensive studies, he synthesized and isolated missing, and sometimes rare, members of this class of compounds so that ultimately a complete set of data was obtained, and the configurations of the hexoses, including glucose, were finally solved. This seminal body of work demystified a major chemical enigma of the time. Alongside his experimental accomplishments, Fischer also introduced a new graphical way to represent carbohydrate structures. Known as the Fischer projection, this style of molecular depiction is still in use today.

Fischer's frequent use of enzymatic resolution in his sugar research had been inspired by a long-standing interest he held in the chemistry of yeasts, which had originated from his father's wish to instill in his son a thorough understanding of fermentation. His father's interest arose from a desire to capitalize on his involvement in the newly founded, and highly profitable, beer industry in Dortmund (Germany). This aspect of Fischer's multifaceted scientific life led to another of his groundbreaking theories, one that constitutes a fundamental principle in biochemistry today, namely the "lock and key" analogy for the specificity of enzyme action. In addition, his accomplishments remind us how chemistry in general, and these enzymatic reactions in particular, have been central to our lives since the dawn of civilization, when our ancestors first began to leaven bread and ferment fruits to produce wines and other spirits. However, it is only in more recent times that these pivotal biotransformations have been understood at the molecular level (Box 4).

D-glucose

Fischer projection

β-D-fructose

Depiction of the "lock and key" analogy for the specificity of enzyme action

Box 4 Yeast and D-glucose in baking and fermentation

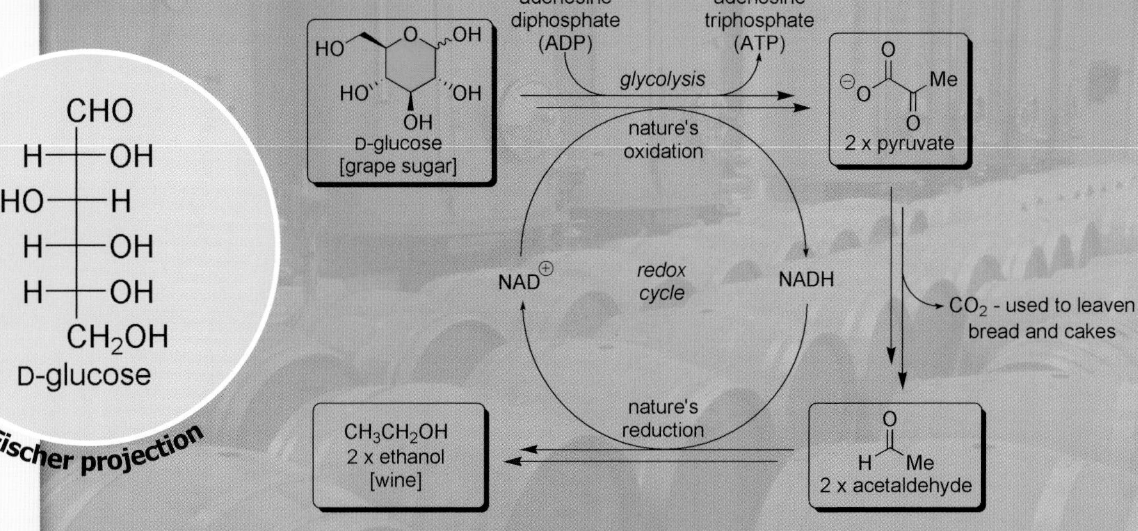

Box 5 Important natural products synthesized by Emil Fischer

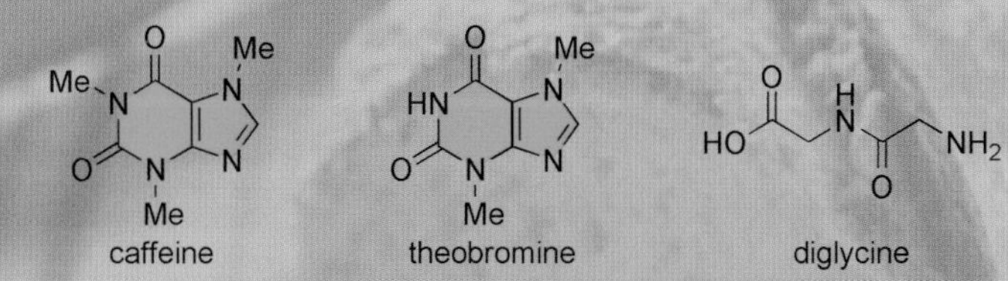

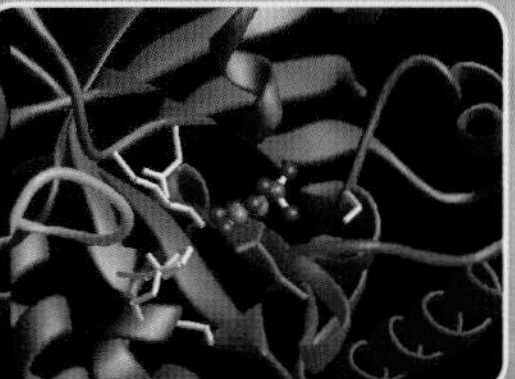

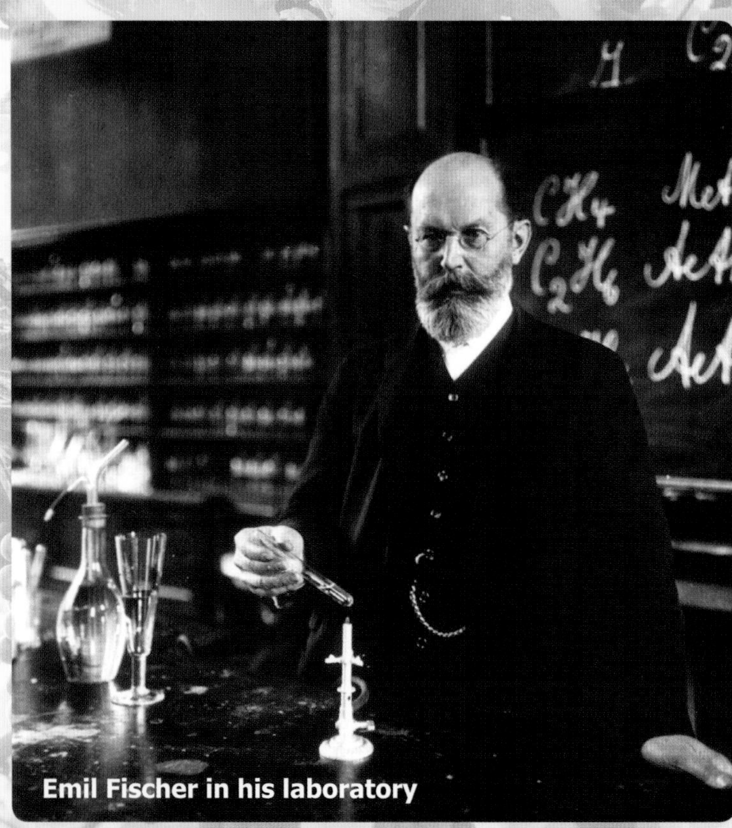

Emil Fischer in his laboratory

Further Reading

Hermann Emil Fischer's Nobel Lecture. The transcript of this lecture can be accessed via http://nobelprize.org/chemistry/laureates/1902/index.html.

J. K. N. Jones, W. A. Szarek, in *The Total Synthesis of Natural Products, Vol. 1* (Ed. J. ApSimon), Wiley, New York, **1973**, pp. 1–80.

F. W. Lichtenthaler, Emil Fischer's Proof of the Configuration of Sugars: A Centennial Tribute, *Angew. Chem. Int. Ed. Engl.* **1992**, *31*, 1541–1556.

F. W. Lichtenthaler, 100 Years "Schlüssel-Schloss-Prinzip": What Made Emil Fischer Use this Analogy?, *Angew. Chem. Int. Ed. Engl.* **1994**, *33*, 2364–2374.

K. C. Nicolaou, E. J. Sorensen, *Classics in Total Synthesis*, Wiley-VCH Publishers, Weinheim, **1996**, pp. 293–315.

D. L. Nelson, M. M. Cox, *Lehninger Principles of Biochemistry*, 4th Ed., W. H. Freeman, New York, **2005**.

As mentioned above, Emil Fischer worked on many projects besides sugars. In the early years of his brilliant career he analyzed and synthesized a number of purines, including the active constituents of coffee and cocoa (caffeine and theobromine, respectively, Box 5). Later, he followed these investigations with the initial forays into the important and highly influential field of peptide synthesis. With the French chemist Ernest Fourneau he took the first tentative steps when they synthesized the amino acid dimer, glycyl glycine (also called diglycine). You will have noticed that, with all these apparently disparate topics of research, Fischer was probing all of the most important biological macromolecules; indeed, he succeeded in defining several major classes and in giving initial clues as to their function. He uncovered some of the first secrets of proteins, carbohydrates, and the genetic code. His vision was to continue to learn and appreciate both the beauty and complexity of nature, and to recognize the awesome power that such comprehension could bring to humanity. Fischer was not only a distinguished experimentalist and visionary chemist, but also an insightful biologist. A sagacious physiologist, he derived much inspiration and pleasure from watching the workings of nature. His innate talent for seeing beyond the confines of his day was impressive and is aptly illustrated by the story of how, only a few weeks before he travelled to Stockholm to accept the Nobel Prize in 1902, he synthesized glucosamine and was excited by how this compound bridged the worlds of carbohydrates and proteins. His discovery prompted him to include in his Nobel lecture the remark that, "...nevertheless, the chemical enigma of Life will not be solved until organic chemistry has mastered another, even more difficult subject, the proteins, in the same way as it has mastered carbohydrates." With these prophetic words he clearly defined a set of objectives that the scientific community at large would grapple with for many years to come, and indeed continues to study (under the newly coined term for this sub-discipline of chemistry and biology, proteomics). However, the confidence he placed in science's mastery of carbohydrates may have been premature, for such molecules are still occupying the minds and time of many investigators today.

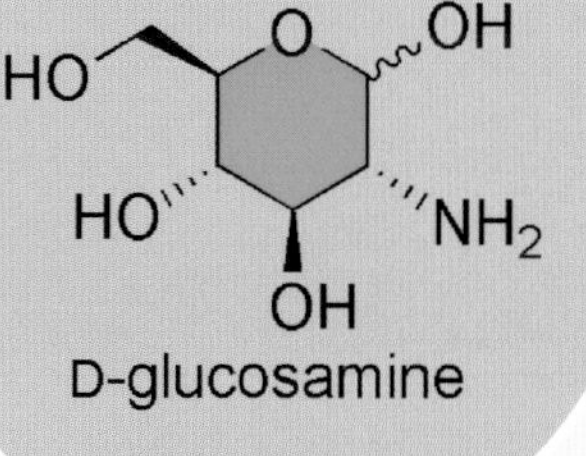

D-glucosamine

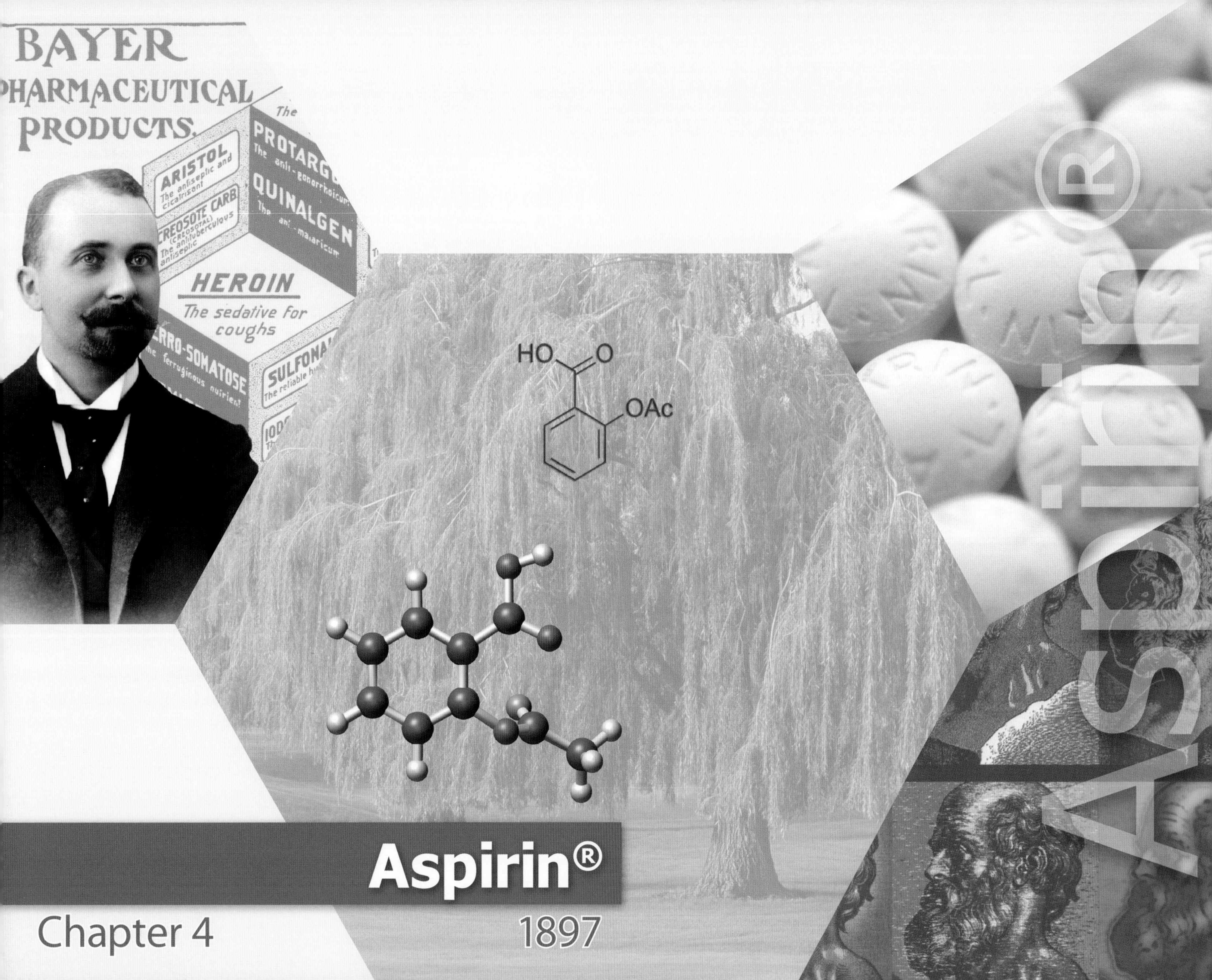

Aspirin®

Chapter 4

1897

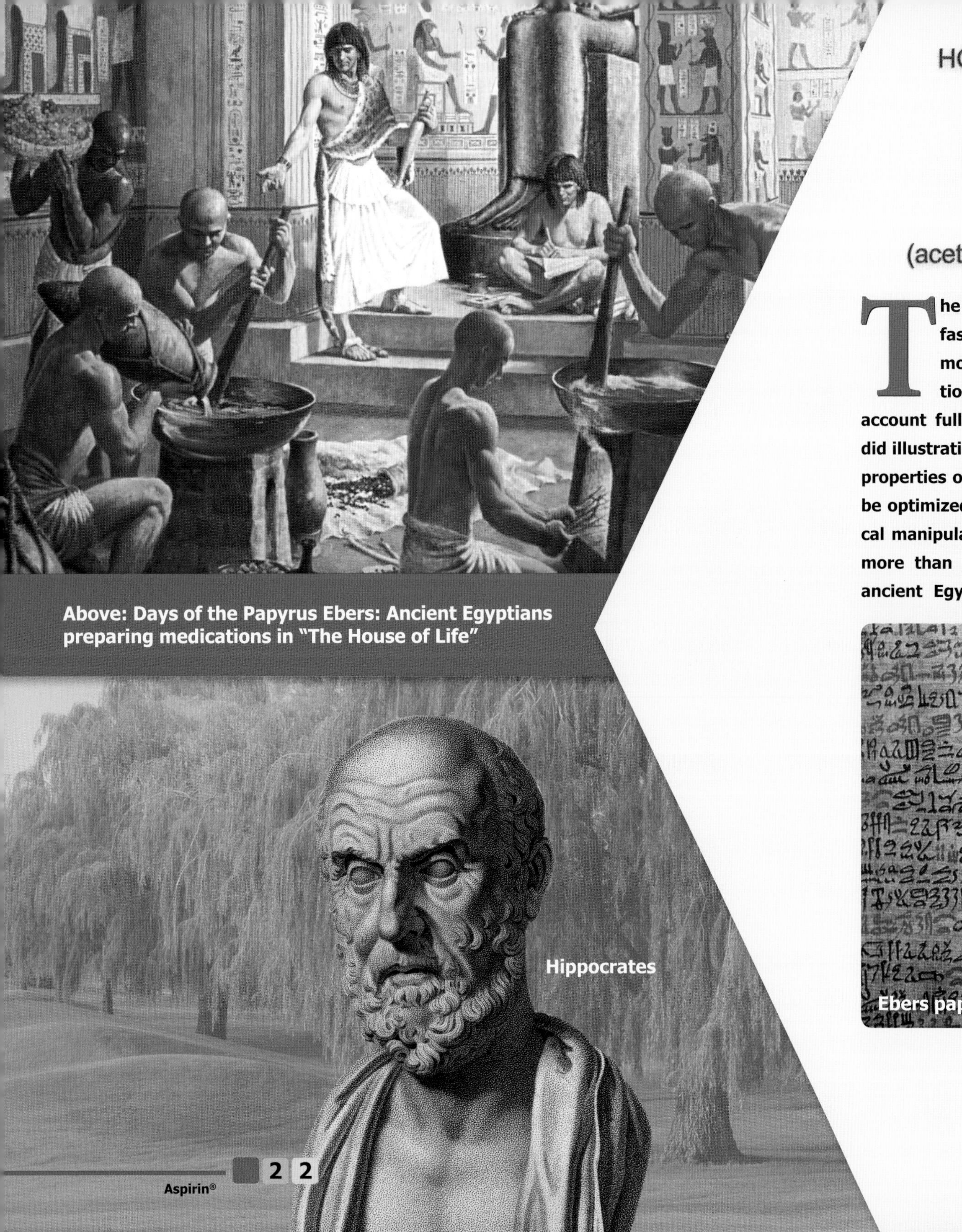

Above: Days of the Papyrus Ebers: Ancient Egyptians preparing medications in "The House of Life"

Aspirin®
(acetylsalicylic acid)

The story of Aspirin® is a fascinating tale about the most successful medication in history. It is an account full of firsts, and a splendid illustration of how the medicinal properties of a natural product may be optimized through subtle chemical manipulations. The story begins more than 3,500 years ago when ancient Egyptian physicians advocated salicin, in the form of herbal preparations, as a remedy for rheumatism and back pain. The Ebers papyrus, a collection of 877 medicinal recipes dating from this early period, describes the preparation of a crude infusion of myrtle bark. Of course, these ancient people had no knowledge of the active ingredient of their remedy, or even an understanding that its effects could be attributed to a single component. This situation was set to prevail until the nineteenth century, when chemistry came of age. One thousand years after the initial report, the father of all doctors, Hippocrates of Kos, prescribed the use of willow bark extracts for pain, fever, and childbirth. Subsequently, the willow tree remained the favored natural source for this tonic; indeed, the name salicin is derived from the Latin name for the willow family, *Salicaceae*. The beneficial effects of salicin were prized in Europe and throughout Asia (particularly China), as well as amongst the North American Indians and the Hottentot people of South Africa. All these cultures have passed on folklore formulae for the relief of pain and fever using specimens now known to contain salicin (Box 1).

Ebers papyrus

In 1763, the Royal Society of London published a landmark report by Edward Stone, an English clergyman. He described how he had successfully administered an extract of pulverized dried willow bark in water, tea, or beer to fifty patients, to treat fever. This example constitutes the first recorded clinical trial (see also Chapter 16, Vitamin B_{12}). By the early nineteenth century, chemistry had developed into a laboratory-based science and was ready to tackle the problem of identifying the active component of such medicines. In 1828, Johann Andreas Buchner, a professor of pharmacy in Munich, prepared an extract of willow bark and, by meticulously removing tannins and other impurities, obtained a relatively pure sample of a yellow substance, which he called salicin. Ten years later at the Sorbonne in Paris, Italian chemist Raffaele Piria made the next significant advance. He successfully hydrolyzed salicin, splitting it into its sugar and phenolic components, and later he succeeded in oxidizing the hydroxymethyl group of the phenolic fragment to afford salicylic acid (Box 1). In 1853, Charles Frédéric Gerhardt from Strasbourg became the first person to prepare acetylsalicylic acid (later named Aspirin®). However, he neither knew the structure of his product, nor how to purify (and thereby stabilize) it. The German chemist Karl-Johann Kraut later managed to crystallize acetylsalicylic acid, but did not apply his material as an analgesic. Thus, for some years to come, salicin-based medicines continued to be derived from natural sources.

Crystals of acetylsalicylic acid

By this time the harvest of salicin now came from a variety of plants, including American teaberry (*Gaultheria procumbens*) and meadowsweet (*Filipendula ulmaria*), as well as the original favorite, willow (*Salix alba*).

Hermann Kolbe, the renowned nineteenth century German chemist (and co-star of our second chapter), was the first to introduce the term 'synthesis' into common chemical usage, and also played a decisive role in the history of Aspirin®. His writings reveal the strength of character that prompted him to redefine many of the prevailing chemical theories of the time, and, not infrequently, to severely criticize his fellow practitioners of chemistry for what he saw as their many shortcomings. Kolbe was a student of Friedrich Wöhler, the first person to transform an inorganic substance to an organic compound (Chapter 2), which may explain why he so keenly rejected vitalism. He described this pseudo-metaphysical explanation for the behavior of matter as "fanciful nonsense." Kolbe preferred to place his trust in mathematically precise compound formulae and new structural theories. Much of his residual fame relates to his synthesis of acetic acid (Chapter 2) and his work involving the reductive electrolytic coupling of carboxylic acids, but

Hermann Kolbe

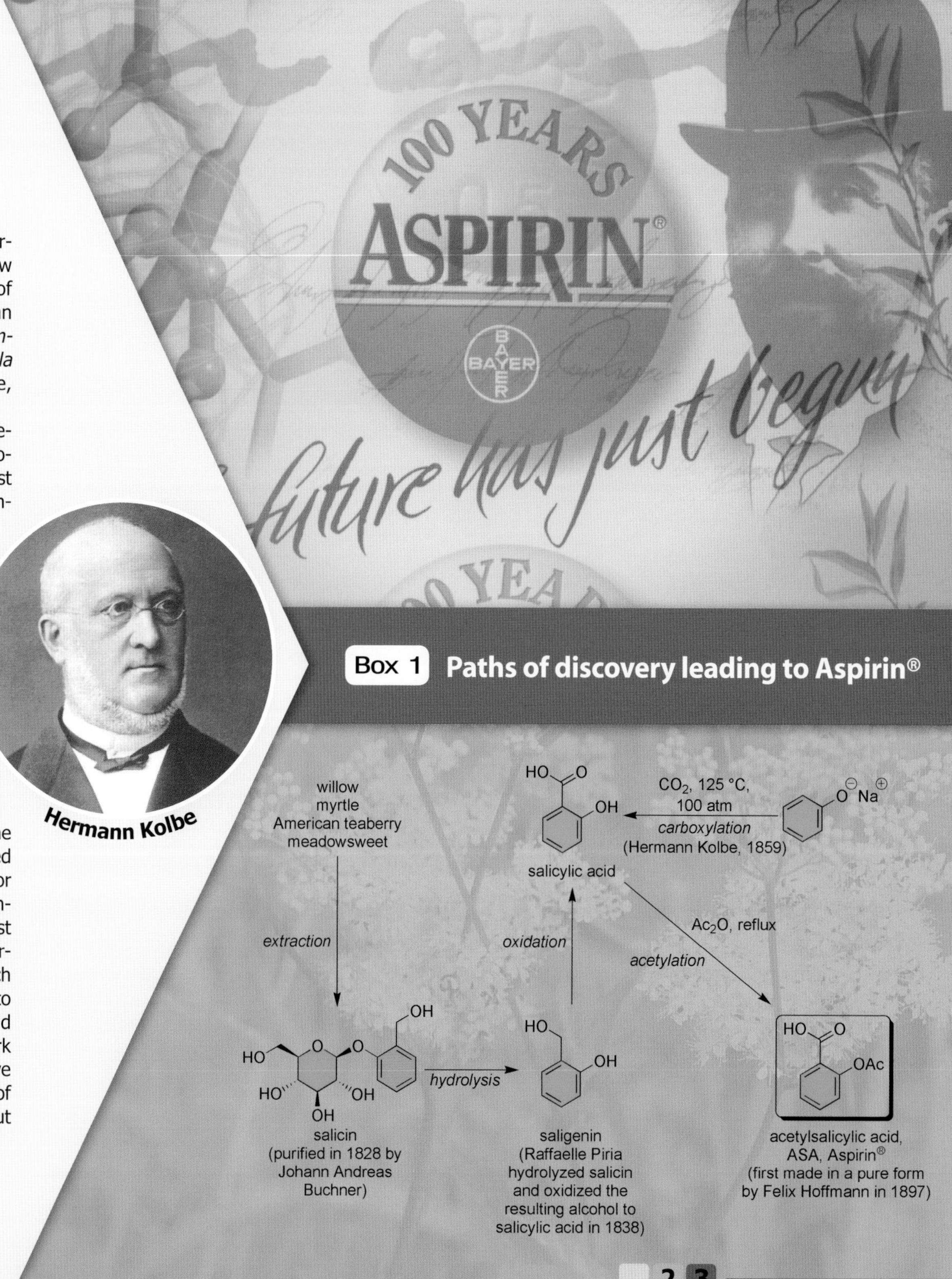

Box 2 The arachidonic acid cascade

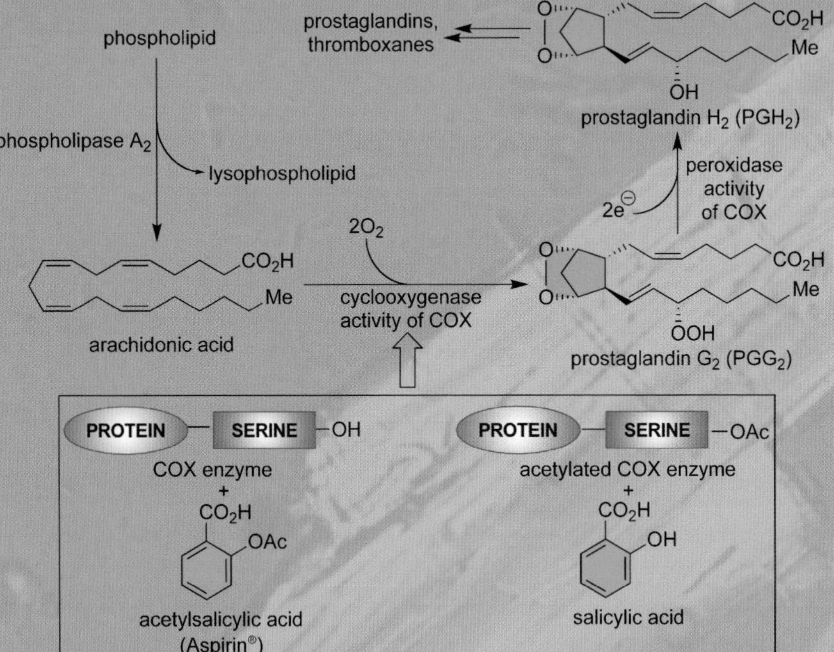

perhaps his most lasting contribution to chemistry was his preparation of salicylic acid in 1859. Kolbe accomplished this landmark synthesis by heating the sodium salt of phenol in the presence of carbon dioxide under pressure to form the crucial carbon–carbon bond. This feat also enabled him to deduce the correct structure of salicylic acid (Box 1).

Kolbe, along with the student of a colleague, Friedrich von Heyden, initiated the industrial production of salicylic acid by what came to be known as the Kolbe Process. This process allowed the production of salicylic acid at a tenth of the price of natural extracts, laying the foundation of the powerful pharmaceutical industry of today. With the increased availability of salicylic acid, physicians were able to add toothache and migraine to the long list of complaints alleviated by this medication, leading to a six-fold increase in annual salicylic acid production at the Heyden plant to 24,000 kilograms in the four years to 1878.

Salicylic acid, however, was not quite the panacea that it had been portrayed up to this point due to a number of side effects, including a particularly foul taste. It also caused irritation to the mucosal membranes of the digestive tract, leading to vomiting and ulceration in some patients. At that time, the relatively youthful company Bayer, which had its beginnings in the production of dyes on an industrial scale (see Chapter 8), had expanded to include agrochemical and pharmaceutical interests. These early medicinal chemists sought to modify the structure of salicylic acid in order to obtain a derivative that might be devoid of undesirable side effects. A number of such compounds were made and sold to patients, but the real breakthrough occurred in 1897, when Felix Hoffmann, a synthetic organic chemist at Bayer, synthesized acetylsalicylic acid (ASA) and was skillful enough to crystallize it, so-obtaining pure material suitable for therapeutic use. Hoffmann's ASA was shown to be free from the side effects that had plagued its predecessors, and Bayer was afforded a 'miracle' drug, which they named Aspirin®. It is said that this accomplishment was also a personal victory for Felix Hoffmann, whose severely rheumatic father had suffered from the discomfort of vomiting and a perforated stomach lining as a result of his prolonged use of sodium salicylate. Of course, there are still some side effects related to the use of Aspirin®, including some gut irritation when used for long periods. Aspirin® has also been linked to Reye's syndrome, a rare condition in children, therefore it is no longer recommended for the treatment of fever in children and teenagers.

On February 1, 1899, Aspirin® was registered as the trade name for acetylsalicylic acid in Germany, and a year later a crucial United States patent was issued. Patent applications filed by Bayer in the United Kingdom and Germany were denied after a vigorous campaign from the Heyden chemical factory, claiming that they had manufactured acetylsalicylic acid first. This proprietary dispute failed to stop Aspirin® from gaining common acceptance and ultimately the Heyden factory became the main supplier of salicylic acid to Bayer during the early twentieth century. Arthur Eichengrün, the head of pharmaceutical and chemical research, and Heinrich Dreser, director of Bayer's Pharmacological Institute, also played significant roles in the discovery and development of Aspirin®.

Aspirin® soon became a standard item in every household medicine cabinet. When Bayer's American plants were sold in 1919 as part of the reparations exacted from Germany after the First World War, Sterling Products of Wheeling, West Virginia, was willing to invest the unprecedented sum of $3 million for Bayer's US drug interests. Aspirin®, however, quickly became a widely available over-the-counter drug as Sterling failed to protect its newly acquired patent. In a strange twist of fate, the pharmaceutical company SmithKline-Beecham later acquired Sterling and sold its over-the-counter business back to Bayer for $1 billion. However, the enormous success of Aspirin® did not signal an end to research surrounding this simple, but powerful drug. Throughout the remainder of the twentieth century research into the mode of action of Aspirin® and increased knowledge of pain and inflammation mechanisms in the body led to an ever-increasing number of conditions treated by this remarkable drug.

In 1935, the Swedish physiologist Ulf Svante von Euler-Chelpin identified a new class of signaling substances that he isolated from sperm obtained from the prostate glands of sheep, a source that inspired the name he coined for these compounds, prostaglandins. Initially, he believed that there was only one prostaglandin; however, we now know of over thirty members of this class of hormones, and it is understood that almost every nucleated cell in our body produces these important chemical mediators (see Chapter 15). A further quarter of a century elapsed before von Euler-Chelpin's compatriot, Sune K. Bergstrøm, was able to formulate the correct structures for the first prostaglandins in 1962. The difficulties encountered in their isolation and structural elucidation arose from the fact that these substances are local hormones that exert their physiological effects near the site of formation and then rapidly decompose. Local emergencies within the body, caused by chemical or physical trauma, elicit the recruitment of enzymes called phospholipases. These lipases hydrolyze

Ulf Svante von Euler-Chelpin

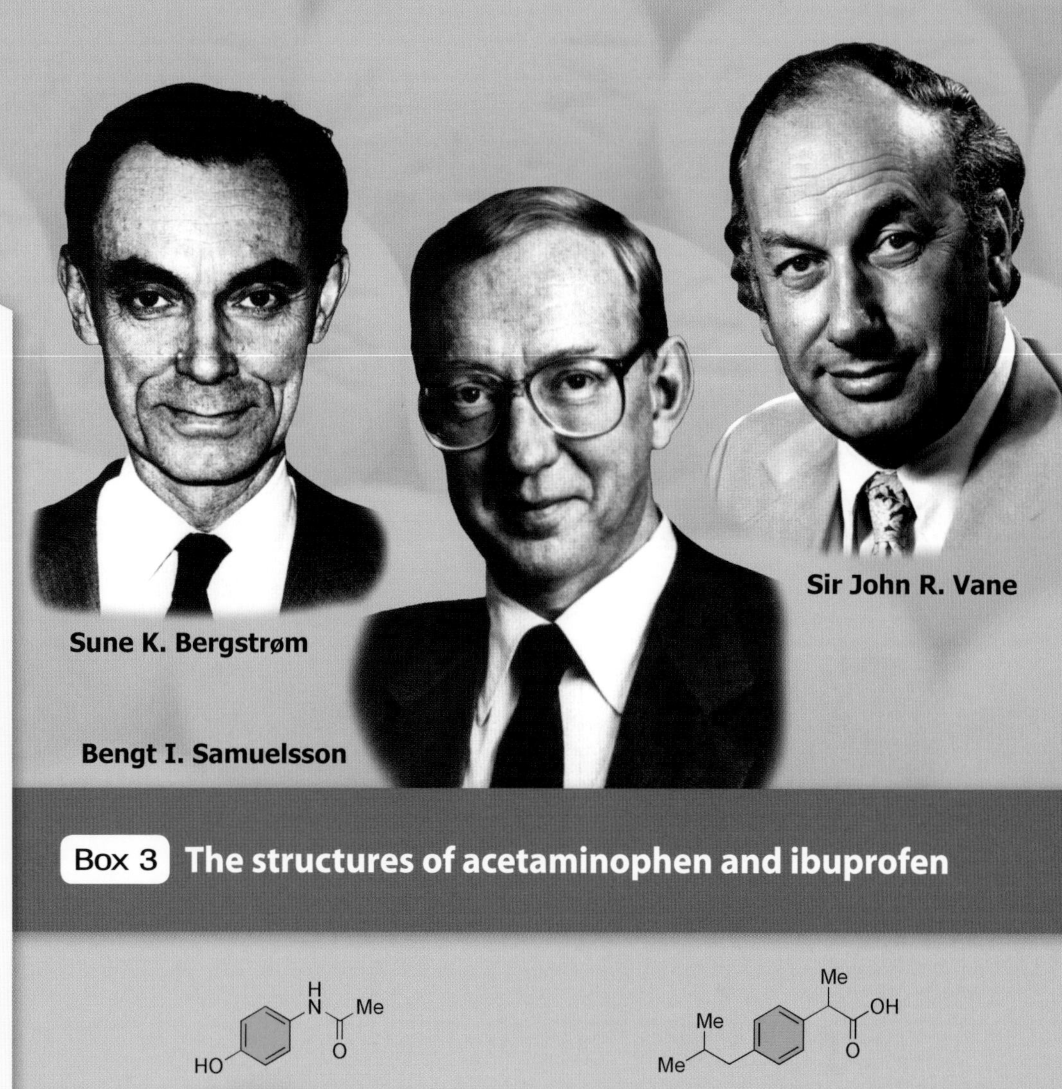

Sune K. Bergstrøm

Bengt I. Samuelsson

Sir John R. Vane

Box 3 The structures of acetaminophen and ibuprofen

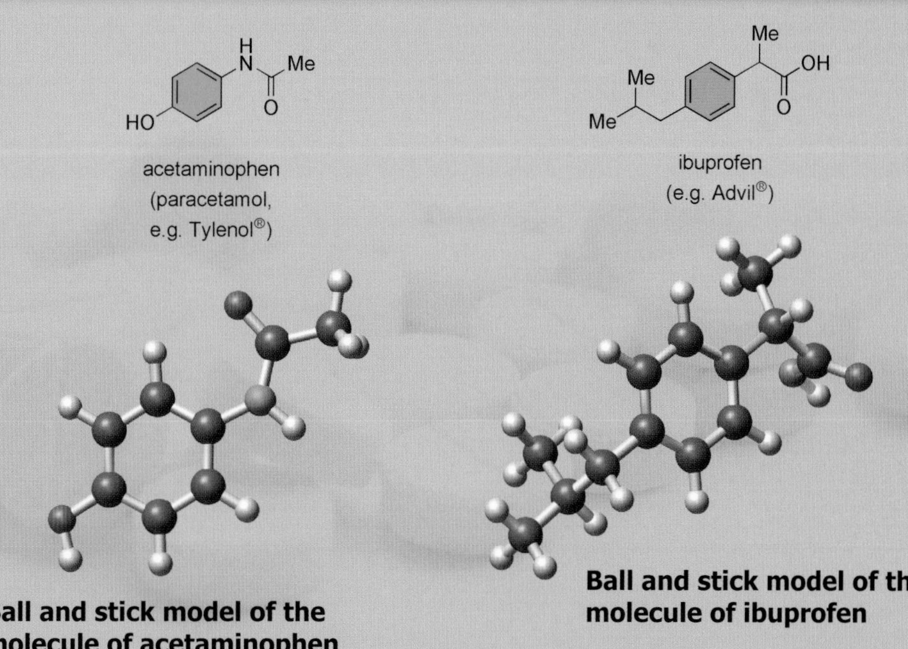

acetaminophen (paracetamol, e.g. Tylenol®)

ibuprofen (e.g. Advil®)

Ball and stick model of the molecule of acetaminophen

Ball and stick model of the molecule of ibuprofen

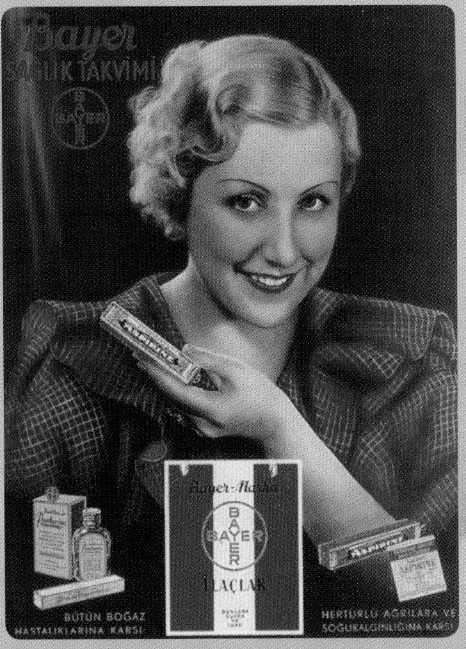

Box 4 The structures of the COX-2 inhibitors Vioxx® and Celebrex®

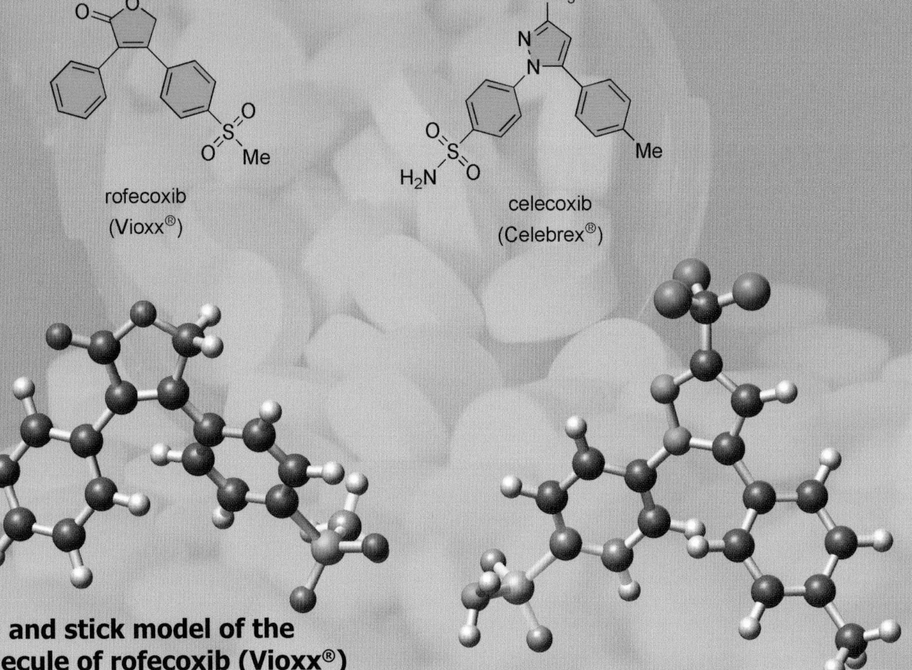

Ball and stick model of the molecule of rofecoxib (Vioxx®)

Ball and stick model of the molecule of celecoxib (Celebrex®)

phospholipids, releasing arachidonic acid, a 20-carbon fatty acid containing four double bonds (Box 2). In only a fraction of a second, the newly liberated arachidonic acid is converted into prostaglandins via a biochemical cascade. Bengt I. Samuelsson and his group at the Karolinska Institute in Stockholm elucidated most of these biosynthetic pathways. The sequence of events involves initial oxidation of arachidonic acid followed by cyclization to furnish a highly reactive cyclic peroxide intermediate, PGG_2. Subsequently, this compound is reduced to another peroxide, PGH_2, prior to further transformation into various prostaglandins and thromboxanes. A single 'super-enzyme' called cyclooxygenase (COX) catalyzes both the oxidation and cyclization steps involved in this essential process. In landmark papers published in 1971, Sir John R. Vane, J. Bryan Smith, and Anthony L. Willis linked the inhibition of this enzyme, at the cyclization step, to acetylsalicylic acid. Acetaminophen (or paracetamol, e.g. Tylenol®, Box 3) and ibuprofen (e.g. Advil®) also work by inhibiting a form of the COX enzyme complex. Later, some of the finer details were clarified as it was shown that the inhibition of this enzyme by Aspirin® occurs through acyl transfer from acetylsalicylic acid to a serine residue within the enzyme active site (Box 2). It was also found that there are at least three variants of the enzyme, COX-1, COX-2, and COX-3.

The identification of the COX-1 and COX-2 subclasses proved to be instrumental in the development of a new generation of 'super-aspirin' drugs with apparently improved properties. The rational design process used to develop these drugs provides an example of the drug discovery process as practiced in modern times. COX-1, a ubiquitous enzyme found throughout our body, is continuously on guard in order to ensure that the normal life cycle of each cell proceeds smoothly. COX-2 plays a different role, springing into action only in response to a specific need at a site of inflammation or cellular damage. Aspirin® irreversibly acetylates both COX-1 and COX-2 at a serine residue found at position 529 and 516, respectively, of the amino acid chain of the protein. This event distorts the docking site for arachidonic acid, thus disabling both these enzymes and preventing them from undertaking their catalytic action. Small differences in the gross structures of the two COX enzymes means that this acetylation has a more profound effect on COX-1 than on COX-2, with the former being inhibited much more strongly than the latter. This unfortunate bias exacerbates some of the side effects of Aspirin®. One such negative side effect is the prevention of the biosynthesis of the protective prostaglandin E_2 in the cells that line the stomach, leading to ulceration and other gastric disturbances when the drug is administered over a long period. The new generation of 'super-analgesic' drugs, represented by

Blood platelets and thrombus formation

Celebrex® and Vioxx® (Box 4), act as specific inhibitors of COX-2 and are, therefore, devoid of some of Aspirin®'s side effects, allowing their use in the long-term management of pain in conditions such as rheumatoid arthritis.

However, as with any new medication, unexpected side effects can appear, especially with long-term use. Recently, clinicians and scientists began to notice a link between Vioxx® and an increased incidence of cardiovascular emergencies (heart attacks and strokes) in patients taking the drug for periods of longer than 18 months. As a result, its manufacturer, Merck, voluntarily removed this popular drug (sales had reached $2.5 billion by 2003) from the world market for unconditional use in September 2004. The implications of this problem with Vioxx® for the safe use of the other COX-2 inhibitors are not yet clear, and several other drugs have also been voluntarily withdrawn as a precaution, underscoring the severe challenges and safety issues associated with the introduction of new medications.

In contrast, Aspirin® is now frequently used to reduce the incidence of myocardial infarctions and strokes, which are associated with arterial occlusions. This advantageous effect, first noted in the 1950's, comes about by inhibition of the biosynthesis of another product from the arachidonic acid cascade, namely thromboxane A_2 (TXA_2), a hormone that was first discovered by Bengt I. Samuelsson in 1975 (Box 5). TXA_2 is a transient chemical messenger, formed within our blood by platelets, whose role is to participate in the complex cascade of events that coagulate blood. In a normal healthy individual this hormone is balanced by prostacyclin (PGI_2, Box 5), another important arachidonic acid metabolite, discovered by Sir John R. Vane in 1977. PGI_2 is a potent vasodilator which prevents platelets from adhering to the arterial wall. In vessels where there is damage from atherosclerosis, PGI_2 biosynthesis is reduced, leading to a preponderance of TXA_2 with the consequence that clot formation is promoted, increasing the risk of stroke or heart attack. The inhibitory effect of Aspirin® on the production of TXA_2 negates the potentially lethal disparity between PGI_2 and TXA_2. The prostaglandin pioneers, Bergstrøm, Samuelsson, and Vane, shared the Nobel Prize in Physiology or Medicine in 1982 for their work on the arachidonic acid metabolites, collectively known today as the eicosanoids. The latter term was coined by E. J. Corey, another pioneer in this field and also a Nobel Laureate, whom we will encounter several times throughout this book.

Aspirin® and its predecessors have saved lives and alleviated suffering for over 3,500 years. What makes this discovery all the more remarkable is that with each year that passes this wonder-drug seems to become even more popular, finding new avenues by which to promote good health. For example, Aspirin® has recently been advocated as an ally in the fight against cancer, for use in treating diabetes, and as an agent to combat pre-eclampsia (a complication of pregnancy). The story of Aspirin® provides a classic example of the discovery and optimization of a drug based on a lead from nature (a natural product), which is as enduring as the drug itself.

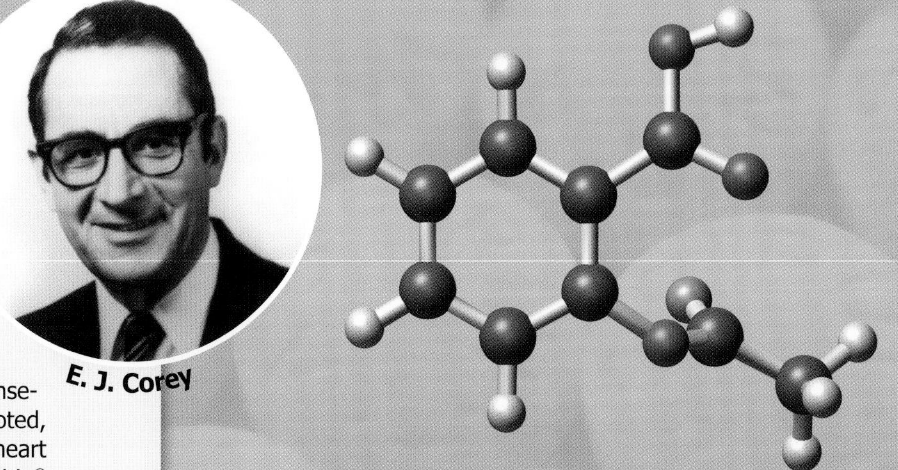

E. J. Corey

Ball and stick model of the molecule of acetylsalicylic acid (Aspirin®)

Box 5 The structures of thromboxane A_2 and prostacyclin

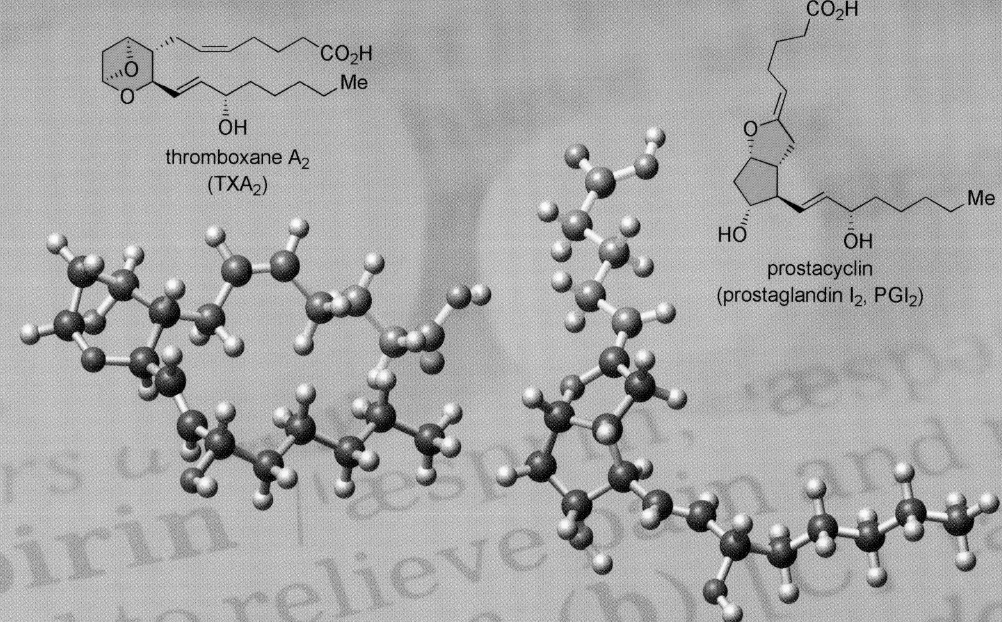

Ball and stick model of the molecule of thromboxane A_2

Ball and stick model of the molecule of prostacyclin

Further Reading

U. Zündorf, *100 Years of Aspirin. The Future Has Just Begun*, Bayer AG Consumer Care Business Group: Germany, **1997**.

G. Weissmann, Aspirin, *Sci. Am.* **1991**, *264*, 84–90.

K. D. Rainsford, *Aspirin and the Salicylates*, Butterworths, London, **1984**.

Me
Me
Me
O
Camphor
Chapter 5
1903

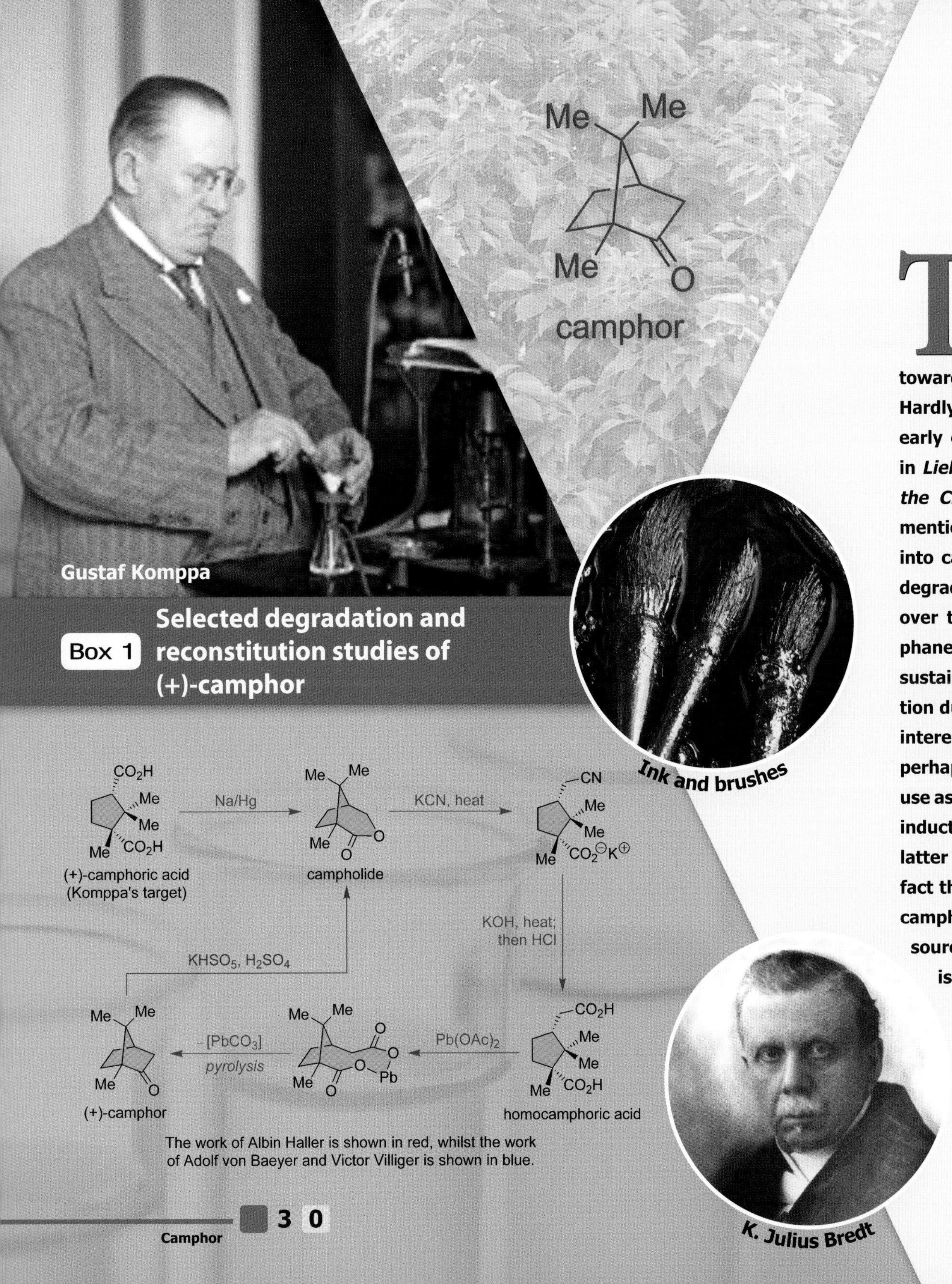

The years surrounding the start of the twentieth century saw a frenzy of activity amongst chemists directed towards the study of camphor. Hardly a page could be turned in the early chemical literature, published in *Liebigs Annalen* or the *Journal of the Chemical Society*, without the mention of some new investigation into camphor, its derivatives, or its degradation products. Strikingly, over the intervening century, camphane derivatives have generally sustained this high level of attention due not only to their varied and interesting physical properties, but, perhaps more importantly, to their use as powerful tools for asymmetric induction in organic synthesis. This latter application is helped by the fact that both of the enantiomers of camphor can be isolated from natural sources (D-(+)-camphor, however, is more widely distributed than L-(−)-camphor). Additionally, camphor has a very rigid skeleton, with faces that can be differentiated easily. The high profile of camphor is exemplified by work stemming from the 1920s in which K. Julius Bredt employed camphane and pinane substrates in the elimination reactions that enabled him to develop his famous rule forbidding olefinic bonds at the bridgehead positions of small bicyclic systems. After this research, camphor and its relatives continued to captivate chemists through their fascinating carbocationic rearrangements, interesting nuclear magnetic resonance (NMR) spectroscopic phenomena, and potential as starting materials for the synthesis of more complex molecular targets (see Taxol®, Chapter 25).

Camphor is a colorless, crystalline solid obtained from certain laurel trees, such as *Cinnamomum camphora*. This intriguing monoterpene has been known since the dawn of civilization, primarily because of the strong smell it imparts to the woods and oils that contain it. The ancient Chinese collected it from its natural sources for medicinal purposes as recounted in the thirteenth century by the great explorer of the orient Marco Polo. In 1571, the exiled Portuguese poet Luis de Cameons described it as the "balsam of disease." In Japan, it was used in scented oils and also in inks, since camphor oil can dissolve many pigments and resins that turpentine, the other common carrier, cannot. In Europe its uses were, and to some extent still are, equally diverse; camphor has had many common household functions from wood preservation and insect deterrence (moth balls), to its application as a therapy for a range of illnesses, including rheumatoid arthritis, sepsis, and hysteria.

Marco Polo

With such a rich history, it is not surprising that the young Finnish chemist Gustaf Komppa chose to spend much of his early career working towards the synthesis of camphor. Degradation and reconstitution studies undertaken in the 1890s, primarily by Albin Haller in France, had established that a certain camphoric acid (or its anhydride) could be converted to camphor by means of a sequence ending with *moderate pyrolysis* of the lead salt of homocamphoric acid (Box 1). Based on this knowledge, Komppa targeted this particular camphoric acid (at that time, different chemists named various compounds camphoric acid, which causes some confusion when reading old literature reports) since he knew that reaching this sub-target would constitute a formal synthesis of camphor. A formal synthesis is one that delivers an advanced intermediate whose conversion to the final target has previously been demonstrated. In 1901, with characteristic humility, Komppa admitted that after eight years of arduous toil directed towards his chosen target, "...these laborious and time-consuming experiments have produced few positive results so far. However now, I have synthesized an acid, that is identical to Bredt's apocamphoric acid, at last." Bredt's symmetrical acid is the targeted camphoric acid lacking the methyl group that is adjacent to one of its carboxylic acids, and thus all Komppa now had to do was to repeat this success with a compound bearing the missing methyl group. His dedication and perseverance were rewarded, as just two years later he was in a position to publish what has become his much celebrated synthesis of camphor. Komppa's solution to one of the era's most taxing synthetic challenges drew a comment from the famous English chemist William Henry Perkin, Jr. (see Chapter 6) voicing admiration and respect for the synthesis (Box 2). Such gracious admissions of defeat have been seen only very rarely in the highly competitive world of synthesis.

Komppa's synthesis, shown in Box 3, began with the fusion of a glutaric acid derivative with an unspecified oxalate ester to afford a five-membered cyclic dienol. Upon treatment of this newly formed substrate with sodium and then iodomethane, he obtained the *C*-methylated adduct. This claim was disputed for several decades because a number of chemists believed that Komppa had affected *O*-methylation instead of *C*-methylation, thus casting doubt on the entire synthetic sequence. The dispute was

Bredt's Rule: Forbids Certain Olefinic Bonds in Small Bicyclic Systems

Cinnamomum camphora

Box 2 Perkin's words regarding Komppa's synthesis of camphoric acid

"In the meantime, Komppa has published his brilliant synthesis of camphoric acid, which, once and for all, establishes the correctness of Bredt's formula, and it is therefore quite unnecessary to investigate our much less satisfactory process any further."

William Henry Perkin, Jr.

Box 3 Komppa's total synthesis of (±)-camphor (1900–1903)

oxalate ester + 3,3-dimethylglutaric acid dimethyl ester → (fusion) → dienol diester → (Na, MeI, EtOH) → methyl diketocamphorate (Komppa's structural assignment confirmed in 1968) → (Na/Hg, CO_2) → diol → (HI, cat. P) → unsaturated acids (or) → (HBr, AcOH) → bromo acid → (Zn, AcOH) → (±)-camphoric acid → (see Box 1) → (±)-camphor

Box 4 Modern industrial synthesis of (–)-camphor

(+)-α-pinene (from turpentine) —catalyst, heat→ camphene —AcOH→ isobornyl acetate —MeOH→ isoborneol —catalyst→ (–)-camphor

•Moth Balls
•Inks •Scents
•Wood Preservation
•Rheumatoid Arthritis
•Sepsis •Hysteria
•Embalming

Further Reading

A. Haller, Sur la transformation de l'acide camphorique droit en camphre droit; synthèse partielle du camphre, *Comptes rendus* **1896**, *122*, 446–452.

A. Haller, G. Blanc, Sur la transformation de l'acide camphorique au moyen de l'acide camphorique, *Comptes rendus* **1900**, *130*, 376–378.

G. Komppa, Die vollständige Synthese der Apocamphersäure resp. Camphopyrsäure, *Chem. Ber.* **1901**, *34*, 2472–2475.

G. Komppa, Die vollständige Synthese der Camphersäure und Dehydrocamphersäure, *Chem. Ber.* **1903**, *36*, 4332–4335.

K. Aghoramurthy, P. M. Lewis, The Komppa Synthesis of Camphoric Acid, *Tetrahedron Lett.* **1968**, *9*, 1415–1417.

J. ApSimon, *The Total Synthesis of Natural Products*, Vol. 2, Wiley-Interscience, New York, **1973**.

not finally settled until NMR studies, carried out in 1968, confirmed the original assignment of exclusive *C*-methylation, thereby vindicating Komppa, whose synthesis now stands unchallenged. In the next step of the synthesis, Komppa employed a double reduction, mediated by sodium amalgam under a stream of carbon dioxide, to convert the diketone into the corresponding diol with concomitant hydrolysis of the methyl esters. The newly acquired dihydroxy diacid was treated with hydroiodic acid and catalytic amounts of phosphorus to furnish an unsaturated diacid. Komppa was not clear which of the two possible regioisomers he had synthesized at this point; it was of little consequence, since both isomers would converge on a single product upon reduction. Thus, the precious camphoric acid was obtained, completing the formal synthesis of camphor.

Today, thousands of tons of camphor are produced synthetically from the more readily available monoterpene, α-pinene, which is a constituent of the turpentine collected from pine tar (Box 4). Although some of the vast amount of camphor produced today is still used medicinally, the bulk of this material is used as a preservative, an embalming agent, or a scent in religious rituals. The modern industrial synthesis of camphor is much less laborious and much more efficient than Komppa's original route, making it a viable alternative to harvesting camphor from its natural sources. However, we should not place too much stock in such comparisons, for that would mean ignoring how much we owe to the early chemical pioneers who laid the foundations of our current knowledge.

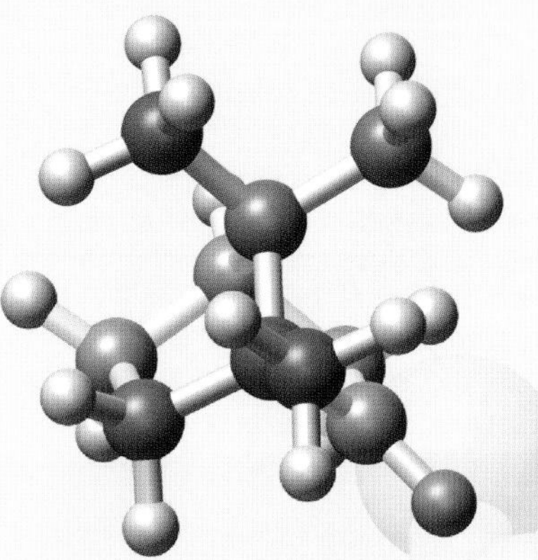

Ball and stick model of the molecule of (+)-camphor

Gustaf Komppa in classroom

Chapter 6 Terpineol

1904

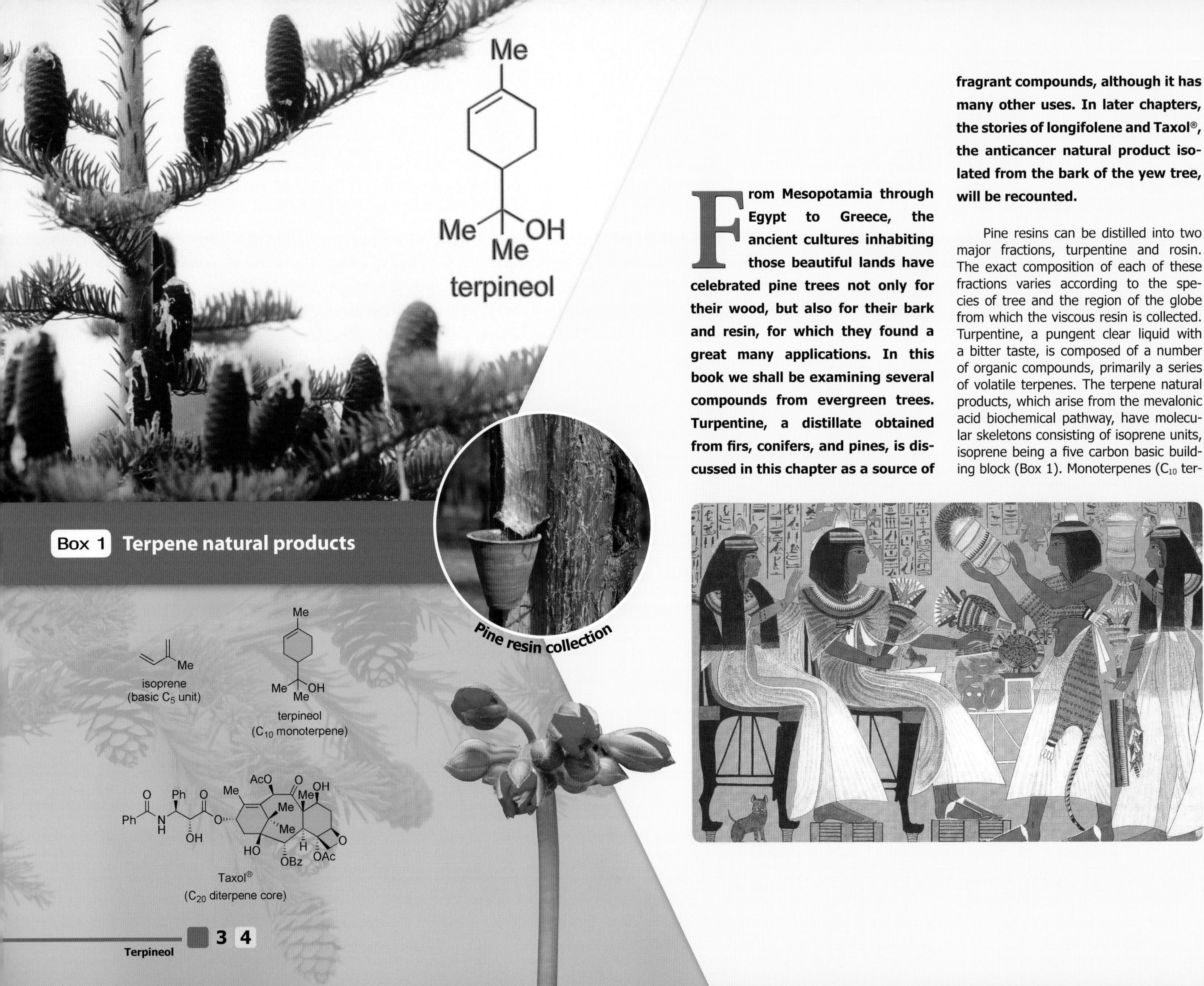

From Mesopotamia through Egypt to Greece, the ancient cultures inhabiting those beautiful lands have celebrated pine trees not only for their wood, but also for their bark and resin, for which they found a great many applications. In this book we shall be examining several compounds from evergreen trees. Turpentine, a distillate obtained from firs, conifers, and pines, is discussed in this chapter as a source of fragrant compounds, although it has many other uses. In later chapters, the stories of longifolene and Taxol®, the anticancer natural product isolated from the bark of the yew tree, will be recounted.

Pine resins can be distilled into two major fractions, turpentine and rosin. The exact composition of each of these fractions varies according to the species of tree and the region of the globe from which the viscous resin is collected. Turpentine, a pungent clear liquid with a bitter taste, is composed of a number of organic compounds, primarily a series of volatile terpenes. The terpene natural products, which arise from the mevalonic acid biochemical pathway, have molecular skeletons consisting of isoprene units, isoprene being a five carbon basic building block (Box 1). Monoterpenes (C_{10} ter-

Box 1 Terpene natural products

penes), such as terpineol, are unsaturated light oils, generally with heavy scents, that have been used since time immemorial as ingredients in paints, disinfectants, medicines, materials for religious rituals, and perfumes.

For thousands of years our ancestors have appreciated and experimented with perfumes in what is a continuing fascination with this luxury commodity. The art of blending fragrances was first cultivated in Mesopotamia and Egypt. Subsequently the knowledge spread to Greece and from there on to Rome. These traditions continued to be practiced and honed during the Middle Ages as the Arabs became experts at crafting exotic blends of perfumes. With the advent of distillation and other techniques, perfumery flourished in Renaissance Italy. However, as chemistry advanced and organic synthesis was born in the nineteenth century, the burgeoning chemical industry of Germany became a center for scent production while France became a hub for perfume blending. Today the coveted crown is shared between fierce rivals on opposing sides of the Atlantic; France, Switzerland, and Germany vie for superiority with the relative newcomer to the art, the United States. The Japanese are also fast becoming serious players in the art of perfumery.

The esteem in which perfumes were held throughout history is illustrated by the magnificent craftsmanship lavished on the vessels that contained these precious essences. Uncovering these treasures of artistry and skill has allowed historians to trace a vivid record of the appreciation of fine fragrances down through the ages and across cultures and civilizations. It is perhaps surprising that while these artifacts have changed dramatically over the centuries, the basic ingredients of the scents themselves have remained essentially the same. Thus, the aroma from the Lebanese cedar (*Cedrus libani*), so highly prized by the Mesopotamians and enjoyed by Cleopatra and Marcus Antony, is still a key component in some stylish scents of today. Various fashions in perfumery can be traced to different periods of history or geographic origins.

But what makes a perfume? In old times, before ethanol could be purified by distillation, crude fatty extracts were prepared by a technique known as enfleurage, wherein delicate floral oils were drawn out from petals. These scented oils were used in cosmetics and medicine. Modern fragrances are mixtures of essential oils and synthetic aroma chem-

Ancient Greek cosmetic box (pyxis)

Perfume bottles from the Middle Ages

Terpineol
- Paint Ingredients
- Oils for Religious Rituals
- Perfumes
- Scents
- Disinfectants
- Medicines

Terpenes

Marcus Antony and Cleopatra as portrayed by Richard Burton and Elizabeth Taylor

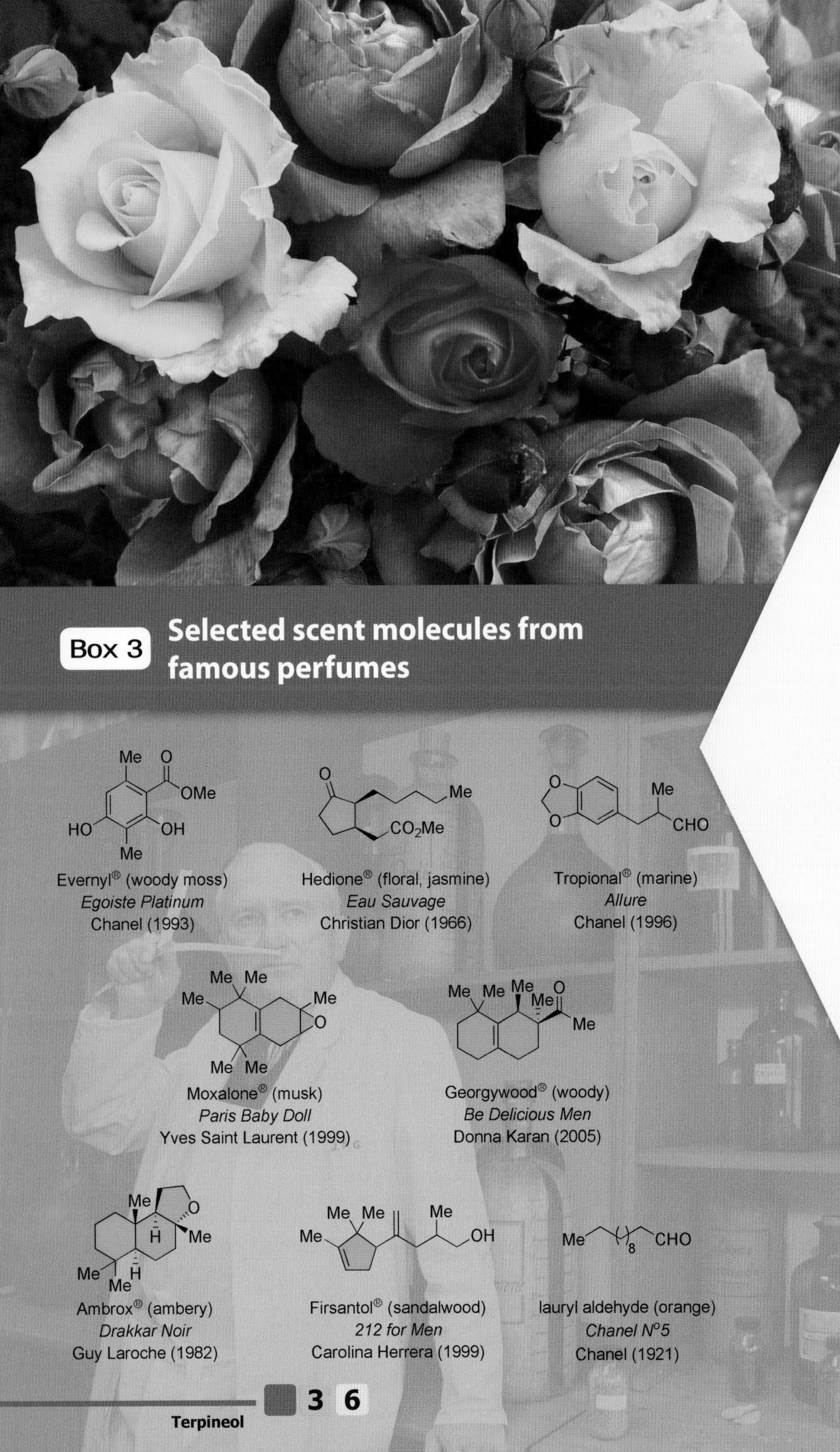

Box 3 Selected scent molecules from famous perfumes

Evernyl® (woody moss)
Egoiste Platinum
Chanel (1993)

Hedione® (floral, jasmine)
Eau Sauvage
Christian Dior (1966)

Tropional® (marine)
Allure
Chanel (1996)

Moxalone® (musk)
Paris Baby Doll
Yves Saint Laurent (1999)

Georgywood® (woody)
Be Delicious Men
Donna Karan (2005)

Ambrox® (ambery)
Drakkar Noir
Guy Laroche (1982)

Firsantol® (sandalwood)
212 for Men
Carolina Herrera (1999)

lauryl aldehyde (orange)
Chanel N°5
Chanel (1921)

icals diluted with ethanol and water. The description of the final perfume depends on its essential oil content. Thus, *eau de parfum* contains 15–30 percent of essential oils, *eau de toilette* has only 4–8 percent, while *eau de cologne* contains less than 5 percent of the expensive scented oils. The additional load (and thus expense) of an *eau de parfum* preparation provides a stronger and longer lasting bouquet by virtue of the higher concentrations of essential oil. Creating a perfume can be likened to the art of composing music, and the finest examples contain an intricate medley of chemical compounds that vary in molecular structure and, therefore, in aroma and volatility. The perfect design yields a layered blend that ensures the development of the scent on the skin with time.

Machines distilling perfume

The modern era for perfumery was ushered in by developments in chemistry beginning in the nineteenth century. Improvements in techniques allowed the fragile active ingredients to be extracted from their natural sources using organic solvents at ambient temperature for the first time, thus the selection of essential oils available was greatly enriched (Box 2). This advance was surpassed only when, in the second half of the nineteenth century, chemists first began to explore organic synthesis as a means to reproduce alluring natural scents. In 1868, British chemist William Henry Perkin, Sr., made coumarin (which smells of hay and woodruff) in the laboratory for the first time. This was followed by the syntheses of other naturally occurring fragrances: musk (1888), vanilla (1890), and violet (1893). Later, camphor (see Chapter 5) and terpineol were added to the growing list of synthetic accomplishments. The 1920s and 1930s fostered yet another creative surge, the influence of which still pervades today. Thus, fragrances such as the legendary Chanel Nº 5 (1921) and Joy (1931) set a new standard, ushering in an era of large scale production, distribution, and consumption of fine perfumes.

Today, these items constitute an intimate part of our lives and as we enjoy them we must not forget that their exquisite bouquets are a result of the symbiotic application of scientific and artistic talents. Synthetic chemists can recreate almost any known natural scent, and in addition they have designed and synthesized a whole host of new molecules with wonderful odors to complement those provided by nature, thereby furnishing an amazing library of pleasing smells. Chemistry, therefore, has been pivotal to the creation of perfumes, and has assisted in elevating them to their current popular and accessible status as items for everyone to enjoy (Box 3).

Simple monoterpenes extracted from natural sources are responsible for the characteristic smells of many plants and fruits, including citrus fruits, pine, spearmint, and geranium. Recognition of this feature spurred an intense interest in these molecules amongst chemists, who began to study them over one hundred years ago. Terpin hydrate captured the imagination of chemists in the last decades of the nineteenth century because not only was its constitution quite a puzzle, but also it was readily crystallized from crude turpentine in a very pure form. For many years chemists (especially William A. Tilden, Otto Wallach, and Georg Wagner) toiled and argued, seeking a solution to the conundrum of the correct formulae to represent this and other simple related terpenes. (Box 4).

William Henry Perkin, Sr.

William Henry Perkin, Jr., had experimental chemistry in his blood. Along with the synthesis of coumarin, his father, William, Sr., had discovered mauveine, the purple dye that revolutionized the fashion world and started an industry in the late nineteenth century (see Chapter 8). William, Sr., passed his enthusiasm for the unknown on to his son, leading to William, Jr.'s, ambitious attempt to synthesize and confirm the structure of various terpenes (Box 4).

In a style resembling that still in use today for reporting a total synthesis, William Perkin, Jr., related his progress along his synthetic path to terpineol, step by step, starting from ethyl cyanoacetate and ethyl 3-iodopropionate (Box 5). Perkin's synthesis made use of the then recently discovered Grignard reaction (Box 6) on two occasions, each time using the Grignard reagent methylmagnesium iodide. The first such reaction involved the addition of this reagent to a ketone, forming a tertiary alcohol, which served as a precursor for the endocyclic alkene of terpineol. The second application of the Grignard chemistry converted an ester group to the tertiary alcohol of the target, thus completing the synthesis. In the same paper, Perkin reported the conversion of terpineol into two other monoterpenes. Dehydration of terpineol gave dipentene (the racemic form of limonene – a key component in many citrus oils), while hydration of terpineol gave the crystalline terpin hydrate (Box 5); thus he had conclusively determined the structures of all of these related terpenes. Throughout the sequence, Perkin described the intermediates in great detail, including such physical properties as melting points, crystal forms, color and smell. For example, he noted that synthetic terpineol had the same distinct smell of lilac characteristic of the natural material. This synthetic achievement is all the more impressive when one considers the relatively

Lilac

Box 4 William Henry Perkin, Jr., explains his motives in pursuing the synthesis of terpineol

"The present investigation was undertaken with the object of synthesising terpin, terpineol, and dipentene, not only on account of the interest which always attaches to syntheses of this kind, but also in the hope that a method of synthesis might be devised of such a simple kind that there would no longer be room for doubt as to the constitution of these important substances."

William Henry Perkin, Jr.

Box 5 Perkin's synthesis of terpineol, terpin hydrate and limonene

EtO CN
ethyl cyanoacetate
+
EtO I
ethyl 3-iodopropionate
CO_2Et
(1) MeMgI
(2) aq. HCl
Grignard reaction
HO Me
CO_2H
aq. HBr
Br Me
CO_2H
Me
CO_2Et
(1) MeMgI
(2) aq. HCl
Grignard reaction
HO Me
• H_2O
Me OH Me
terpin hydrate (crystals)
aq. H_2SO_4
Me
Me OH Me
terpineol (oil)
$KHSO_4$
Me
Me
dipentene (limonene, oil)

Box 6 The Grignard Reaction and the total synthesis of zeaxanthin

Victor Grignard

In 1901, François Auguste Victor Grignard published his groundbreaking doctoral thesis describing his studies on organomagnesium reagents. Grignard, born in Cherbourg (France) in 1871, was part of a generation of French chemists who made a significant imprint on chemical synthesis with the invention of a number of organometallic methods, many of which are still in use today. For example, Paul Sabatier, who would share the 1912 Nobel Prize in Chemistry with Grignard, developed the process of hydrogenation at around this time (see Chapter 28).

Grignard reagents are formed from alkyl, alkenyl, or aryl halides, by reaction with metallic magnesium in ethereal solvents. Although to this day the exact nature of the species formed is poorly understood, they react as nucleophiles, and are commonly employed in additions to carbonyl groups. In the century since the invention of the Grignard reagent, a number of similar reagents have been developed, notably organolithium compounds, which are also often used as strong bases. Grignard and organolithium reagents have also been employed as precursors to other organometallic species (for example, organocopper reagents), each of which has subtly different reactivity.

$$\mathrm{R{-}X} \xrightarrow[\text{ether}]{\text{Mg}} [\mathrm{R{-}MgX}] \xrightarrow{\mathrm{R^1C(O)R^2}} \mathrm{R{-}C(OH)R^1R^2}$$

R = alkyl, alkenyl, aryl; X = Cl, Br, I; [R–MgX] = *Grignard reagent*

Today, the Grignard reaction is applied, in its various guises, to the synthesis of many natural products, and industrial materials. The example below shows the application of a Grignard reagent (vinylmagnesium chloride) in BASF's synthesis of the carotenoid zeaxanthin, a compound used as a nutritional supplement.

(Scheme: epoxy ketone + vinyl-MgCl, *Grignard reaction* → tertiary allylic alcohol ⇒ zeaxanthin)

primitive analytical tools available in Perkin's time.

The importance of the structures established by Perkin's total synthesis may not have been fully appreciated until much later, when it became apparent just how many diverse terpenoids exist in nature and how many important roles they fulfill. The chemistry of terpenes has been greatly advanced since the early 1900s by many scientists, including Duilio Arigoni, whose brilliant synthetic and biosynthetic contributions have helped shape the field. Examples of this ubiquitous class of natural products include all the steroids, many vitamins, and miscellaneous compounds such as the anticancer agent Taxol®, mentioned above. In this book you will, time and time again, come across increasingly complex descendants of isoprene, the syntheses of which require sophisticated and ingenious strategies to master their much more complex structures. Perkin's synthesis delivered racemic terpineol (an equal mixture of both enantiomers). Chemists would later learn tricks allowing them to synthesize selectively one enantiomeric form of a given substance at will. So remember that, with Perkin's synthesis of terpineol, we are still at the very beginning of our journey through the synthesis archives.

Duilio Arigoni

Spindle bottle,
Late Cypriot IIA:2
(1410–1375 B.C.)

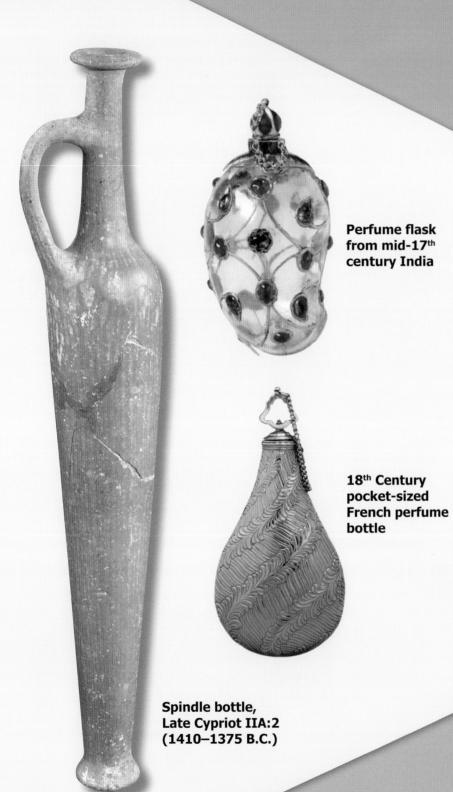

Perfume flask from mid-17th century India

18th Century pocket-sized French perfume bottle

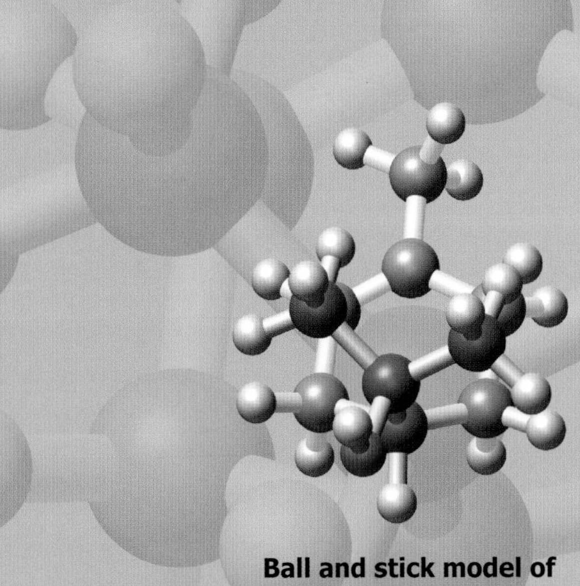

Ball and stick model of the molecule of terpineol

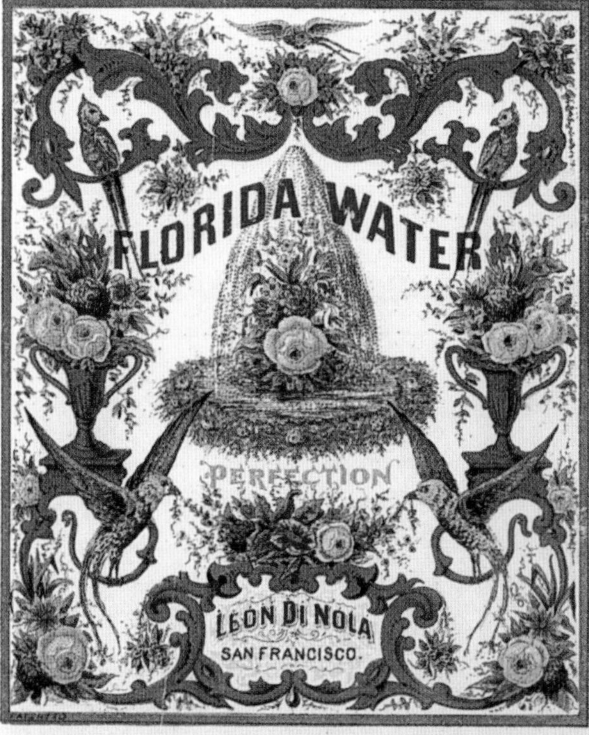

Further Reading

W. H. Perkin, Jr., LXVI. Experiments on the Synthesis of the Terpenes. Part I. Synthesis of Terpin, Inactive Terpineol, and Diterpene, *J. Chem. Soc.* **1904**, *85*, 654–671.

K. B. G. Torssell, *Natural Product Chemistry*, 2nd Ed., Apotekarsocieteten, Stockholm, **1997**, pp. 251–312.

E. T. Morris, *Scents of Time. Perfume from Ancient Egypt to the 21st Century*, Bulfinch Press, New York, **1999**.

P. Kraft, J. A. Bajgrowicz, C. Denis, G. Fráter, Odds and Trends: Recent Developments in the Chemistry of Odorants, *Angew. Chem. Int. Ed.* **2000**, *39*, 2980–3010.

Tropinone

Chapter 7

1917

Me–N

tropinone

Box 1 Sir Robert Robinson's wise words on the value of basic research

"The synthesis of brazilin would have no industrial value; its biological importance is problematical, but it is worth while to attempt it for the sufficient reason that we have no idea how to accomplish the task. There is a close analogy between organic chemistry in its relation to biochemistry and pure mathematics in its relation to physics. In both disciplines it is in the course of attack of the most difficult problems, without consideration of eventual applications, that new fundamental knowledge is most certainly garnered."

Sir Robert Robinson

Brazilwood sawdust provides dye pigment

Born in 1886 in the north of England, Sir Robert Robinson was one of the finest synthetic organic chemists ever known. Robinson was also a keen chess player and mountaineer, pursuits that reflected his innate desire to take on the toughest mental and physical challenges. He played an essential role in founding the discipline as we know it today and, quite apart from his countless successful syntheses of diverse natural products, he provided insights into mechanistic and electronic chemical theory, which contributed to our current appreciation of molecular structure and the course of reactions. In this chapter we shall explore, in addition to some of his contributions to the theory of organic chemistry, the total synthesis of tropinone, one of his many synthetic achievements. This synthesis ably illustrates Robinson's penetrating analysis of molecular structure. His mechanistic perceptions allowed him to design a beautiful and elegant cascade sequence to solve one of the most taxing synthetic challenges of his era.

Sir Robert Robinson's career is characterized not only by his incisive thought and brilliance (Box 1), but also by the tremendous energy that he directed towards his research. This creativity and vigor was probably garnered from his father. Robinson senior was a highly successful business owner, manufacturing surgical dressings in the north of England, and was described by his son as "... a tireless inventor who went every day to his mechanics shop...". Indeed, in 1902, when he entered the University of Manchester for his undergraduate studies, his father steered the young Robert towards the practical choice of chemistry, which could assist the family business, and away from his natural aptitude for physics and mathematics. At the University of Manchester,

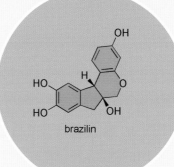

brazilin

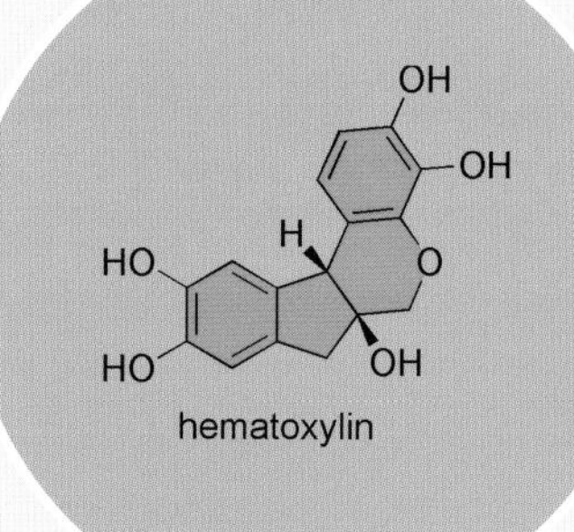

hematoxylin

Gilbert N. Lewis

Irving Langmuir

Sir Christopher Ingold

Robinson was mentored by William Henry Perkin, Jr. (see Chapter 6), who passed on superb experimental skills and intuition to his young protégé and ignited Robinson's interest in the synthesis of natural products. During the period spanning 1905 to 1909 Robinson produced the work of his doctoral thesis, concerned with the chemistry of brazilin, in Perkin's private laboratory. Brazilin was Robinson's own choice, since at this time Perkin had just succeeded in synthesizing limonene (Chapter 6) and wanted to move on to the synthesis of the rather closely related molecule, sylvestrene. Robinson, however, dismissed this project, saying, "Such work involved long and laborious preparations and would not have suited my case." Brazilin and the related compound hematoxylin are the precursors to the quinone species responsible for the color of brazilwood, and their selection for study is a reminder that during this period the residual effect of the dye industry boom, which had driven the popularization of organic chemistry during the latter decades of the preceding century, was still governing the research interests of synthetic chemists. It is also interesting that these two compounds, brazilin and hematoxylin, continued to intrigue Robinson for over half a century; he published a final paper, at last communicating their synthesis, in 1970!

A second influential individual in Robinson's early career was Arthur Lapworth, also a faculty member at the University of Manchester, for he was responsible for the development of Robinson's interest in the more theoretical aspects of organic chemistry. Lapworth used the undertaking of natural product syntheses as a vehicle to progress his objective of devising a rationale for why organic reactions occur as they do. He used physico-chemical techniques, such as reaction kinetics, to probe such issues, knowledge of which, it must be stressed, was conceptually very immature at that point in time. Indeed, it was only in 1916 that the seminal paper, "The Atom and the Molecule" by Gilbert N. Lewis, was published in the *Journal of the American Chemical Society*, detailing the concept of electron pairs and octets (the beginning of what we now know as Valence Bond Theory). These ideas took some time to proliferate and were not rapidly assimilated into general chemical understanding. Eventually, the Valence Bond Theory, developed by Gilbert N. Lewis, Irving Langmuir (Nobel Prize in Chemistry, 1932), and Sir Christopher Ingold, and cham-

Box 2 Examples of Robinson's theoretical contributions to organic chemistry and Kekulé's structure of benzene

Curly arrows, shown below, were introduced by Robinson to denote the movement of electrons during chemical reactions. This representation is still used today in scientific papers, and in the teaching of organic reaction mechanisms.

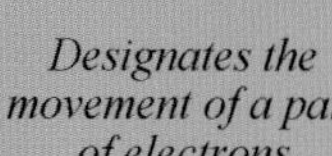
Designates the movement of a pair of electrons
Designates the movement of a single electron

"...These are, of course, the chief characteristics of benzenoid systems, and here the explanation is obviously that six electrons are able to form a group which resists disruption and may be called the aromatic sextet....Thus the symbols are each considered to express a view of one aspect of the problem...

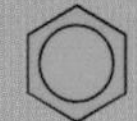

The circle in the ring symbolizes the view that the six electrons in the benzene molecule produce a stable association which is responsible for the aromatic character of the substance."
— Sir Robert Robinson

Sir Robert Robinson

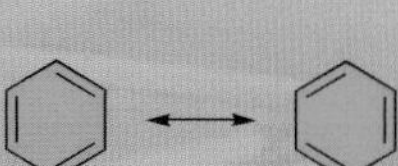

Kekulé's representation of the structure of benzene

Space filling model of benzene

Friedrich August Kekulé

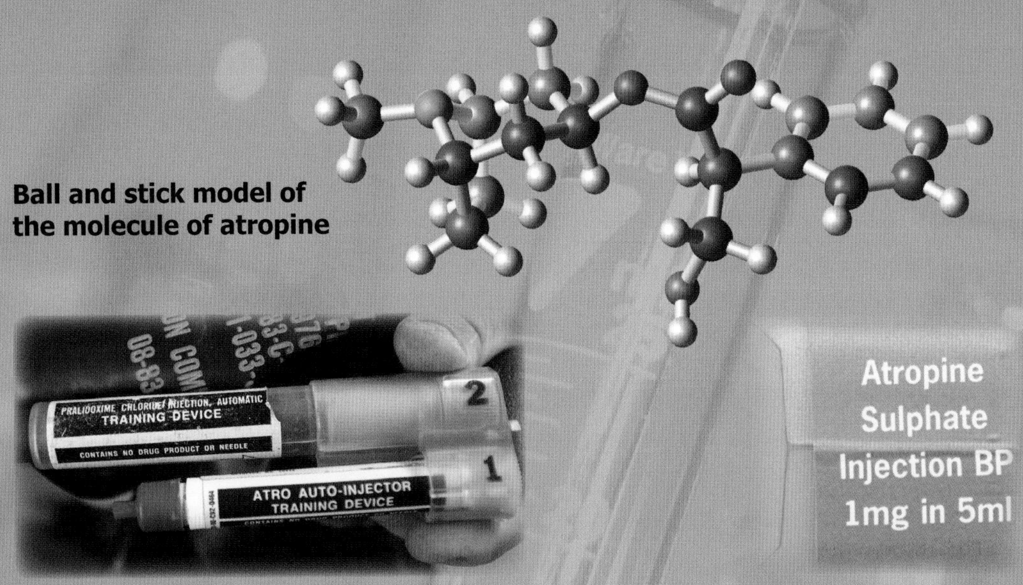

Ball and stick model of the molecule of atropine

Auto-injectors containing atropine and pralidoxime chloride, compounds that counter the effects of certain chemical weapons.

Box 3 Tropinone and its relationship to some other famous alkaloids

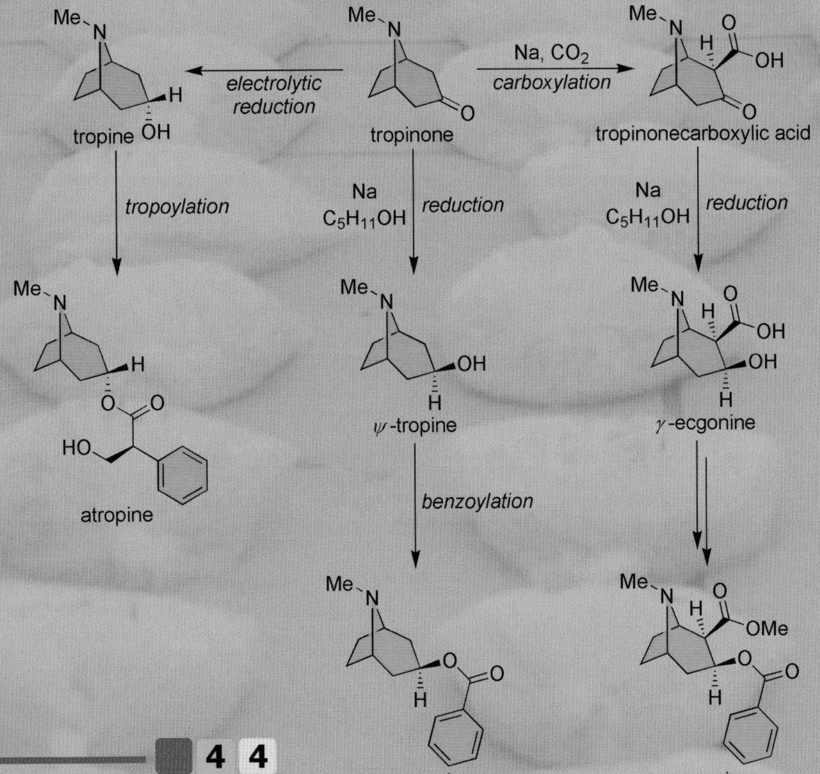

Linus Pauling

Erich Hückel

Robert S. Mulliken

pioned by Linus Pauling (Nobel Prizes in Chemistry, 1954, and Peace, 1962), gained widespread acceptance. Indeed, it was only overhauled in the 1960s and 1970s with the advent of Molecular Orbital (MO) Theory and quantum mechanics, theories to which other early theoretical chemists, such as Erich Hückel and Robert S. Mulliken (Nobel Prize in Chemistry, 1966), contributed. These complementary new theories offered a more complex, but also more accurate picture of molecular structure. In the late 1920s, Robinson withdrew from the heated discussions on bonding theory, which were filling the chemical literature, to concentrate on his primary passion of synthesis, but not before making his mark on this advancing field.

To today's student, perhaps the most recognizable of Robinson's theoretical contributions is the introduction of the curly arrow into common usage as a chemical notation to describe the course of a reaction (Box 2). Here an arrow represents the movement of a pair of electrons from a bond or atom to a new location during a chemical transformation. Later this notation was adapted for radical reactions, wherein a half arrow serves to impart the movement of a single electron. Thus, chemists can now write the likely progress of a reaction on paper, predicting and communicating the reactivity of specific molecules with ease. Robinson also understood aromaticity at a level well beyond that of many of his contemporaries. Several years before Erich Hückel had formulated his famous rule for discerning when a molecule is stabilized by 'aromatic' character (planar systems with cyclically conjugated π electrons will be aromatic when the number of these electrons can be represented by the formula $4n + 2$, where n is an integer), and before the extension of Valence Bond Theory to include ideas of resonance, Robinson had defined the unusual properties of benzene in electronic terms. His imagination went beyond Kekulé's benzene structure (Box 2), proposed in the eighteenth century as a result of the contributions of several researchers. Robinson had no difficulty in extending these ideas to include the sextet of electrons of pyridine, thiophene, pyrrole, and furan. He fully appreciated how the lone pair of electrons in the latter three compounds could interact to produce aromatic character in these heterocycles (see Chapter 8). However, it is for his respected synthetic triumphs that Robinson is best remembered.

Ball and stick model of the molecule of tropinone

The 1947 Nobel Prize in Chemistry was awarded to Robinson in recognition of his elegant investigations on plant products of biological importance, especially the alkaloids. In the words of Arne Fredga, then the chairman of the Nobel Prize Committee, "Among synthetic chemists you [Robinson] are today acknowledged as a leader and a teacher, and second to none." Due to the prevailing influence of the dye industry at the turn of the century, his interest had begun in the realm of plant dyestuffs, such as the anthocyanins, a group of red, blue, or violet pigments appearing throughout the plant world, and giving color to many specimens from cornflowers to beetroot. Subsequently, Robinson had become interested in a more diverse set of natural products, particularly those with medicinal properties. The isolation of alkaloids (alkaloid being a general name for natural products containing a basic nitrogen atom) had begun in the late nineteenth century and curiosity about their chemistry and properties has continued to fascinate and beguile such that, even today, many synthetic chemists still revel in the construction of the more complex members of this broad family of compounds. Robinson was responsible for solving the riddle of the structure of an infamous member of this family, morphine (Chapter 10), which at one stage was the subject of more than twenty structural proposals. After spending years clarifying many of the salient features related to the exact structure of the poison strychnine, he also elucidated its true structure in 1946 (Chapter 12).

One of Robinson's most brilliant total syntheses was that of tropinone, published in 1917 (see Boxes 4 and 5). Tropinone is a key member of the tropane alkaloid class of natural products (Box 3), isolated from various plants of the family Solanaceae, including *Atropa belladonna* (deadly nightshade) and *Datura stramonium* (jimsonweed) (Box 6). Tropinone was considered a pivotal synthetic target because from it one could make numerous other congeners of this class of alkaloids (Box 3), whose most famous member is probably cocaine, a molecule that certainly engendered interest due to its stimulant and analgesic properties. Atropine, another well known tropane alkaloid isolated from *Atropa belladonna*, is a deadly poison, but also has a number of medical applications. For example, it is used as a muscle relaxant in surgery

***Atropa belladonna* (deadly nightshade)**

***Datura stramonium* (jimsonweed)**

Richard Willstätter

Box 4 Robinson's analysis of the tropinone structure

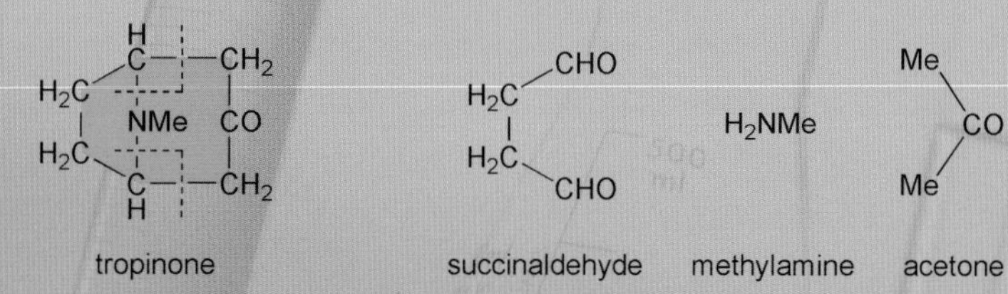

Box 5 Robinson's total synthesis of tropinone

Carl Wilhelm Scheele as an apprentice

Friedrich Wilhelm Adam Sertürner at work

and eye examinations, and as an antidote to mushroom, morphine, and nerve gas poisoning.

Among the earliest explorers in chemistry was a brilliant experimentalist from Scandinavia. Carl Wilhelm Scheele, a Swedish apothecary-pharmacist-chemist of the eighteenth century, contributed enormously to society through his pioneering discoveries that included the elements oxygen (O_2), chlorine (Cl_2), tungsten (W), and molybdenum (Mo), and numerous natural products such as the fruit acids prussic and tartaric, and also glycerin, and nitroglycerin.

If Scheele was the father of acidic substances from nature, Friedrich Wilhelm Adam Sertürner, a German apothecary - chemist can be considered as the father of the alkaloids, including morphine, naturally occurring substances that exhibit alkali-like properties such as forming salts with carboxylic and inorganic acids.

Richard Willstätter (the winner of the 1915 Nobel Prize in Chemistry) had previously accomplished some noteworthy achievements in connection with unraveling the chemistry of the atropine class, including a long synthesis of tropinone. Robinson, however, clearly considered that he could improve upon Willstätter's rather complex and arduous synthesis. The paper that Robinson published on the synthesis of tropinone gives us a glimpse into the prophetic mind of this chemical genius, for from his analysis of tropinone (Box 4) he concludes "... an inspection of the formula of tropinone (I) discloses a degree of symmetry and an architecture which justify the hope that the base may ultimately be obtained in good yield as the product of some simple reaction and from accessible materials. By imaginary hydrolysis at the points indicated by the dotted lines, the substance may be resolved into succindialdehyde, methylamine, and acetone, and this observation suggested a line of attack of the problem which has resulted in a

D-(−)-tartaric acid

L-(+)-tartaric acid

$H{-}C{\equiv}N$

prussic acid

HO–CH2–CH(OH)–CH2–OH

glycerin

O_2NO–CH2–CH(ONO_2)–CH2–ONO_2

nitroglycerin

Box 6 Datura metel, by artist Eudoxia Woodward

In this genus (Datura) there are fifteen to twenty species, annuals and perennials, shrubs and trees, with trumpet-shaped flowers. All are hallucinogenic. Datura was adapted to Latin by Linnaeus from the Sanscrit Dhatura. Early Sanscrit and Chinese writings mention Datura metel. The metel was reported by the Arabian Doctor Avicenna, 11th Century. In China it was considered sacred. Datura metel (or fastuosa), an annual herb, can grow to five feet. The flower can be seven inches long, white inside, violet and white outside, with a purple calyx. The fruit (capsule) starts out green and purple, becoming beige when ripe. The flowers can be double-flowered (some are blue, red and yellow). It is native to India and has been widely naturalized. The Daturas contain certain alkaloids, principally hyoscyamine and scopolamine. This painting of Datura metel shows its progressive growth from bud to full blossom to seed pod.

The artist has included the phonetic sanscrit at the top center of the painting. In the lower right, the Indian origin of the plant is shown and as the plant's lifecycle is revealed counterclockwise the path to introducing the plant to China (near seed pod) is visualized. Images of death and hallucinations are included to reflect its use and abuse in ancient Chinese and Indian medicine. The table is Chinese. The molecular structures (space filling on the table; ball and stick, upper right) are of scopolamine and were rendered by Robert M. Williams using Chem 3D and re-painted by the artist revealing the chemical basis for the hallucinogenic properties of Datura metel. This plant is also known as Devil's trumpet, metel, and downy thorn apple.

Eudoxia Woodward was Robert B. Woodward's second wife and mother to his second set of children, Crystal and Eric Woodward. She is an artist residing in Belmont, Massachusetts.

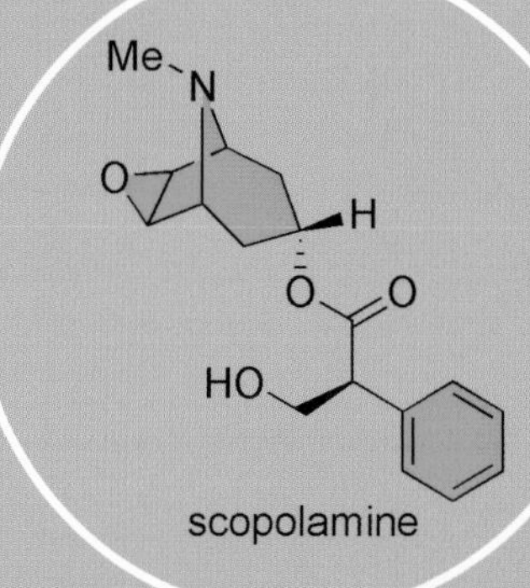

scopolamine

Me–N, H, O, O, HO

hyoscyamine

Robert B. Woodward

direct synthesis." So the seed was planted for the concept of what would later become known as retrosynthetic analysis. This theory was proposed and then greatly advanced and systematically formalized by another great chemist, E. J. Corey, who was awarded the 1990 Nobel Prize in Chemistry in recognition of his enormous impact on organic synthesis (see Chapter 14).

In a single reaction, using succindialdehyde, methylamine, and acetone, Robinson was able to reduce his theoretical analysis to practice, albeit in low yield. Despite the elegance and enviable beauty of this cascade sequence of events, Robinson was not yet content and he experimented further until he discovered an improved method in which acetone was replaced with calcium acetonedicarboxylate. In this case, the initial product was tropinonedicarboxylic acid. This could be induced to lose two molecules of carbon dioxide when basified and subjected to heat, giving tropinone in improved overall yield (Box 5).

Time can never diminish the exquisite beauty of Robinson's total synthesis of tropinone. The simplicity of this remarkable synthesis, the splendor of the cascade, and the stunning analysis that led to its conception will always stand out, marking this endeavor as a true classic in total synthesis, and inspiring new generations of chemists.

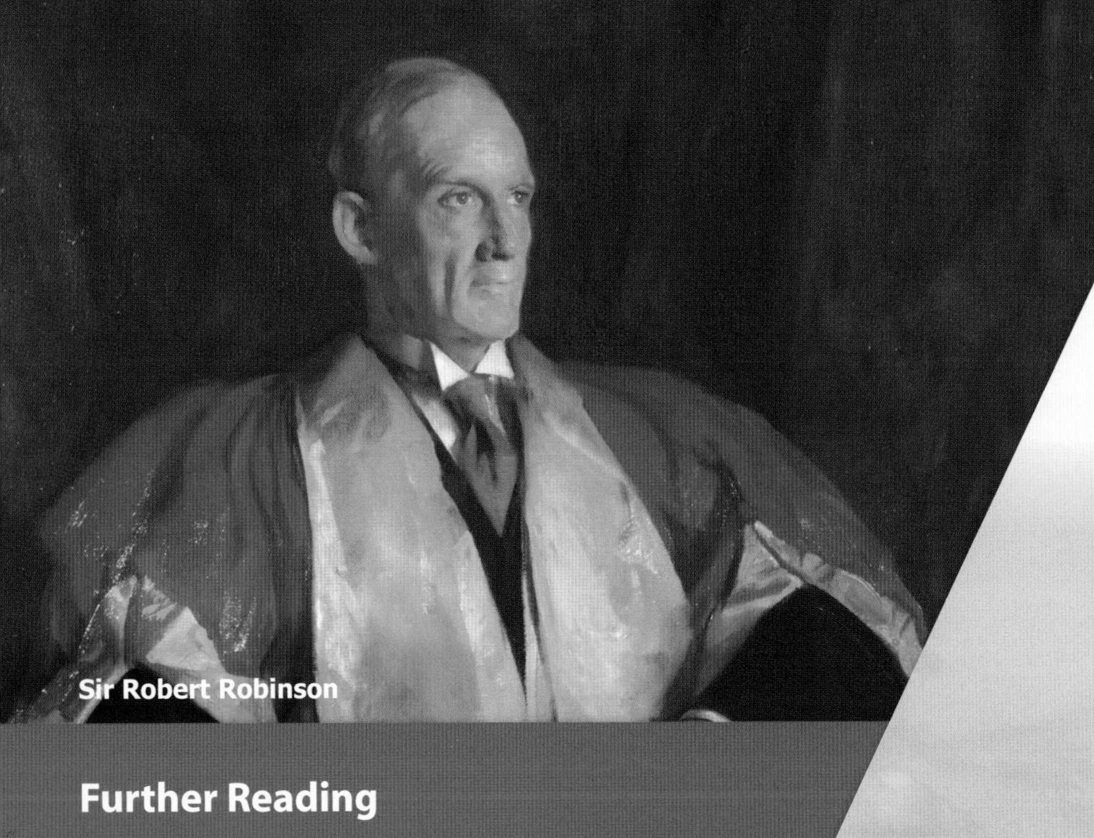
Sir Robert Robinson

Further Reading

G. N. Lewis, The Atom and the Molecule, *J. Am. Chem. Soc.* **1916**, *38*, 762–785.

R. Robinson, A Synthesis of Tropinone, *J. Chem. Soc.* **1917**, *111*, 762–768.

R. Robinson, Some Polycyclic Natural Products, *Nobel Lecture*, December 12, **1947**. The transcript of this lecture can be accessed via http://www.nobelprize.org/nobel_prizes/chemistry/laureates/1947/robinson-lecture.html.

M. D. Saltzmann, Sir Robert Robinson – A Centennial Tribute, *Chem. Ber.* **1986**, *22*, 545–548.

E. J. Corey, X.-M. Cheng, *The Logic of Chemical Synthesis*, Wiley-Interscience, New York, **1989**.

K. C. Nicolaou, T. Montagnon, S. A. Snyder, Tandem Reactions, Cascade Sequences, and Biomimetic Strategies in Total Synthesis, *Chem. Commun.* **2003**, *5*, 551–564.

K. C. Nicolaou, D. J. Edmonds, P. G. Bulger, Cascade Reactions in Total Synthesis, *Angew. Chem. Int. Ed.* **2006**, *45*, 7134–7186.

University of Oxford

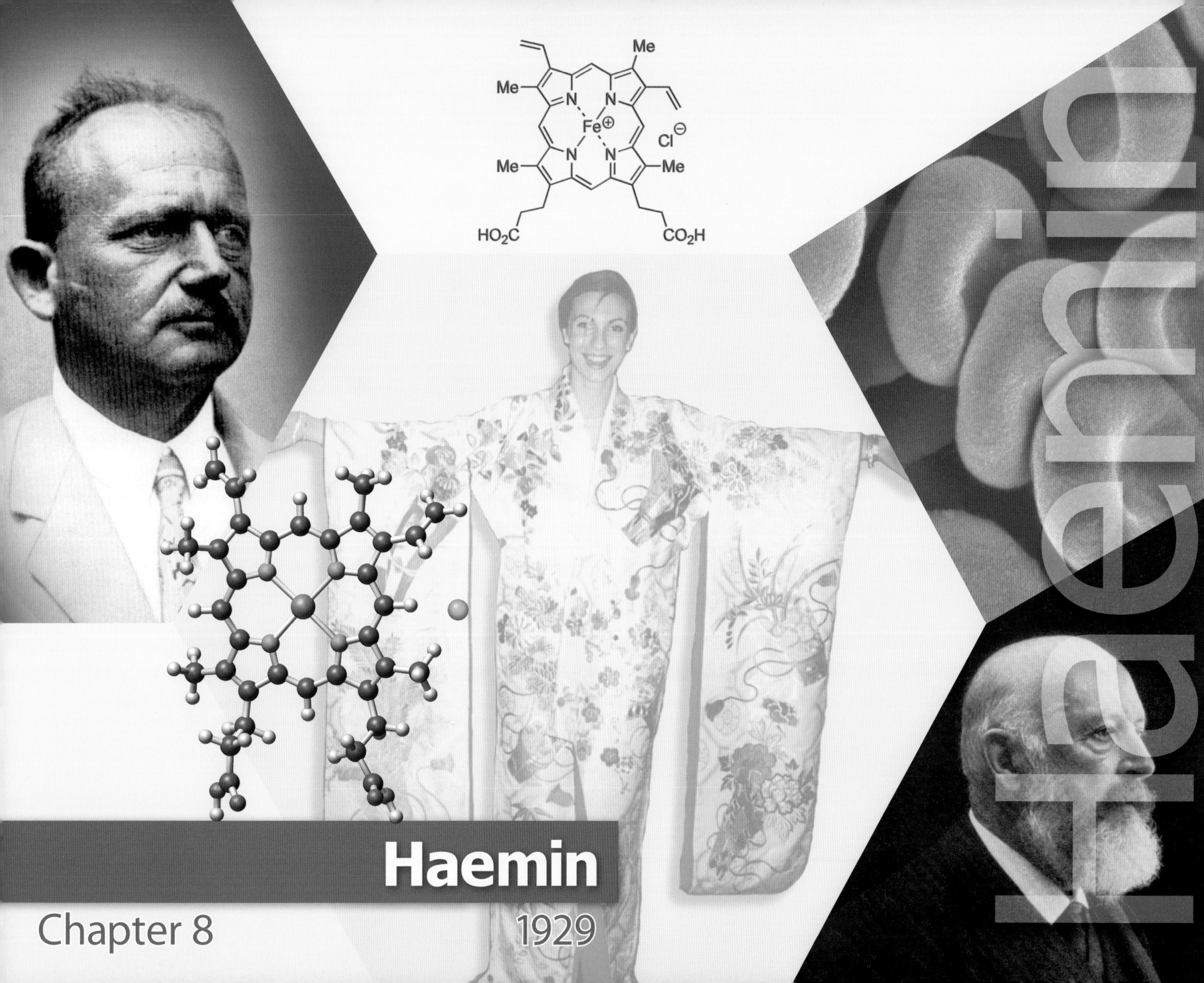
Me
Me
N
N
Fe⊕
N
N
Cl⊖
Me
Me
HO_2C
CO_2H
Haemin
Chapter 8
1929
Hemin

Hans Fischer

Box 1 Hans Fischer's synthesis of haemin

HBr

a

a: fusion in succinic acid

macrocyclization

haemin

The distinct red color of blood, with all its emotional connotations, is due to the pigment haemin. In our bodies, haemin is found wrapped up in the protein globin, to form haemoglobin, which resides in our red blood cells (or erythrocytes). Haemin itself consists of an aromatic porphyrin macrocycle, comprised of four pyrrole units, which encapsulate a central iron cation. The metal absorbs oxygen reversibly so that it can be carried around the body from the lungs to the site where it is needed (in order to release energy to fuel all cellular processes). The enveloping protein is not just a vehicle for this transportation; it also serves to protect the metal site from the surrounding environment.

Porphyrins are widely distributed throughout the living world where they play key roles in many of the important electron transfer reactions in animals, plants, and single cell organisms. Their electrochemical function is mediated by a number of different central metal atoms such as iron, magnesium, and copper. For example, cytochrome P450 is another haemin-containing protein, one that plays an important role in the metabolism of drugs and other foreign chemicals, facilitating their removal from the body by oxidation to a more water-soluble form. Chlorophyll, the green pigment found in plants, is another example of a porphyrin. Its role is to absorb energy from sunlight, using it to supply electrons for a pivotal reduction process in the photosynthesis pathway. This brief survey of a few porphyrins merely skims the surface of this large group of biomolecules, whose diversity and importance captured the hearts and occupied the minds of many of the early synthetic chemists, particularly one called Hans Fischer.

Like his namesake, Emil Fischer (whose investigations into sugars were described in Chapter 3), Hans Fischer was fascinated by the chemistry of living systems. Indeed, he studied chemistry and medicine simultaneously, and held degrees in both subjects. His independent academic career began in Munich where he lectured first on internal medicine and subsequently on physiology. Later, he moved to Innsbruck where he became a Professor of Medicinal Chemistry. In 1921, Hans Fischer returned to Munich to take up the position of Professor of Organic Chemistry at the Technical University. He remained there until his death in 1945. The biological significance of the porphyrins did not escape Fischer's attention, prompting him to dedicate many years to their structural determination, a goal that he accomplished in the case of haemin through chemical synthesis. His synthesis was founded on the insightful clues provided by the work of William Küster and others on haematinic acid and various other degradation products of haemin. Küster proposed a structure for haemin as early as 1912, which Fischer later described as "on the whole presenting an accurate picture of haemin."

However, uncertainties remained and the all important details were only definitively established when Fischer completed the synthesis of haemin in 1929 (Box 1). Fischer's synthesis of haemin exploits the greater nucleophilicity of the 2-position of pyrrole relative to that of the 3-position in order to assemble the porphyrin macrocycle from three different pyrrole building blocks. Perhaps the most striking feature of this synthesis is the fusion of two different dipyrrole sub-units in the macrocyclization step. It is also interesting to note that the reagents used by Fischer and his contemporaries are quite harsh and they were frequently employed under brutal conditions. Of particular relevance to this feature are the circumstances under which, approximately thirty years later, Robert B. Woodward synthesized chlorophyll a, a substance structurally related to haemin. Woodward had sharper and more selective chemical tools at his disposal, allowing the efficient execution of his brilliant strategy. Despite the hurdles faced by Fischer, he was able to synthesize several members of the porphyrin class of biomolecules and illustrate many of the relationships that exist between them. Hans Fischer held a holistic view of chemistry, biology, and medicine, and the interface between these sciences fascinated him. Fischer was awarded the 1930 Nobel Prize in Chemistry for "his researches into the constitution of haemin and chlorophyll and especially for his synthesis of haemin." The lecture he delivered on receipt of the award described his belief in the common origins and shared evolutionary paths that unite all these life-giving molecules. He remained dedicated to probing this hypothesis by means of experiment to the end of his life.

3
2
N 1
H
pyrrole

The rich and deep coloring of the pigment haemin is a secondary feature when compared to its vital oxygen transporting function, but its color mesmerized mankind long before its physiological role was understood. The human fascination with color has a long history; the earliest known people painted on the walls of caves, and the staining of fabrics with natural dyes has a long and proud tradition. The roots of the word porphyrin aptly illustrate the link between our appreciation of colors and the development of cultures, industries, and sciences. The ancient Greeks used the word *porphura* to describe the color purple, having borrowed the term from an earlier Semitic word used by the Phoenicians to refer to certain molluscs from the Purpura and Murex families. These molluscs were the source of the rare and expensive pigment, Tyrian purple, which gives fibers an intense purple color. It took approximately ten thousand molluscs to produce just a single gram of the precious dye, which meant that the dye was frequently worth more than its weight in gold. As a result, purple became associated with aristocracy and wealth; indeed, the Roman Emperor Nero decreed that only royalty could wear purple, a tradition that thrived well beyond his limited lifetime. Tyrian purple is exceptional because it is so colorfast. The Greek writer Plutarch tells of how Alexander the Great was able to recover textiles from the Persians that had been dyed purple 190 years earlier in Greece. Alexander's for-

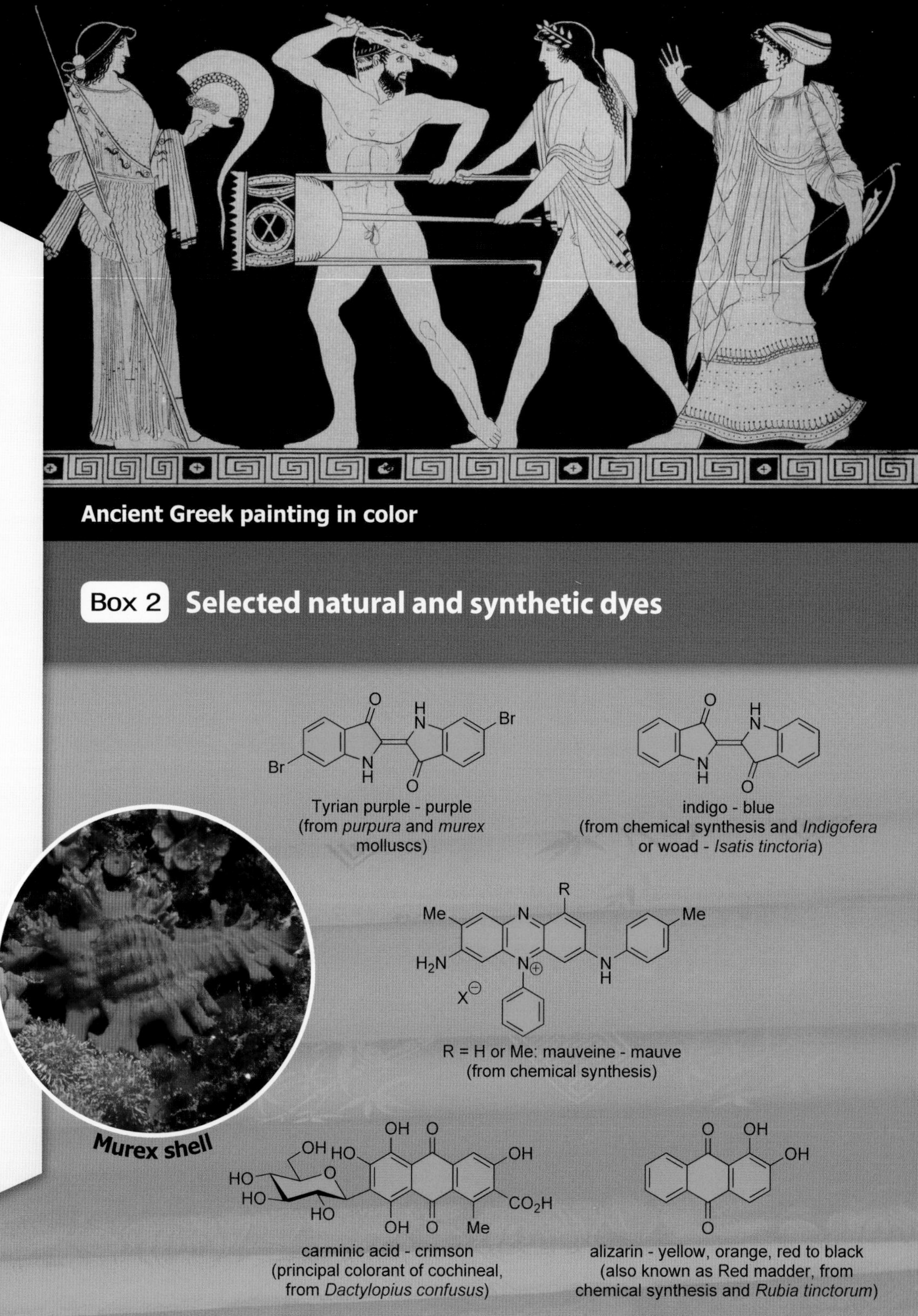

Johann Friedrich Wilhelm Adolf von Baeyer

Indigo jeans

Silk cloth

Box 3 The first commercial synthesis of indigo, devised by K. Heumann in 1890

anthranilic acid (CO_2H, NH_2) → (Cl$\diagdown$$CO_2H$; *alkylation*) → phenylglycine *o*-carboxylic acid (CO_2H, N(H)CO_2H)

phenylglycine *o*-carboxylic acid → (base, heat; *cyclization*) → indoxyl (OH, NH)

indoxyl → (*aerial oxidation*) → indigo

Cochineal insects feeding on Nopal cacti

ays farther east continued to promote innovations in dye practices as new technical knowledge and methods were spread as he traveled. This process included the spread of rich materials and expertise from the artisans of India, where he found particularly exquisite and beautiful cloths.

China and India provide some of the oldest written records of dyestuffs, and the skills developed in these nations have profoundly influenced the world of color for millennia. A five-thousand year old text from China describes recipes used to impart red, black, and yellow coloring to silks, whilst in India and Pakistan archaeological investigations have revealed that the production and dyeing of cotton has existed in these lands for a similar length of time. Indeed, India donated its own name to one of the most enduring of natural dyes, indigo (Box 2). This famous blue dye is the same as that which is obtained from woad upon exposure to air and treatment with dilute alkali, and is also closely related to a legume dye used by the Navajo Indian tribes in North America. The use of woad in Europe dates back to Neolithic times and indigo was employed even earlier throughout Asia. Early inhabitants of the United Kingdom used woad blue as a body paint to frighten away the Roman invaders; later, Queen Elizabeth I of England forbade woad processing within five miles of her residences because of the foul smell associated with this productive cottage industry. During the sixteenth and seventeenth centuries, indigo imported from India began to replace woad, and by the nineteenth century the world consumption of this natural blue dye was skyrocketing. The huge popularity of indigo inspired the celebrated chemist Adolf von Baeyer (full name Johann Friedrich Wilhelm Adolf von Baeyer) to study its constitution, and in 1866 he established its structure. This accomplishment soon led to a commercial synthesis of the dye by Karl Heumann, which has been hugely successful, in particular as the original color for denim (Box 3). The 1905 Nobel Prize in Chemistry was awarded to Baeyer in recognition of his contributions to the chemical industry and to the study of organic dyes and aromatic compounds (see Box 5).

Perhaps the most beguiling natural dye is cochineal (Box 2). Cochineal is a deep red dye that was used by the native people of South and Central America. It is obtained from a scale insect that lives on the pads of prickly pear cacti. The insect produces the pigment as a defensive chemical since carminic acid, the principal colorant of cochineal, is a bitter astringent. The Mixtex Indians cultivated the cacti in order to farm the insects, going so far as to light fires at night to keep the bugs warm. In the sixteenth century, the Spanish, who had colonized Latin America, were so impressed with the intensity of the scarlet produced by the indigenous people that they transported it back to Europe, where it became a luxurious and sought after commodity. The Spanish refused to allow any cochineal insects to leave their colonies and thereby maintained a monopoly in the supply of this valuable dye for almost two hundred fifty years. The cochineal industry declined somewhat with the advent of the synthetic aniline dyes that revolutionized the industry in the 1900s, although niche markets remained. For instance, the guards around

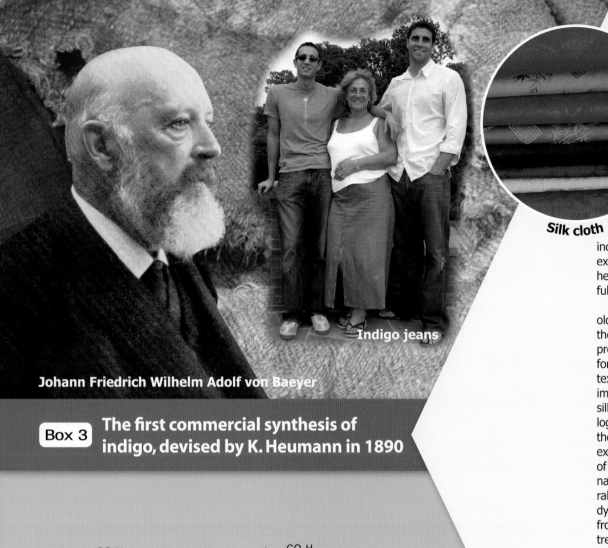

Buckingham Palace in London still wear tunics dyed with cochineal red.

In the mid-nineteenth century a new ethos in chemistry was being embraced and promoted, especially by August Wilhelm von Hofmann, who had, in turn, been mentored by the legendary Justus von Liebig. These two great men believed in empirical chemistry and in the teaching of chemistry by laboratory instruction. During this period, Europe was in the grip of the industrial revolution. Along with the generation of substantial wealth, the industrial revolution inspired a heady confidence that Man, the scientist and engineer, could accomplish almost anything. Together, these influences ensured that chemistry entered a new phase that would see a massive growth in its direct application.

William Henry Perkin, Sr., was born in 1838, the son of a builder. Any hopes his father harbored that his eldest son might follow his trade were dashed when, as a young boy, William was mesmerized by the chemicals and processes used in photography. His curiosity led him, at the age of thirteen, to enroll at the City of London School, where he attended weekly lectures on various topics in chemistry given by Thomas Hall. Hall recognized the talent and enthusiasm of his young pupil and persuaded William's reluctant father to purchase him some glassware so that he could conduct experiments at home. From here, and with the support of his former teacher, Perkin cultivated his interest further by attending Hofmann's lectures at the Royal College of Science in London. Two years later, aged seventeen, Perkin became Hofmann's assistant. Hofmann had two main areas of research, which he believed could be connected by synthesis: coal tar (and its derivatives) and quinine. The professor had completed his doctoral thesis on the topic of coal tar chemistry and its numerous aromatic constituents and he was now motivated to find a synthetic means of making quinine (Chapter 9). Quinine, used as a tonic to treat the deadly disease of malaria and other fevers, was imported in large quantities and at huge cost into Europe from its natural sources in South America. Perkin set to work in the makeshift laboratory that he had created in his home, determined to make quinine from coal tar during the Easter break of 1856. Quinine was known to have the molecular formula $C_{20}H_{24}N_2O_2$, a fact that led Perkin to believe, naïvely, that it could be made from two molecules of allyl toluidine ($C_{10}H_{13}N$) upon oxidation. However, when he treated impure allyl toluidine with potassium dichromate he made a solid brown mess. Ever the optimist, he considered this result as an indication that he might have found a general test for aromatic bases. To test his new hypothesis, he attempted the same procedure on impure aniline sulfate. This time the precipitate was black, but it stained the cloth that he kept for cleaning spills a deep purple color (Box 4). In tune with the entrepreneurial spirit of the time, Perkin rushed to consolidate

August Wilhelm von Hofmann

Buckingham Palace guards in cochineal red

William Henry Perkin, Sr.

Box 4 Perkin's synthesis of mauveine

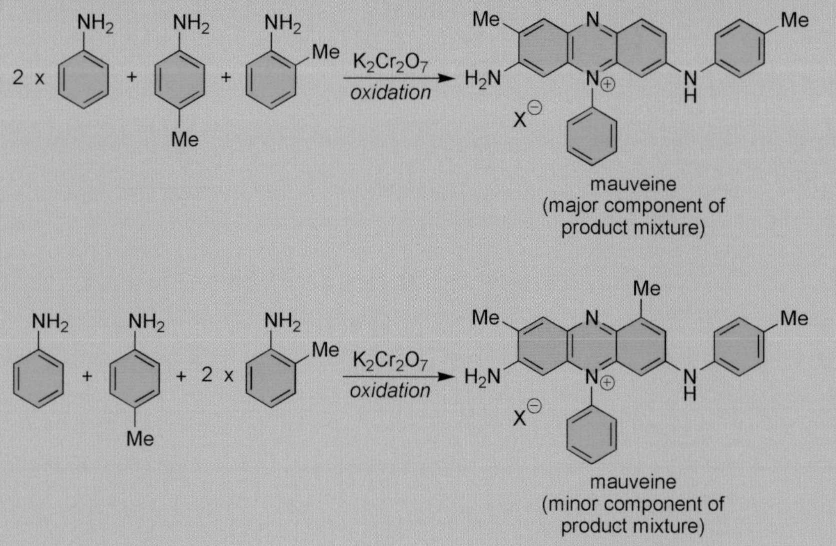

In 1825, Michael Faraday, better remembered as the pioneer of electricity and magnetism, isolated a liquid by distillation from the sticky condensate that was plaguing London's newly laid gas pipes. He determined that this liquid, which he called "bicarburet of hydrogen", was a hydrocarbon and that it had a molecular formula that could be expressed as a multiple of C_2H. Whilst the latter deduction was in error, Faraday had discovered the core member of a very important class of organic molecules, the so-called aromatic molecules. The term aromatic has been employed ever since the early nineteenth century when chemists began studying these compounds due to the sweet or spicy fragrances that appeared to be the common link between them. The name has stuck; however, in the modern definition, not all aromatic molecules have a noticeable or pleasant smell, and not all fragrant molecules have an aromatic group as part of their structure.

Professor Michael Faraday lecturing at the Royal Institution in London

Justus von Liebig

Friedrich Wöhler

Jöns Jakob Berzelius

Justus von Liebig later rechristened Faraday's bicarburet of hydrogen, benzene, a name he derived from that of the already known benzoic acid, which was the principal ingredient of church incenses at the time. Benzoic acid had, since 1557, been obtained as the major constituent of the Javan incense, gum benzoin. Liebig and his friend, Friedrich Wöhler, became fascinated by these aromatic compounds because they exhibited unusual chemical stability. Together, they published several groundbreaking papers describing a series of reactions undertaken on an aromatic extract of bitter almonds (now known to be benzaldehyde) and its derivatives. They identified a core part of the molecule that remained unchanged through the entire reaction series. They named this motif the "benzoyl radical". In this context, the term radical was used to describe a group of atoms mimicking a single element in their indivisible behavior. Their results were hailed as "the most important that have yet been obtained in vegetable chemistry" by the great Swedish chemist Jöns Jakob Berzelius. However, soon afterwards, Eilhard Mitscherlich corrected their work by carrying out the decarboxylation of benzoic acid to give benzene, proving that this was actually the true indivisible portion. Liebig's consternation over this event raged on for years, and he frequently hounded the frail Mitscherlich by publicly attacking his character and worth.

Around this time, August Wilhelm von Hofmann and William Henry Perkin, Sr., took the lead in the study of aromatic chemistry through their discovery of new reactions and the aniline dyes. However, none of these researchers had any structural information about benzene and its derivatives. Following the groundbreaking work of Italian chemist Stanislao Cannizzaro, the true molecular formula of benzene (C_6H_6) was determined, based on the newly corrected relative atomic mass of carbon. With the formula of benzene clarified, chemists began to concentrate on how its constituent atoms were joined, and a raft of weird and wonderful suggestions resulted.

August Wilhelm von Hofmann

Thomas J. Katz

While none of these structures were ultimately proved correct, they captured the interest of organic chemists, who sought to prepare isomers of benzene. Thus Dewar benzene was prepared in 1963 by the group of Eugene van Tamelen, and prismane (Ladenburg's benzene) was synthesized by Thomas J. Katz and colleagues in 1973. The Katz group had previously devised a practical synthesis of another benzene isomer, benzvalene, which had been synthesized and isolated earlier in only tiny amounts by Kenneth E. Wilzbach.

$Me(H)C=C=C=C=CH_2$

Crum Brown's benzene (1864)

Dewar's benzene (1867) (van Tamelen, 1963)

Ladenburg's benzene (1868) (prismane; Katz, 1973)

benzvalene (Wilzbach, 1967; Katz, 1971)

Kekule's resonance structures of benzene

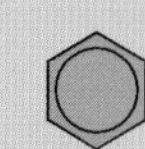

Robinson's depiction of the structure of benzene

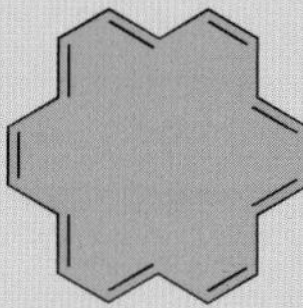

Sondheimer's [18]-annulene

Friedrich August Kekulé

Josef Loschmidt

Erich Hückel

Popular history credits the German chemist Friedrich August Kekulé with ultimately identifying the true structure of benzene as a hexagonal ring of six CH units, an idea he later claimed was inspired by a dream in which a snake appeared eating its own tail. In the recording of scientific history, it is often more convenient to attribute discoveries to an individual, however, this may not always give an accurate picture of events, where prevailing scientific opinion moves forward incrementally, and more than one individual can contribute to the shift. In this particular case, a few more details are needed to complete the picture. Kekulé's structure did not appear in print until 1865, although he claimed to have known the true structure as early as 1861. The Kekulé structures we know today with resonating single and double bonds did not appear until 1872. In 1854, the French chemist August Laurent, without explanation, depicted the core of benzoyl chloride as a hexagon. More importantly, in 1861, a little known Austrian chemist called Josef Loschmidt wrote a small book titled *Chemische Studien* (Chemical Studies) in which he drew structures for over 300 molecules, including several good representations for a number of benzene derivatives. Furthermore, correspondence dating from January 1862 indicates that Kekulé had seen this book. Kekulé certainly was responsible for popularizing the hexagonal structure of benzene and, thereby, ending a long period of confusion. Kekulé's structure was later adapted by Robinson, who drew a hexagon containing a circle, which indicated the delocalized nature of the electrons. This can also been shown in a three-dimensional image, representing the delocalized molecular orbital of benzene. However, several questions still remained over the true nature of aromaticity that would have to wait until the 1930s when innovations in quantum theory unearthed the solutions.

In 1931, a German physicist, Erich Armand Arthur Joseph Hückel, working from the Technical Institute in Stuttgart (Germany), carried out a series of theoretical calculations using the newly emerging quantum theory to find the molecular orbitals (the specific paths the electrons of the molecule follow) for selected aromatic compounds. He developed a rule describing the requirements for 'aromaticity', which stated that a molecule must be planar and contain 4n + 2 (where n is an integer) π-electrons within a cyclically conjugated (i.e. bearing a π orbital on each atom) system. This theory explained the unusual reactivity and stability of aromatic molecules. It could also be used to describe the aromatic character of other compounds, such as pyrrole, furan, thiophene, pyridine and the porphyrins. Higher analogues of benzene, such as [18]-annulene, first reported by Franz Sondheimer in 1959, have also been synthesized and studied, confirming Hückel's rule. The work of Franz Sondheimer, Emanuel Vogel, and Peter J. Garratt, together with that of many others, has done much to advance our understanding of aromaticity in the second half of the twentieth century.

Aromatic compounds have their own unique chemical characteristics and patterns of reactivity, and the study of their synthesis and properties has become a major field in its own right. Aromatic groups are represented in many natural products and designed drugs, as demonstrated throughout this book.

Franz Sondheimer

Peter J. Garratt

Emanuel Vogel

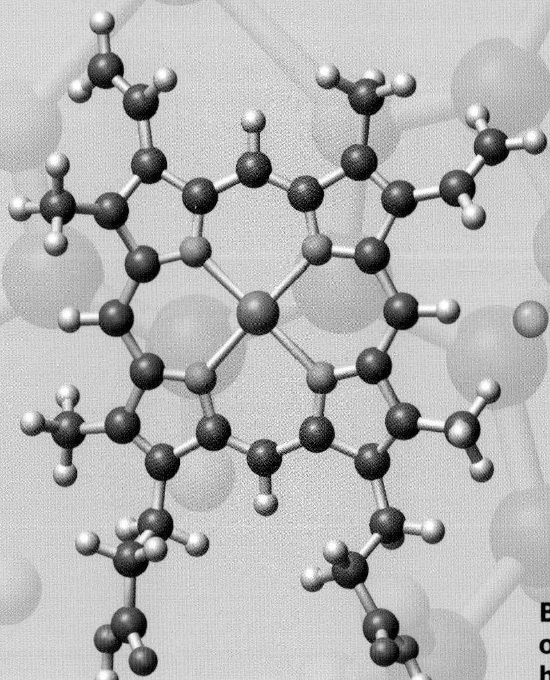

Ball and stick model of the molecule of haemin

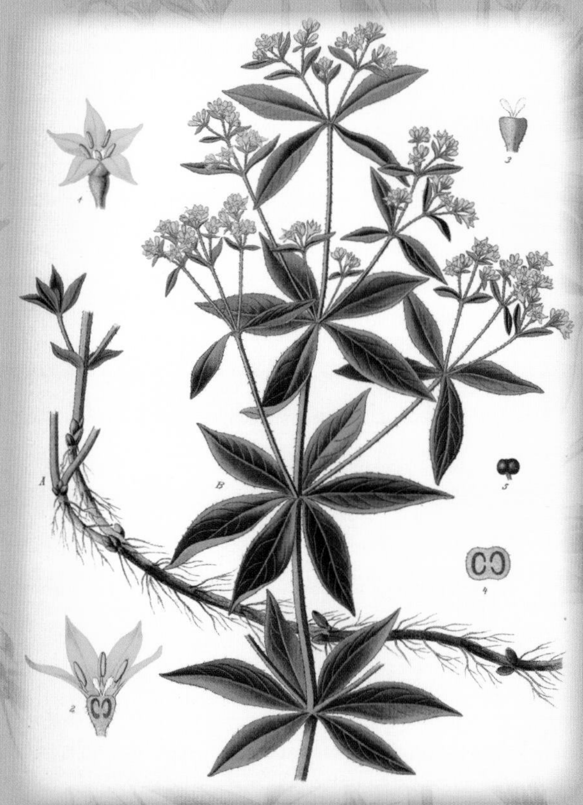

Red madder plant (*Rubia tinctorum*)

Further Reading

W. H. Brock, *The Norton History of Chemistry*, W. W. Norton, New York, **1992**.

L. R. Milgrom, *The Colours of Life*, Oxford University Press, Oxford, **1997**.

S. Garfield, *Mauve*, Faber and Faber, London, **2000**.

O. Meth-Cohn, M. Smith, What Did W. H. Perkin Actually Make When He Oxidised Aniline to Obtain Mauveine?, *J. Chem. Soc., Perkin Trans. 1* **1994**, 5–7.

A. Bader, A Chemist Turns Detective, *Chem. Brit.* **1996**, *32*, 41–42.

Empress Eugenie

this discovery by enlisting the assistance of the Scottish textile firm Pullars. Pullars tested Perkin's purple dye and reported that it could be successfully fixed to silks and exhibited reasonable color-fastness in response to sunlight. In Britain at this time, 75,000 tons of natural dyes were imported and, with the ever-growing middle classes boosting consumerism, fashion and home decor were becoming big businesses. This led William, still only eighteen years old, and his brother, Thomas, to build a factory to manufacture 'mauve', as they called their new purple dye.

The success of mauve netted the brothers a fortune and proved that industrial chemistry was the way of the future, spawning a host of other dye factories in Britain and continental Europe, especially in Germany, where two friends, Adolf von Baeyer and Friedrich Westkott (a paint merchant and a master dyer, respectively), built a factory to make coal tar dyes. Baeyer's dye factory represents the tentative early steps of a company that developed into a bastion of today's dye, agrochemical and pharmaceutical industries (see Chapter 4, Aspirin®). The excitement surrounding the production of aniline-based dyes contributed to a huge body of research into the chemistry of aromatic amines, and became of great importance in the early chemical industry. The influence of mauve is sometimes exaggerated, since, after achieving an initial high profile as a very fashionable color (Empress Eugenie, the wife of Napoleon III of France, took a particular liking to the color, and Queen Victoria of England wore it when officially in mourning after the death of her beloved husband, Prince Albert), its popularity was surpassed by other more practical dyes that were cheaper to manufacture, such as safranine. What Perkin did establish beyond doubt, however, was the principle of science-based industries as not just viable, but also highly profitable. In addition, he specifically proved the concept that natural dyes could be replaced by synthetic alternatives whose properties might be improved and fine-tuned through chemical research.

Perkin narrowly missed out on the next lucrative innovation to be developed and exploited by this emerging industry, one that ultimately proved a far greater success than mauve. In 1869, Perkin and Heinrich Caro, a German calico printer, filed patents describing a synthesis of alizarin within twenty-four hours of each other. Their independent processes differed only in small details, each beginning with the oxidation of anthracene to anthraquinone. Alizarin (Box 2) is itself a natural product, the main constituent of the madder plants (e.g. *Rubia tinctorum*) that have been used for dyeing cloth since ancient times, and which give a variety of colors from oranges and reds, to violet blue and black, depending on the pH of the dye solution. The Egyptians were fond of using this dye 3,500 years ago and it has been found in Pompeii and at numerous archeological sites throughout Asia. The synthesis of alizarin (or madder red) was amongst the first examples (see also Aspirin®, Chapter 4) of the large-scale industrial synthesis of a natural product. The synthesis provided alizarin so efficiently and cheaply that the long-standing process to exploit the natural source was abandoned. As such, the success of synthetic alizarin provided an important proof-of-principle and helped to launch a new approach to manufacturing through chemical synthesis. Today, such industrial chemical processes enrich not only the world of fashion, but also many other aspects of our lives. We have already described the role of chemical synthesis in the production of food, wine, medicines and perfumes, and we will encounter many more applications as the history of total synthesis unfolds.

Queen Victoria

OMe
N
OH
H
N
Quinine
Quinine
Chapter 9
1944, 2001

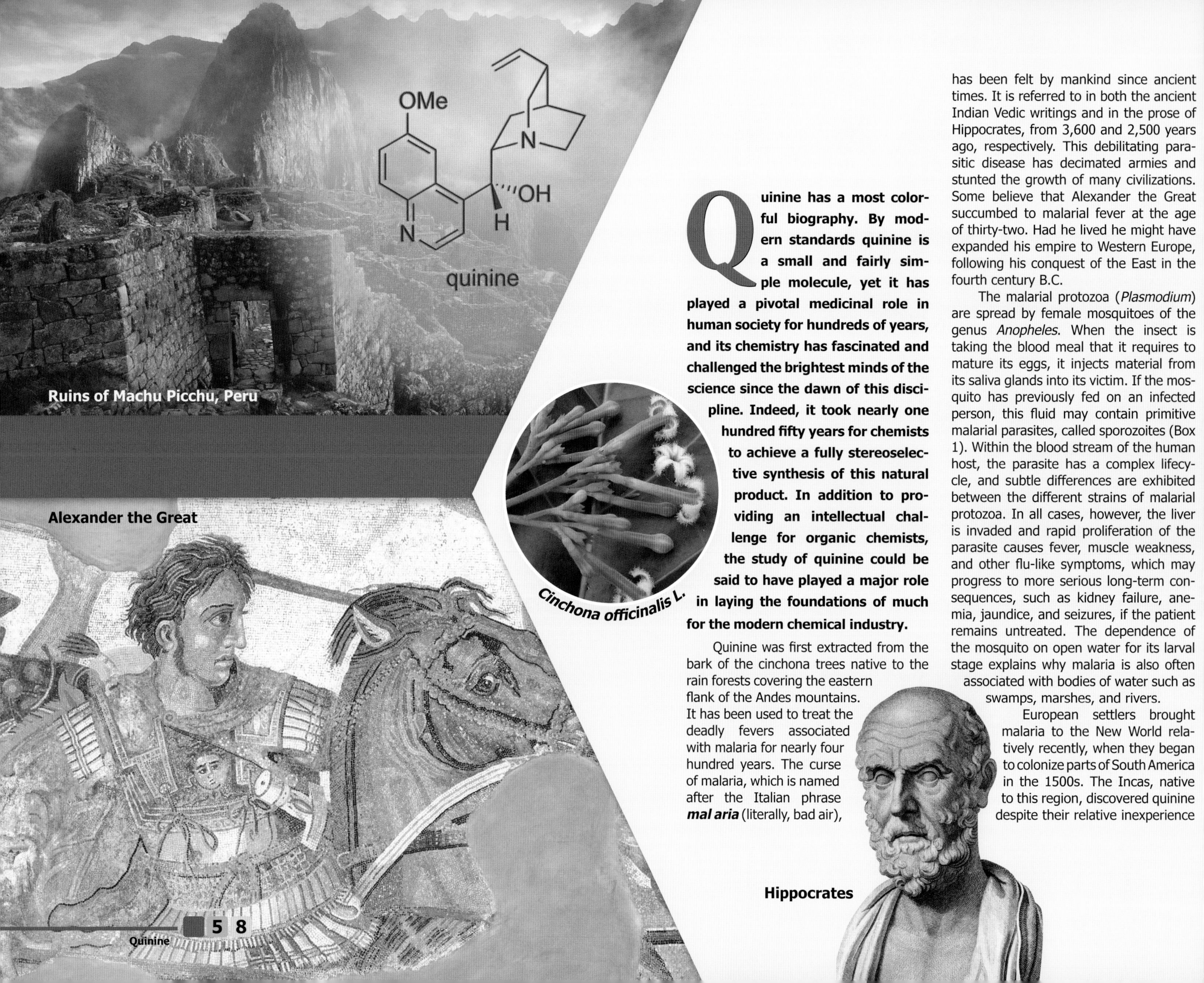

Ruins of Machu Picchu, Peru

Alexander the Great

Cinchona officinalis L.

Hippocrates

Quinine has a most colorful biography. By modern standards quinine is a small and fairly simple molecule, yet it has played a pivotal medicinal role in human society for hundreds of years, and its chemistry has fascinated and challenged the brightest minds of the science since the dawn of this discipline. Indeed, it took nearly one hundred fifty years for chemists to achieve a fully stereoselective synthesis of this natural product. In addition to providing an intellectual challenge for organic chemists, the study of quinine could be said to have played a major role in laying the foundations of much for the modern chemical industry.

Quinine was first extracted from the bark of the cinchona trees native to the rain forests covering the eastern flank of the Andes mountains. It has been used to treat the deadly fevers associated with malaria for nearly four hundred years. The curse of malaria, which is named after the Italian phrase ***mal aria*** (literally, bad air), has been felt by mankind since ancient times. It is referred to in both the ancient Indian Vedic writings and in the prose of Hippocrates, from 3,600 and 2,500 years ago, respectively. This debilitating parasitic disease has decimated armies and stunted the growth of many civilizations. Some believe that Alexander the Great succumbed to malarial fever at the age of thirty-two. Had he lived he might have expanded his empire to Western Europe, following his conquest of the East in the fourth century B.C.

The malarial protozoa (*Plasmodium*) are spread by female mosquitoes of the genus *Anopheles*. When the insect is taking the blood meal that it requires to mature its eggs, it injects material from its saliva glands into its victim. If the mosquito has previously fed on an infected person, this fluid may contain primitive malarial parasites, called sporozoites (Box 1). Within the blood stream of the human host, the parasite has a complex lifecycle, and subtle differences are exhibited between the different strains of malarial protozoa. In all cases, however, the liver is invaded and rapid proliferation of the parasite causes fever, muscle weakness, and other flu-like symptoms, which may progress to more serious long-term consequences, such as kidney failure, anemia, jaundice, and seizures, if the patient remains untreated. The dependence of the mosquito on open water for its larval stage explains why malaria is also often associated with bodies of water such as swamps, marshes, and rivers.

European settlers brought malaria to the New World relatively recently, when they began to colonize parts of South America in the 1500s. The Incas, native to this region, discovered quinine despite their relative inexperience with malaria, providing the world with the first and most enduring drug to combat this terrible disease. In the early 1600s, the Countess of Chinchon, consort to the Spanish Viceroy of Peru, was saved from near death by administration of a native medicinal antipyretic tonic. The infusion was made from the bark of a tree known as *quina-quina* by the local people. Details of her miraculous recovery were transmitted rapidly. Jesuit missionaries returning to the courts of Europe (especially to Rome, then the malaria capital of Europe) spread word of a new miracle cure for malaria. Quinine quickly became a popular medication and soon quina bark (later known as cinchona bark) was being exported across the Atlantic Ocean in considerable quantities. Those who were wealthy enough began to benefit from its curative powers. There was, however, some initial reluctance to accept the new treatment. The Protestants of Northern Europe, in particular in England, would not imbibe a tonic so closely associated with the papacy, and as a result suffered the deadly consequences of malaria. At that time, such preparations were also known as Cardinal's powder or Pope's powder. Oliver Cromwell, the Puritan leader who overthrew King Charles I in the seventeenth century English civil war, preferred to treat his fevers with bloodletting and consumption of mercury rather than take a Jesuit tonic containing quinine. Needless to say, Cromwell soon died, but whether his death can be attributed to malaria or the brutal treatment regime employed in an attempt to cure his disease is harder to appraise. As the European nations expanded their empires into parts of the world where malaria was prevalent, such as the Indian subcontinent, quinine became more widely embraced, and since then it has been used to treat malarial fever. Although it is

National coat of arms of Peru (displaying a cinchona tree)

The Countess of Chinchon

19th Century Print, A Soldier Grasps Atahnalpa, King of the Incas

MALARIA

Box 1 Transmission of malaria by mosquitoes

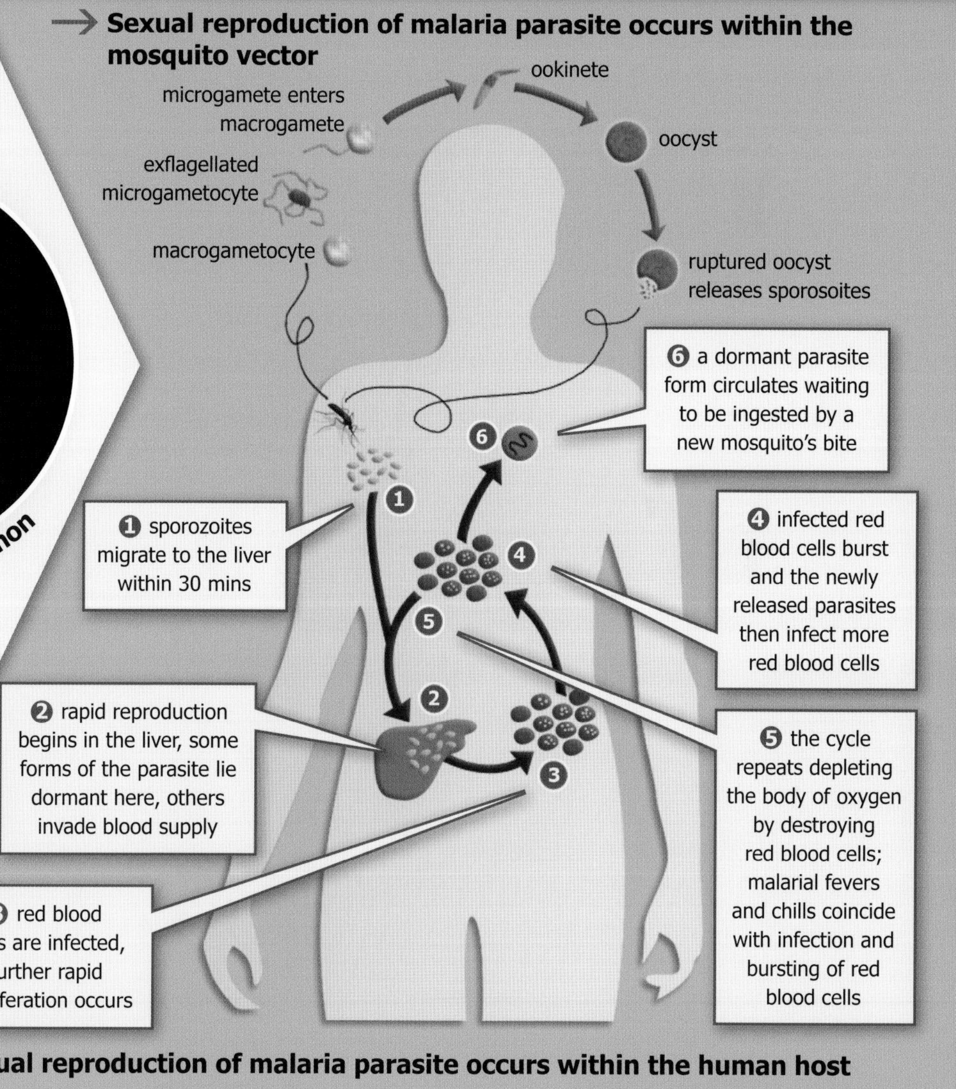

→ Sexual reproduction of malaria parasite occurs within the mosquito vector

microgamete enters macrogamete

ookinete

oocyst

exflagellated microgametocyte

macrogametocyte

ruptured oocyst releases sporosoites

❻ a dormant parasite form circulates waiting to be ingested by a new mosquito's bite

❶ sporozoites migrate to the liver within 30 mins

❹ infected red blood cells burst and the newly released parasites then infect more red blood cells

❷ rapid reproduction begins in the liver, some forms of the parasite lie dormant here, others invade blood supply

❺ the cycle repeats depleting the body of oxygen by destroying red blood cells; malarial fevers and chills coincide with infection and bursting of red blood cells

❸ red blood cells are infected, further rapid proliferation occurs

→ Asexual reproduction of malaria parasite occurs within the human host

August Wilhelm von Hofmann   William Henry Perkin, Sr.

Box 2 Examples of coal tar chemistry

The "mathematical approach" to quinine by Sir William Henry Perkin, Sr. (1856)

2 x *N*-allyltoluidine ($C_{10}H_{13}N$) (coal tar product) + 3[O] --- ? / – H_2O ---> quinine ($C_{20}H_{24}N_2O_2$)

aniline + *p*-toluidine + *o*-toluidine --- $K_2Cr_2O_7$ / *oxidation* ---> mauveine (R = H or Me) (see Chapter 8)

Ira Remsen and Constantin Fahlberg's synthesis of saccharin (1879)

o-tolylsulfonamide (coal tar derivative) --- [O] ---> saccharin

difficult to estimate the true figures, quinine has certainly saved millions of lives around the world.

Against this background, the pioneering chemist August Wilhelm von Hofmann considered the laboratory synthesis of quinine to be one of his most pressing tasks and cherished ambitions. In 1849, while heading the new Royal College of Chemistry in London, he published the idea that quinine might be accessed from coal tar. This material was readily available as a by-product from the production of coal gas, then a major fuel source, from coal. He suggested that two equivalents of naphthylamine might be transformed into quinine by the addition of two equivalents of water through, "...the discovery of an appropriate metamorphic process...". Hofmann was by no means alone in his quest for a route to quinine; with enormous sums being offered as the prize for a successful synthesis, competition to find a solution to the conundrum was intense. In 1854, the German chemist Adolf Strecker identified the correct molecular formula for quinine ($C_{20}H_{24}N_2O_2$). Eighteen year old William Henry Perkin, Sr., a precocious student of Hofmann's, attempted to prepare quinine in his homemade laboratory in 1856 during his Easter vacation. Following his mentor's arithmetical approach, he considered that it might be possible to obtain quinine by combining two molecules of *N*-allyltoluidine ($2 \times C_{10}H_{13}N$) with three atoms of oxygen ($3 \times O$) following the loss of water ($–H_2O$). His experiments failed to give quinine, but instead furnished mauveine (Box 2), a purple dye that took the fashion world by storm, making Perkin both famous and wealthy (see Haemin, Chapter 8). As a result, the coal tar dye industry boomed, and from this innovation many more developments followed in quick succession. In another serendipitous discovery, Ira Remsen and Constantin Fahlberg, at Johns Hopkins University in Baltimore (USA), first synthesized saccharin, in 1879, from a substance isolated from coal tar (Box 2). Arguably the most important upshot of coal tar chemistry was that it gave birth to a fledgling pharmaceutical industry (see also Aspirin®, Chapter 4).

Adolf Strecker

Although Perkin's equation was, in a sense, mathematically correct, his attempt to synthesize quinine was obviously futile when regarded in the light of modern chemical knowledge. Inspection of the chemical structures of quinine and his aromatic amine starting material reveals the folly of this approach. However, at that time, the structure of quinine was unknown, and modern structural theory was yet to be developed.

The power of quinine to cure malaria led a great many scientists and entrepreneurs to

Ira Remsen

Oliver Cromwell in the midst of battle

expend their energies on the isolation of the "sel essential de quina-quina" (the essential salt of quina-quina), beginning with Count Claude Toussaint de la Garaye in 1746. He isolated a crystalline substance which he believed to be the active constituent, but which later turned out to be a salt of quinic acid. In 1790, Antoine François Fourcroy isolated a red resin, which he called chinchona red, following a systematic investigation of cinchona bark. Again, he believed he had succeeded in preparing the active component of the bark preparations, but his hopes were dashed when chinchona red was shown to be inactive against malaria. A solution to this vexing problem would have to wait until 1820. By that time, the French pharmacists Pierre Joseph Pelletier and Joseph Bienaimé Caventou had already isolated a number of important alkaloids, including strychnine (Chapter 12). Their careful and skilled investigation of cinchona bark eventually led to the isolation of quinine. This time, their material was shown to be effective for the treatment of malarial fevers, thus ending the decades-long search for the active component of quina-quina.

Using modern analytical techniques, such as mass spectrometry, NMR spectroscopy, and x-ray crystallography, chemists today would need perhaps only a few days to determine the structure of a compound such as quinine. However, in the nineteenth century, none of these methods were

Box 3 The degradation of quinine to quinotoxine and Rabe's reconstitution of quinine

Louis Pasteur's acid-catalyzed rearrangement of quinine to quinotoxine

quinine —H_2SO_4→ [ ] —$-H^{\oplus}$→ ⇌ quinotoxine

Paul Rabe's 1918 reconstitution of quinine from quinotoxine

quinotoxine —NaOBr→ —NaOEt, EtOH→ [ ] —$-Br^{\ominus}$→ ⇌ (solvent) —Al powder→ quinine (+ three other stereoisomers)

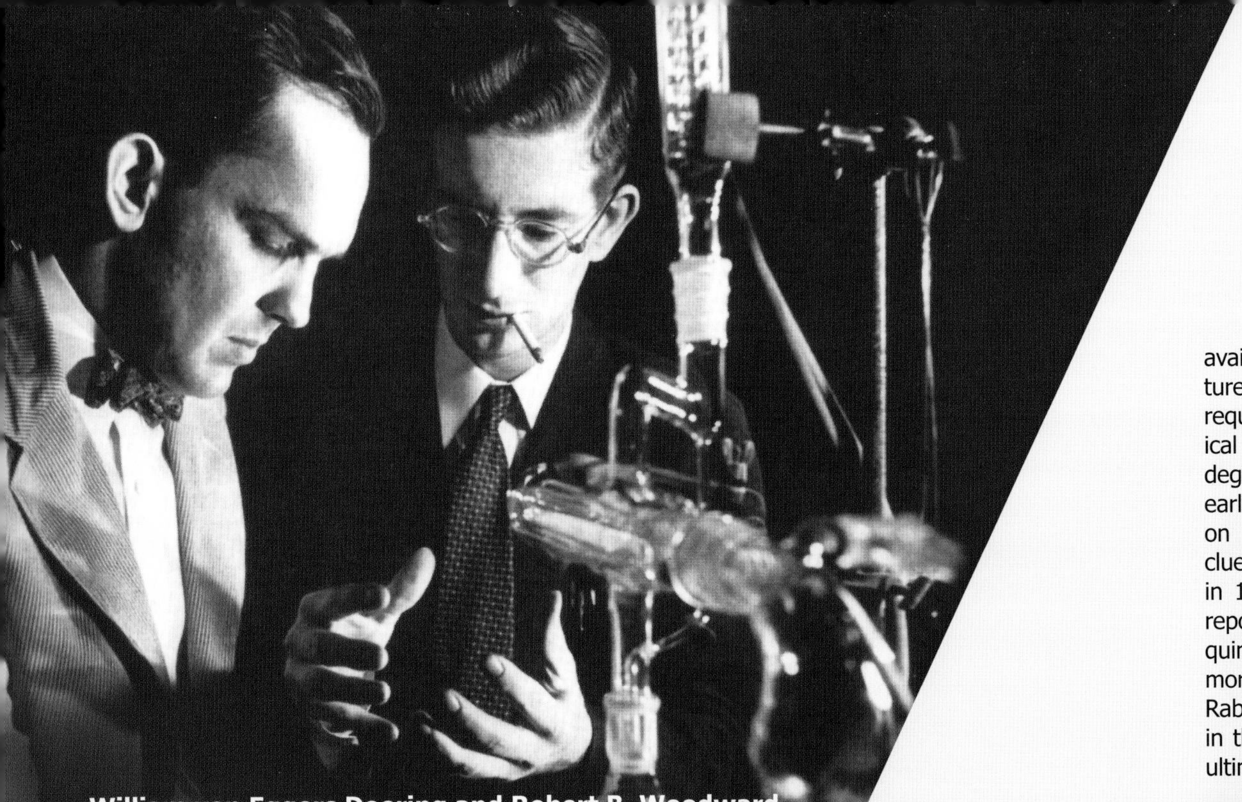

William von Eggers Doering and Robert B. Woodward

available and structure determinations required arduous chemical investigations and degradation studies. Many early organic chemists worked on the quinine problem, providing clues to the true structure. Eventually, in 1908, the German chemist Paul Rabe reported the correct atom connectivity of quinine. This marked the beginning of a more rational approach to its synthesis. Rabe spent the next quarter of a century in the dedicated pursuit of this goal, and ultimately achieved a partial synthesis of quinine. Louis Pasteur had come across quinine during his work on tartaric acid (see Chapter 13). In the 1850s, he had been searching for chiral bases to resolve racemic acids, and during the course of this investigation he treated quinine with sulfuric acid, forming quinotoxine (Box 3). Pasteur never deconvoluted the molecular basis of this transformation; nevertheless, quinotoxine has since played an important role in many attempts to access quinine by total synthesis. In 1918, Rabe reported the reconstitution of quinine, amongst a mixture of other stereoisomers, from quinotoxine (Box 3).

Vladimir Prelog

In 1944, the brilliant chemist Robert B. Woodward and his student, William

Box 4 Woodward's formal synthesis of quinine (1944)

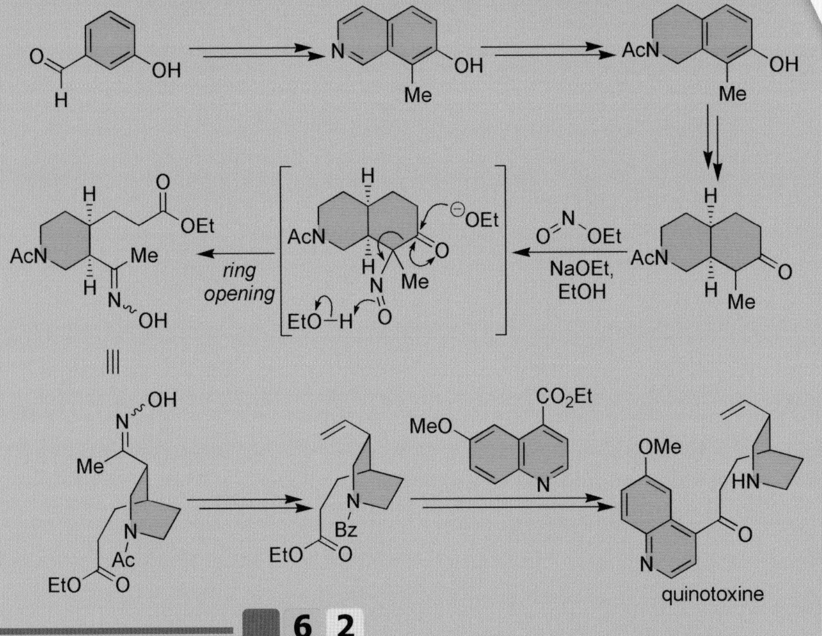

ring opening

NaOEt, EtOH

quinotoxine

Joseph Bienaimé Caventou (left) and Pierre Joseph Pelletier

von Eggers Doering, at Harvard University (USA), achieved a formal synthesis of quinine by synthesizing quinotoxine (Box 4). In the interim, Vladimir Prelog, who would later win the 1975 Nobel Prize in Chemistry for his contributions to stereochemistry, demonstrated that quinotoxine could be prepared from the relatively simple, monocyclic degradation product homomeroquinene. Woodward devised an ingenious route to this intermediate, proceeding via an isoquinoline. He used the second ring of this bicyclic structure to control the stereochemistry of the substituents in homomeroquinene. The strategy of using temporary ring systems to control stereochemistry served Woodward well in subsequent years, and was instrumental in many syntheses, including that of erythromycin A (Chapter 17). Following the synthesis of quinine, Woodward would go on to revolutionize total synthesis before his death in 1979.

Woodward's synthesis of quinine occurred during World War II. Japanese and German forces had seized control of much of the world's quinine supplies, having captured the cinchona plantations in the Dutch colonies in Java and the quinine stocks in Amsterdam, Netherlands. This led to a focused research effort to find alternative sources of this drug, which was vital to the war effort. Despite the fact that their synthesis would never provide meaningful quantities of quinine, Woodward and Doering were thrust into the limelight when they published their work, and their synthetic triumph was reported in the national press.

The Woodward–Doering formal synthesis of quinine successfully assembled the molecular framework of quinotoxine, but the now outdated Rabe protocol for the conversion of this material to quinine (Box 3) left much to be desired with regard

Milan Usković

Gilbert Stork

Box 5 The Hoffmann-La Roche total synthesis of quinine (1970)

NaN_3, PPA, heat
Schmidt rearrangement
Bz
CO_2Me
MeO
Me
OMe
AcO
AcOH, NaOAc
heat
KO*t*-Bu, O_2
8
OH
quinine
desoxyquinine

Box 6 Stork's stereocontrolled total synthesis of quinine (2001)

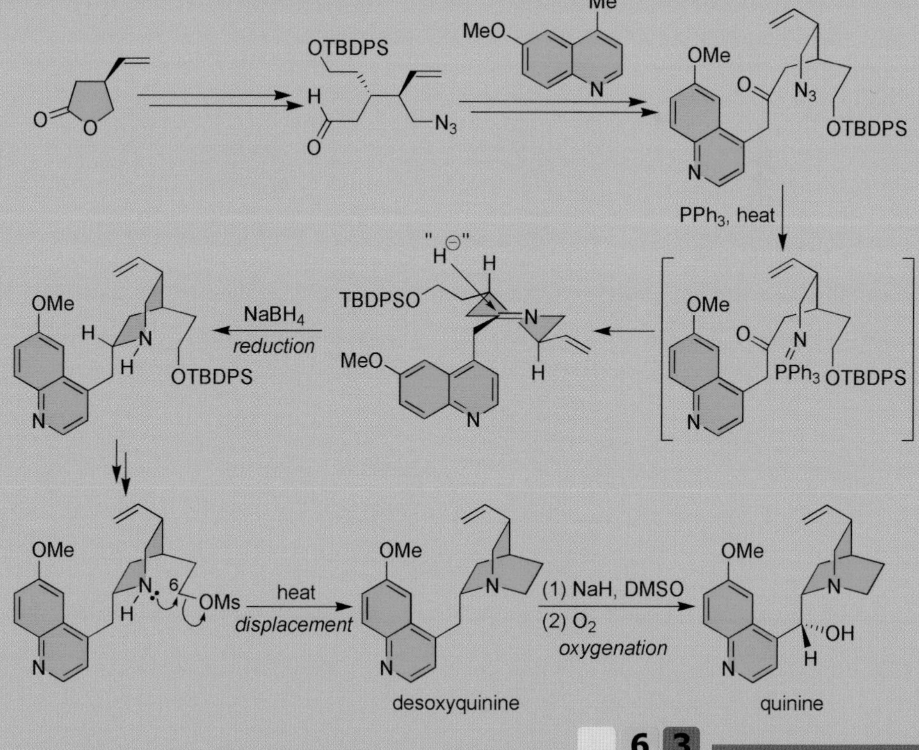

Cinchona officinalis L.

Box 7 Modern anti-malarial drugs

chloroquine (marketed under a variety of names)

proguanil hydrochloride

atovaquone

combination of proguanil and atovaquone (Malarone®)

artemisinin (natural product, used in Chinese medicine called qinghaosu)

mefloquine hydrochloride (Lariam®)

sulfadoxine

pyrimethamine

combination of sulfadoxine and pyrimethamine (Fansidar®)

to stereocontrol. This problem was revisited with some measure of success by a group of scientists at the pharmaceutical company Hoffmann-La Roche in the United States headed by Milan Usković (Box 5). The Usković team prepared desoxyquinine using a similar temporary ring strategy to control the stereochemistry of the bicyclic system. The key step in their approach involved oxygenation of the benzylic position, using molecular oxygen to introduce the secondary alcohol. This impressive reaction proceeds via a single electron transfer process, to give the desired isomer with good selectivity. However, the new route still afforded a mixture of isomers at the C8 stereocenter due to lack of selectivity during formation of the bicyclic quinuclidine structure.

The Rabe and Hoffmann-La Roche syntheses of quinine share a common feature in that both routes construct the bicyclic system through the formation of the C8–N bond, which leads to the poor stereoselectivity in the generation of the C8 chiral center in each case. The problem of C8 stereocontrol, also found in other related approaches, was recognized by Gilbert Stork at Columbia University (USA). Stork used this crucial information in his own retrosynthetic analysis of quinine, choosing to disconnect the quinuclidine ring system at an alternative site. This design allowed him to successfully synthesize (–)-quinine, the naturally occurring enantiomeric form of quinine, in 2001 (Box 6), in a fully stereoselective manner for the first time, nearly one hundred fifty years after Perkin's original attempt. Stork's approach used modern reactions to generate an acyclic precursor to the quinuclidine ring. Reduction of the azide group led to ring closure to produce an imine, which underwent reduction with high selectivity for the desired C8 stereochemistry, as anticipated in the planning stages. The quinuclidine system could then be completed by formation of the C6–N bond, and the hydroxyl group generated through a modified Usković procedure to give quinine.

Cinchona officinalis

Eric N. Jacobsen

Recently, the C8–N approach has been revived by two groups, using modern chemical reactions to provide a high level of stereocontrol. Groups led by Eric N. Jacobsen at Harvard University and Yuichi Kobayashi at the Tokyo Institute of Technology (Japan) have both prepared quinine via the intramolecular opening of an epoxide to form the C8–N bond. In each case, the epoxide was installed in a highly selective manner using the Sharpless asymmetric dihydroxylation reaction (see Chapter 26). Interestingly, this reaction is controlled by chiral ligands related to quinine itself.

The chemistry of quinine is fascinating and the story surrounding the development of its various syntheses is both thrilling and controversial, as chronicled in two recent review articles (see Further Reading). However, malaria continues to be a devastating disease, particularly in the developing world, despite the discovery of quinine and the many newer drugs available for its treatment (for selected examples, see Box 7). The relatively sparse access to such drugs, amongst those in most need, and a growing problem of protozoan resistance contribute to a staggering global death toll of more than one million individuals annually. Alongside the spread of HIV/AIDS, this terrible burden hinders the progress of nations desperately trying to escape the trap of poverty. Only a worldwide effort, supported by developed countries, can overcome this scourge to humanity, and it is encouraging that such efforts have been intensified in recent years. The Bill and Melinda Gates Foundation has been instrumental in reviving the campaign to eradicate this killer disease.

Cinchona officinalis L.

The habitat of *Cinchona officinalis L.*

Cinchona officinalis L.

Melinda and Bill Gates in Mozambique

Gilbert Stork

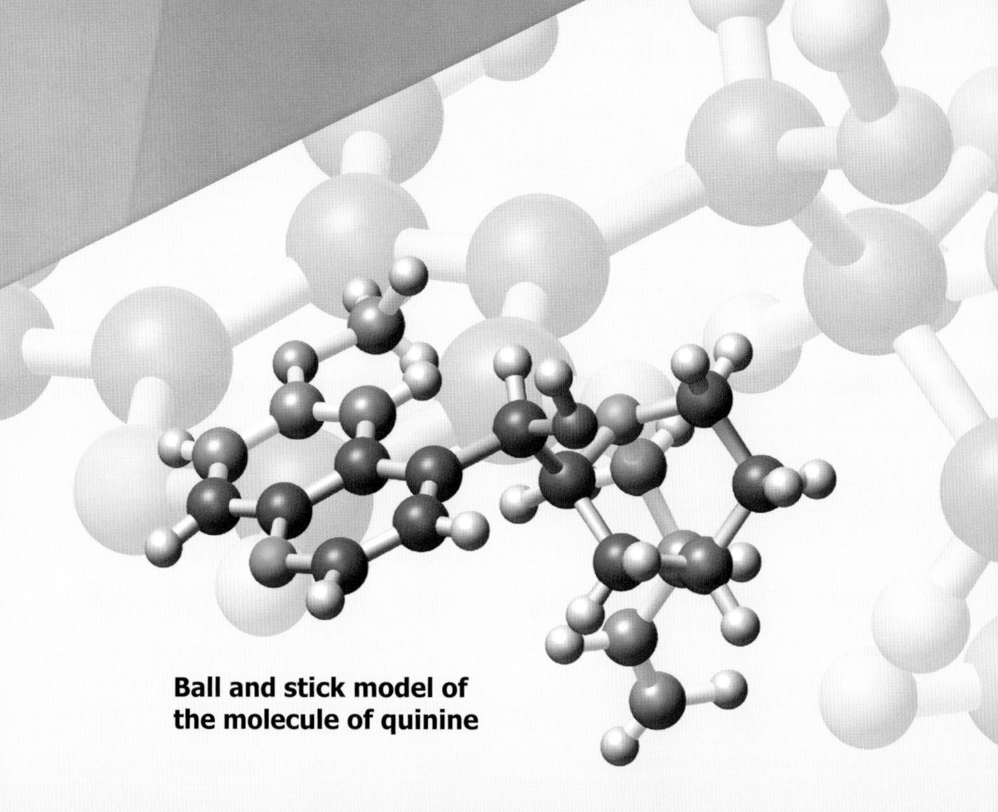

Ball and stick model of the molecule of quinine

Drying cinchona bark

Further Reading

G. Stork, D. Niu, R. A. Fujimoto, E. R. Koft, J. M. Balkovec, J. R. Tata, G. R. Drake, The First Stereoselective Total Synthesis of Quinine, *J. Am. Chem. Soc.* **2001**, *123*, 3239–3242.

S. M. Weinreb, Synthetic Lessons from Quinine, *Nature* **2001**, *411*, 429–431.

K. C. Nicolaou, S. A. Snyder, *Classics in Total Synthesis II*, Wiley-VCH, Weinheim, **2003**, pp. 443–463.

R. G. Ridley, Medical Need, Scientific Opportunity and the Drive for Antimalarial Drugs, *Nature* **2002**, *415*, 686–693.

M. Honigsbaum, *The Fever Trail: In Search of a Cure for Malaria*, Farrar, Straus and Giroux, New York, **2002**.

T. S. Kaufman, E. A. Rúveda, The Quest for Quinine: Those Who Won the Battles and Those Who Won the War, *Angew. Chem. Int. Ed.* **2005**, *44*, 854–885.

J. I. Seeman, The Woodward–Doering/Rabe–Kindler Total Synthesis of Quinine: Setting the Record Straight, *Angew. Chem. Int. Ed.* **2007**, *46*, 1378–1413.

HO
O
H
HO
H
H
N
Me
Morphine
Chapter 10
1952
Morphine

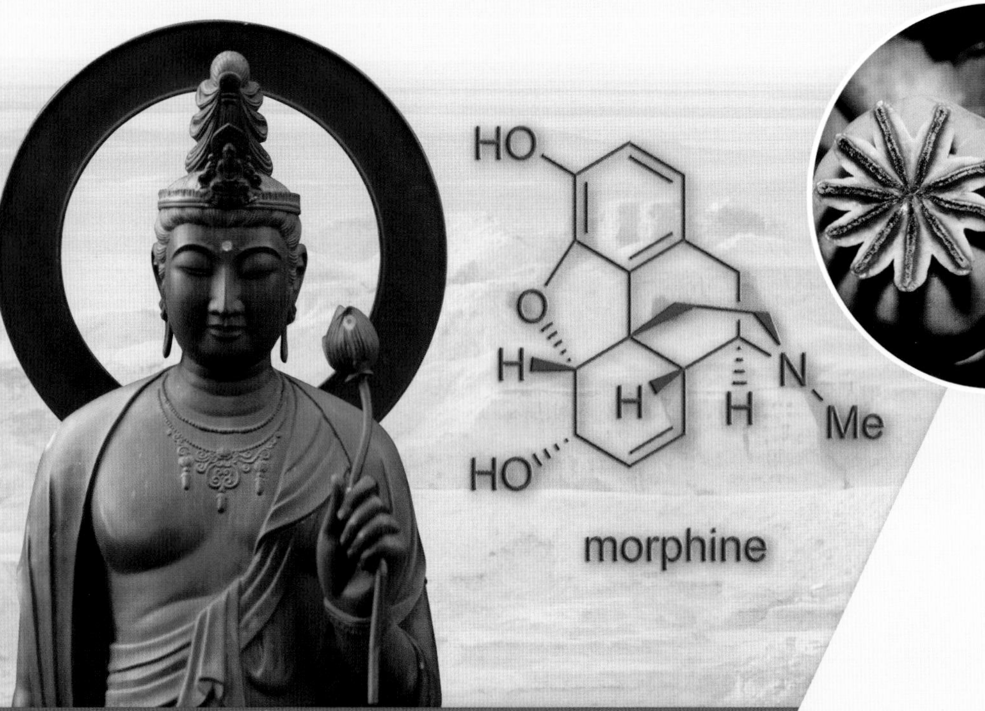

A long time ago, in Asia, a legend began: *Buddha had cut off his eyelids in order to prevent sleep overtaking him, and where his eyelids fell to the earth a herb grew, which blossomed bearing a beautiful nodding violet flower that gave sleep and tortured dreams to mankind.* The flower being described here is the opium poppy, and the natural product this plant produces is morphine.

The dichotomy of this myth, in which Buddha, a powerful symbol of good, gives rise to a flower that taunts mankind with disturbed sleep, is reflected in the contrasting biological properties of morphine. This natural product not only has the power to alleviate intense pain, but also rapidly induces dependence and addiction. In this chapter, we shall explore some of the history, biology, chemistry, and medicine of morphine, and some related analgesics. We shall also take a brief look at a collection of other natural and synthetic compounds that have become drugs of abuse, reflecting the illicit face of organic chemistry. As with many of the compounds described in this book, the molecule of morphine has an intricate structure which has a long record of challenging the skills of organic chemists.

Morphine is the main constituent of opium, the dark gum obtained when the milky latex exuded from the unripe seed head of the opium poppy (*Papaver somniferum*) is dried. It constitutes 4–21 % of opium by mass, with codeine being the second most abundant component (0.8–2.5 %). Opium also contains a smooth muscle relaxant called papaverine, which shares distant biosynthetic precursors with morphine and codeine (Box 2). While morphine is still one of the most powerful analgesics known, it is also a narcotic. Narcotics, in the medical sense, are drugs that suppress the central nervous system (CNS), inducing drowsiness or stupor. People have consumed opium since the very beginnings of civilization. The ancient Sumerians used it 6,000 years ago, although their stone tablets do not reveal its precise purpose. In the Ebers papyrus, an ancient Egyptian medical text dating back more than 3,500 years, opium is promoted as a tonic for colicky children or for those that cry incessantly. In the intervening millennia, knowledge of the effects of opium has spread around the globe. Dioscorides, the notable first-century Greek botanist-physician, provides a vivid description of how to collect opium (Box 1). By the Middle Ages, opium was becoming a standard apothecary's item from China and India in the Orient, to Europe and the New World in the West. During this period, the eccentric alchemist and physician Paracelsus (more properly known as Theophrastus Phillippus Aureolus Bombastus von Hohenheim) contributed to the popularization of opium by encourag-

Paracelsus

Box 1 Dioscorides' description of how to collect opium from the poppy seed head

Dioscorides

"it behoves them that make opium...to scarify about the asterisk with a knife... and from the sides of the head make straight incisions in the outside, and to wipe off the tear that comes out with the finger into a spoon, and again to return not long after, for there is found another thickened, and also on the day after..."

ing its use as a general analgesic as he toured Europe preaching and teaching. He also developed recipes for opium solutions, which he named laudanum (from the Latin, *laudare*, meaning to praise). These preparations were used widely as soporific painkillers for many centuries to follow. However, opium gives its users a sense of euphoria, and is also highly addictive, so the boundary between medicinal and recreational use was constantly eroded. Eventually, the problems associated with its addictive properties led to opium being regarded as a drug of abuse, capable of damaging society, but this was not fully appreciated until the use of opium was well established around the world. In the late eighteenth and early nineteenth centuries, drinking tinctures of laudanum or consuming a grain or two of opium held a certain social cachet. At that time, the romantic poets were busily creating some of the best known poetry and prose in western literature. Lord Byron, Samuel Taylor Coleridge, and Mary and Percy Shelley, to name but a few of these talented writers, moved around Europe socializing with an elite group of fashionable intellectuals; in these circles the consumption of both opium and laudanum for pseudo-medicinal purposes, or even just for pleasure, was commonplace (Box 3). The habit of using opium and laudanum recreationally continued unabated for many years, ensnaring a tremendous number of well-known historical figures, including Louisa May Alcott, George Washington, and Florence Nightingale. There was a period when laudanum was cheaper than hard liquor, and many of the poorer people of cities such as London celebrated the end of the working week by drinking excessive quantities of laudanum as a cheap alternative to gin or other spirits. It was a long time before rumbles of general dissent across the world drove authorities to reclassify opium use as a serious social malaise, and much longer still before the number of addicts saw any meaningful decline.

George Washington

The opium trade even led to wars between China and several European powers, most notably Great Britain, in the mid 1800s. Until the early part of the seventeenth century, Chinese opium consumption was following the European model in that its use was mainly medicinal, and limited to treating rheumatic pains and diarrhea, although it is said that it was sometimes added to cakes and sweets to satisfy guests on special occasions. However, in 1644, when the Emperor Tsung Chen banned the popular habit of smoking tobacco, the people of China seemed to turn gradually to opium as a convenient alternative. Demand was soon far in excess of the local production capacity, and foreign merchants were eager to profit from the lucrative and untapped market of China. Indeed, by the end of the 1830s, the British East India Company was supplying more than one million pounds (£) worth of Indian opium

Florence Nightingale

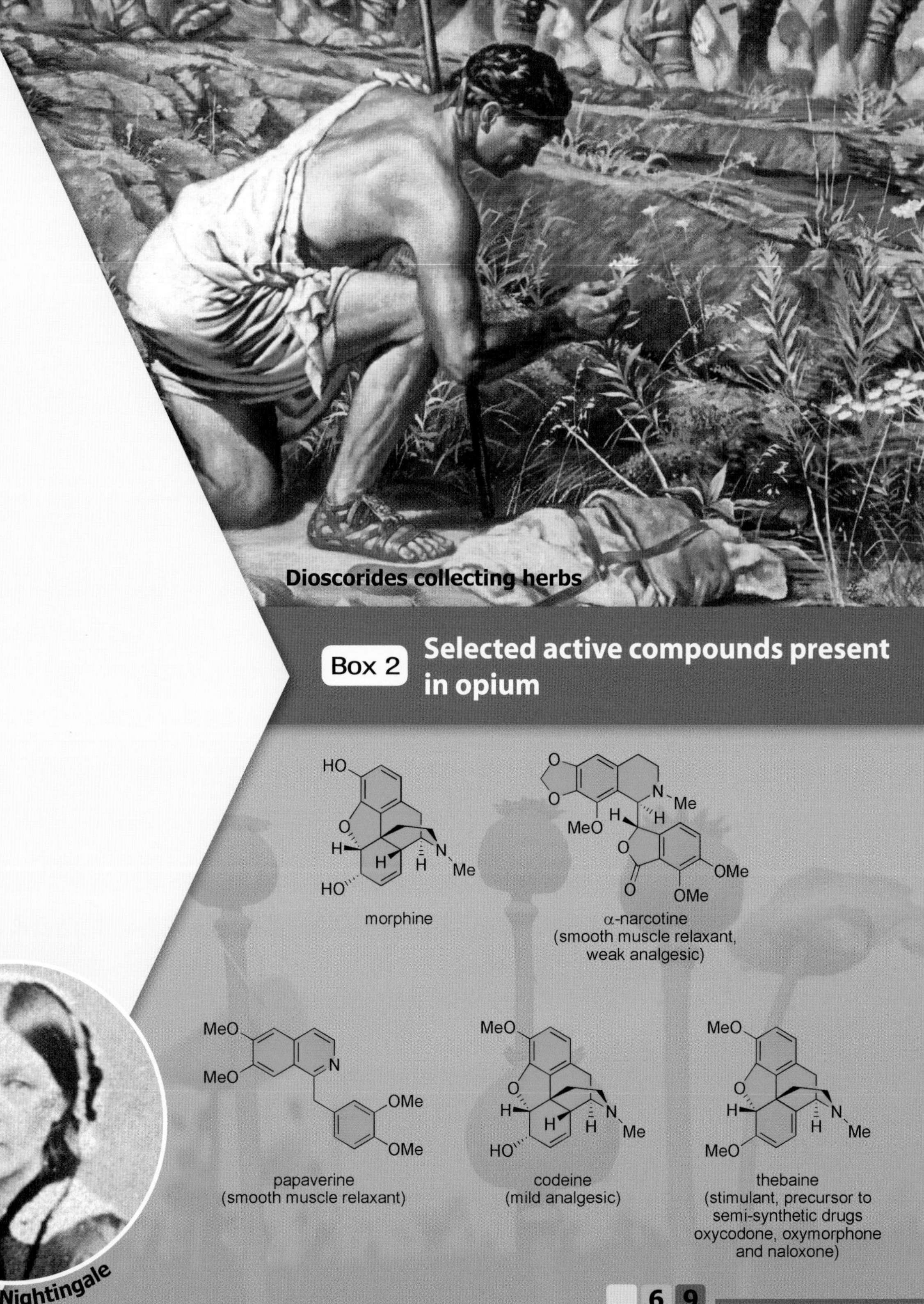

Dioscorides collecting herbs

Opium smoking in France in 1903

Box 3 The first verse of Samuel Taylor Coleridge's poem "Kubla Khan" written while under the influence of opium

In Xanadu did Kubla Khan
A stately pleasure-dome decree:
Where Alph the sacred river, ran
Through caverns measureless to man
Down to a sunless sea.
So twice five miles of fertile ground
With walls and towers were girdled round:
And there were gardens bright with sinuous rills,
Where blossomed many an incense-bearing tree;
And here were forests ancient as the hills,
Enfolding sunny spots of greenery.

Samuel Taylor Coleridge

per annum to China, a vast sum in those days. The Chinese authorities appointed an official, Lin Tse-Hsu, with the specific remit to curb this trade, which was crippling Chinese society. One of his first actions was to close the port of Canton to the British, provoking a fight that had escalated into a full-blown war by the beginning of November 1839. The war was to last until 1842, at which point the 10,000-strong combined British forces had succeeded in taking control of enough ports and strategic positions to force the Chinese to capitulate and sign a treaty which, amongst other things, ceded Hong Kong to Great Britain and gave British merchants free and unrestricted trading rights. Hong Kong was not passed back into Chinese rule until the treaty expired in 1997, proof of the longevity of the opium legacy.

Thus, opium played a crucial role in shaping our world over innumerable generations. However, the details of its constituents and their respective properties could not be unraveled and investigated until the science of chemistry began to come of age. Historical accounts differ as to who was responsible for the first isolation of morphine from opium, and exactly when this milestone event occurred. Most proclaim that a German apothecary by the name of Friedrich Wilhelm Adam Sertürner isolated it sometime between 1803 and 1806, while others say a lesser-known scientist, Armand Séquin, should receive the recognition for its isolation in 1803. It is quite possible that both men could independently lay claim to this prize, having isolated morphine at around the same time. The exact date of the first isolation is of little consequence, as it was to be some time before this knowledge was consolidated upon and expanded. This white crystalline substance was originally known as morphium, after the Greek god of dreams, Morpheus, who was the son of Hypnos, the god of sleep. After its isolation, purified morphine was frequently used to treat pain, particularly after the advent of the hypodermic syringe in 1853. Clinicians, however, were not blind to its addictive properties and they launched a search for a non-addictive derivative or substitute. Codeine (methylated morphine) was identified as a constituent of opium in 1832. This analgesic is approximately one tenth as powerful as morphine, and it has since found widespread use as a mild pain reliever, such that today it is sold over the counter in pharmacies the world over. Codeine, when ingested, is partially demethylated to regenerate morphine (Box 4).

At St. Mary's Hospital Medical School in London in 1874, British scientist C. R. Alder Wright was experimenting by combining morphine (a base) with various acids. He boiled anhydrous morphine with acetic anhydride and acetic acid over a stove for several hours, producing a compound that he named tetraacetylmorphine. We now know

Greek god Hypnos, father of Morpheus — God of Dreams

this substance as diacetylmorphine or, more commonly, heroin (Box 4). Wright, although first, was not alone in pursuing this line of investigation. The natural product salicylic acid was acetylated in a similar manner to afford Aspirin® (Chapter 4). Felix Hoffmann, working at the then fledgling pharmaceutical company of Bayer, was credited with discovering this enormously valuable analgesic. It is not hard to understand why, therefore, within eleven days of his first triumph, Hoffmann took an interest in acetylating morphine with the hope of obtaining yet another useful and valuable drug. It might be said that Bayer had a forewarning of the problems associated with their compound as they searched for a name under which to market this discovery. Employees who worked with the drug by testing it on themselves reported that it was a heroic and dangerously pleasing drug, so it was named heroin. Bayer first promoted it as a treatment for severe dry coughs and tuberculosis.

Heroin is amongst the earliest examples of an important category of drugs called prodrugs. A prodrug is an inactive compound that is converted to an active drug within the body. They are usually designed because the desired active drug has poor bioavailability or low stability. For example, it may not be lipophilic (lipid soluble) enough to be absorbed efficiently across the body's lipid membranes, such as those that line the walls of the gut. Lipophilicity is the key to the extraordinary activity of heroin. It is lipophilic enough to cross the blood-brain barrier (a very non-polar barrier that protects the CNS from blood-borne substances) so that it reaches the core of the CNS in higher concentrations than the more water soluble morphine. This leads to a much more intense response within the body. Bayer misguidedly prided itself on selling heroin as a non-addictive morphine substitute. In 1900, another giant pharmaceutical company, Eli Lilly, began selling heroin over the counter in the United States. It is bizarre to browse through pharmaceutical advertisements from this period and see heroin and Aspirin® advocated side by side as treatments for coughs and flu. Later, it was important in military medicine in World War I where it was administered intravenously to grievously wounded and suffering troops on a regular basis. Heroin was even championed, albeit briefly, as a treatment for morphine addiction. Beginning in the 1920s, however, it began to disappear from medicine cabinets throughout the world as clinicians recognized the serious problems associated with heroin. Today it tops the list of illegal drugs, and governments across the world unremittingly seek to stamp out the illicit heroin trade. Such is the potency of heroin that it remains available by prescription in some countries, including Great Britain (under the name diamorphine), where it is used to treat severe pain.

The fertile research era of the twentieth century saw a horde of synthetic opioids discovered and developed. Given the constraints of space we shall mention only the two most famous synthetic examples, meperidine and methadone (Box 5). In 1939, in the pharmaceutical laboratories

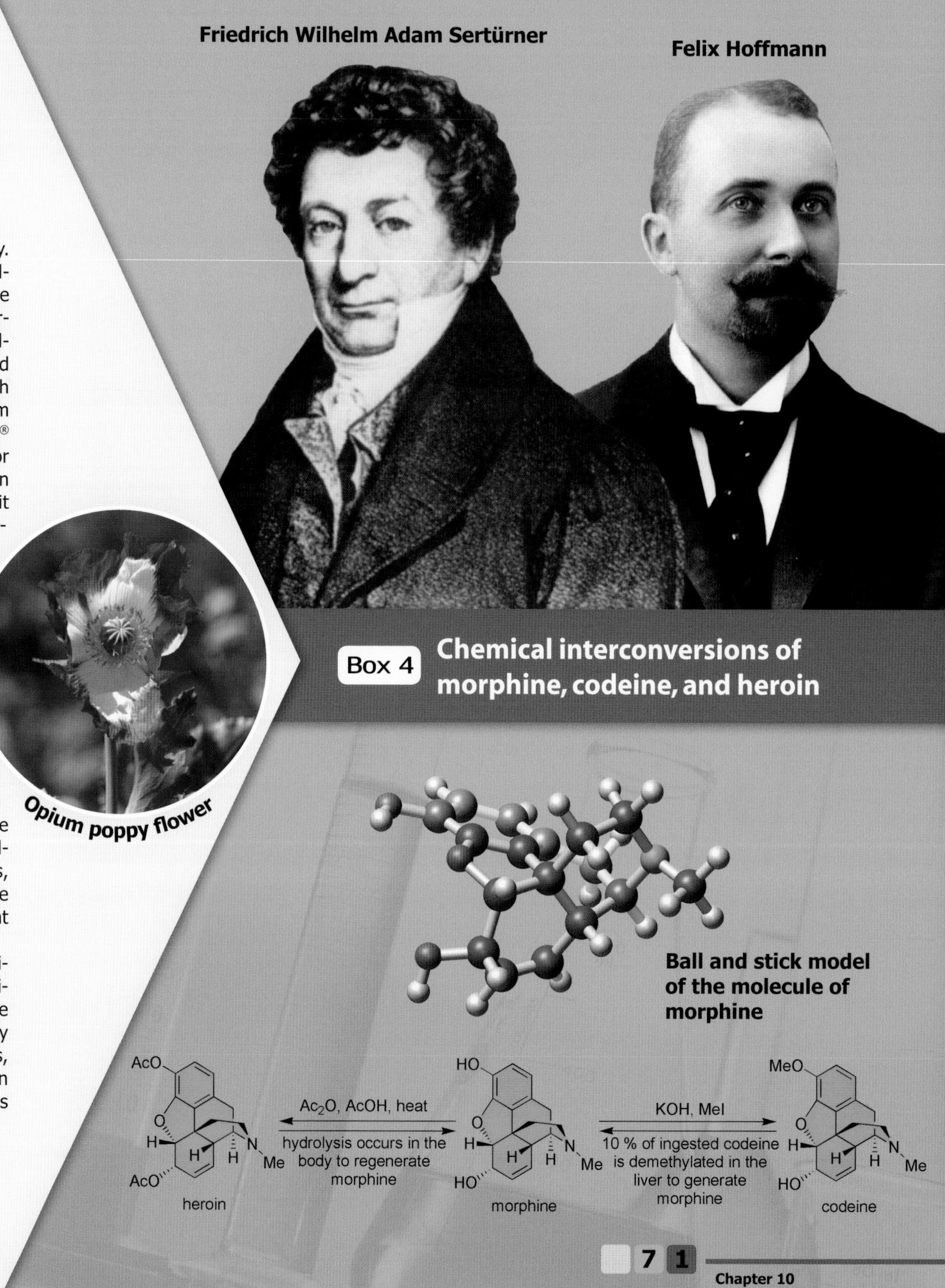

Opium poppies

Box 5 Selected opioid drugs

Synthetic drugs:

naloxone (opioid antagonist, treats heroin overdose)

meperidine (also called pethidine or demerol, analgesic)

methadone (used to manage heroin addiction)

Endogenous opioid:

met-enkephalin (neuropeptide)

of the company I. G. Farbenindustrie (Germany), Otto Eisleb and his colleagues were continuing the century-old search for a non-addictive opioid drug, one which they hoped would serve as a water soluble hypnotic for sedation purposes (a means of slowing the gastrointestinal tract to make surgery easier) and a strong non-addictive painkiller. The latter of these three criteria took on ever-increasing importance due to the need to treat large numbers of soldiers wounded in the war that was raging throughout Europe. Their first success came with the discovery of dolantin. This compound had been synthesized originally as part of a program to develop atropine-like drugs, but was found to be a useful alternative to morphine. By 1944, 1,600 kilograms of dolantin were being produced annually, indicating the importance that the war gave to research into new medicines. Today, this drug, which is milder than morphine, is better known by the clinical names meperidine, demerol, or pethidine. It is used to alleviate post-operative pain or to relieve pain during childbirth, although its use is in decline because of undesirable side effects. When the structure of meperidine is viewed as a two-dimensional drawing, it appears to be structurally disparate from morphine; however, later studies indicated that the three-dimensional conformation adopted by this molecule in solution bears key overlaps with morphine, and it is thought to interact with the opioid receptors in an analogous fashion.

Back at I. G. Farbenindustrie, the newly discovered dolantin had spurred further research activity; it acted as the template for a new series of compounds. On September 25, 1941, a German patent was filed that credits Eisleb's close colleagues, Max Bockmühl and Gustav Ehrhart, with the discovery of polamidon, the drug we now call methadone. The complications of war mean that we know little else regarding their plans for this new analgesic, other than it was tested by the military under the code name Amidon. The development of this drug gained momentum after the war ended, when the US Foreign Economic Management Department requisitioned German intellectual property. Eli Lilly led a field comprised of a number of other American pharmaceutical companies in commercializing this new drug. Methadone is at least as powerful as morphine pharmacologically, but it is also highly addictive and quite toxic. Today, it is used in many countries, somewhat controversially, as a prescribed replacement for heroin in attempts to establish a maintenance regime for those suffering of heroin addiction, with the goal of eventually weaning them off opiates altogether.

The sensation of pain is a very complex phenomenon which takes on many guises. Its perceived intensity is not always proportional to the type or extent of tissue damage, and, within the CNS, psychological, emotional, and hormonal chemical mediators may all become involved. Hence, an inordinate number of different pathways and chemicals play a role in a cascade of responses, depending on the type of pain, its cause, and a plethora of other variables. The situation is so complex that our current knowledge is considered by the experts in the field to be relatively immature. However, pain is also the number one reason for Americans to visit a physician and in the United States alone its economic impact currently approaches \$100 billion annually. Therefore, one can begin to appreciate why analgesics, new and old, are such important medicines. Before we expand further on the biological activity of morphine, it is necessary to clarify two slightly confusing definitions. Opioids are drugs, both natural and synthetic, that exhibit similar effects to morphine because they act at the same receptors in the body (named the opioid receptors); structurally, this group is diverse, including neuropeptides (endorphins), small synthetic molecules, and the opiates themselves (Box 5). Opiates, meanwhile, are all close structural relatives of morphine (Box 2). Opioid drugs act directly on the CNS to produce their analgesic effects. This is in stark contrast to the non-steroidal anti-inflammatory drugs (NSAIDs) such as Aspirin® (Chapter 4), which act at the periphery, as do local anesthetics. A rash of research into identifying the receptors and deconvoluting the pathways involved in both opioid analgesia and the euphoria associated with it was carried out in the early 1970s using radiolabeled opiates. In this technique a radioactive isotope of an element is incorporated into a compound using special reagents during synthesis. The fate of these labeled molecules can then be followed as they pass through, and act upon, an animal model or tissue culture by scanning for the emissions from the radioactive atom. In this instance, the label used to mark selected opiates was tritium, an isotope of hydrogen, which gradually decays, emitting traceable β-particles with a half-life (the time taken for half the tritium to decay) of approximately twelve years. Thus, a number of different types of opioid receptors (subsequently named μ, κ, and δ) were identified, primarily in the gut and the CNS. Later work to elucidate the exact sites of these receptors within the brain was to shed much light on our perception of pain. Solomon H. Snyder and his student, Candace Pert, working at Johns Hopkins University Medical School (Baltimore, USA), were first to publish their work identifying these opioid receptors. They confirmed the early hypothesis and used special techniques pioneered by another great pharmacologist, Avram Goldstein from Stanford University (California, USA), who had suggested opioid receptors must exist in the CNS. The biological effects of opioid drugs are summarized in Box 6.

The investigations of these researchers, and those that followed, were greatly assisted by the discovery of a synthetic antagonist of opioid receptors, called naloxone (Box 5), which could also be used to probe opioid agonist interactions with their receptors. An agonist mimics the action of a natural ligand, stimulating a receptor, while an antagonist blocks a receptor, thus preventing stimulation. Naloxone had been discovered as the by-product of the persistent search for a non-addictive, but powerful, analgesic. Naloxone is now used to reverse the effects of heroin overdose in patients because it displaces the drug from the opioid receptor and blocks the receptor site without stimulating it.

Prior to the opioid receptor identification, it had long been suspected that our bodies manufacture their own powerful painkillers in order to subdue panic and numb pain, thus helping us to deal with situations of stress and injury. In the 1960s, the opinion that morphine might share the opioid receptors with these endogenous analgesics was rife, but before this theory could be subjected to rigorous analysis the natural ligands had first to be found. In 1964, Choh Hao Li (University of California at Berkeley, USA) isolated a pituitary gland hormone that he named β-lipotropin; and, in a crucial observation, he noted that one portion of this molecule appeared to exhibit analgesic properties. Because Li's primary research interest was in fat metabolism,

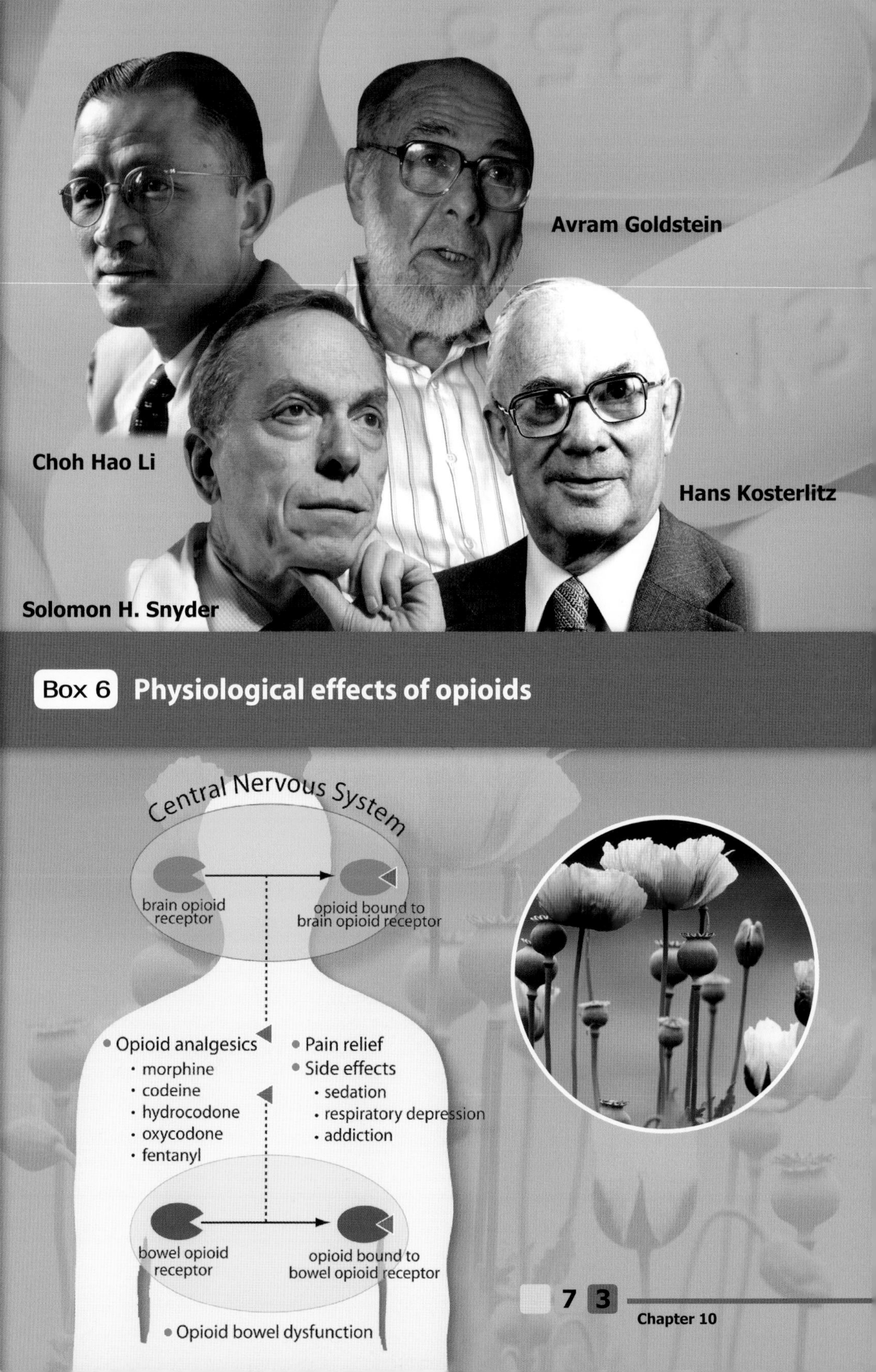

Box 6 Physiological effects of opioids

Unharvested coffee beans

Coca plant

Cola nuts

he did not at first pursue this key finding. As a result, the trail then fades for over a decade until, one year after the triumph of the Johns Hopkins team in locating the elusive opioid receptors, a group from Aberdeen (Scotland) made a significant breakthrough. Hans Kosterlitz, a quiet and thoughtful leader in this field, had been joined in his laboratories by an energetic student named John Hughes. Together, this brilliant pair isolated two hitherto unknown peptides from the brains of pigs, which were shown to share certain physiological properties with morphine. They named these short amino acid chains the enkephalins (from the Greek phrase for in-the-head). Again, despite having two-dimensional structures that are very different to that of morphine, the enkephalins have been shown to share key three-dimensional features with the opiates. Interestingly, the amino acid sequence of one of these compounds was contained within Li's β-lipotropin. Following on shortly from this pivotal discovery, researchers from a number of additional laboratories isolated similar small peptides, some of which were also sub-units of β-lipotropin, that fell under the general umbrella category of endorphins (the name being derived from ***endo***genously produced mo***rphin***e analogue). Endorphins have since received a great deal of attention, and most people have become familiar with their properties. As expected from molecules that stimulate the opioid receptors, they not only act as analgesics and sedatives, but they can induce feelings of euphoria and pleasure. We now know that eating chocolate, laughing, or strenuous exercise all cause the body to release endorphins. It is this process and the subsequent action of the endorphins that constitute or explain the comforting role of chocolate to stressed individuals, the compelling appeal and pleasure attained from smiling, or the so-called 'athlete's high.'

In addition to opioid drugs, several other natural and synthetic compounds are commonly misused and abused. Starting from the mild end of the spectrum we have caffeine, an ingredient in coffee, tea, soft drinks, and chocolate (Box 7). This compound, which acts as a psychomotor stimulant, is present in the coffee beans, cola nuts, tea leaves, and cocoa nuts used to make these beverages and sweets. Cocaine (Box 7) is another psychomotor stimulant, which acts by inhibiting catecholamine uptake by nerve terminals. It is found in the leaves of the coca plant. Cocaine is usually isolated and then supplied as its hydrochloride salt. When the free base is liberated by treatment with sodium bicarbonate, crack (or free-base) cocaine is produced. This latter form of the drug can be smoked rather than injected or sniffed, and is much more potent and dangerous. Cocaine still finds occasional use in medicine as a local anesthetic; however, the quest to find a non-addictive alternative has been much more successful in this case. As early as 1905, procaine (marketed under the name novocaine) was introduced as a non-addictive local anesthetic, and is still in use today. We end our discussion on cocaine with a curious set of facts; it was not until 1929 that Coca-Cola® became totally cocaine-free, when caffeine was added in order to compensate for its withdrawal. Indeed, the name Coca-Cola® was derived from the two principal additives of the original drink, extracts from coca leaves and cola nuts.

Cannabis, marijuana, and hashish are just a few of the plethora of terms used to describe products of the hemp plant, *Cannabis sativa*. The first two refer to the dried leaves and flower heads, the third to the extracted resin. When these products are smoked or ingested, an active ingredient, Δ^9-tetrahydrocannabinol (THC, Box 7), depresses CNS activity and has psychotomimetic (hallucinogenic) effects. In other words, users report feelings of euphoria and relaxation, combined with sharpened sensory perception. Cannabis is perhaps the recreational drug whose negative effects are most underestimated; objective tests have shown that it impairs learning, memory, and motor performance. It may also have a slew of other negative long-term effects, including induction of depression, although the full extent of these after-effects remains to be determined.

Ecstasy, or MDMA (methylenedioxymethamphetamine) (Box 7), is a recently emerged synthetic drug frequently associated with a particular dance and music culture popular amongst teenagers and young adults. When taken, ecstasy produces a relaxed upbeat mood, high energy levels, and heightened social sensitivity. It works by affecting serotonin levels in the brain. Worryingly, this drug has become increasingly fashionable and common just as scientists are discovering the serious long-term consequences of ecstasy use. It has been found that ecstasy permanently damages serotonergic neurons throughout the brain with, as yet not fully understood, negative consequences. Mescaline (Box 7), a compound structurally related to ecstasy, is obtained from the Mexican peyote cactus (*Lophophora williamsii*), whose effects have been written about for centuries. The famous British author Aldous Huxley was an advocate of mescaline intoxication. When the dried crown of this spineless cactus is ingested it causes powerful hallucinations. LSD (lysergic acid diethylamide, Box 7) is a drug that elicits similar effects to mescaline, but with much higher and more frightening potency. LSD is actually a chemical derivative of lysergic acid, which is produced by an ergot fungus that blooms on cereal crops under damp conditions. LSD produces visual, auditory, tactile, and/or olfactory hallucinations, which can be nightmarish and menacing. These so-called trips can reoccur without prior warning many years after the drug has been taken.

In Chapter 7, we were introduced to the elegant biomimetic synthesis of tropinone accomplished by the great chemist and Nobel Laureate (1947) Sir Robert Robinson, in 1917. Tropinone belongs to a class of compounds called alkaloids, which were first identified by the German apothecary K. F. Wilhelm Meissner. Meissner's 1818 definition was broad and included all basic molecules emanating from plants. With time this rather general characterization evolved to mean natural products bearing at least one basic nitrogen functionality. Under both these classifications, morphine and the other opiates are prime

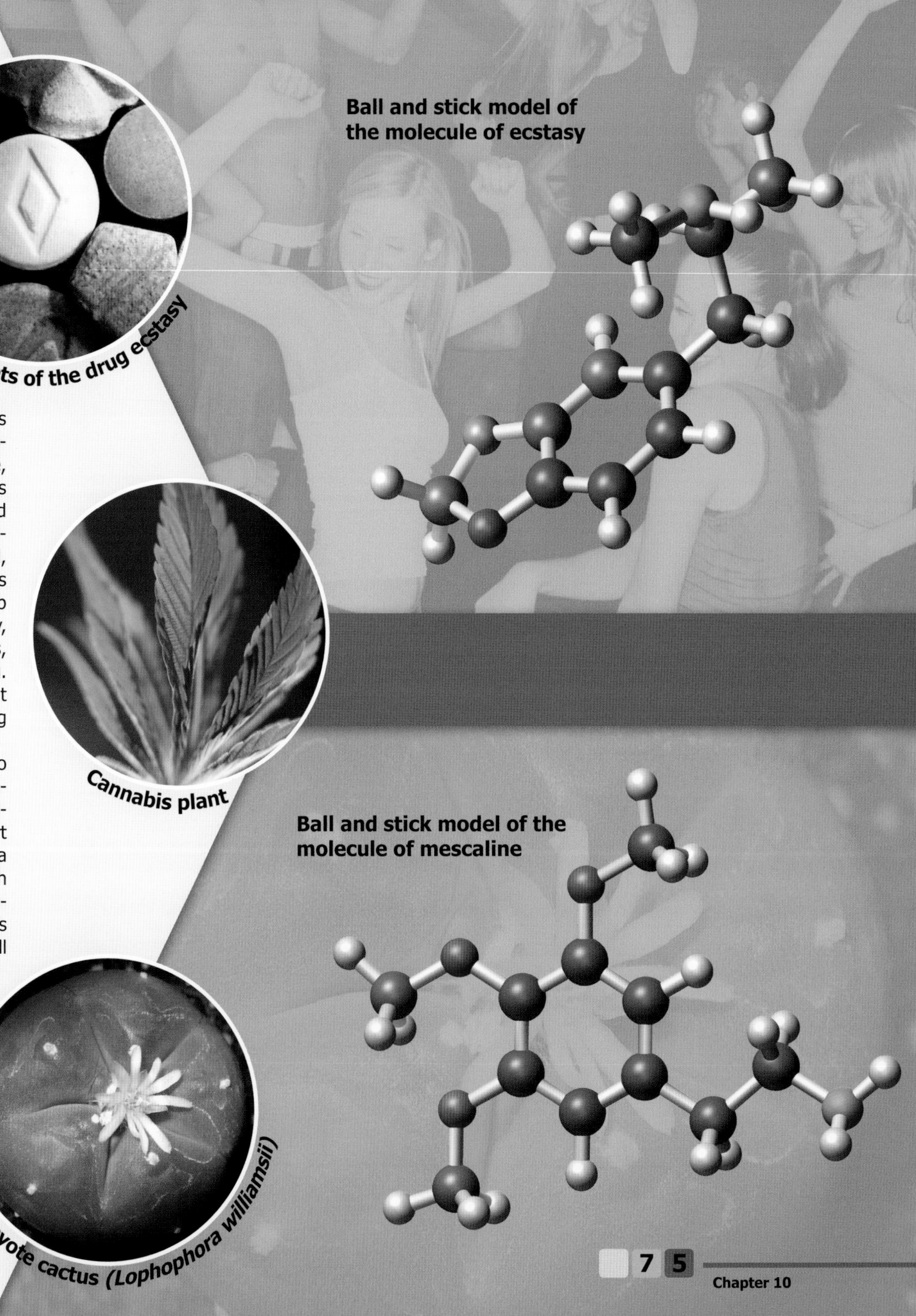

Various tablets of the drug ecstasy

Ball and stick model of the molecule of ecstasy

Cannabis plant

Ball and stick model of the molecule of mescaline

Mexican peyote cactus (*Lophophora williamsii*)

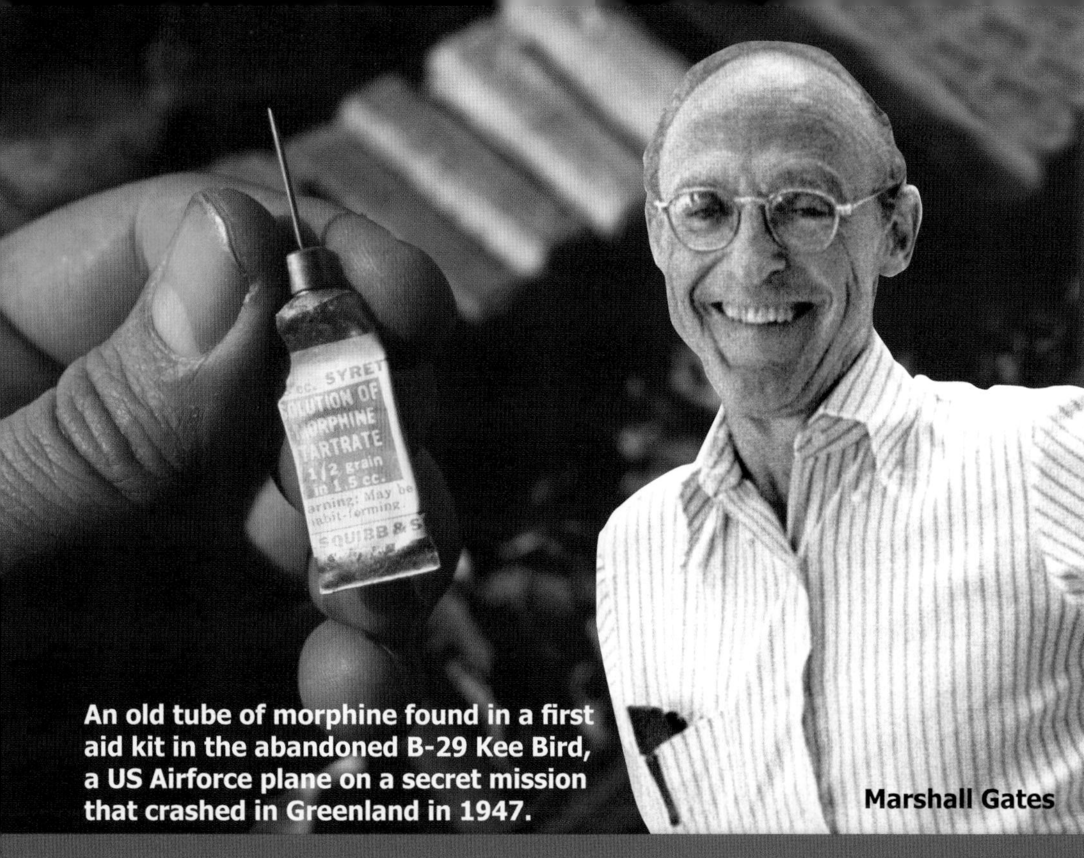

An old tube of morphine found in a first aid kit in the abandoned B-29 Kee Bird, a US Airforce plane on a secret mission that crashed in Greenland in 1947.

Marshall Gates

Box 8 Summary of Gates' total synthesis of (±)-morphine

2,6-dihydroxynaphthalene → → high pressure *Diels–Alder reaction* → → → → → → Rapoport's demethylation conditions → (±)-codeine → (±)-morphine

Sir Robert Robinson

examples of alkaloids. The alkaloids, molecules like reserpine, strychnine, and quinine, mesmerized isolation chemists throughout the nineteenth century and, as the twentieth century dawned, they retained the attention of organic chemists, who were eager to deconvolute their complicated and tangled structures. Sir Robert Robinson built a substantial part of his chemical reputation upon his alkaloid expertise, although not the entirety, for he was an intellectually agile man of many talents. He loved to try and piece together the intricate jigsaw of how the individual atoms of an unknown alkaloid were knotted together, taking into consideration every known detail of their three-dimensional structures, and he greatly enjoyed synthesizing them. Clearly he was not going to be able to resist investigating perhaps the most important alkaloid of his era, morphine. Given that morphine had first been isolated in the opening decade of the nineteenth century, it is quite amazing that it was not until Robinson's work that an accurate structure was proposed. Furthermore, Robinson's 1925 structural proposal, based upon extensive degradation studies, could not be confirmed until the first total synthesis of morphine was completed by Marshall Gates twenty-seven years later. Robinson's success represented the culmination of a century's worth of investigation, during which time many of the names synonymous with the development of organic chemistry (Justus von Liebig, Pierre Joseph Pelletier, and Heinrich Otto Wieland, to name but a few) played a role in analyzing the structure of morphine. It is just speculation, but anyone familiar with Robinson's love of alkaloids might suspect that he would have reveled in the challenge of synthesizing morphine. However, synthetic chemistry was simply not equipped for the rigors of a target like morphine at the time.

During World War II, members of the synthetic chemistry community focused their attention on so-called wartime drugs; the minds of the most illustrious chemists grappled unrelentingly with the extreme challenges of just a few crucial natural product drugs. Chapter 13 relates how the triumphs of the Penicillin Project came about under exactly these circumstances, following a desperate search for a reliable treatment for the infected wounds of soldiers. At that time, morphine also captivated synthetic chemists on both sides of the Atlantic. These investigators were spurred on by the goal of finding creative new approaches to making this drug on an industrial scale or to find new related opiates so that the war wounded need not suffer unnecessary pain. The champion of the morphine race was Marshall Gates, a young professor, just thirty-five years of age, based initially at Bryn Mawr College (Pennsylvania, USA) and then at the University of Rochester (New York, USA). He and his student, Gilg Tschudi, announced the completion of the first total synthesis of morphine in 1952. Central to their synthesis (summarized in Box 8) is the use of a Diels–Alder reaction, although in 1948, when they first published their early reports on how they used this process in the synthesis of a fragment of morphine, the reaction had not yet been given its title (see Box 10). Gates' synthesis began from a convenient dihydroxynaphthalene (Box 8), which was known in the literature, having been constructed more than forty years earlier by Richard Willstätter and Jacob Parnas. One should realize that, while today's chemist would simply have to reach for a chemicals catalog and place an order to obtain such an uncomplicated starting material, in Gates' time chemists had far fewer such luxuries and they had to prepare many of their own materials and reagents. The naphthalene was converted to an *ortho*-quinone, which was, in turn, transformed into the dienophile for the Diels–Alder reaction; the diene partner was simply butadiene. After the rather sluggish [4+2]-cycloaddition (as this process is also known), Gates was in possession of an adduct with three of the required five rings now in place. A complex series of manipulations, including a host of oxidations, reductions, and isomerizations, allowed the synthesis to crawl forward. Hindered by the intricacy and unpredictability of the three-dimensional structure of the growing molecule, Gates and Tschudi persevered and, after closing the fifth and final ring of morphine, they were just a few steps away from their goal. Finally, a reduction afforded codeine, which was readily demethylated using Henry Rapoport's previously reported conditions to give morphine itself. This was a magnificent achievement. Gates, however, remained unaffected by his success and shunned attention, preferring to quietly continue his work as a much-respected teacher and researcher.

Morphine, although conquered, did not stop luring chemists towards its deceptive web of atoms. In each decade since Gates' triumph at least one group of chemists has successfully completed a total synthesis of morphine, and many more have published accounts of their struggles with this molecule. Most recently, world-renowned chemist Barry M. Trost and his research group at Stanford University completed an enantioselective synthesis of (–)-codeine, and thus (–)-morphine, fifty years after Gates' original accomplishment. Among Trost's many talents, the use of palladium-catalyzed cross-coupling reactions (see Chapter 22) stands out, and reactions of this type were fundamental to his conquest of morphine (Box 9). This elegant total synthesis of morphine provides us with an aesthetically pleasing end to the story of this amazing, but fearful, molecule which has impacted the lives of people for so long.

Poppy field

Pd

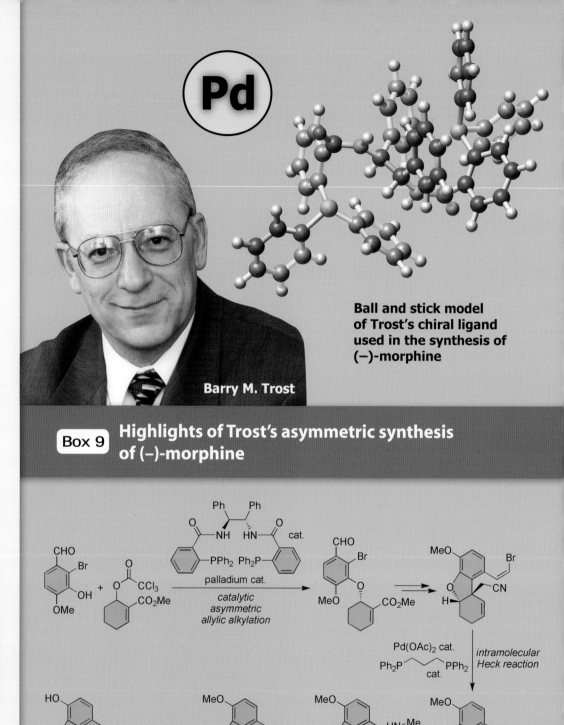

Ball and stick model of Trost's chiral ligand used in the synthesis of (–)-morphine

Barry M. Trost

Box 9 Highlights of Trost's asymmetric synthesis of (–)-morphine

palladium cat.

catalytic asymmetric allylic alkylation

$Pd(OAc)_2$ cat.

Ph_2P PPh_2 cat.

intramolecular Heck reaction

Rice's demethylation conditions

(–)-codeine

(–)-morphine

Box 10 The Diels–Alder Reaction

Early in the twentieth century, a number of chemists began to notice some strange reactions involving conjugated dienes (compounds containing two adjacent alkene groups); however, no explanation for these reactions was offered, and the true identity of the products was often unknown. This situation was to change, in 1928, with a seminal publication from Otto Diels and his student, Kurt Alder. They studied the reaction, shown below, between *para*-quinone (dienophile) and cyclopentadiene (diene), to generate a new six-membered ring. They immediately recognized the utility of their process, stating in their first paper,

"Thus it appears to us that the possibility of synthesis of complex compounds related to or identical with natural products such as terpenes, sesquiterpenes, perhaps even alkaloids, has been moved to the near prospect."

diene + dienophile → [transition state (*endo*)] —Diels–Alder reaction→ monoadduct —Diels–Alder reaction→ diadduct

Diels and Alder were awarded the 1950 Nobel Prize in Chemistry for "their discovery and development of the diene synthesis," as this reaction was known up to that time. Usually, one would expect such a useful process to be applied almost immediately by others in their own research (see, for example, the use of the Grignard reaction by Perkin, Jr., Chapter 6); however, this was delayed in the case of the Diels–Alder reaction, due, in part, to a proprietary warning issued by the discoverers in their first paper,

"We explicitly reserve for ourselves the application of the reaction developed by us to the solution of such [natural product synthesis] problems".

Thus, the diene synthesis did not contribute to any total synthesis campaign until the syntheses of cantharidin (1951) and morphine (1952), reported by Gilbert Stork and Marshall Gates, respectively. Following this rather slow uptake, the Diels–Alder reaction has become one of the most important transformations in modern organic chemistry, and has been extended to include enantioselective reactions promoted by metal or organic catalysts, as well as variants involving heteroatoms (oxygen, nitrogen, or sulfur) in either the diene or dienophile component.

Part of the attraction of this process, along with the efficient generation of new bonds, is its predictability. The Diels–Alder reaction is an example of a concerted pericyclic process (meaning that all the bond-forming and bond-breaking steps take place concurrently, and that the reaction proceeds through a cyclically conjugated transition state), therefore stereochemical information in the reactants is faithfully transcribed in the products. In the late 1960s, Kenichi Fukui and colleagues developed Frontier Molecular Orbital (FMO) Theory, which led to greater understanding of the progress of chemical reactions. This theory provides an explanation for the occurrence and selectivity of the Diels–Alder reaction, allowing chemists to predict the outcomes accurately. The Diels–Alder reaction is, in fact, one of the most predictable in the chemist's arsenal, allowing an almost unshakable confidence in its application within a synthetic plan. Some of the amazing developments and applications of this famous transformation can be found in two reviews, written to coincide with the centenary of Alder's birth (one written by E. J. Corey, and one by K. C. Nicolaou and his students) (see Further Reading).

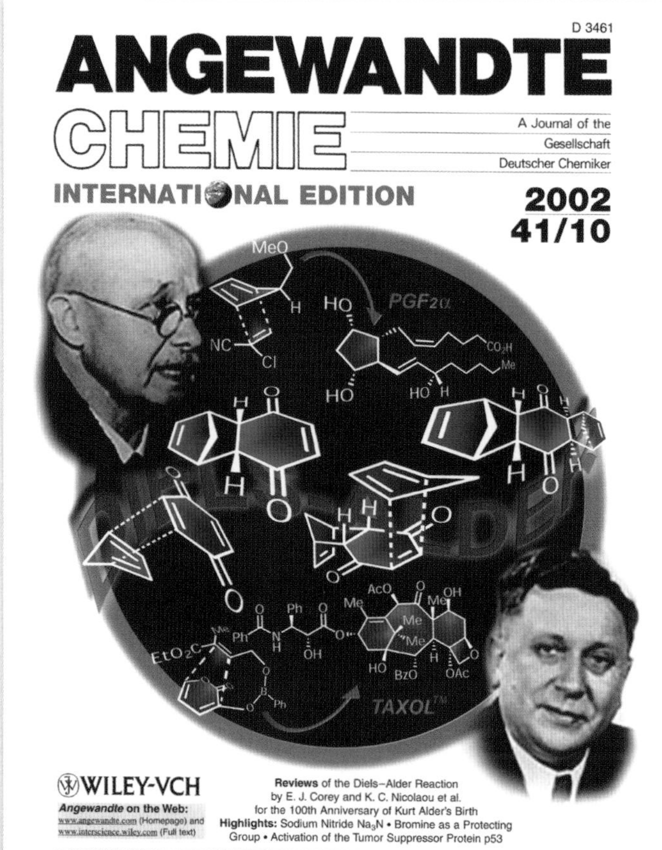
D 3461

ANGEWANDTE CHEMIE

A Journal of the Gesellschaft Deutscher Chemiker

INTERNATIONAL EDITION

2002 41/10

WILEY-VCH

Angewandte on the Web: www.angewandte.com (Homepage) and www.interscience.wiley.com (Full text)

Reviews of the Diels–Alder Reaction by E. J. Corey and K. C. Nicolaou et al. for the 100th Anniversary of Kurt Alder's Birth

Highlights: Sodium Nitride Na_3N • Bromine as a Protecting Group • Activation of the Tumor Suppressor Protein p53

ACIEF5 41 (10) 1629–1808 (2002) · ISSN 1433-7851 · Vol. 41 · No. 10 · May 17, 2002

Further Reading

M. Gates, G. Tschudi, The Synthesis of Morphine, *J. Am. Chem. Soc.* **1952**, *74*, 1109–1110.

R. Schmitz, Friedrich Wilhelm Sertürner and the Discovery of Morphine, *Pharmacy in History* **1985**, *27*, 61–74.

J. Mann, *Murder, Magic and Medicine*, Oxford University Press, Oxford, **1992**.

B. M. Trost, W. Tang, Enantioselective Synthesis of (–)-Codeine and (–)-Morphine, *J. Am. Chem. Soc.* **2002**, *124*, 14542–14543.

E. J. Corey, Catalytic Enantioselective Diels–Alder Reactions: Methods, Mechanistic Fundamentals, and Applications, *Angew. Chem. Int. Ed.* **2002**, *41*, 1650–1667.

K. C. Nicolaou, S. A. Snyder, T. Montagnon, G. Vassilikogiannakis, The Diels–Alder Reaction in Total Synthesis, *Angew. Chem. Int. Ed.* **2002**, *41*, 1668–1698.

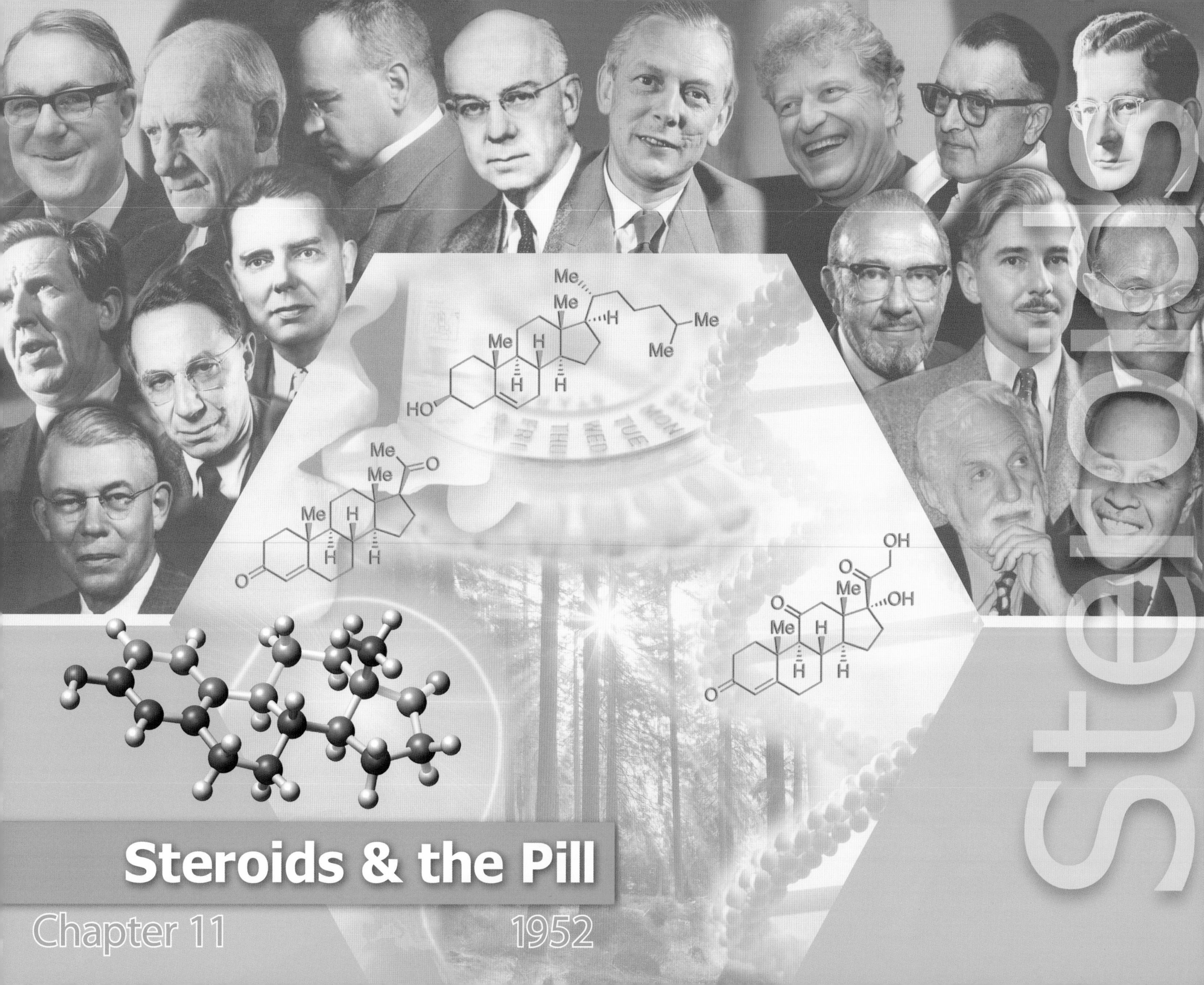
Steroids & the Pill
Chapter 11
1952
Steroids

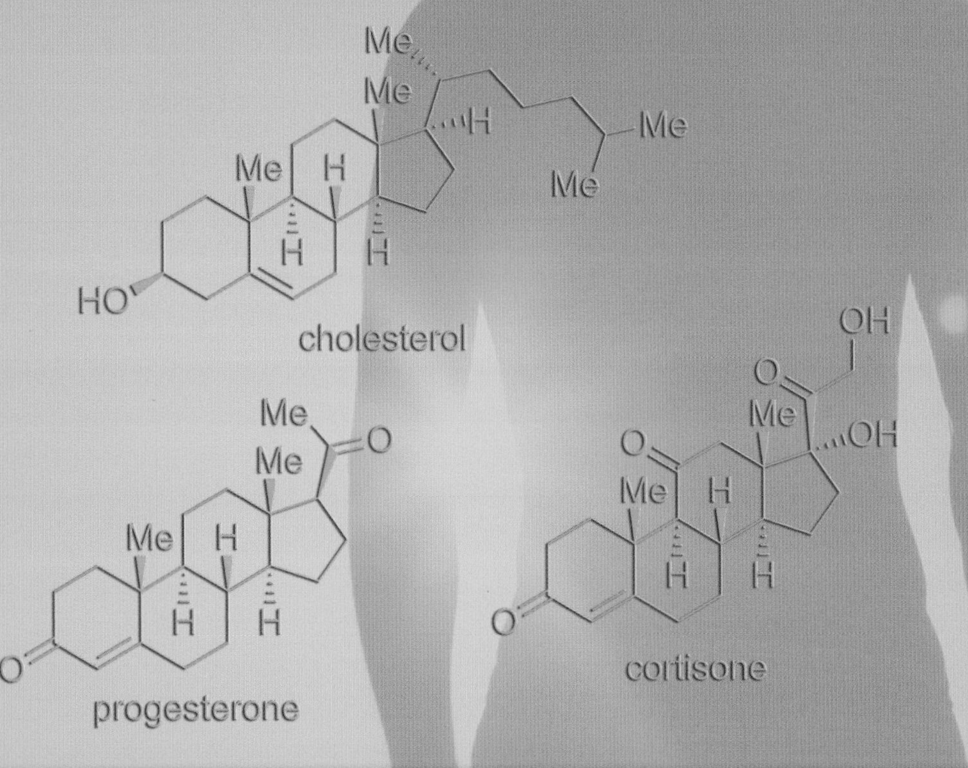

Steroids are found in all the corners of the living world, from plants and fungi to human beings. Within our bodies, steroids serve a host of vital functions, with some acting as hormones. Hormones, which encompass a wide range of biological molecules, are chemical messengers and regulators. They are produced by a specific cell or tissue, and cause a physiological change, or initiate a new activity, within a cell or tissue located elsewhere within the organism. Following their discovery and characterization during the 1930s, the steroid hormones provided an opportunity for the development of medicines that can modulate the chemistry of the body. Today there are a large number of steroid-based drugs on the market with uses ranging from birth control to management of conditions such as asthma and skin disorders. Steroid hormones exert their effects on tissues even when present in exceedingly small amounts, making them very powerful molecules indeed.

These hormones are made within the body from cholesterol, which has a bad public image because high blood levels of this molecule have been linked to cardiovascular disease (see Chapter 26). When correctly balanced, however, this lipid fulfills a number of crucial roles, such as contributing to membrane structure and as the precursor to bile acids. Indeed, 17 % of the solid matter of our brains is cholesterol. Cholesterol is synthesized in sufficient quantities in the majority of our cells from acetyl coenzyme A (as are many other essential fatty acids) through the mevalonic acid pathway (Box 1). The problem stems from the fact that, despite this self-sufficiency, the modern western diet tends to be rather cholesterol-rich, such that many people in the industrialized nations of the world suffer from high blood cholesterol levels. This excess causes vascular damage and contributes to potentially lethal cardiovascular diseases.

In 1927, German chemist Heinrich O. Wieland was awarded the Nobel Prize in Chemistry for his work on bile acids and sterols (cholesterol derivatives) (see timeline of landmark accomplishments on pages 82 and 83). In the following year his compatriot, Adolf O. R. Winhaus, also received the Nobel Prize in Chemistry in recognition of his work, over quarter of a century, on the constitution of sterols. He was christened the "father of steroid chemistry." Despite these honors, this respected pair had made a rather serious mistake at the core of their research, misassigning the structure of cholesterol. In 1932, English scientist John D. Bernal used the then new and groundbreaking technique of x-ray crystallography to correct their error by solving the structure of ergosterol (albeit

Mexican farmer holding a Mexican yam, a source of steroids

with a minor error). In the decades that followed, the steroids became an obsessive subject of study for scientists, who began to realize the full extent of the control over human biology exerted by these pivotal molecules, particularly in relation to male and female reproductive mechanisms. As a consequence of this drive for new understanding, extraordinary and exceedingly laborious measures were taken to isolate and study new steroids. For example, in 1936, Edward A. Doisy of Washington University (St. Louis, USA) processed four tons of fresh sow ovaries in order to obtain a meager 25 mg of the female sex hormone estradiol. Doisy had earlier led a team that isolated estrone, the first steroid hormone, from the urine of pregnant women (1929). A second team, led by Adolf F. Butenandt in Germany, also isolated estrone. Neither of these groups could establish the correct structure for their prized discovery at the time, however, both Doisy and Butenandt would be awarded Nobel Prizes for their work on the steroid hormones. Butenandt's was for Chemistry in 1939 (which he declined due to pressure from the Nazi regime) and Doisy's was for Physiology or Medicine in 1943. Leopold Ružička, a famous Swiss chemist who had also made significant contributions to the deconvolution of steroid structure and biochemistry, was also honored with the Nobel Prize in Chemistry in 1939. During this frenetic era, a revolution was taking place in chemistry. Until the 1920s, all structural information about a molecule had to be garnered from its chemical reactions; molecules were degraded and derivatized to yield structural clues, or a synthesis was undertaken in order to prove the proposed constitution. With the development of x-ray crystallography and other techniques, the physical properties of a molecule took on a new importance, and chemists instigated studies into a new sub-discipline encompassing analytical chemistry and spectroscopy. In 1939, Willard Allen, in the United States, isolated thirty milligrams of progestin (from nearly twenty kilograms of ovaries) and used UV (ultraviolet) spectroscopy to identify key groups within its structure. This example is one of the first cases where instrumental techniques complemented chemical methods in structural elucidation.

Mexican yam (*Dioscorea macrostachya*)

Steroids are terpenes (see Chapter 6) harboring a complex, all-carbon tetracyclic skeleton, that eluded early synthetic attempts until 1939, when Werner E. Bachmann at the University of Michigan (USA) succeeded in preparing the sex hormone equilenin. This sterling accomplishment was followed, in 1951, by a total synthesis of the more complex steroid, cortisone, by Robert Burns Woodward (Box 3). Woodward was a brilliant synthetic organic chemist who was destined to change the face of total synthesis forever through his insight and the elegant solutions he found to so many synthetic problems. Cortisone belongs to a sub-class of steroids originating from the adrenal gland. The isolation, characterization, and investigation of the biological effects of these adrenal cortex steroids earned the Americans, Edward C. Kendall and Philip S. Hench, and the Swiss, Tadeus Reichstein, the 1950 Nobel Prize in Physiology or Medicine. Also during the 1950s, researchers at the pharmaceutical firm Merck in Rahway (New Jersey, USA) including Lewis Sarett and

Mexican farmer digging for Mexican yams

Box 1 The mevalonic acid biosynthetic pathway to cholesterol

acetyl coenzyme A [Ac CoA] → hydroxymethylglutaryl coenzyme A [HMG CoA] → (HMG CoA reductase) → mevalonic acid → farnesyl pyrophosphate (FPP) → many other lipid biomolecules

farnesyl pyrophosphate (FPP) → squalene → (squalene epoxidase) → squalene oxide → (enzymatic cyclization) → lanosterol → cholesterol

Box 2 Landmarks in steroid chemistry

1927

Heinrich O. Wieland wins the Nobel Prize in Chemistry for his work on sterols and bile acids.

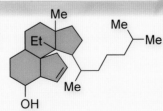

1928

Adolf O. R. Winhaus wins the Nobel Prize in Chemistry for his work on the constitution of the sterols, however, the structure for cholesterol is incorrect.

Wieland–Winhaus cholesterol structure

1929

Edward A. Doisy and Adolf F. Butenandt independently report the first isolation of a steroid hormone, estrone, from the urine of pregnant women, structure unknown.

1930

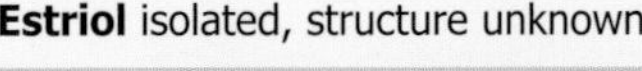

Estriol isolated, structure unknown.

1931

Androsterone isolated, structure unknown.

1932

John D. Bernal publishes x-ray structure of the cholesterol derivative ergosterol which is only slightly mistaken in its positioning of the pendant hydroxyl group.

Accumulated evidence from many sources allows correct structure of cholesterol to be deduced.

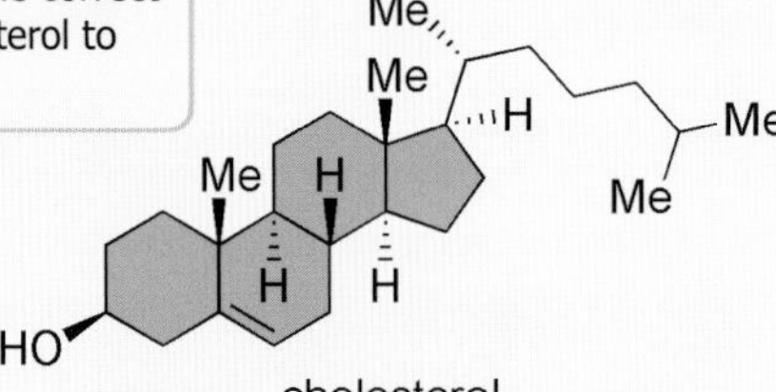

cholesterol

1933

Equilenin and equilin isolated.

1934

Progesterone isolated.

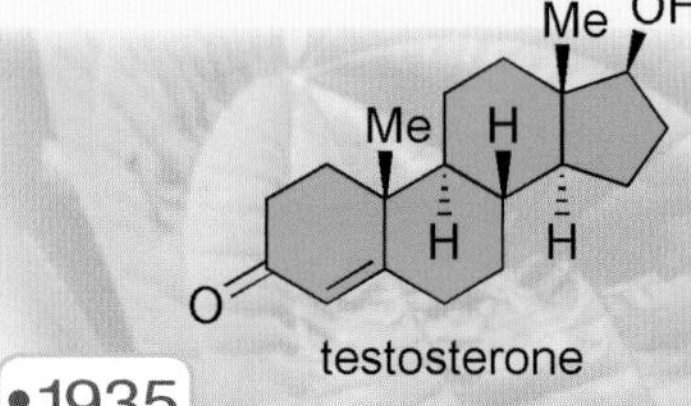

testosterone

1935

Estradiol and testosterone isolated.

1936

Edward A. Doisy isolates 25 mg of estradiol from four tons of sow ovaries.

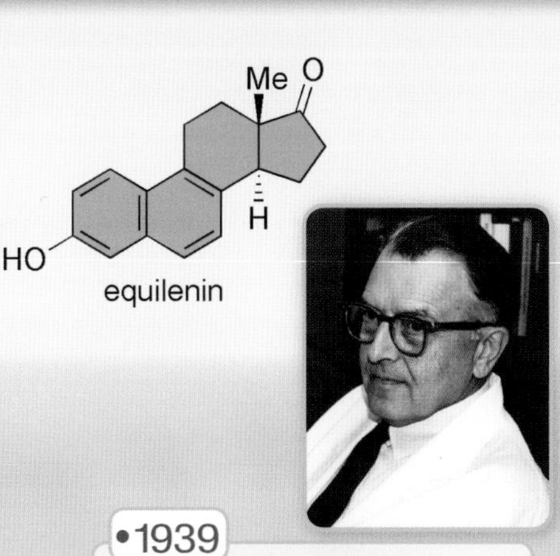

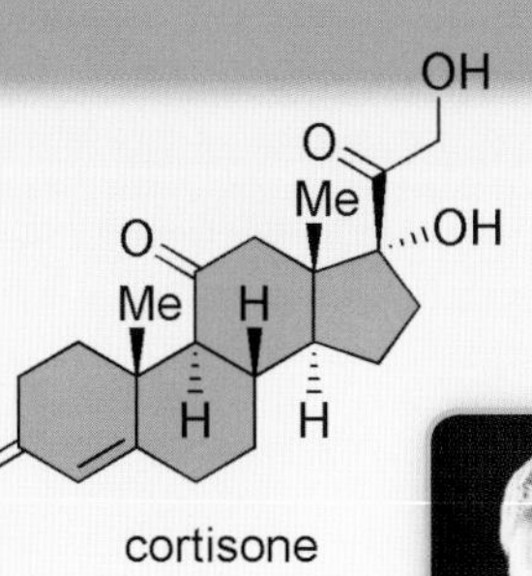

•1939

Willard Allen isolates 30 mg of progestin from 19.7 kg of ovaries and establishes its structure using a combination of methods, including UV spectrometry.

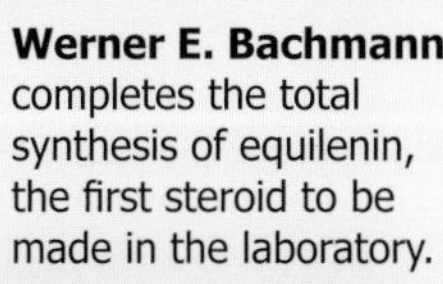

Werner E. Bachmann completes the total synthesis of equilenin, the first steroid to be made in the laboratory.

•1950

Edward C. Kendall, Tadeus Reichstein, and Philip S. Hench win the Nobel Prize in Physiology or Medicine for "their discoveries relating to the hormones of the adrenal cortex, their structure and biological function."

•1951

Robert B. Woodward completes the synthesis of cortisone and cholesterol.

•1953

Sir Robert Robinson and Sir John W. Cornforth complete the synthesis of *epi*androsterone.

•1971

William S. Johnson completes the biomimetic synthesis of progesterone.

progesterone

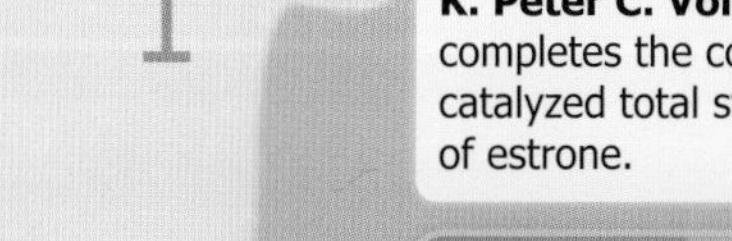

•1977

K. Peter C. Vollhardt completes the cobalt catalyzed total synthesis of estrone.

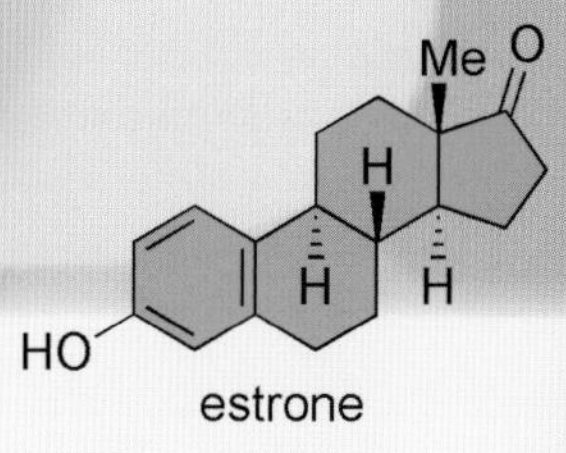

Steroid Chemistry

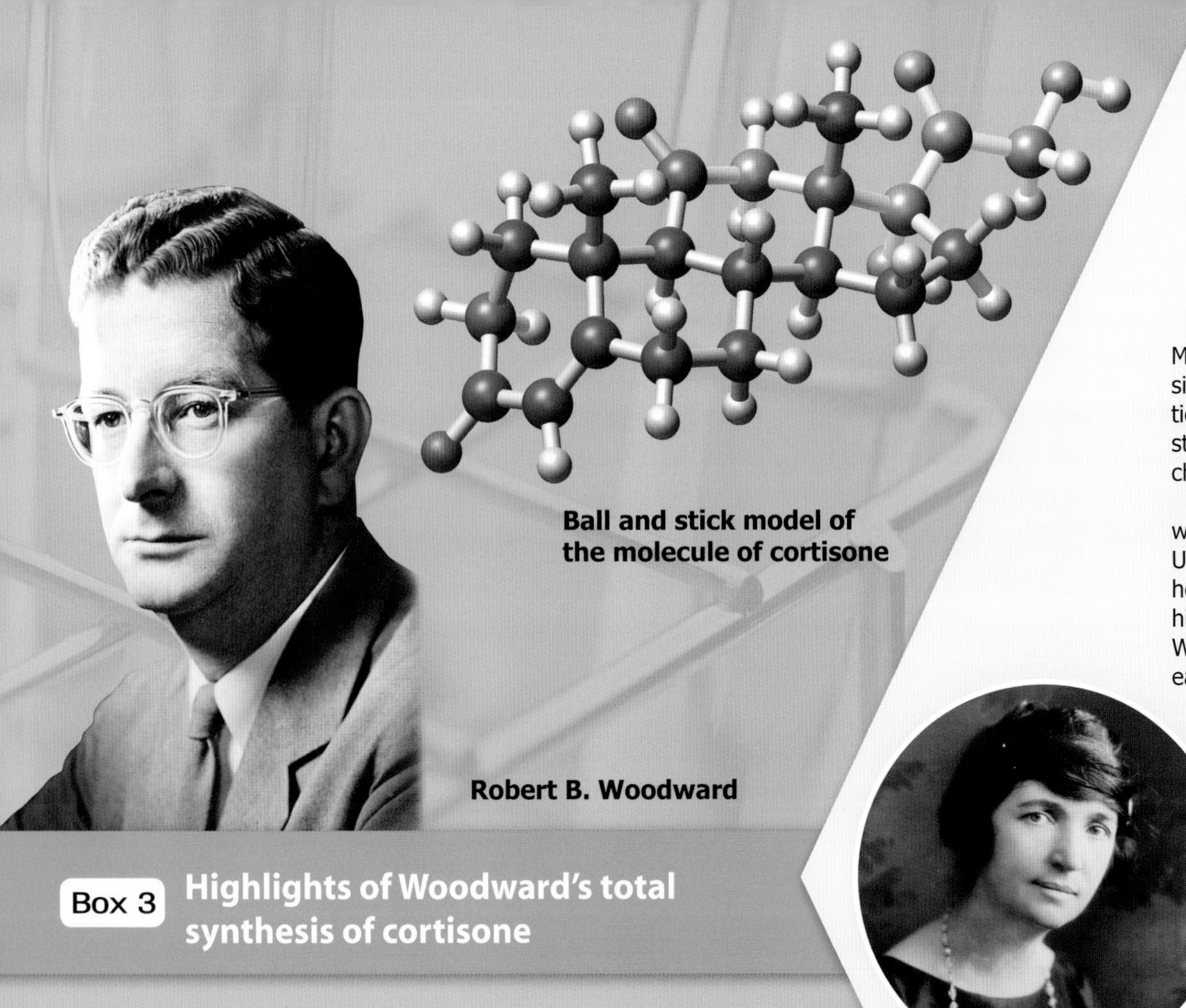

Ball and stick model of the molecule of cortisone

Robert B. Woodward

Margaret Sanger

Gregory Pincus

Box 3 Highlights of Woodward's total synthesis of cortisone

Max Tishler, made significant contributions to the field of steroid synthesis and chemical manipulation.

Born in Boston, Woodward was the son of British immigrants to the USA. Unfortunately, his father died when he was less than two years old, leaving his mother to raise him single-handed. Woodward's academic ability was revealed early in his childhood when he methodically reproduced a series of reactions at home equivalent to that of a typical university undergraduate practical course, using a chemistry set bought for him by his mother. During his high school education he skipped grades, such that by the age of nineteen he was poised to graduate from the Massachusetts Institute of Technology (MIT, USA). His bachelor's degree was awarded despite the fact that he had been expelled in the second year for taking final exams without attending the corresponding classes, in order to leave more time for library and laboratory work in his schedule. He had made up for the time lost following this disciplinary incident by successfully condensing two years worth of courses into a single year. The thirty year period following his graduation in 1936 was a most dramatic time in the history of the art and science of total synthesis, primarily due to Woodward's amazing collection of accomplishments in the field (see also Chapters 12, 16, and 17). He masterminded a relentless attack on the boundaries of possibility within synthetic organic chemistry, tackling molecules whose diversity and complexity were considered unattainable in the laboratory at the time. In each case, the ideas and concepts deployed bear the stamp of his exquisite artistry and sheer genius. These features were recognized by the Royal Swedish Academy of Sciences who awarded him the Nobel Prize in Chemistry in 1965 "for his outstanding achievements in the art of organic synthesis."

In the same year that Woodward synthesized cortisone, an important meeting took place in New York City. The veteran birth control campaigner Margaret Sanger, who had trained as a nurse in New York's poverty stricken Lower East Side, was introduced to Gregory Pincus at a dinner party. Pincus was a physiologist who had been vilified by the scientific establishment some years earlier after he accomplished the *in vitro* fertilization of rabbits. Many believed this work had violated the sanctity of life. Sanger described to him her lifelong dream of a "magic pill" which would prevent unwanted pregnancies and give the poor working-class women, whose staunch advocate she had become, more control over their bleak lives. Pincus told her of recent investigations into fertility that had suggested the use of steroid hormones may result in just such a treatment. Sanger decided there and then to fund Pincus, through the auspices of her charitable foundation (Planned Parenthood), to build on the work already done in this area. In the 1930s and 1940s, scientists had established that a woman cannot become pregnant for a second time during a pregnancy because her ovaries secrete estrogen (also known as oestrogen) and progesterone. The estrogen acts on the pituitary gland so that it withholds the hormones necessary to promote ovulation, and the progesterone reinforces this message by suppressing the production of leutenizing hormone (LH). It was clear from these findings that manipulation of steroids might allow artificial control of a woman's fertility. Schering Laboratories, a German pharmaceutical company who showed early interest in commercial applications of this emerging science, isolated 20 mg of natural progesterone from 625 kg of ovaries obtained from 50,000 sows in 1934. Impressive as it was, this feat demonstrated just how difficult, and therefore economically unfeasible, this isolation technique was at the time (progesterone sold for around $80 per gram in the 1930s, an astronomical sum for the time). The difficulties in extraction and purification of natural progesterone delayed further work in the field until the 1940s.

In 1941, chemist Russell E. Marker, from the Pennsylvania State University (USA), devised a solution to this urgent supply problem. He found a way to synthesize progesterone from a substance called diosgenin, which could be extracted from a plant. His persistent investigations also led to the discovery of a new and more plentiful source of diosgenin, the Mexican yam (*Dioscorea macrostachya*), known colloquially as *cabeza de negro*. Marker, who was collaborating with the American pharmaceutical company Parke-Davis, could not persuade his colleagues to apply for foreign patents for his process, so he severed his ties in the United States and left for Mexico. There he set up his own company, which was named Syntex (from ***synt***hesis and M***ex***ico), with an eclectic set of business partners, Emerik Somolo (a Hungarian immigrant to Mexico) and Federico Lehmann (a German-trained scientist), to produce progesterone and other steroids. He also published a number of papers in the chemical literature at this time which stated his working address as being Hotel Geneve, Mexico City, a curious fact, but one which was in keeping with his rather eccentric personality. Unfortunately, although a great scientist, Marker was something of a maverick, and following a dispute over profit sharing he left Syntex in 1945, taking the crucial details of his process with him. In the 1940s and '50s, chemist Percy L. Julian, already notable for his 1935 total synthesis of physostigmine, an alkaloid used in the treatment of glaucoma, also made a number of advances in the practical production of steroids. Working for the Glidden Company, he developed techniques for the large-scale isolation of sterols from soy bean oil and went on to devise methods for their conversion into several therapeutically useful steroids, including progesterone and hydrocortisone.

In 1939, Carl Djerassi, an Austrian who fled from war-torn Europe to the USA, joined the Swiss pharmaceutical company Ciba at their New Jersey facility. He left Ciba temporarily to pursue a doctoral degree at the University of Wisconsin

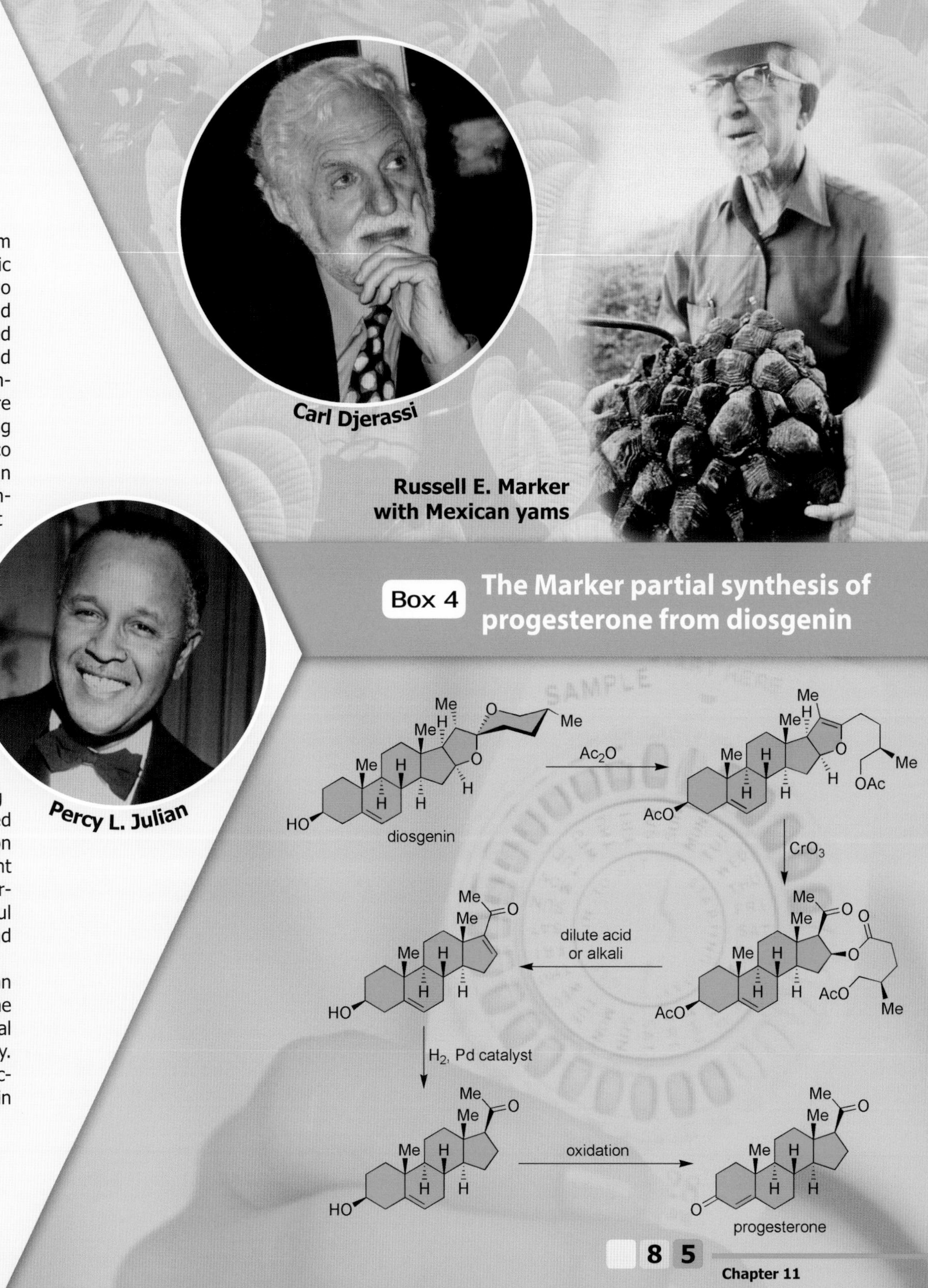

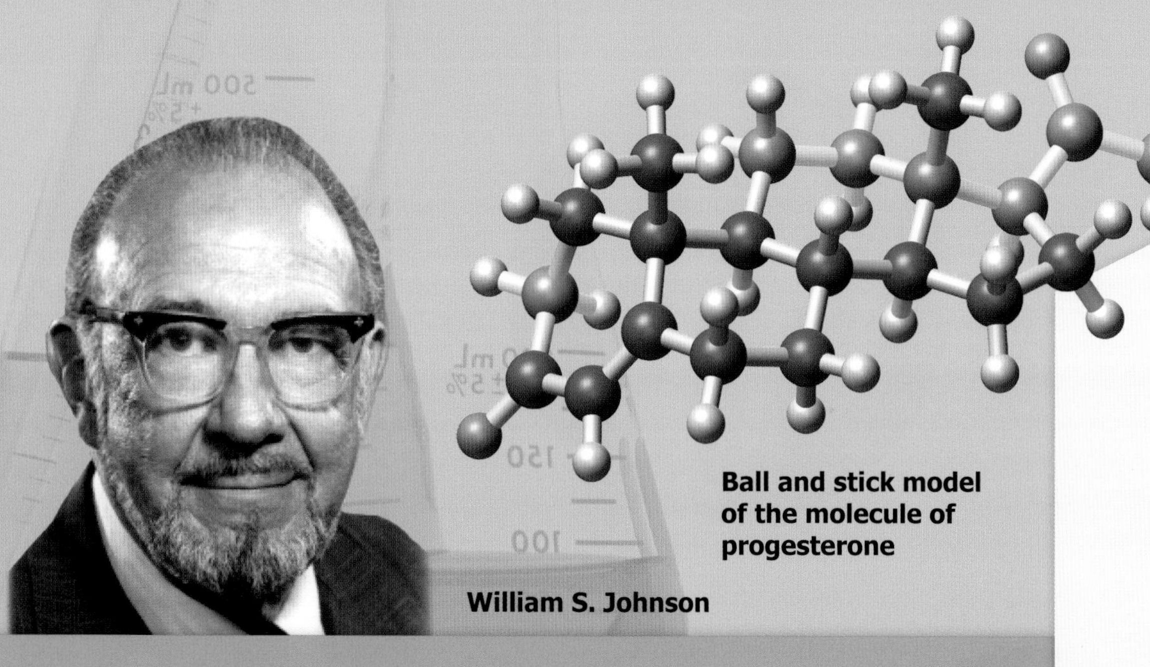

Ball and stick model of the molecule of progesterone

William S. Johnson

Box 5 Johnson's biomimetic total synthesis of progesterone

CF_3CO_2H

cation π-cyclization

aq. K_2CO_3

ozone

aq. KOH

(±)-progesterone

(USA). His dissertation was written on the topic of steroids, in particular the transformation of the male sex hormone testosterone into the female sex hormone estradiol through an innovative sequence of reactions. Inspired by these investigations, Djerassi returned to the company and argued fiercely with his managers because he desperately wanted to work on the synthesis of cortisone, the burning steroid problem of the day. However, this field of research was reserved for a team of scientists at the corporate headquarters in Switzerland. Disappointed, Djerassi left Ciba to take up a position with Syntex, thus filling the void left by Marker's departure. In 1951, coinciding with Woodward's total synthesis of cortisone, Djerassi succeeded in synthesizing the same steroid from diosgenin. His endeavors also culminated in the synthesis of a designed steroid, norethindrone, which was eight times more potent than natural progesterone and, most importantly, orally active. Syntex sent this compound to a biologist, Elva Shipley, at Endocrine Laboratories, Inc., in Wisconsin for further testing. Meanwhile, Frank Colton, the chief chemist at the American pharmaceutical company G. D. Searle, developed another synthetic progesterone. Gregory Pincus, however, beat both companies in the race to what became known as *the pill*, by undertaking the first clinical trials of progesterone as a contraceptive, in collaboration with the respected Harvard gynecologist John Rock. Their trial was a resounding success, establishing the pill regimen as an effective birth control strategy. After further investigations, and based on comparative results, Rock favored Searle's formulation, named Enovid®, so it was this drug that was submitted to the Food and Drug Administration (FDA) for approval in the United States. Initially, the FDA would only permit its prescription for menstrual disorders, as contraception itself was controversial and, in certain states, still illegal. By 1960, however, the situation had changed, mainly because half a million American women were already taking Enovid® for 'menstrual disorders.' Over the years since these groundbreaking days, the steroid dosage has been reduced dramatically and so-called combination pills, using synthetic progesterones and estrogens, have almost completely replaced the progesterone-only pill. Today, hundreds of millions of women worldwide take these contraceptive pills every day. They have radically changed the face of many societies around the world and have contributed enormously to the emancipation of women.

Enovid® bottle

From the standpoint of organic synthesis, the 1970s was a resplendent era, and this period saw a renaissance in the steroid field. In 1971, William S. Johnson synthesized progesterone in the laboratory employing a stunning biomimetic cascade, an event that has remained as a milestone in the history of total synthesis (Box 5). This result was followed by K. Peter C. Vollhardt's brilliant synthesis of estrone in 1977, in which an imaginative cobalt-catalyzed strategy was enlisted to obtain this highly prized natural substance (Box 6).

The Johnson synthesis (Box 5) was exquisite, not only because the impressive cascade at its heart formed three new carbon-carbon bonds and three additional rings stereoselectively in a single step, but also because it faithfully followed the natural synthesis. It had been established prior to this synthetic triumph that the linear molecule squalene oxide is converted to lanosterol (a biogenic precursor of cholesterol, Box 1) through an enzyme-catalyzed process. Johnson's daring vision was to mimic this sequence in the laboratory. In so doing, he also proved that enzymes were not essential to obtain the desired stereochemical outcome, and confirmed the hypothesis for the stereochemical outcome of such cyclizations put forward by Gilbert Stork and Albert Eschenmoser. Crucial to the success of this strategy was the development of a practical means by which to initiate and terminate the key polyolefin cyclization event. Since this achievement there has been a growing recognition that the analysis of how nature assembles a molecule can inspire synthetic plans for secondary metabolite construction in the laboratory. Today, so-called biomimetic syntheses are very popular, and many are amongst the most elegant examples of modern syntheses. Johnson's biomimetic cascade reaction to prepare the steroid skeleton is a striking example of the use of carbocations in organic synthesis. Carbocations (see Box 7) are reactive species in which a carbon atom has only three bonds and bears a positive charge. These species are usually highly reactive, unstable, and short-lived. Thus, in Johnson's cascade, the initially generated carbocation reacts with the nearby double bond, generating a second carbocation. The sequence is repeated to form two further rings before the final cation is trapped by a nucleophilic solvent molecule.

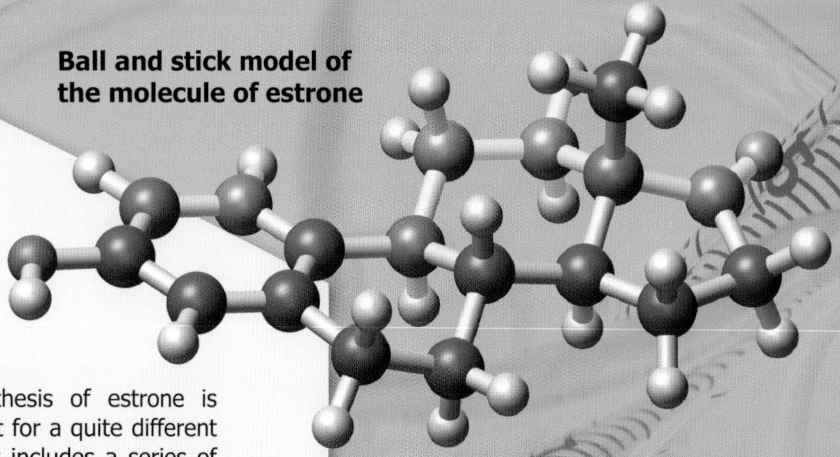

Vollhardt's synthesis of estrone is also remarkable, but for a quite different reason. His strategy includes a series of splendid complexity-building reactions proceeding in a domino fashion (Box 6). Thus, implementation of a specially developed cobalt-mediated cycloaddition was followed by application of the venerable Diels–Alder reaction, thereby installing the full tetracyclic carbon framework of estrone in a rapid and efficient cascade.

The study of steroids has continued from both a chemical and biological perspective, leading to new medicines and the development of elegant chemical synthetic strategies and tactics. The use of both natural and designed steroids has led to treatments for a wide range of diseases (Box 10), along with a more sinister application in the enhancement of athletic performance. Over the years, the steroids have proved tempting synthetic targets, with several more brilliant syntheses to be found in the chemical literature. The steroid framework has also proved to be a mine of theoretical information due to its particular layout. Studies on steroid chemistry have provided valuable insight into the fundamental nature of chemical reactivity, carbocations (Box 7), and radicals (Box 9), as well as the concept of conformational analysis (Box 8). The discovery and investigation of steroids have left a lasting imprint on many facets of chemistry, biology, and medicine.

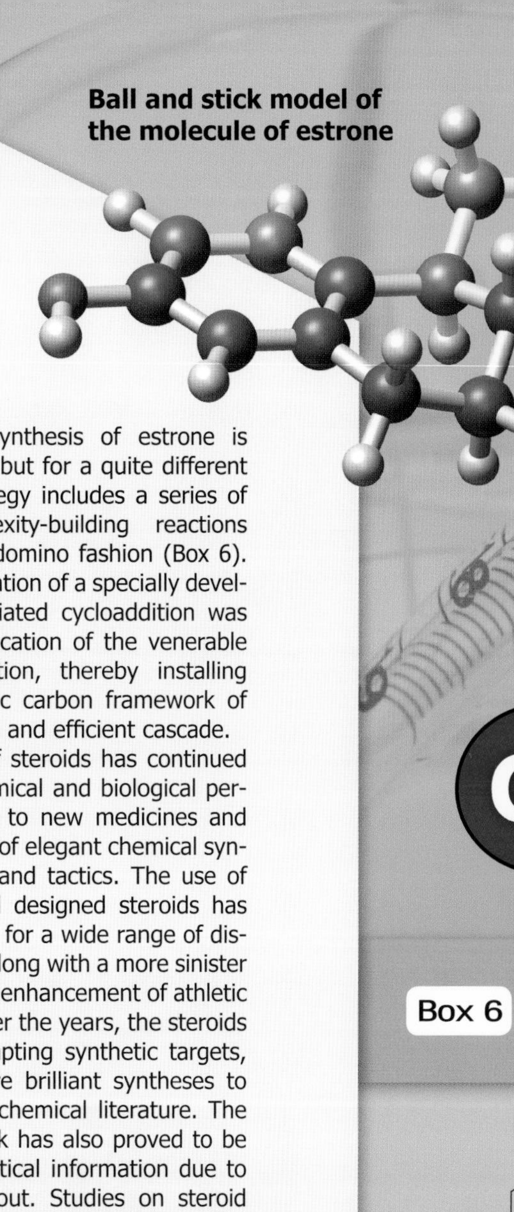
Ball and stick model of the molecule of estrone

Co

K. Peter C. Vollhardt

Box 6 Vollhardt's cobalt-catalyzed total synthesis of estrone

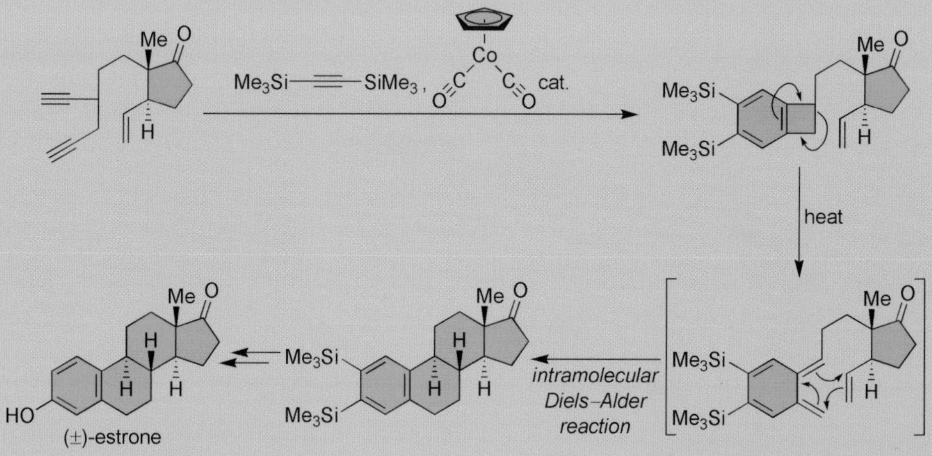

Me O
Me_3Si—≡—$SiMe_3$, Co, O≡C, C≡O, cat.
heat
intramolecular Diels–Alder reaction
HO
(±)-estrone

Box 7 Carbocations

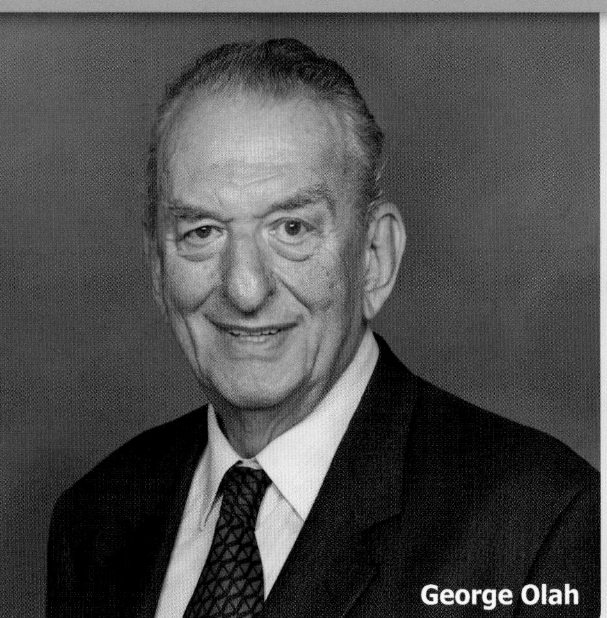
George Olah

tert-butyl carbocation

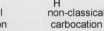

iso-propyl carbocation

non-classical carbocation

Carbocations are of fundamental importance to modern organic synthesis and many other areas of chemistry, including the petrochemical industry. A good understanding of how these species form and react allows chemists to design elegant reaction cascades, new types of chemical transformations, new reagents, and new industrial processes. The existence of carbocations had been known for some time, having been inferred from a range of chemical observations. However, carbocations are high energy, reactive species with only fleeting lifetimes under normal circumstances, and their study was a major challenge. This challenge was met by the Hungarian-born chemist George Olah, of the University of Southern California (Los Angeles, USA), who developed techniques for preparing long-lived carbocations in solution, allowing them to be studied by a range of techniques. Olah used very powerful acids (or superacids), such as triflic acid (CF_3SO_3H) or $HF•BF_3$, to prepare a wide range of carbocations, and observe them by NMR spectroscopy and other techniques. This work proved the planar nature of trivalent carbocations. In his landmark publication in 1972, Olah proposed the now widely used system for naming these species. He won the Nobel Prize in Chemistry in 1994 "for his contributions to carbocation chemistry".

Saul Winstein

Olah's work also contributed to the conclusion of a major controversy of the era. In the 1950s, Saul Winstein reported a short-lived carbocation derived from norbornane, to which he assigned the so-called non-classical penta-coordinate structure shown on the left. Other prominent chemists of the day believed Winstein's structure to be incorrect, and despite many efforts by leading physical chemists the mystery remained. Olah's method for preparing long-lived carbocations allowed the study of Winstein's species by NMR spectroscopy, proving the penta-coordinate structure to be correct. Amongst Olah's many other contributions are a number of widely used reagents for organic synthesis including HF•pyridine, as well as the development of certain reactions of hydrocarbons used in the environmentally benign production of high-octane fuels.

Prior to moving to California, Olah worked in the more basic surroundings of his native Hungary and published some of his early successes in relatively obscure Hungarian journals. Fortunately, Hans Meerwein of the University of Marburg (Germany), another prominent figure in the field of cation research, read these journals and wrote to Olah offering his encouragement. In addition to a string of correspondence, he also sent Olah a gift of a cylinder of boron trifluoride, an expensive chemical that was not commercially available in Hungary. When he was awarded the Nobel Prize in Chemistry in 1994, Olah recalled Meerwein's generosity during his early career, crediting the German chemist for both his moral and material support, the former often as important as the latter in science. In a similar style, George Olah himself extended his generosity and support to his many students and colleagues.

Box 8 Conformational Analysis

Sir Derek H. R. Barton

hydrocortisone

cyclohexane - chair conformation

cyclohexane - boat conformation

Odd Hassel

The flat drawings of organic molecules used to represent their structures on paper can contain a great deal of information regarding the spatial arrangement of the constituent atoms, however, they often fail to give a good representation of the true three-dimensional shape of the compound. In reality, most organic molecules are far from static, and are constantly rotating, vibrating, and flexing. Of the almost infinite number of possible arrangements (conformations), only those lowest in energy will be populated to a significant extent. Understanding the factors that govern the preferred conformation of a given molecule allows chemists to predict the likely shape of the compounds they study. In 1950, British chemist Sir Derek H. R. Barton published what was to become one of the most influential papers in organic chemistry, entitled "The Conformation of the Steroid Nucleus." In this seminal work, Barton described his theory of conformational analysis, and every organic chemist who has followed has, in some way or another, employed these principles in their work. Barton's work was inspired by the fundamental and methodical work of Odd Hassel from Oslo (Norway) on the three-dimensional shapes ('chair' and 'boat') of cyclohexane rings. The importance of these scientific contributions was recognized with the awarding of the 1969 Nobel Prize in Chemistry to Barton and Hassel, "for their contributions to the development of the concept of conformation and its application in chemistry." The rigid conformation of the steroid skeleton is illustrated here in a perspective drawing of hydrocortisone (left).

Box 9 Radicals

A reactive species that we have not yet discussed is the carbon-centered radical. Unlike carbocations (Box 7) and carbanions (see Chapter 6), radicals are usually uncharged but possess an unpaired electron, represented in chemical drawings by a single dot. The unpairing of electrons is generally energetically unfavorable, and radicals are, therefore, highly reactive species. This reactivity can be exploited to effect challenging chemical transformations; however, this power has only been harnessed relatively recently (the 1980s saw an explosion in the use of radicals in synthesis). Radicals can be used in chain reactions and can also react in cascade sequences. These reactions are usually initiated by a high-energy species designed to compensate for the energy required to generate a radical.

Sir Derek H. R. Barton

The steroid skeleton has been constructed in the laboratory by a number of radical processes, and, at one time, many influential chemists believed that a radical cascade might be responsible for the biosynthesis of steroids. In addition to his work on conformational analysis (Box 8), Sir Derek H. R. Barton was a pioneer of the use of radical reactions in organic chemistry. He developed a number of methods for generating radicals and discovered several reactions, such as the Barton deoxygenation procedure for the reductive removal of unwanted hydroxyl groups. Many of his methods are still used frequently today. Barton employed his nitrite ester method for radical generation to prepare aldosterone 21-acetate from corticosterone acetate as shown below.

ClNO: R = H (corticosterone acetate) → R = N=O; $h\nu$, $-{}^{\bullet}NO$

oxygen-centered radical → carbon-centered radical; ${}^{\bullet}NO$

aldosterone acetate oxime; $H_3O^{\oplus}$ → aldosterone 21-acetate

Another pioneer in the field of radical chemistry was Gilbert Stork, who developed many powerful tools for carbon–carbon bond formation in chemical synthesis. Over his long career, Stork published many elegant syntheses involving radical reactions in key steps. In the example shown below, Stork made use of a radical cascade involving a stereoselective cyclization followed by an intramolecular radical addition en route to prostaglandin $F_{2\alpha}$ (see Chapter 15). In recent years, many other natural products have been synthesized using radical-based strategies.

Gilbert Stork

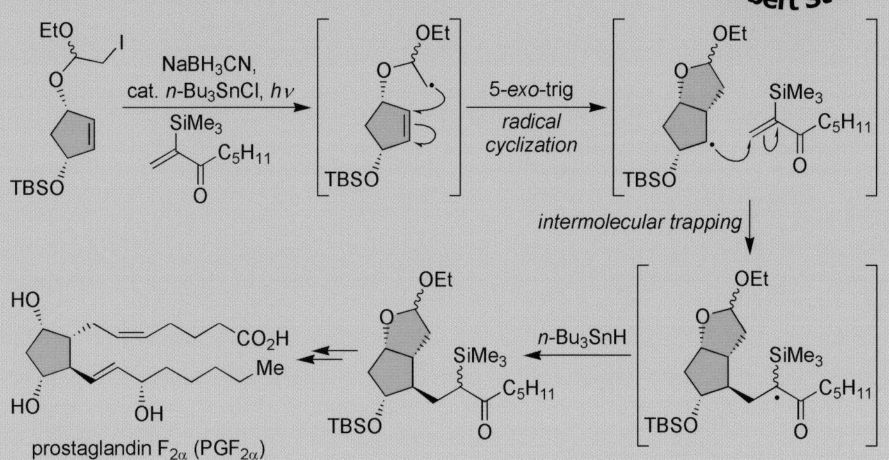

Gerald Pattenden

An impressive synthesis of the steroid skeleton developed by Gerald Pattenden and his group at Nottingham University (UK) is shown below. This process makes use of a common recipe used for radical reactions. The AIBN initiator generates radicals through the loss of nitrogen gas, and the tin hydride species acts as a chain carrier, propagating the chain reaction. This elegant cascade sequence assembles the entire steroid framework from an acyclic precursor, forming four rings and seven stereocenters in a single step.

n-Bu_3SnH, AIBN, heat → radical cascade

Box 10 Selected steroids used in medicine

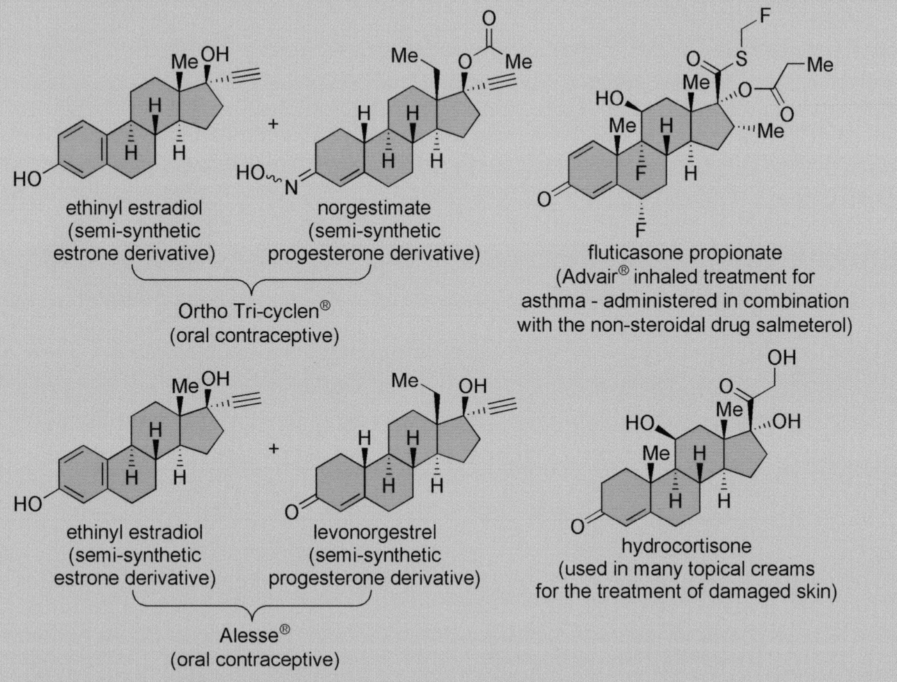

ethinyl estradiol (semi-synthetic estrone derivative) + norgestimate (semi-synthetic progesterone derivative)

Ortho Tri-cyclen® (oral contraceptive)

fluticasone propionate (Advair® inhaled treatment for asthma - administered in combination with the non-steroidal drug salmeterol)

ethinyl estradiol (semi-synthetic estrone derivative) + levonorgestrel (semi-synthetic progesterone derivative)

Alesse® (oral contraceptive)

hydrocortisone (used in many topical creams for the treatment of damaged skin)

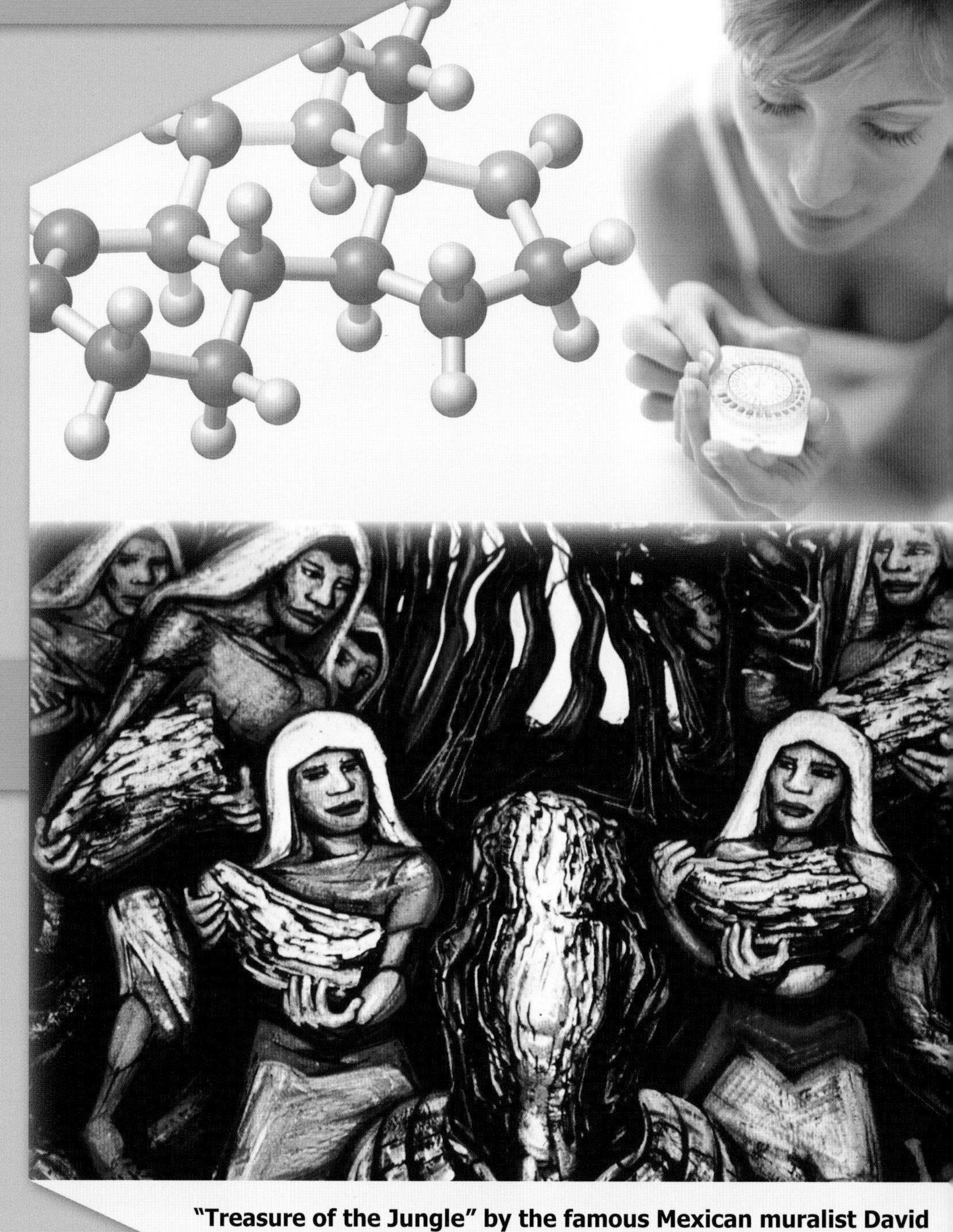

"Treasure of the Jungle" by the famous Mexican muralist David Siqueiros, depicting local workers harvesting Mexican yams. Commissioned by Syntex nearly fifty years ago, this image was provided by Carl Djerassi as were the other images in this chapter showing Mexican yams.

Further Reading

R. B. Woodward, F. Sondheimer, D. Taub, The Total Synthesis of Cortisone, *J. Am. Chem. Soc.* **1951**, *73*, 4057.

B. Asbell, *The Pill: A Biography of the Drug That Changed the World*, Random House, New York, **1995**.

K. C. Nicolaou, E. J. Sorensen, in *Classics in Total Synthesis*, Wiley-VCH, Weinheim, **1996**, pp. 83–94 and 153–166.

C. Djerassi, *This Man's Pill*, Oxford University Press, Oxford, **2001**.

The history of the pill online: http://www.pbs.org/wgbh/amex/pill/.

G. A. Olah, Stable Carbocations. CXVIII. General Concept and Structure of Carbocations Based on Differentiation of Trivalent ("Classical") Carbenium Ions from Three-Center Bound Penta- or Tetracoordinated ("Nonclassical") Carbonium Ions. The Role of Carbocations in Electrophilic Reactions, *J. Am. Chem. Soc.* **1972**, *94*, 808–820.

D. H. R. Barton, The Conformation of the Steroid Nucleus, *Experientia* **1950**, 316–329.

POISON
POI
CHNINE NITRATE
S. P. IX—CRYSTALS
Distributors
and Guarantors
N
N
H
H
H
O
O
H
Strychnine
Strychnine
Chapter 12
1954, 1993

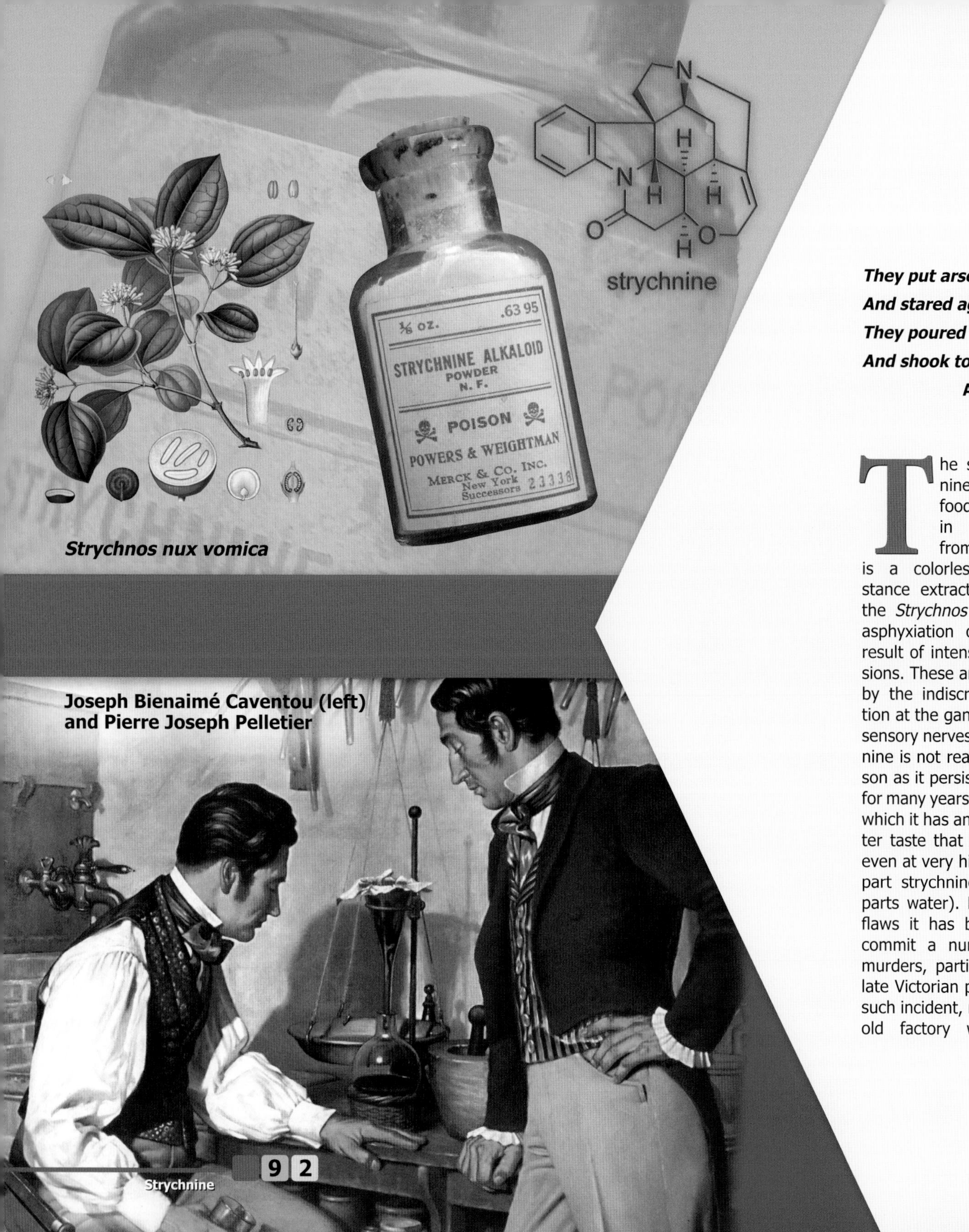

Strychnos nux vomica

Joseph Bienaimé Caventou (left) and Pierre Joseph Pelletier

They put arsenic in his meat
And stared aghast to watch him eat;
They poured strychnine in his cup
And shook to see him drink it up"

Alfred Edward Housman (British poet)

The secret addition of strychnine powder to a character's food or drink is a pivotal event in many murderous plots from fiction. Strychnine is a colorless crystalline substance extracted from plants of the *Strychnos* genus that causes asphyxiation of the victim as a result of intense muscle convulsions. These are brought about by the indiscriminate stimulation at the ganglia of motor and sensory nerves. In reality, strychnine is not really the ideal poison as it persists in the body for many years, in addition to which it has an intensely bitter taste that is perceptible even at very high dilution (1 part strychnine to 700,000 parts water). Despite these flaws it has been used to commit a number of real murders, particularly in the late Victorian period. In one such incident, nineteen year old factory worker Ellen Donworth was poisoned in Waterloo, London, in 1891. It took three people to hold down her twitching and trembling body before she lost consciousness and finally died. The manner of her death suggested strychnine poisoning, and, nine days later, the coroner confirmed this diagnosis. Donworth had fallen pregnant and, having no husband to support her, turned to prostitution in order to supplement her meager income. Her murderer, Thomas Neil Cream, a Scottish born doctor turned serial killer, was convicted and hanged for his crimes in 1892. Cream targeted prostitutes, whom he killed with strychnine pills after offering to perform illegal abortions. Of his known crimes, the first took place in Chicago and more followed in London.

As a qualified doctor, Cream had been able to purchase strychnine with ease from pharmacists because the poison had many legitimate medicinal applications at the time. Medical practitioners believed it had value in the treatment of many disorders, from the disposition to grieve to wandering pelvic pains and congestive headaches. In addition to the dubious medicinal properties now confined to history, strychnine was, and continues to be, used as a pesticide. Two Frenchmen, Pierre Joseph Pelletier and Joseph Bienaimé Caventou, isolated strychnine in 1818 from the Saint Ignatius bean, the seed of the *Strychnos ignatii* tree (named by Jesuits who used the plant medicinally). The long tapering branches of this tree, indigenous to the Philippines, bear beautiful white flowers that have a fragrance reminiscent of jasmine. It is likely that Pelletier and Caventou's investigation of this Asiatic plant was prompted by the numerous folklore uses for its beans, which had been traded around the world. Strychnine has also been found in a range of other plants from the genus *Strychnos*, which are distributed around the warmer climes of the world, including *Strychnos nux vomica* from southeast Asia.

The two French chemists had no idea of the molecular structure of their extract because chemistry was still in its infancy. Indeed, the search for the structure of strychnine did not come to an end until one hundred twenty years later. Two brilliant and astute chemists working independently, Sir Robert Robinson (see Chapter 10) and Hermann Leuchs, had spent several decades securing much of the information regarding the exact connectivity of this particularly complex alkaloid. Their combined results allowed Robinson to propose the correct structure, with his work appearing in print in 1946. The unraveling of the structure of strychnine is frequently attributed to Robert B. Woodward, who independently confirmed Robinson's structural analysis through his own research. However, his results were not published until 1947 and 1948. The structure determination, a feat enough in itself, is all the more remarkable because it was conducted entirely using classical chemistry, before the advent of modern spectroscopic techniques, which greatly facilitate such analyses.

Sir Robert Robinson

Many plants have evolved to make poisonous alkaloids (Box 1) as a chemical defense against herbivores. The ancient Greeks used hemlock, a potion containing the alkaloid coniine, for state executions, Socrates being the most famous victim of this system. Curiously, despite the amino acid-based biogenesis of all these alkaloids, they are rare in the animal kingdom and rarer still in the marine environment. We have already encountered several famous alkaloids; tropinone was the subject of Chapter 7, the story of quinine was the focus of Chapter 9, and the famous opiates morphine and heroin were featured in Chapter 10.

In the decade following the structural elucidation of strychnine, x-ray crystallographic data not only reconfirmed Robinson's structure, but also assigned the absolute stereochemistry of this intricate heptacyclic molecule. In 1952, with this complete information in hand, Robert Robinson remarked, "for its molecular size it is the most complex substance known." The striking molecular architecture that had elicited this famous comment includes seven rings and six contiguous asymmetric centers, concentrated within a molecule that has only twenty-four skeletal atoms. Despite Robinson's assertion, in a

The Death of Socrates by Jacques-Louis David

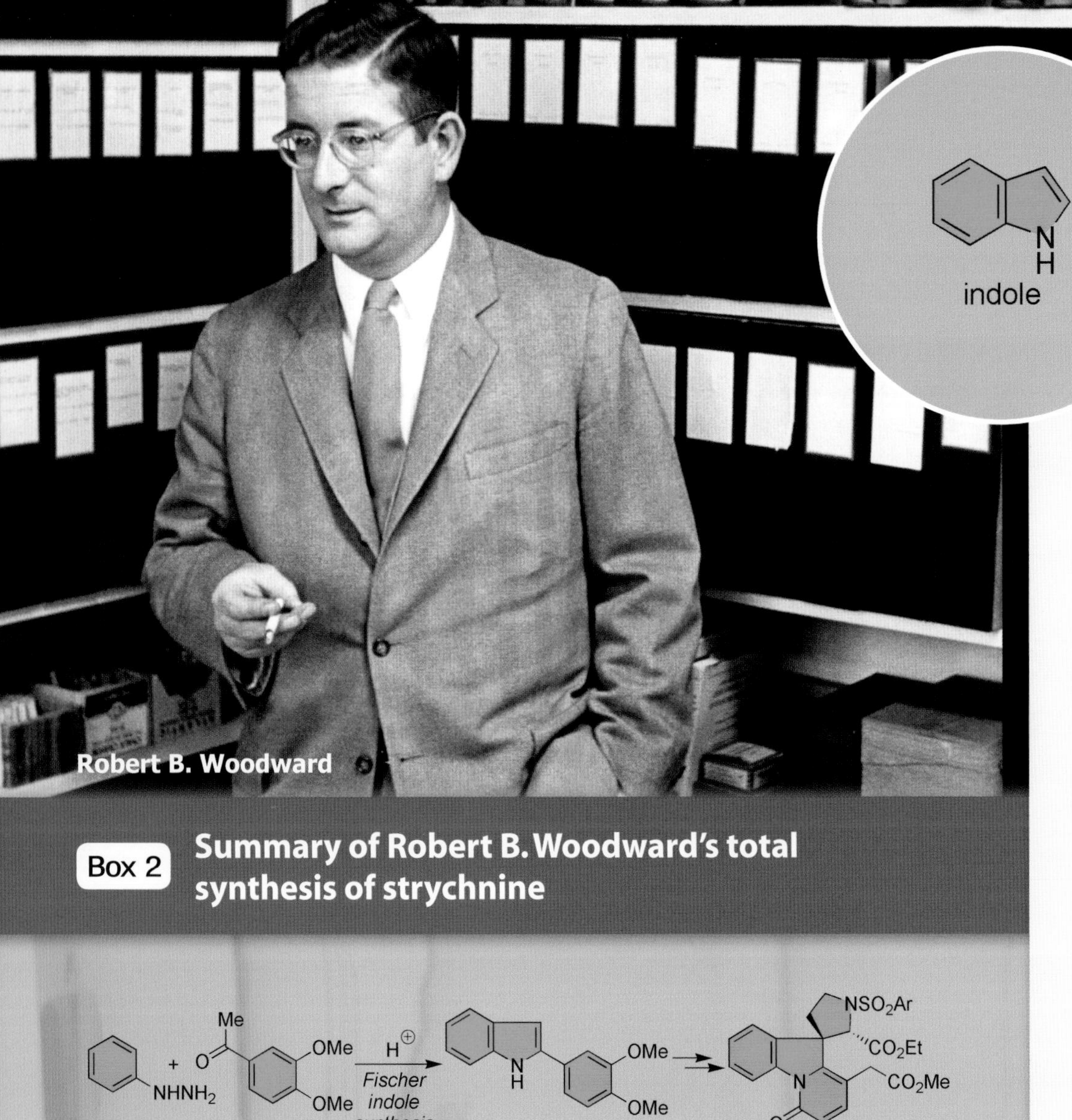

further two years the total synthesis of strychnine was completed by R. B. Woodward, one of the century's finest chemists (Box 2).

Woodward was a legend in the field of total synthesis, as recognized in 1965 with the awarding of the Nobel Prize in Chemistry, "for his outstanding achievements in the art of organic synthesis." The style in which he tackled the very toughest problems of the era had become famous, as it combined subtlety in planning with an unremitting ability to overcome even the most challenging of obstacles en route to his natural product targets. Amongst his many stunning accomplishments, the total synthesis of strychnine is perhaps his most illustrious, along with his conquest (together with A. Eschenmoser) of vitamin B_{12} (Chapter 16).

Woodward began the arduous journey toward "...the tangled skein of atoms..." that constitutes strychnine with the assembly of the indole core, which he achieved by employing a famous reaction developed by another great chemist, Emil Fischer (see Chapter 3). From a modern standpoint, Woodward's construction of the remaining rings and stereocenters may seem a little laborious; however, with the technology available at the time it required all the skills and talent of this master of organic chemistry to overcome the challenge. The conquest of strychnine by synthetic organic chemistry also proved to many who doubted it that chemists could make compounds of this striking complexity in the laboratory.

Emil Fischer

Following Woodward's synthesis in 1954, strychnine continued to captivate further generations of synthetic chemists and new approaches to this fascinating target, as well as novel theories regarding its biosynthesis, appeared in the chemical literature. Throughout the twentieth century, the prevailing opinion had always been that strychnine should be approached through one of two degradation products, isostrychnine or the so-called Wieland–Gumlich aldehyde (Box 3). Isostrychnine was Woodward's target in his successful synthesis, however, its conversion to strychnine has only ever been achieved in 5–20 % yield, despite thorough investigation by many chemists. Alternative approaches held the possibility of much higher efficiency and, thus, seemed attractive. In the 1990s, strychnine received an unprecedented renewal of interest as a synthetic target and several new total syntheses were reported. Of all these reports, the one originating from the laboratories of Larry E. Overman (University of California, Irvine, USA) was particularly impressive and concise (Box 4). Not only did it represent the first asymmetric synthesis of (–)-strychnine, but, by proceding through the Wieland–Gumlich aldehyde rather than isostrychnine, it also afforded strychnine in a sequence of only twenty-four steps with an overall yield of

Box 2 Summary of Robert B. Woodward's total synthesis of strychnine

3 %. This is approximately 100,000 times higher than Woodward's original yield! The power of synthetic chemistry had certainly expanded at a rapid pace over the forty year span between these two milestone publications. Overman's approach to the strychnine problem invoked the use of an impressive cationic aza-Cope/Mannich domino sequence and an enzymatic desymmetrization protocol to provide the key chiral building block. Amongst the other total syntheses reported, that of Masakatsu Shibasaki (University of Tokyo, Japan) stands out as a demonstration of the power and utility of the asymmetric Michael reaction developed in that laboratory. Using only a miniscule quantity of a chiral catalyst, over one kilogram of a key building block was prepared in enantiomerically pure form. In addition to this triumph, Professor Shibasaki's work on the asymmetric construction of carbon-carbon bonds has had a major impact on the art of modern organic synthesis.

Ball and stick model of the molecule of strychnine

Masakatsu Shibasaki

Larry E. Overman

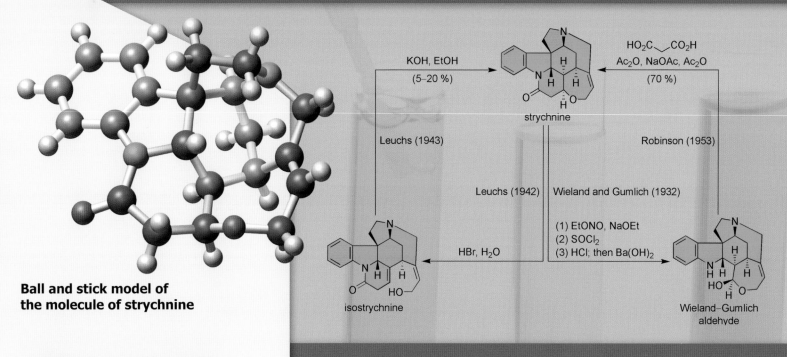

Box 3 Degradation and reconstruction of strychnine

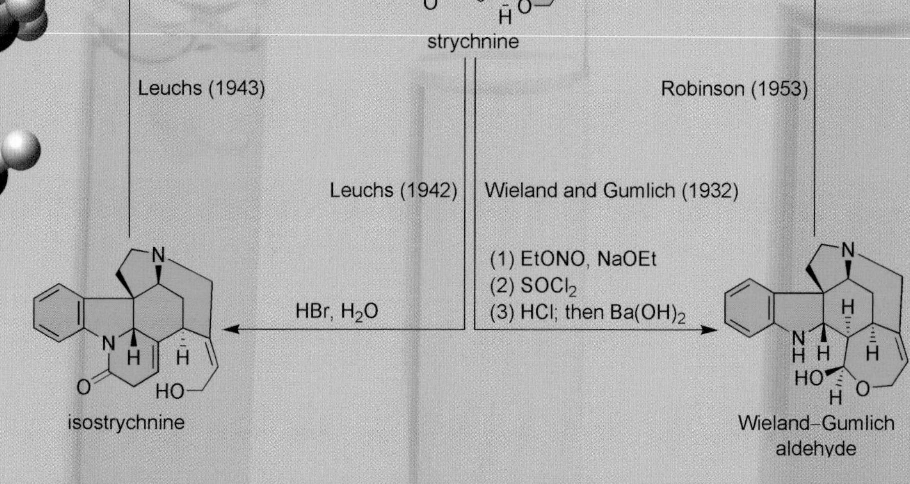

Box 4 Summary of Overman's synthesis of strychnine

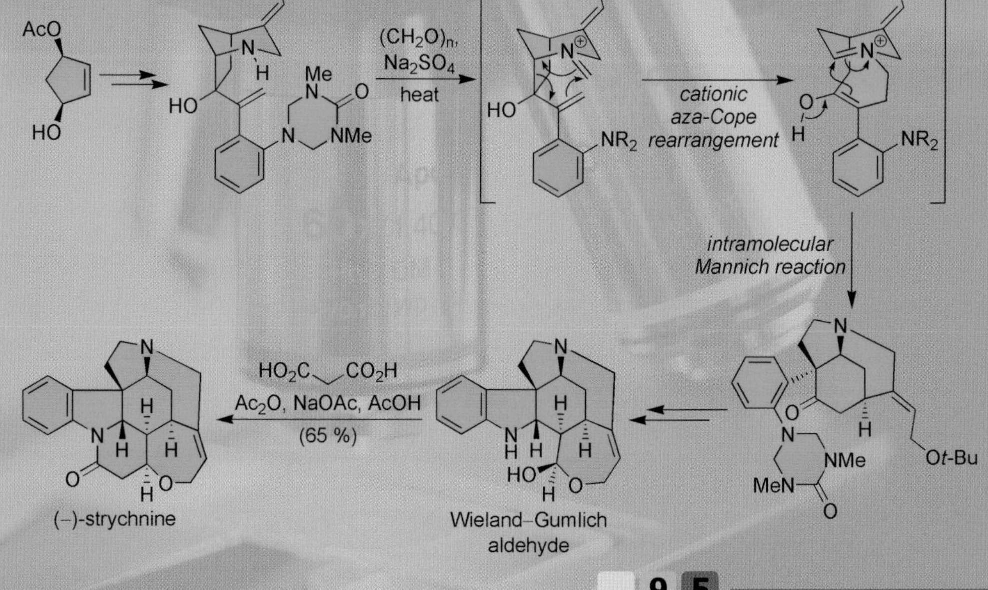

Strychnos ignatii

Further Reading

P. J. Pelletier, J. B. Caventou, *Note sur un Nouvel Alcali* (Note on a New Alkaloid), *Annales de chimie et de physique* **1818**, 323–336.

R. B. Woodward, M. P. Cava, W. D. Ollis, A. Hunger, H. U. Daeniker, K. Schenker, The Total Synthesis of Strychnine, *J. Am. Chem. Soc.* **1954**, *76*, 4749–4751.

S. D. Knight, L. E. Overman, G. Pairaudeau, Asymmetric Total Syntheses of (–)- and (+)-Strychnine and the Wieland–Gumlich Aldehyde, *J. Am. Chem. Soc.* **1995**, *117*, 5776–5788.

T. Ohshima, Y. Xu, R. Takita, S. Shimizu, D. Zhong, M. Shibasaki, Enantioselective Total Synthesis of (–)-Strychnine Using the Catalytic Asymmetric Michael Reaction and Tandem Cyclization, *J. Am. Chem. Soc.* **2002**, *124*, 14546–14547.

K. C. Nicolaou, E. J. Sorensen, in *Classics in Total Synthesis*, Wiley-VCH, Weinheim, **1996**, pp. 21–40 and 641–653.

J. Mann, *Murder, Magic and Medicine*, Oxford University Press, Oxford, **1992**.

Strychnine has by now beguiled the non-chemist and chemist alike due to its powerful and deadly biological effects for nearly two centuries. Its constant appearances in the chemical literature over such a long period affords this molecule the rare ability to act as a barometer, showing just how the art and science of synthesis has developed in style, depth, and efficiency from its early tentative origins to the modern day. Thus, it appears that throughout the annals of time strychnine has served as an inspiration in both fiction and science.

Strychnos descussata

Apothecary jar

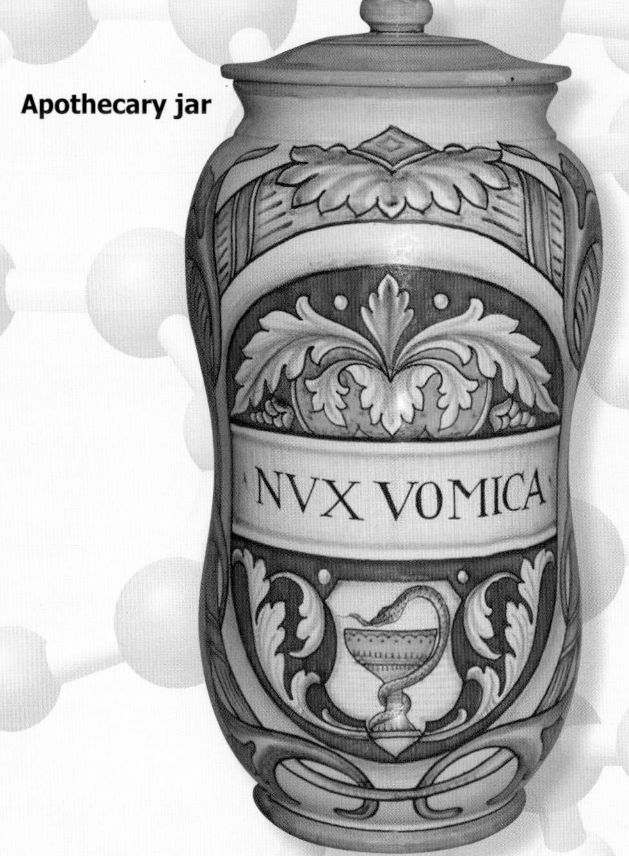

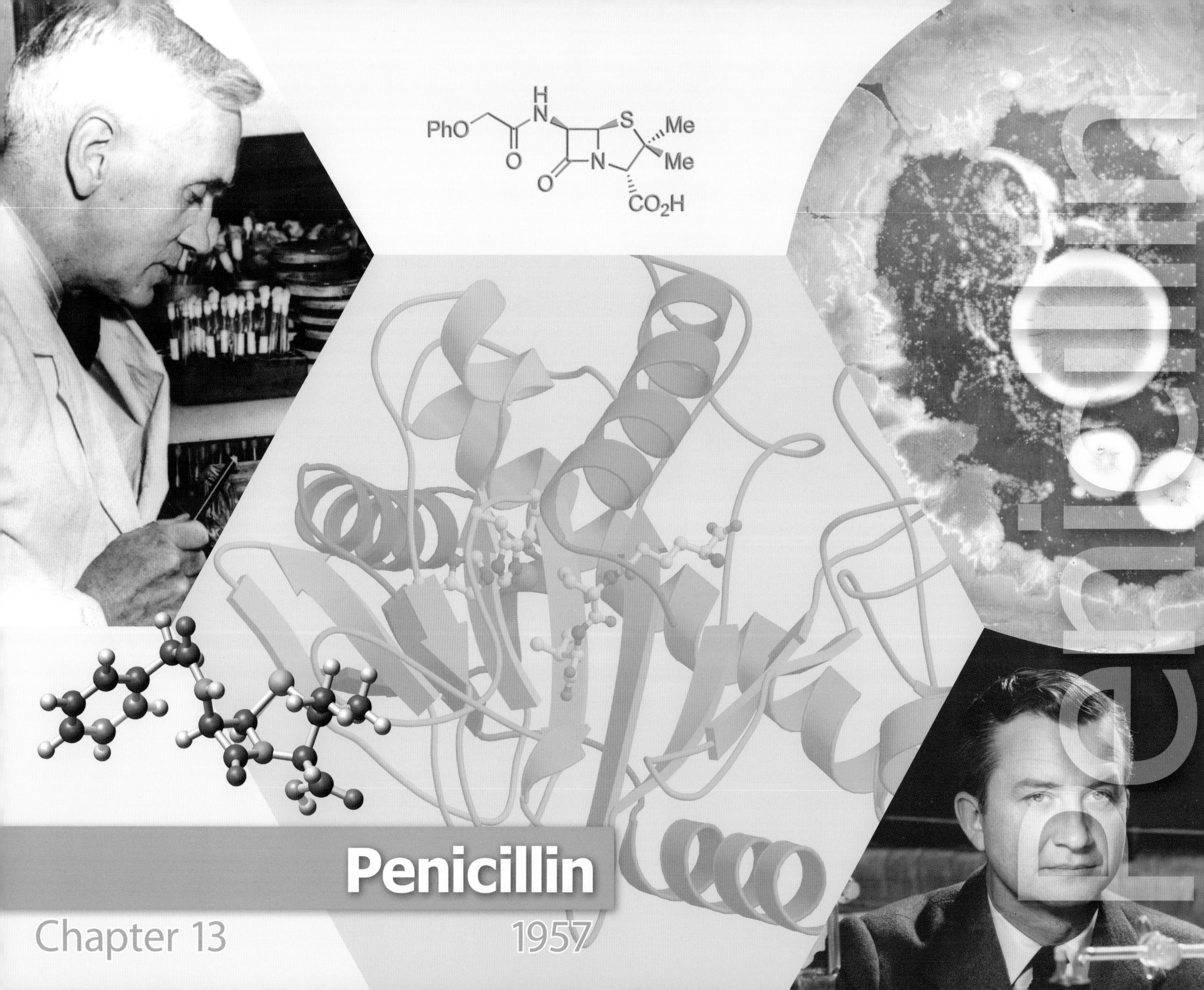
PhO
H
N
O
S
Me
Me
O
N
CO2H
Penicillin
Chapter 13
1957
Penicillin

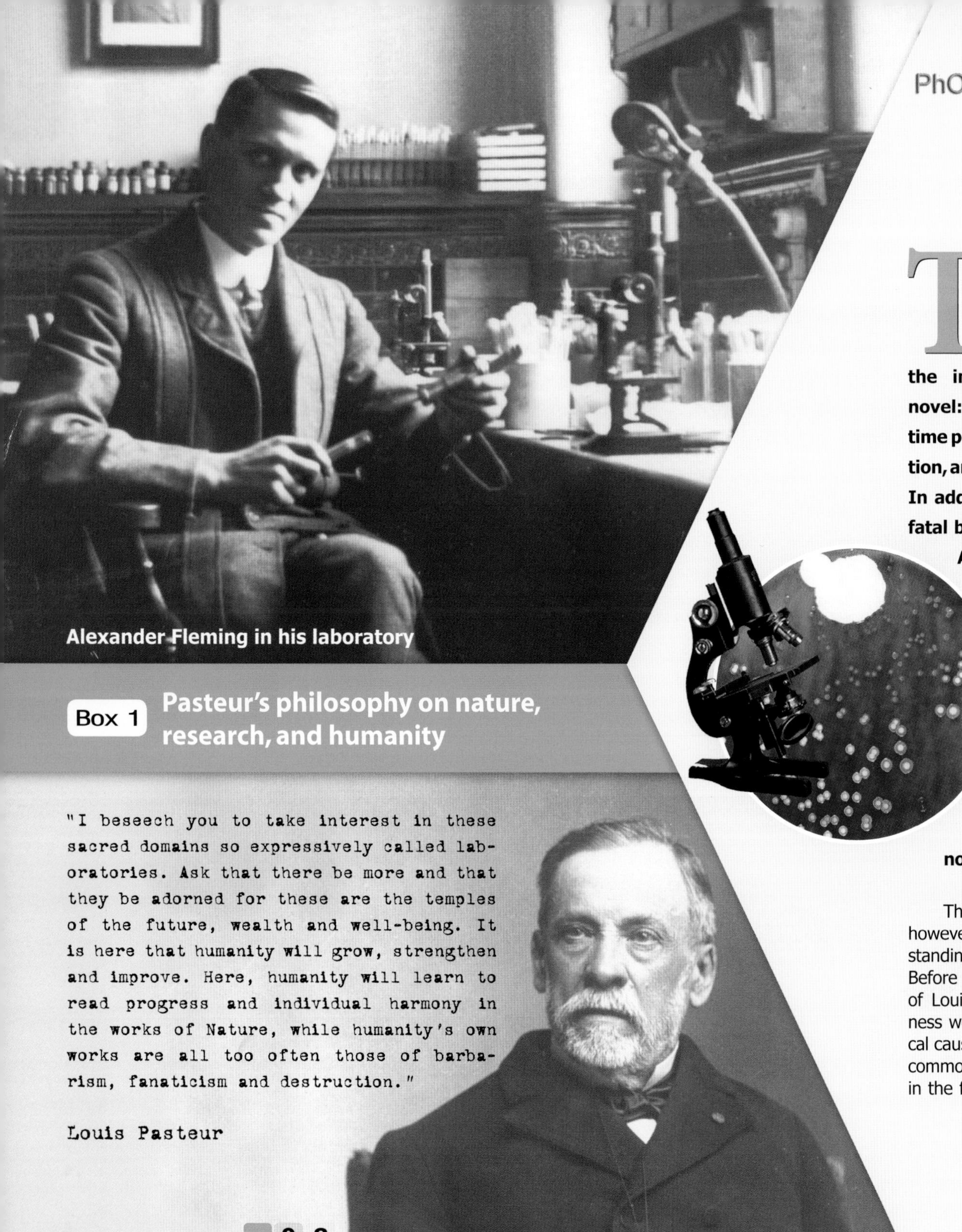

Alexander Fleming in his laboratory

Box 1 Pasteur's philosophy on nature, research, and humanity

"I beseech you to take interest in these sacred domains so expressively called laboratories. Ask that there be more and that they be adorned for these are the temples of the future, wealth and well-being. It is here that humanity will grow, strengthen and improve. Here, humanity will learn to read progress and individual harmony in the works of Nature, while humanity's own works are all too often those of barbarism, fanaticism and destruction."

Louis Pasteur

penicillin V

The narrative surrounding the discovery and development of penicillin is truly remarkable, possessing all the ingredients of a best-selling novel: serendipitous discovery, wartime political intrigue, fierce competition, and an eclectic cast of characters. In addition, the dreadful problem of fatal bacterial infections was solved. All this was a consequence of the extraction of a broad-spectrum antibiotic from an ordinary mold, following an observation made by Alexander Fleming in 1928. Despite its importance and the efforts of many chemists, the total synthesis of penicillin would not be accomplished until 1957.

This epic story begins much earlier, however, with a revolution in our understanding of the underlying cause of disease. Before the magnificent accomplishments of Louis Pasteur, disease, death, and illness were frequently ascribed metaphysical causes. In the western world, this idea commonly meant that sickness, especially in the form of epidemics, was deemed to be the wrath of God, punishing man for his sins. Pasteur's seminal works culminated in the 'germ theory of disease,' which asserted that infectious diseases were caused not by God's vengeance, but by microbes.

Louis Pasteur studied chemistry in Paris, and in 1848 he made his first profound contribution to the advancement of science, aged just twenty-six. He examined tartaric acid crystals under a microscope and noticed that they existed in two distinct forms that were mirror images of each other. He was able to separate these forms and discovered that, in solution, they rotated the plane of polarized light in opposite directions. This investigation represents the inception of the pivotal and fundamental field of stereochemistry. The tartaric acid Pasteur studied came from wine sediments, and his interest in the science of fermentation would lead to more important discoveries. In 1856, Pasteur came to the aid of a student's family, who were experiencing production problems at their fermentation plant; sometimes alcohol was produced as expected, but other batches gave lactic acid instead. Pasteur examined the fermentation mixtures by microscopy and noticed that during normal production the yeast cells were plump and budding, but when lactic acid was being produced, the yeast cells were smaller and accompanied by rod-like microbes. He found that briefly heating the liquid medium before fermentation began would kill the undesirable microbes and lead to reproducible fermentations. This sterilization procedure, known as *pasteurization*, is still in common use today, particularly for dairy products. Pasteur was subsequently able to show, using an ingeniously designed swan-necked flask containing a fermentable solution, that microbes such as these were airborne particles. Next, he tackled a serious disease in silk-producing worms whose malaise was having a devastating economic impact on the buoyant European silk trade. He showed that healthy worms could become infected by nesting on leaves previously occupied by diseased specimens. These early influential studies led Pasteur to develop his idea that germs spread contagious diseases, and that these foreign particles were living micro-organisms. Pasteur went on to do more groundbreaking work. Building on Edward Jenner's discovery of a vaccine for smallpox, Pasteur developed vaccination as a prophylactic strategy to outwit other viral contagions, such as rabies, as well as the bacterial infections of anthrax and cholera. Having sensed the widespread recognition of his work, Pasteur died with the knowledge that he had succeeded in bequeathing France a powerful and vibrant establishment for research, The Pasteur Institute, in Paris. He had pursued this project with passion for he strongly believed in applying experimental science to eradicate humanity's plagues and often spoke emphatically on this subject (Box 1). His words ring true and are surprisingly relevant even in today's much altered world.

Meanwhile, in the United Kingdom, a young Quaker surgeon by the name of Joseph Lister was enthralled by Pasteur's work, prompting him to propose a connection between wound sepsis and microbes from the air. Lister's search for the cause of sepsis had begun because he was troubled by the assertion made earlier by the famous chemist Justus von Liebig that sepsis was a kind of combustion occurring when expanding moist body tissue met with oxygen from the air. In Lister's surgical ward at the Glasgow Royal Infirmary, wound sepsis killed fifty percent of his patients. Lister was further inspired to do something about this dreadful situation when he read that carbolic acid (Box 2) was being used to treat sewage in Carlisle, and that fields treated with the resulting slurry were freed of a contagion that would normally lead to infection in cattle grazing on the same site. Lister began to clean wounds and dressings with a carbolic acid solution, and in 1867 he was in a position to announce to the British Medical Association that his ward had been sepsis-free for an astonishing nine month period due to the implementation of this protocol. Lister had pioneered the use of antiseptic solutions and broadcast the importance of hygiene in operating theatres, thereby saving innumerable patients from a painful gangrenous death.

The German doctor Robert Koch, working alone as the District Medical Officer for Wollstein during the Franco-Prussian War in the 1870s, was finally able to show definitively that the anthrax bacillus directly causes disease in test animals. He also developed techniques for culturing bacteria, thereby illustrating the phenomenal resilience of these specific microbes, while also proving that, in general, a host animal was not necessary for these germs to thrive. He photographed cultures, studied the conditions required for bacterial growth, and developed stains to improve their visibility. It was by employing this knowledge that he was able to identify both the *Vibrio cholerae* and the *Mycobacterium tuberculosis* that cause the deadly diseases of cholera and tuberculosis, respectively, work which earned him the 1905 Nobel Prize in Physiology or Medicine.

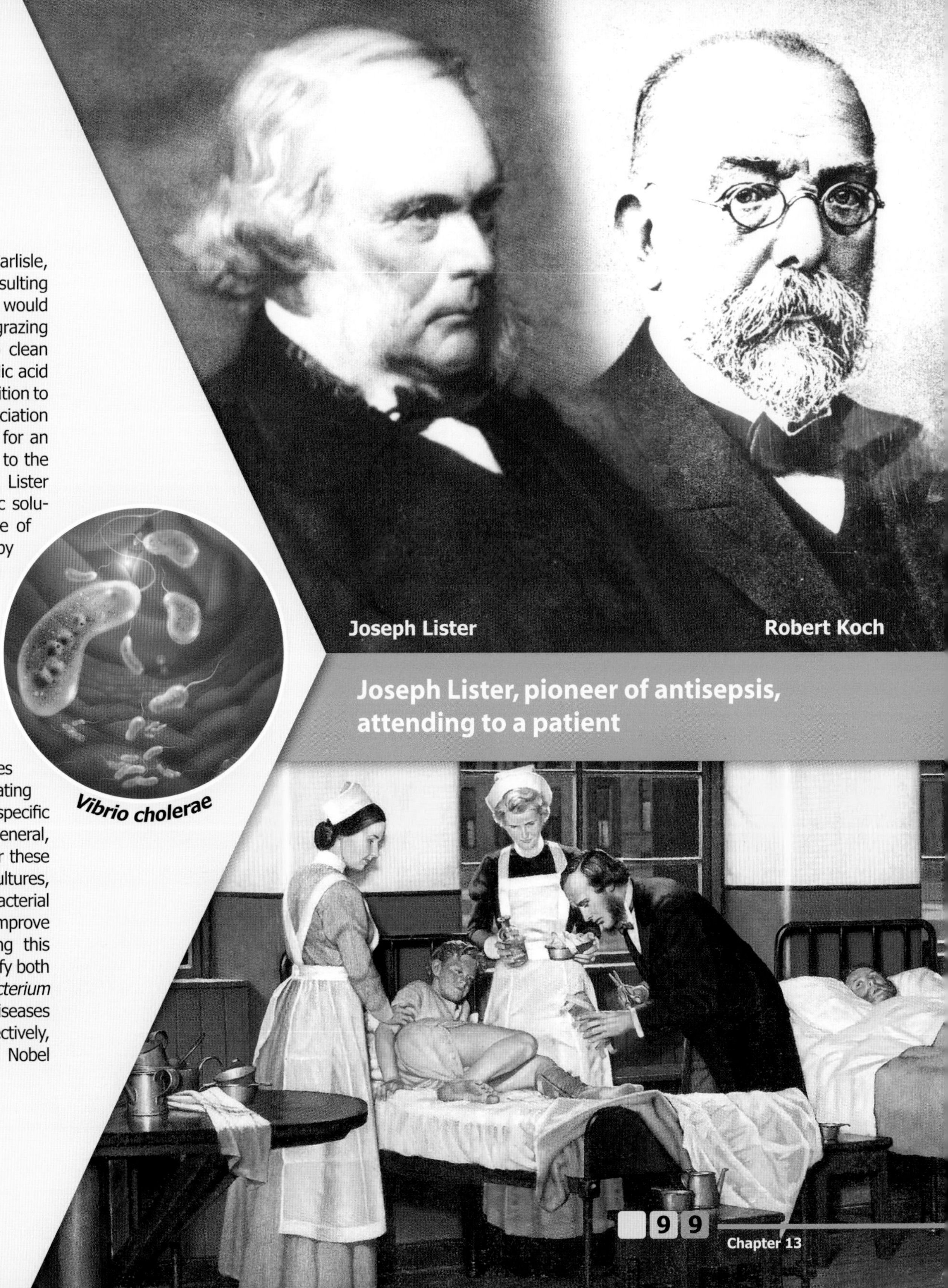

Vibrio cholerae

Joseph Lister

Robert Koch

Joseph Lister, pioneer of antisepsis, attending to a patient

Paul Ehrlich
Gerhard Domagk

Box 2 Pre-penicillin antibacterial agents

trypan red
(red dye)
Paul Ehrlich, 1904

phenol
(carbolic acid)
Joseph Lister, 1867

prontosil rubrum
(red dye)
Gerhard Domagk, 1933
(first sulfonamide drug)

Identifying these disease-causing agents had been a long and arduous battle requiring all the energy of those distinguished scientists whose ingenious work had already revolutionized medical practice, as we have just seen. Each of these men left his own indelible mark on history, immediately contributing to saving lives, yet our story is just in its beginning. At the start of the twentieth century life expectancy in England was just forty-five years, and infant mortality was at a rate of one hundred and fifty per thousand live births. Many of the premature deaths that contributed to these appalling statistics could be attributed to fatal bacterial infections. Bacteria had been identified, routes of transmission had been elucidated, and sterilization and surgical hygiene to prevent infection had been proven, yet still no solution to the problem of successfully treating an already-infected patient was available. People had good reason to dread bacterial infections since they frequently ended with the death of their victim.

Paul Ehrlich, a friend and one-time colleague of Koch, suggested that bacterial infection might be curable by treatment with a drug that was toxic towards the bacteria, whilst being harmless to the patient, thus introducing the so-called 'magic bullet' concept. This idea had been inspired by the selective uptake of dyes into bacterial cells, a technique that had been developed for enhancing microscopy, and Ehrlich was fascinated by what influenced the disparity between various cell types. The large number of dyes produced by the German dye industry at the time afforded Ehrlich with a vast array of chemical candidates for testing. He and his Japanese colleague, Kiyoshi Shiga, eventually found that trypan red (an azo dye, Box 2) could effectively kill the bacteria *Trypanosoma gambiense*, the causative agent of the fatal sleeping sickness transmitted by the tsetse fly in Africa. However, in tests on humans in Uganda, its use resulted in unacceptable side effects ranging from blindness to death. Ehrlich is also famous for introducing the arsenic salts salvarsan (arsephenamine) and neosalvarsan (neoarsephenamine) as drugs to treat syphilis (caused by *Treponema pallidum* infection). Pain at the injection site, side effects, and frequent relapses rapidly consigned these agents to history. Nevertheless, Ehrlich's studies launched the age of chemotherapy. Ehrlich was also a pioneer in the fields of hematology and immunology, with his contributions being recognized by the awarding of the 1908 Nobel Prize in Physiology or Medicine, which he shared with the Russian Ilya Ilyich Mechnikov, "in recognition of their work on immunity."

In 1927, Gerhard Domagk, while continuing investigations in the same vein at the laboratories of I. G. Farbenindustrie, concentrated mainly on finding agents effective against haemolytic *Streptococci* (the bacteria associated with throat infections). He recommended prontosil rubrum (Box 2), a red azo dye, for clinical trials – a fortunate choice, for this compound contains the critical sulfonamide functionality and proved to be very effective against these bacteria. So it was that the first clinically used sulfonamide antibacterial agent was born, to be followed soon by many other members of this same class. The dis-

Petri dish

covery of the antibacterial effects of prontosil won Gerhard Domagk the Nobel Prize in Physiology or Medicine in 1939. His real triumph, however, was much more personal, as it emerged later that Domagk had, in desperation, used prontosil rubrum successfully to treat his infant daughter, who was dying of *Staphylococcal* sceptacemia, long before the drug became available to the public. After an initial period of euphoria over the discovery of the sulfonamides the febrile activity calmed and their shortcomings began to become all too apparent. Although still used in rare cases today, these antibacterials are very narrow in spectrum and essentially obsolete.

Fleming's famous petri dish

Enter Sir Alexander Fleming, the son of a farmer from rural Scotland, whose entire career could be said to have been based on a series of serendipitous happenings. In 1900, the young Alec, as he was called, joined a Scottish regiment in order to fight in the Boer War with two of his brothers. This experience honed his swimming and shooting skills, but never actually took him as far as the war-torn Transvaal region of South Africa (a province lying between the Vaal and Limpopo rivers). On returning to London, and in need of a profession, Fleming decided to follow his older brother into medicine. He obtained top scores in the qualifying examinations, giving him a free hand over which school he might choose for his studies. He elected to enroll at St. Mary's, one reason being that he had played water polo against them in a previous sporting fixture. In 1905 Fleming found himself specializing in surgery, a career choice which would require him to leave St. Mary's in order to take up a position elsewhere. The captain of St. Mary's rifle club, who relied on Fleming's flawless shooting skills, heard of his impending departure and did his best to prevent it by winning him over to his own discipline, bacteriology, thereby maintaining the integrity of his champion team. Alexander Fleming never left St. Mary's, becoming instead the world's most famous bacteriologist.

In the 1920s, Fleming identified lysozyme, an enzyme found in tears, which exhibited a natural and mild antibacterial action. This protein was the first antibiotic to be isolated from the human body, but as it was not powerful enough to attack the most prevalent and aggressive infections, Fleming continued his search. One day, so the legend goes, when he was clearing out petri dishes containing bacterial cultures that had begun to accumulate in one of his sinks, he noticed that one of the containers had a mold growing on the nutritional agar. This rather common occurrence was made fascinating by the fact that Fleming's habit of careful observation also revealed that no bacterial colonies were growing around the periphery of the fungus. Fleming went on to show that not only was bacterial growth inhibited, but that healthy bacteria underwent cell lysis and death when exposed to the mold (*Penicillium notatum*). It should be noted at this point that the full detailed tale of the discovery of penicillin has its roots earlier than Fleming's investigations.

Sir Alexander Fleming

Box 3 Highly contested structures of penicillin

oxazolone thiazolidine structure (incorrect)

β-lactam thiazolidine structure (correct)

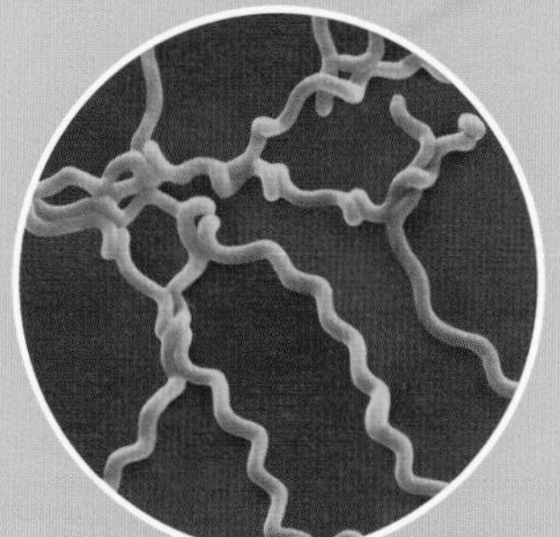

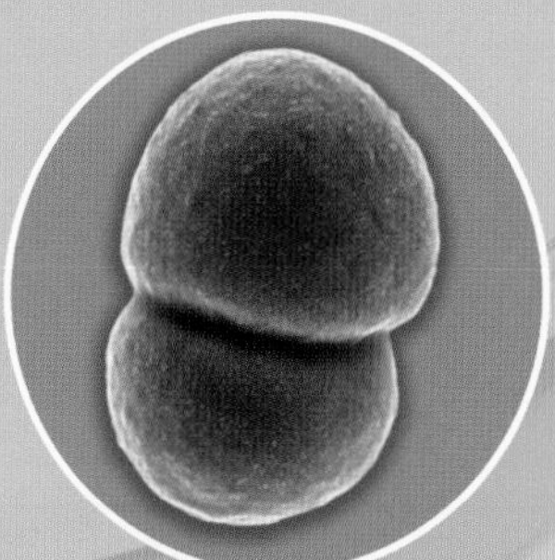

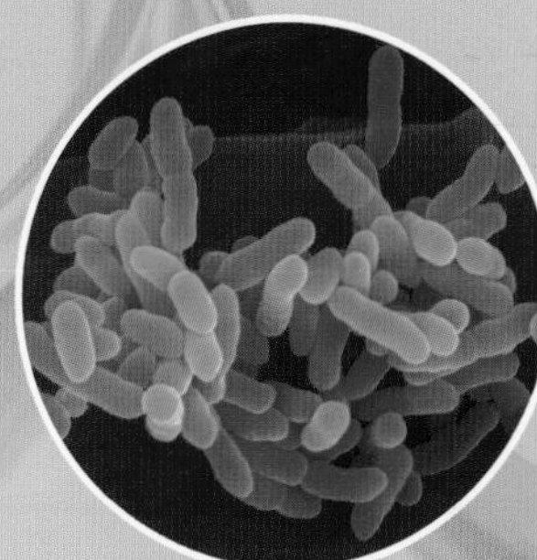

Bacterial strains

Alexander Fleming and the Oxford penicillin team: Fleming, Florey, Chain, Sanders, Abraham (left to right)

Box 4 The mechanism of action of penicillin

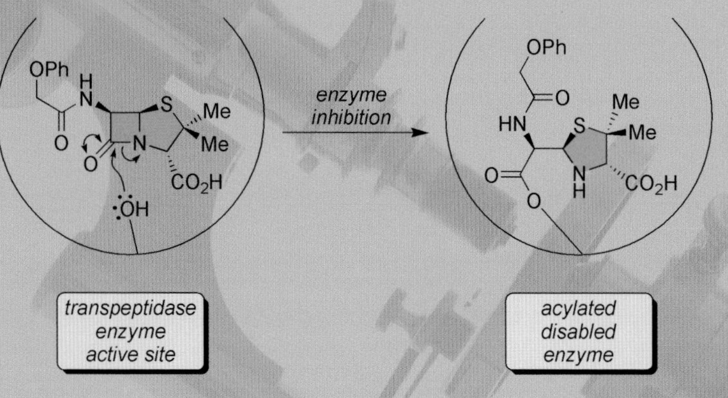

Both Joseph Lister and Ernest Duchesne (a French medical student) independently reported the use of *Penicillium* molds in bandages to treat infected patients at the end of the nineteenth century. These compresses were ignored by the scientific and medical communities due to their low potency until Fleming rekindled interest in them. Furthermore, Chinese medical texts dating back some 3000 years advocated the use of moldy soybean curd to treat skin infections.

At the beginning of the 1930s, frustrated by progress in advancing his discovery to the next stage, Fleming passed on some of his culture to Howard W. Florey and Ernst B. Chain at Oxford University. The Oxford group, which also included Norman G. Heatley and Edward P. Abraham, refined the growth and isolation of the penicillin extracts just enough to facilitate the instigation of clinical trials, which immediately began to deliver very promising results. However, pure penicillin was still in such short supply that it had to be recovered from the urine of patients for reuse. The spectacular success of penicillin as an antibiotic would later earn Fleming, Chain, and Florey the 1945 Nobel Prize in Physiology or Medicine. In the meantime, the elevation of penicillin to its legendary status as a world-changing antibiotic would require the launching and successful execution of one of history's most captivating international scientific adventures, the so-called Penicillin Project.

With the outbreak of World War II, interest in penicillin had intensified. Numerous scientists strove to produce the antibiotic on a large scale in response to an urgent new need for the drug to treat wounded soldiers and civilians who had subsequently contracted infections. The proximity of the battle frontline, the frequent aerial bombardments of the UK, and the need for a rapid solution to the problem led to a huge Anglo-American collaborative project on penicillin. The Rockefeller Foundation in New York arranged for Florey and Heatley to come to America in 1941 to meet with Charles Thom, chief mycologist at the US Department of Agriculture. A two-pronged strategy for the procurement and development of penicillin was immediately formulated. The first approach was directed towards the elucidation of the structure of penicillin, which would make the ultimate goal of its chemical synthesis at least conceivable. Upwards of forty independent laboratories and hundreds of chemists became involved in this labyrinthine task. The second line of attack was directed towards further improving the fermentation process for production of the drug. The latter of these pivotal works was relocated to the US Agriculture Department Laboratories in Peoria, Illinois. Here an intense research program drew on a myriad of sources to find extra momentum; in this eclectic project, progress was even aided by local residents who brought moldy household items to the laboratories for investigation in the search for more productive penicillium strains.

It was the second of these two approaches that paid

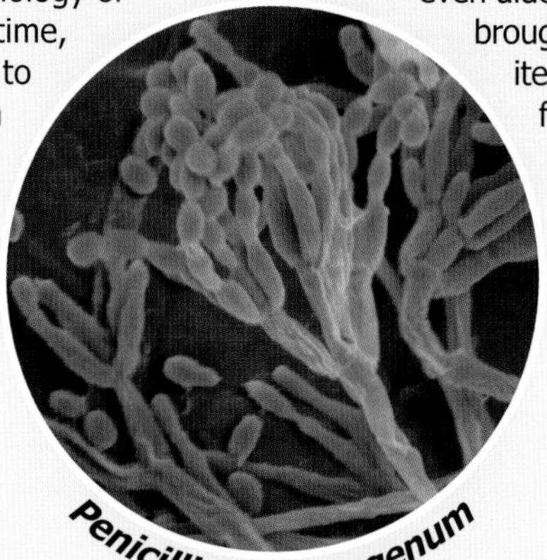

Penicillium chysogenum

the earliest dividends. Thus, a much-improved yield of penicillin was secured in the 1940s from *Penicillium chrysogenum*, a discovery made courtesy of a moldy cantaloupe melon brought into the laboratory by Mary Hunt, an employee at the Peoria laboratories. Within three years, twenty-nine plants were fermenting this high-yielding fungal strain to produce penicillin using a corn-steep liquid medium (a by-product of the massive Mid-West corn production) in an amazing effort organized by an extraordinary conglomeration of scientific establishments. The pharmaceutical companies of Merck, Pfizer, Squibb, and Abbott were all involved, along with leading British and American academic and governmental institutions.

Ripening cantaloupe on a vine

General Dwight D. Eisenhower began the invasion of Europe from the southern shores of England on D-Day supported by some three million doses of penicillin (300 billion units or approximately 180 tonnes), the product of one of the most exciting and lucrative joint ventures in history. Only the notorious Manhattan Project directed towards the development of the atom bomb exceeded it in magnitude during that period. Thus, in less than four years, scientists had gone from recovering penicillin from patients' urine, due to its short supply, to the phenomenal level of production whereby 1,633 billion units of penicillin were produced in 1944 alone. The tremendous effort extended towards the development of penicillin may be less familiar to us than the Manhattan Project, but it is easy to argue that the former collaboration yielded far greater benefits to mankind. Overall, this venture succeeded admirably in optimizing fermentation and production protocols, allowing for a successful and practical supply of the new miracle drug, ultimately saving countless lives. The process, however, still relied on natural biosynthesis of the intact penicillin molecule, a fact that limited investigations into producing analogs with enhanced activity, especially to combat the new demon of bacterial resistance that was just beginning to rear its ugly head. These imperatives stipulated urgent attention be paid to finding a chemical synthesis.

X-Ray crystallography

Before a chemical synthesis could be attempted the non-trivial task of deconvoluting the molecular structure of penicillin had to be accomplished. From the ardent debate amongst many of the most renowned chemists of the time, two possible structures emerged in the early 1940s as leading contenders for the honor of representing the magical molecule of penicillin (see Box 3). The so-called oxazolone-thiazolidine formula was proposed by Sir Robert Robinson (Nobel Prize in Chemistry, 1947, see also Chapter 7) and fiercely defended by him, as well as by a number of other notable chemists such as Sir John Cornforth (Nobel Prize in Chemistry, 1975). Its β-lactam rival was advocated by Merck scientists and by the Oxford axis of Abraham and Chain. Despite

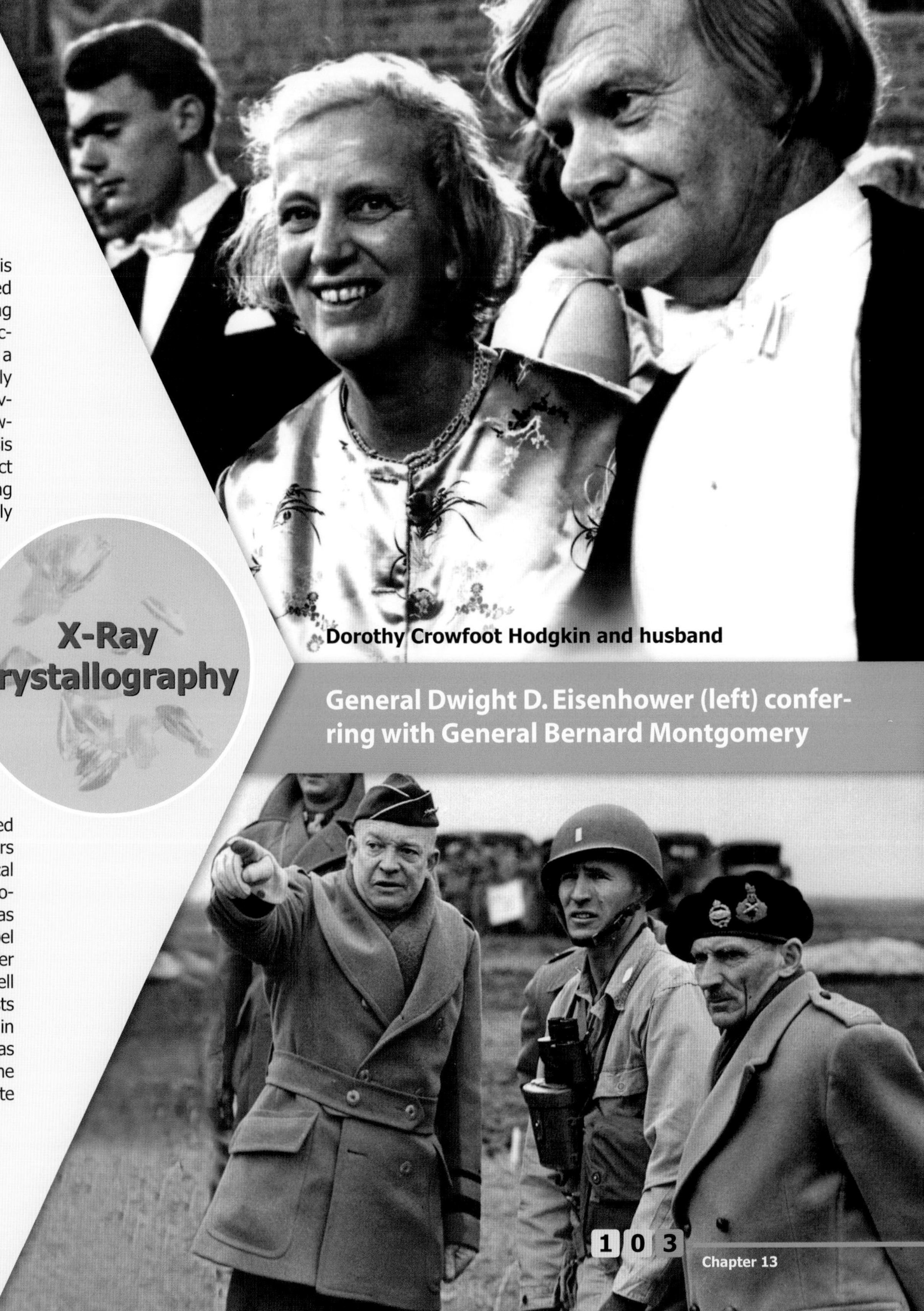

Dorothy Crowfoot Hodgkin and husband

General Dwight D. Eisenhower (left) conferring with General Bernard Montgomery

Box 5 Sheehan's thoughts on the total synthesis of penicillin V

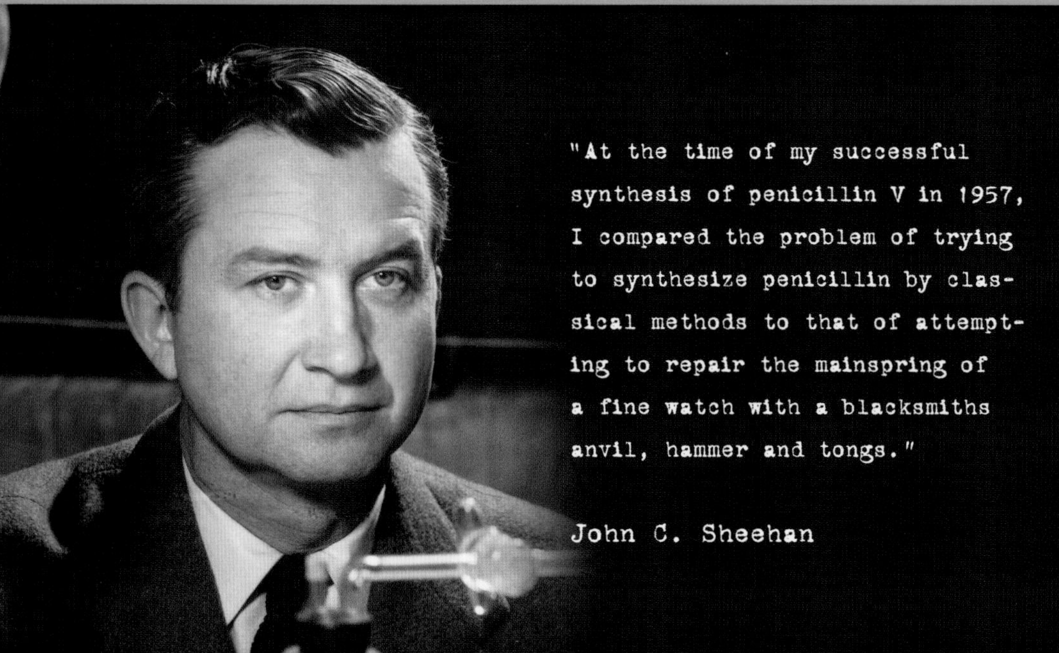

"At the time of my successful synthesis of penicillin V in 1957, I compared the problem of trying to synthesize penicillin by classical methods to that of attempting to repair the mainspring of a fine watch with a blacksmiths anvil, hammer and tongs."
John C. Sheehan

Box 6 Sheehan's total synthesis of penicillin V

(1) KOH

(2) N=C=N (DCC)

penicillin V potassium salt

experimental evidence for the existence of a β-lactam structural motif provided by the Merck scientists, conventional wisdom could not accept the presence of such a strained and reactive feature within a naturally occurring substance. It was only after the brilliant crystallographic work of Dorothy Crowfoot Hodgkin that the dispute was finally settled in favor of the β-lactam structure in 1945, pleasing its backers and winning fame for the unobtrusive crystallographer from Oxford.

The striking four-membered β-lactam ring of penicillin, which was so decisively revealed by Hodgkin's crystallographic analysis, also turned out to be the motif that was responsible for the lethal action of the drug against bacteria. This activity was found to be related to the conformation adopted by penicillin, wherein the fused 4,5-ring system enforces an orthogonal alignment of the nitrogen lone pair and the carbonyl π-bond such that the resonance stabilization exhibited by traditional amides cannot be attained in this case. This feature, in combination with the intrinsic strain of the four-membered ring, creates a situation where the carbonyl functionality of the β-lactam ring acts as a highly effective acylating agent due to its particularly strong electrophilic reactivity. Thus, it is now known that penicillin irreversibly acylates the bacterial transpeptidase enzyme responsible for the cross-linking reaction which unites the terminal glycine residue of a pentaglycine strand with the D-alanine residue of a neighboring pentapeptide, in an indispensable step during the construction of bacterial cell walls (Box 4). The acylation process deactivates this cross-linking enzyme, thereby compromising the integrity of the bacterial cell wall, resulting in rapid cell death. Unlike bacteria, only a phospholipid membrane surrounds mammalian cells, so transpeptidase inhibition is completely selective for bacterial cells. It has been shown that the penicillin molecule adopts an overall conformation that is very similar to the D-alanine-D-alanine residue of the substrate involved in this chain elongation process, thus it gains ready access to the active site of the enzyme where it reacts to disable its host.

The biosynthesis of penicillin, including the unprecedented β-lactam ring, was elucidated through a series of brilliant chemical and biological studies, many of which were carried out by Sir Jack E. Baldwin and his group at Oxford University. In addition to this seminal work, Baldwin is also known for his contributions to biomimetic synthesis, as well as for a set of rules he devised to predict the outcome of certain ring-closing reactions.

Ironically, while the biological activity of penicillin relies on the characteristic lability of the β-lactam ring, it is this same feature that led most synthetic organic chemists of the wartime period to consider penicillin an impossible target to conquer by chemical synthesis. Indeed, despite a huge effort directed towards its synthesis, both during and after World War II, no success was reported until much later. Indeed, it was 1957 before John C. Sheehan at the Massachusetts Institute of Technology (USA) was able to announce triumphantly the total synthesis of penicillin V, the result of a relentless ten-year campaign (Boxes 5 and 6).

Sheehan's celebrated success was due to his innovative and daring approach. He had recognized very early on that the main hurdle to be overcome prior to any total synthesis of penicillin was the construction of the highly strained and sensitive β-lactam ring. In addition, it was clear to him that a suitable method for accomplishing this challenging ring formation had to be developed since none of the existing technologies could be expected to rise to the challenge. It is here that Sheehan's insight and brilliance ensured his team's triumph. He conceived of and developed the *N,N′*-dicyclohexylcarbodiimide (DCC)-mediated coupling of carboxylic acids with amines to afford amides. Applied in its intramolecular version (wherein the amine and acid both belong to the same molecule) on an appropriately functionalized precursor, this method would solve one of the most recalcitrant problems in chemical synthesis of the 1940s and 1950s – the construction of the β-lactam ring of penicillin. The DCC coupling reaction, later extended to provide a solution for ester bond formation, remains a powerful synthetic tool in contemporary organic synthesis, and has been the inspiration for many similar reactions. This innovation was not the only one made by Sheehan during the penicillin synthesis; the phthaloyl protecting group for primary amines was another important and enduring contribution emanating from his group.

With these two innovations at their disposal, the Sheehan group was able to complete the synthesis of this previously impossible target through the sequence briefly outlined in Box 6. Thus, coupling of a phthaloyl-protected amino aldehyde with a suitable amino thiol led to the construction of the thiazolidine ring of penicillin. Further elaboration furnished an advanced amino diacid intermediate which served as the precursor for the β-lactam ring of penicillin V. Indeed, exposure of this precursor to Sheehan's DCC coupling conditions provided penicillin V as its potassium salt. So it was that the first total synthesis of penicillin was accomplished, a milestone event in the history of the art of total synthesis. This feat marked not only the beginning of a highly productive era in the synthesis of β-lactam derivatives, many of which were subsequently synthesized, but also represented the addition of a new dimension to total synthesis endeavors, that of seeking to invent new synthetic technologies along the way. The task of developing this paradigm further, taking it to impressive new heights, would be assumed by one of Sheehan's students, the now highly celebrated master of this science E. J. Corey, about whom we will learn much more later in subsequent chapters.

Penicillin ushered in a new epoch in antibiotic research, one that grew by leaps and bounds over the ensuing decades. A series of new naturally occurring β-lactams such as cephalosporin C, clavulanic acid, and thienamycin were delivered in quick succession for use as drugs (Box 7). As soon as these discoveries had been made in microbiology laboratories, synthetic chemists busily focused on synthesizing the newly discovered natural products and modifying their structures in an effort to discover new antibacterial agents with improved pharmacological profiles. Their work led to an equally impressive collection of synthetic or semisynthetic β-lactam antibiotics including ampicillin, amoxycillin, and methicillin (Box 7). During this golden

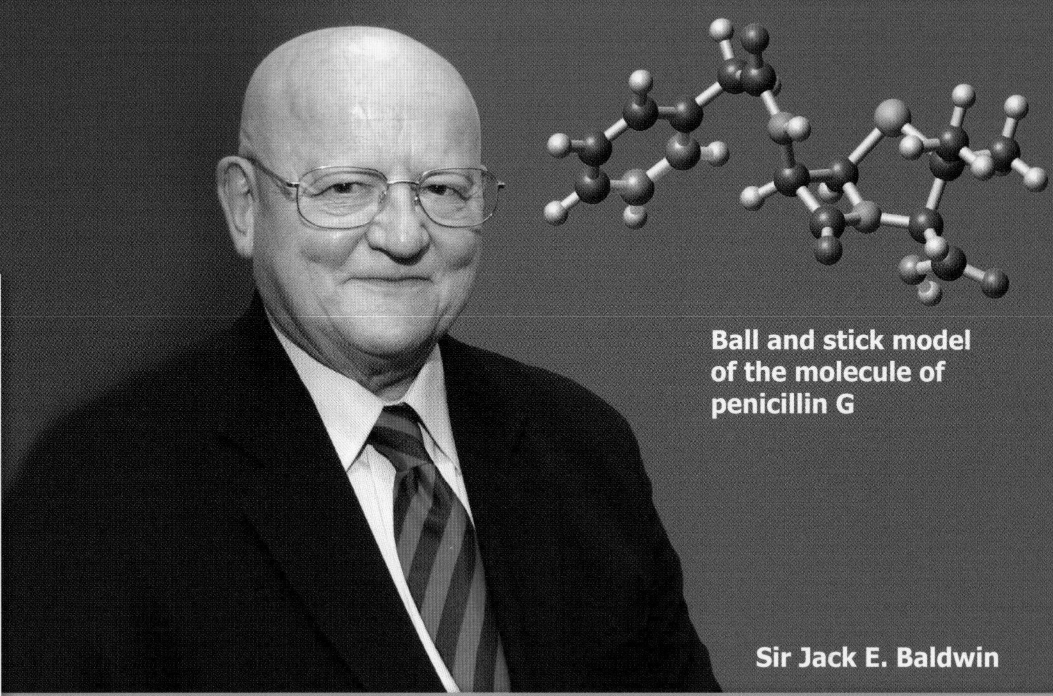

Ball and stick model of the molecule of penicillin G

Sir Jack E. Baldwin

Ribbon model of isopenicillin *N*-synthase

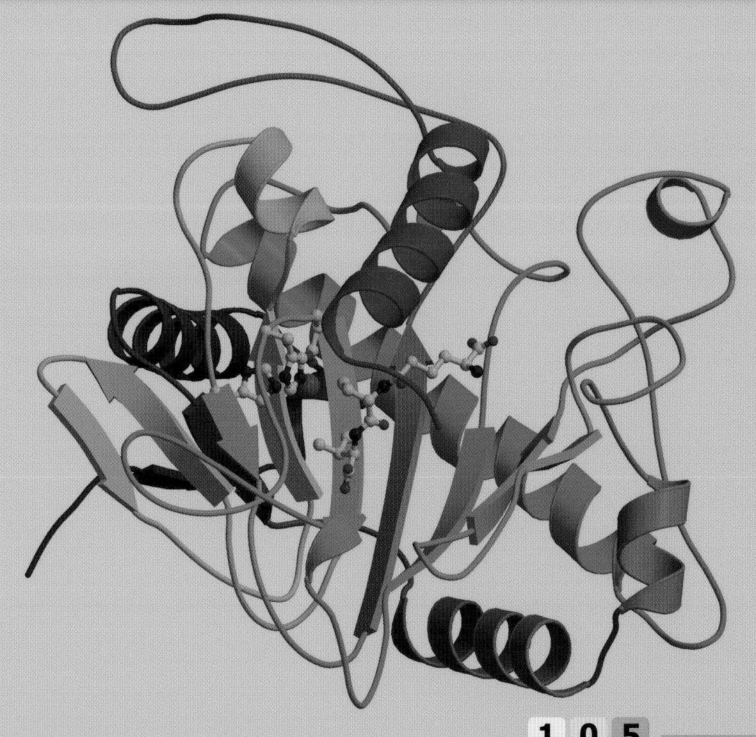

Box 7 Molecular structures of various β-lactam antibiotics

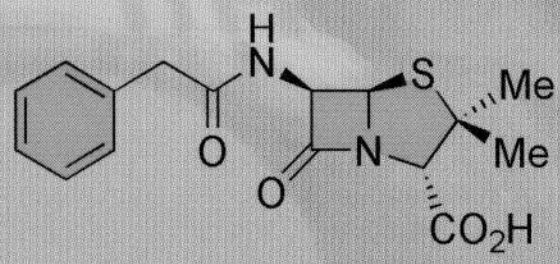

penicillin G
(natural)

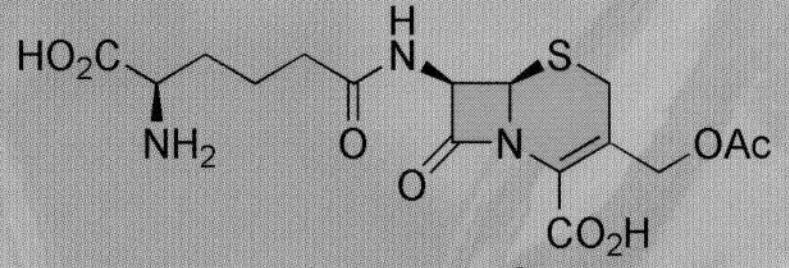

cephalosporin C
(natural)

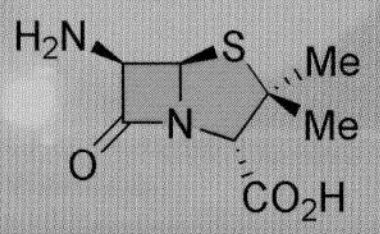

6-aminopenicillanic acid
(natural)
(industrial production from penicillins G and V by the action of an enzyme, penicillin amidase)

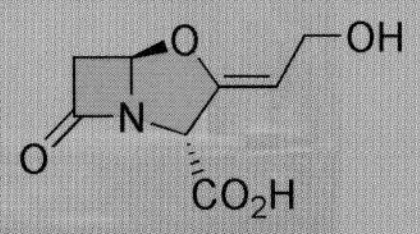

clavulanic acid
(natural)
(weak antibacterial activity, often used with amoxycillin to combat β-lactamases)

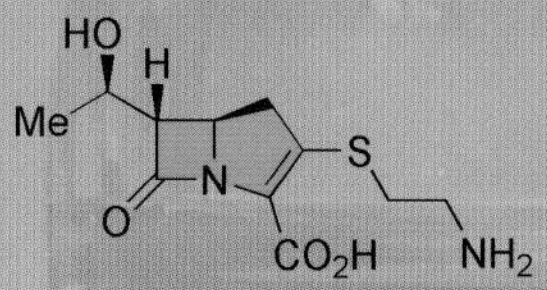

thienamycin
(natural)

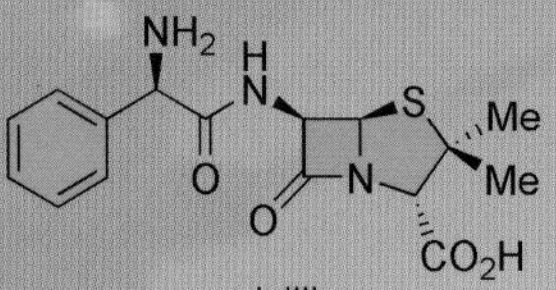

ampicillin
(semi-synthetic)

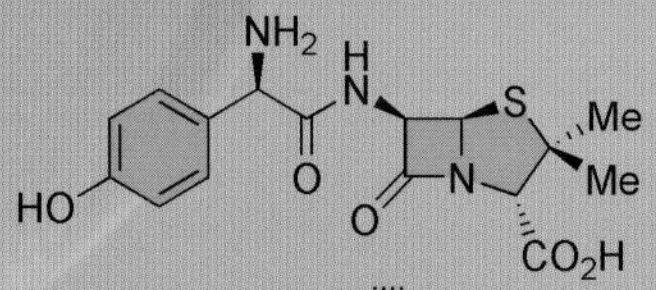

amoxycillin
(semi-synthetic)

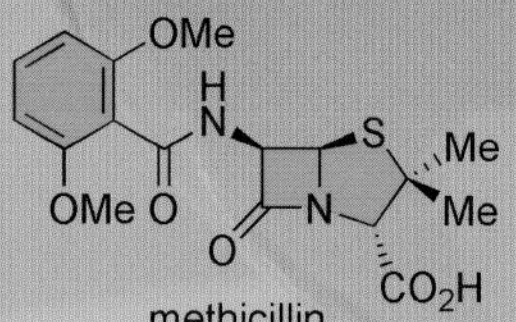

methicillin
(semi-synthetic)

era for antibiotics, however, came some disturbing news. Their widespread use, and sometimes misuse, led to the rapid evolution and spread of antibiotic-resistant bacterial strains. A new menace for humanity was now looming on the horizon!

Bacterial strains have evolved the capability to evade the action of β-lactam antibiotics by producing an enzyme, called β-lactamase, which can cleave the β-lactam ring, thus deactivating the molecules before they reach their site of action. To combat these newly acquired enemies, scientists have since sought out and developed various novel antibacterial agents; some of these powerful new antibiotics came from nature and some emerged from totally manmade designs. The story of drug-resistant bacteria, and of the antibiotics used to defend us against them, will be revisited in the chapter introducing vancomycin (Chapter 31), a natural product that is today the lifesaving drug of last resort in cases of severe infection.

We must conclude this account by underscoring once again the profound significance of the discovery of penicillin. Besides constituting a landmark medical breakthrough that saved lives and alleviated human suffering, this fortunate event also revealed to scientists a new treasure trove of biologically active molecules ripe for exploration. Thus, to the forest, which held the key to the development of Aspirin®, we now add the kingdom of microbes as a rich hunting ground for molecules endowed with healing powers. Indeed, many such molecules have since been isolated from the soil and other habitats where bacteria and fungi hold sway and, from this bountiful harvest, scientists have derived a host of 'magic bullets' and billion dollar drugs, as we shall see in forthcoming chapters.

University of Oxford

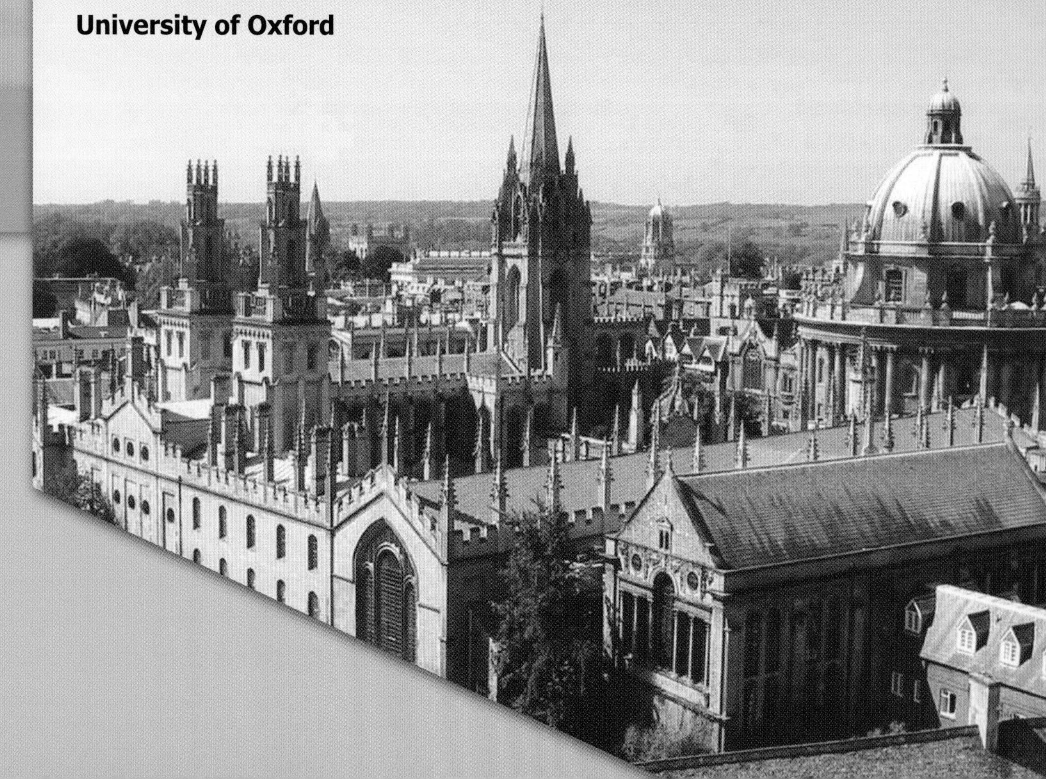

Further Reading

J. C. Sheehan, *The Enchanted Ring: The Untold Story of Penicillin*, The MIT Press, Cambridge, Massachusetts, **1982**.

J. Mann, M. J. C. Crabbe, *Bacteria and Antibacterial Agents*, Oxford University Press, Oxford, **1996**.

K. C. Nicolaou, E. J. Sorensen, *Classics in Total Synthesis*, Wiley-VCH, Weinheim, **1996**, pp. 41–53.

Me
Me
H
Me
LOGIC of CHEMICAL SYNTHESIS
CALCOMP 565
Longifolene
Retrosynthetic Analysis
Longifolene
Chapter 14
1961

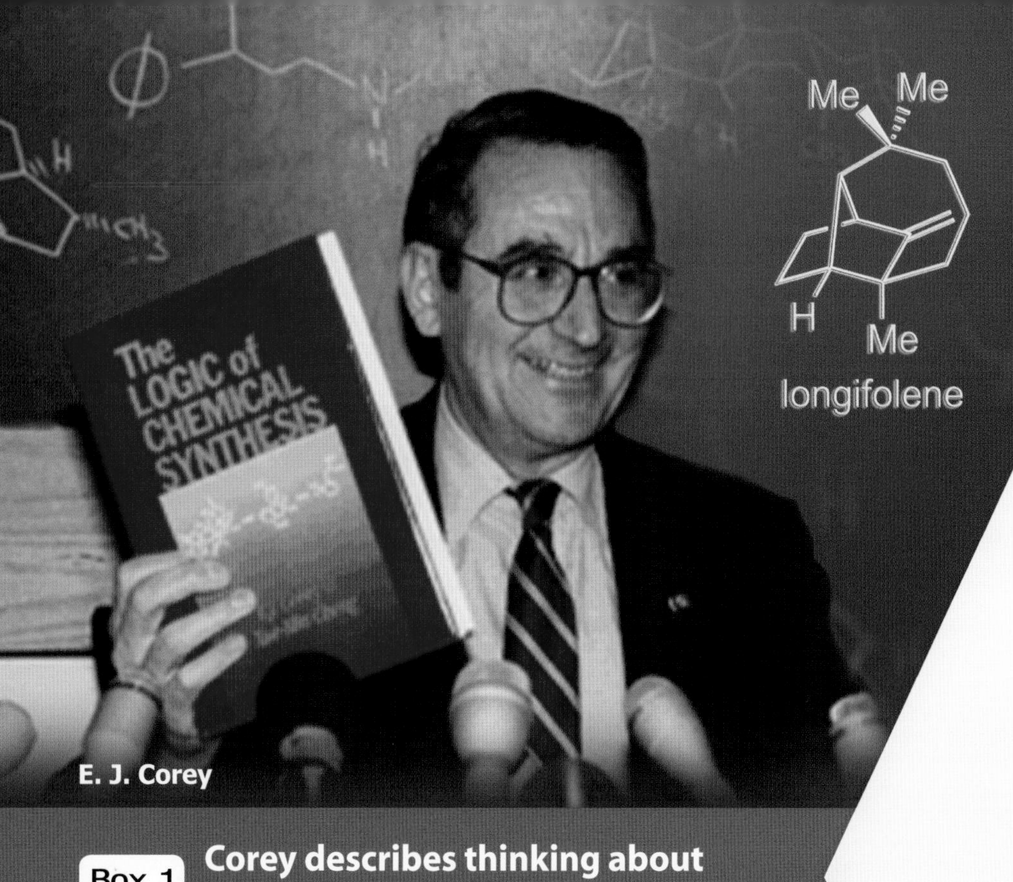

E. J. Corey

Box 1 Corey describes thinking about organic synthesis

"How does a chemist find a pathway for the synthesis of a structurally complex carbogen? The answer depends on the chemist and the problem. It has also changed over time. Thought must begin with perception – the process of extracting information which aids in logical analysis of the problem. Cycles of perception and logical analysis applied reiteratively to a target structure and to the 'data field' of chemistry lead to the development of concepts and ideas for solving a synthetic problem. As the reiterative process is continued, questions are raised and answered, and propositions are formed and evaluated with the result that ever more penetrating insights and more helpful perspectives on the problem emerge. The ideas which are generated can vary from very general 'working notions or hypotheses' to quite sharp or specific concepts."

E. J. Corey

We hope that as we relate each story you will begin to develop an empathy and rapport with the chemists whose dedicated research fills these pages. It is our aspiration that as each new investigation is unveiled you should be feeling the thrills, recognizing the challenges, and comprehending the drive to explore and conquer the molecule at hand in communion with the scientists who undertook the task in the first place.

This congenial ambition to entertain, however, should not overshadow the more philosophical themes and objectives being explored as we delve through the history of our science. As synthetic organic chemists, we would like to share with you just how the science that we feel so passionate about has matured, changing dramatically from its tentative infancy into the confident scientific discipline currently at the heart of so many different aspects of our lives. Today it shapes the world, major industries and other disciplines are built upon its foundations, and it gives us so many essentials, from polymers and fuels to medicines and the materials for bio- and nanotechnologies.

You may be asking yourself what the total synthesis of longifolene, in 1961, has to do with these broad general statements about benefits to society. The answer lies in the fact that the synthesis of this relatively small molecule marked a sharp turning point in the art and science of total synthesis. As we shall see, it ushered in a new era during which the recently acquired muscles of synthetic organic chemistry were flexed. In the period that followed, we shall witness this science charging forth with new found vigor, successfully meeting a myriad of challenges, and accepting every gauntlet thrust its way, to end up in its present position as a powerful, world-shaping discipline. Longifolene was synthesized by E. J. Corey at Harvard University (USA), and it took the mind of this truly great scientist to recognize that chemical synthesis had changed and that it was ripe for new advancements. The breakthrough he pioneered took the form of a unifying theory, named retrosynthetic analysis, which describes how one can rationally design a synthesis of any given complex organic molecule.

E. J. Corey was born in Massachusetts (USA) into a family of first generation Americans of Lebanese descent. He was actually named William for the first eighteen months of his life, but the tragic death of his father prompted his mother to rename him Elias in memory of her late husband. Following the trials of the depression and World War II, this student, already showing exceptional talent, entered the Massachusetts Institute of Technology (MIT) in 1945, to pursue undergraduate studies in mathematics.

Retrosynthetic Analysis

Synthetic Strategy

However, he was soon recruited to the field of chemistry by the "joys of solving problems in a laboratory setting" and by the "intrinsic beauty and great relevance to human health" that he perceived in organic chemistry, in particular. Perhaps it was his earlier love for the rigors of mathematics that guided his thinking, allowing him to pioneer such a logical system as retrosynthetic analysis, during the early days of his career. Whatever the inspiration, this new concept took synthetic organic chemistry by storm and led to the award of the 1990 Nobel Prize in Chemistry to Corey for "his development of the theory and methodology of organic synthesis."

Corey describes the context in which his theory was first conceived as being related to certain stimuli pertinent to the post-war era of chemistry. Synthetic organic chemistry had evolved, leaving behind the days when limits were set simply by the accessibility of starting materials. Such constraints are exemplified by the reliance on coal tar during the late 1800s and early 1900s, which you might perceive if you browse through the early chapters of this book. Studies on the chemistry of coal tar, the sticky black residue produced during the refinement of coal to give coal gas, the prevalent fuel of the day, first afforded a broad range of dyes (see Chapter 8). From these discoveries, fledgling chemical industries were spawned, which quickly branched out, initiating the birth of pharmaceuticals as a distinct scientific discipline and a commercial venture. Even the earliest antibiotics, such as prontosil rubrum and trypan red (discussed in Chapter 13), originated from dyes that owed their discovery to investigations into coal tar chemicals. Later, chemistry expanded rapidly and the knowledge base grew exponentially, so that complex natural products like morphine (Chapter 10), strychnine (Chapter 12), and vitamin B_{12} (Chapter 16) could be tackled by chemical synthesis. Success in these ventures required a supreme knowledge of individual transformations, a dedicated persistence, plenty of creativity, and, of course, a little luck. Concomitant to this progress towards more intricate and challenging molecular structures, the practice of organic synthesis was being revolutionized. The mechanisms of chemical reactions and electronic theory were delineated and chemists were beginning to understand conformational effects and transition state stereochemistry, allowing selective reagents and reactions to be developed. Chromatography, the technique employed to separate and purify organic compounds, was discovered and refined. At the same time, the science and technology of spectroscopy, used to determine molecular structures and the purity of chemical samples, was being established at a rapid pace. Despite a prevailing faith at this time which held that essentially no molecule was beyond the efforts of synthetic organic chemistry, there was a seemingly gaping flaw in that each molecule dictated an

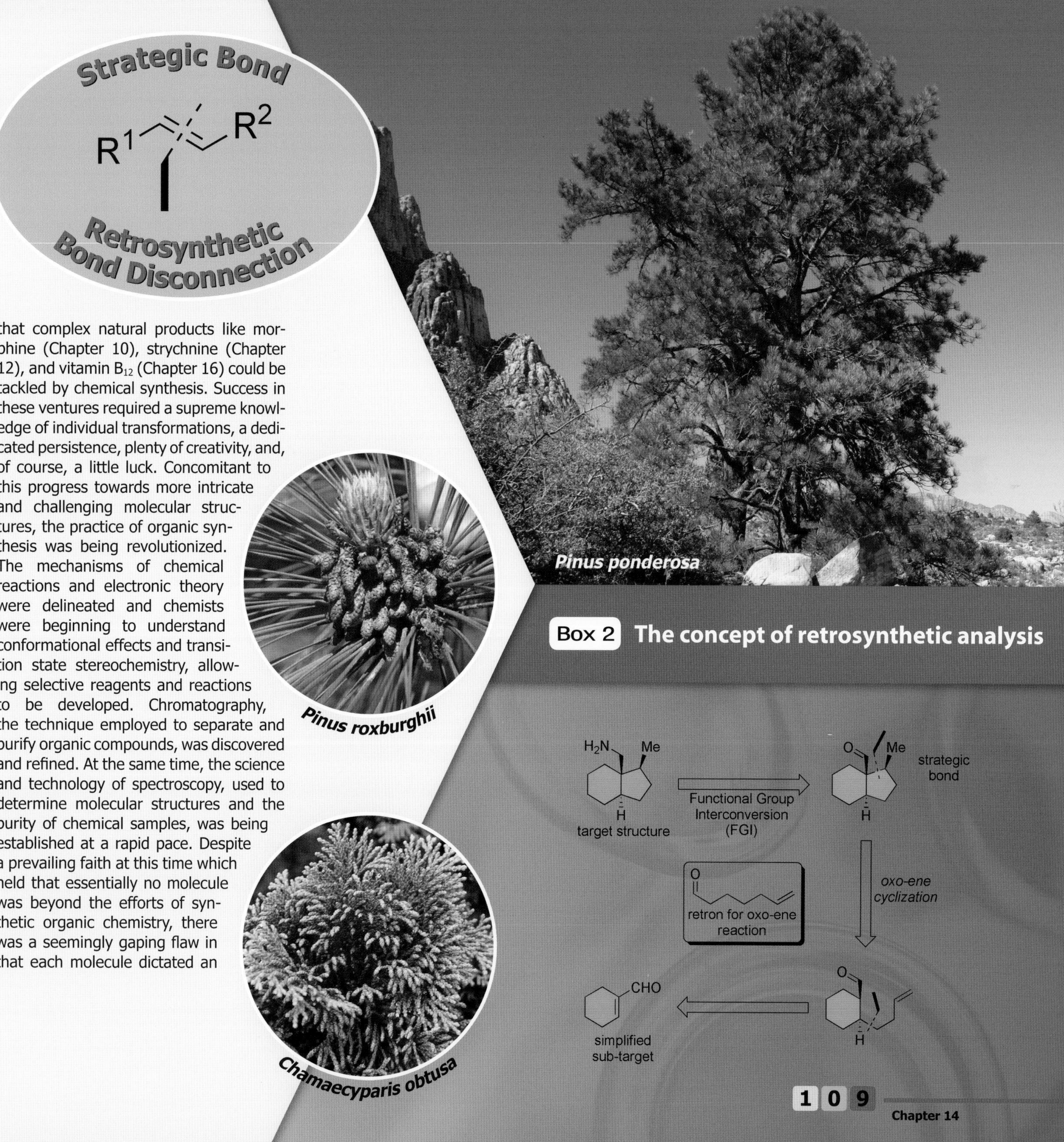

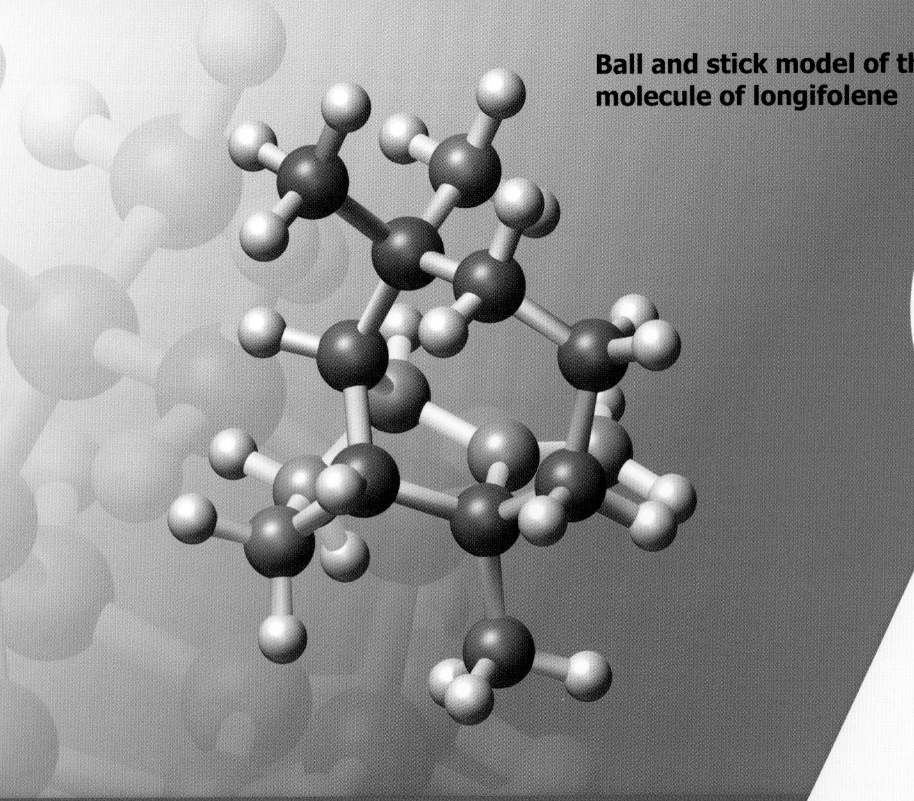

Ball and stick model of the molecule of longifolene

Chamaecyparis obtusa

Box 3 Retrosynthetic analysis of longifolene

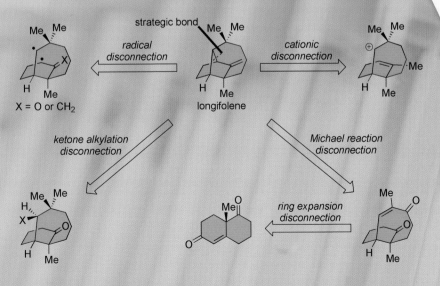

strategic bond

radical disconnection

cationic disconnection

longifolene

X = O or CH_2

ketone alkylation disconnection

Michael reaction disconnection

ring expansion disconnection

Functional Group Interconversion

entirely unique approach. In addition, perhaps many chemical steps after setting forth on a synthetic route, the sword of Damocles would fall and a transformation would fail utterly with disastrous consequences. The chemist would face the awful decision as to whether to start all over again, or work for extended periods, sometimes years, to circumvent this single obstacle, knowing that the next daunting hurdle might be just a few more steps further down the line.

Corey devised an approach that would minimize these issues by injecting a high degree of rationale and predictability into the design strategy for synthesizing a molecule. Thus, the chemist would examine the structure of any given molecule and undertake an analysis that would identify 'strategic bonds.' Working backwards, the so-identified bonds could be broken, or 'disconnected,' in an imaginary manner, to reveal a new series of simplified sub-targets (Box 2). If this process is continued iteratively and elaborated to include steps such as rearrangements or functional group interconversions (FGIs) that do not afford simplification, but that generate handles for subsequent strategic disconnections, then a course through the most reasonable pathways available for building the target molecule can be navigated. The rigor and pervasive logic of this approach, and the ability to identify key building blocks ('retrons') and disconnections independent of gross structure made the theory amenable to computing applications, an avenue that Corey also pioneered in the 1960s. Although an uncreative and inflexible computer can never replace the imagination of a talented chemist, computer programs have been shown to offer valuable assistance in analyzing the options available for modes of molecular 'deconstruction.' Retrosynthetic analysis allows a number of different strategies and tactics to be explored and the merits of each compared, thereby facilitating and improving the design and planning stages of a chemical synthesis. Nevertheless, ingenuity and talent on the part of the chemist still commands a premium because using creativity and imagination in the choices required to define strategic bonds and disconnections can greatly enhance the efficiency and elegance of the resultant synthetic plan. Whether undertaken by a chemist or a computer, the success of retrosynthetic analysis is beyond doubt since, today, no synthesis is undertaken without applying this concept at the outset.

Longifolene, a sesquiterpene isolated from a variety of plant sources (e.g. *Pinus ponderosa*, *Pinus roxburghii*, and *Chamaecyparis obtusa*), was amongst the first molecules on which Corey demonstrated his new theory. In 1957, whilst still a member of the faculty at the University of Illinois (Urbana-Champaign, USA), he took a sabbatical post, sponsored by the Guggenheim Memorial Foundation, within the prestigious chemistry department at Harvard University (USA) by invitation from Robert B. Woodward. There he showed Woodward an illustration of his retrosynthetic idea together with the synthetic plan for longifolene. The elder and highly esteemed chemist was enthused and immediately recognized the sagacity of this young academic, precipitating the offer Corey received shortly thereafter to join the faculty at Harvard permanently as a professor.

Terpenes, independent of size, are excellent molecules on which to test such an analysis because they have stubborn, all carbon-based skeletons frequently arranged in complex polycyclic frameworks. Indeed, Corey states, "the virtuosity of nature in the construction of intricate molecules is nowhere more evident than in those [terpenoid] families." Terpenes are made naturally through the mevalonic acid pathway (for other examples of terpenes, see Chapters 6, 21, and 25) and are defined by containing a number of C5 isoprene units. Akin to many other smaller terpenoids, longifolene has found use in the perfumery industry due to its rich woody odor. The all-carbon nature of terpenoid architectures is quite challenging, with the difficulties stemming from a number of fundamental principles that make carbon–carbon bonds harder to form than corresponding bonds between carbon and one of the heteroatoms, such as oxygen and nitrogen, commonly found in natural products. Furthermore, saturated carbon skeletons have no amenable handles (functional groups) on which to operate in order to carry out the transformations needed to build up molecules in the laboratory.

It was against this background that Corey successfully designed and executed the elegant total synthesis of longifolene. This natural product had been at the center of a fierce debate for several years over its true structure, a dispute that was not settled until the x-ray crystallographic analysis of a derivative was completed by R. H. Moffett and D. Rogers in 1953. Corey's retrosynthetic analysis of this molecule, which, at least initially, afforded several alternative strategies, is shown in Box 3. The first bond selected as a 'strategic bond' for disconnection is the only one whose cleavage takes the molecule back (retrosynthetically) to a simple 6,7-fused bicycle; all other possibilities leave a more complex bridged bicycle, and can therefore be rejected readily. Precedent for particular reactions directed the remaining retrosynthetic analysis and choices. This auspicious blueprint was then smoothly translated, more or less in its original form, into a practical laboratory synthesis of longifolene, attesting to the power of the strategy that had been designed using the retrosynthetic analysis (Box 4).

Corey has continued to synthesize many challenging natural products, a few examples of which are shown in Box 5, using his theory of retrosynthetic analysis to guide him. His vision and mastery of chemical synthesis are legendary. These virtues have allowed him, hand-in-hand with his graceful syntheses, to innovate and discover countless new synthetic methods, which have found widespread applications in chemical synthesis. The theory of retrosynthetic analysis is now taken for granted, its power to assist in the teaching of organic chemistry and the designing of

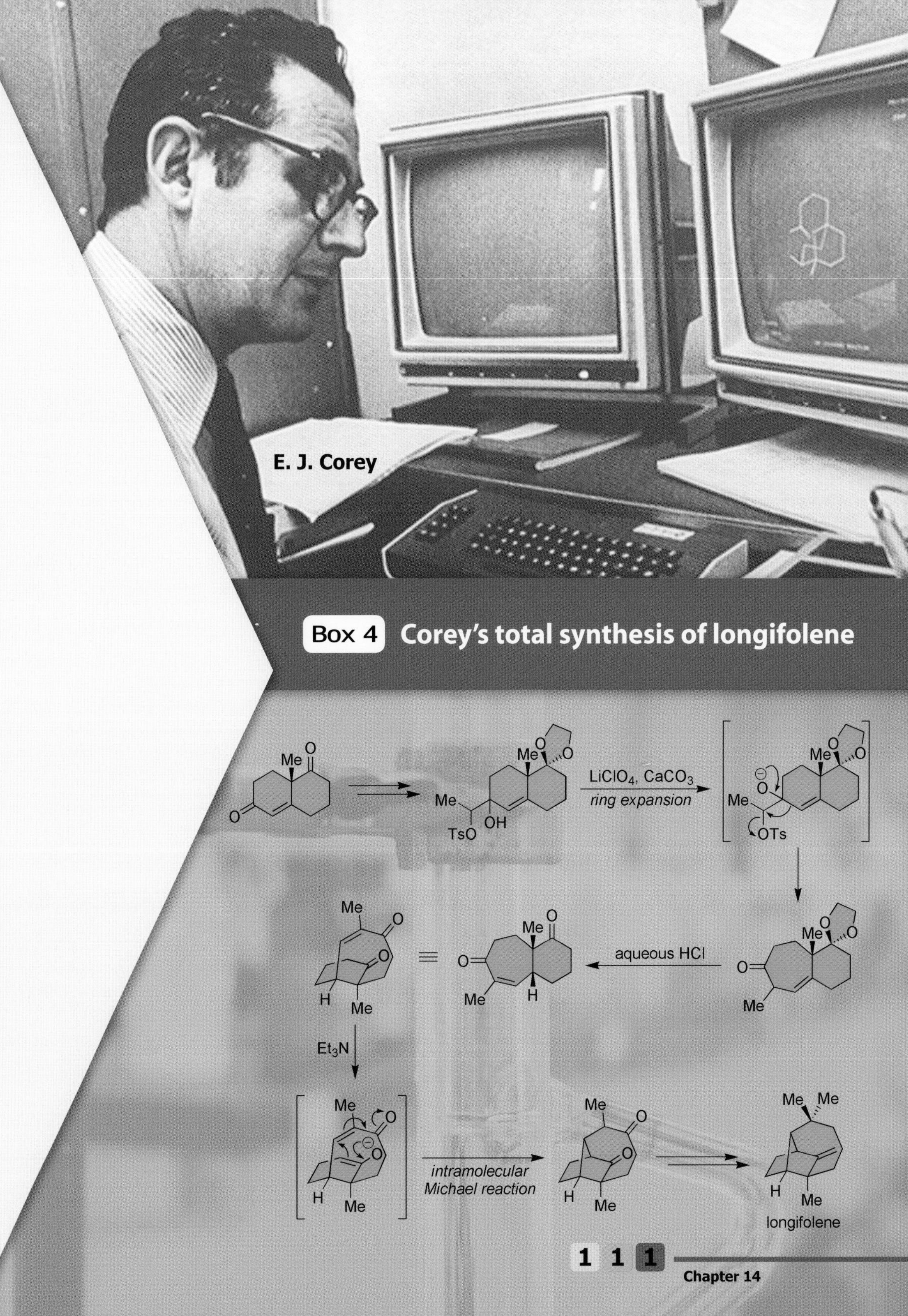

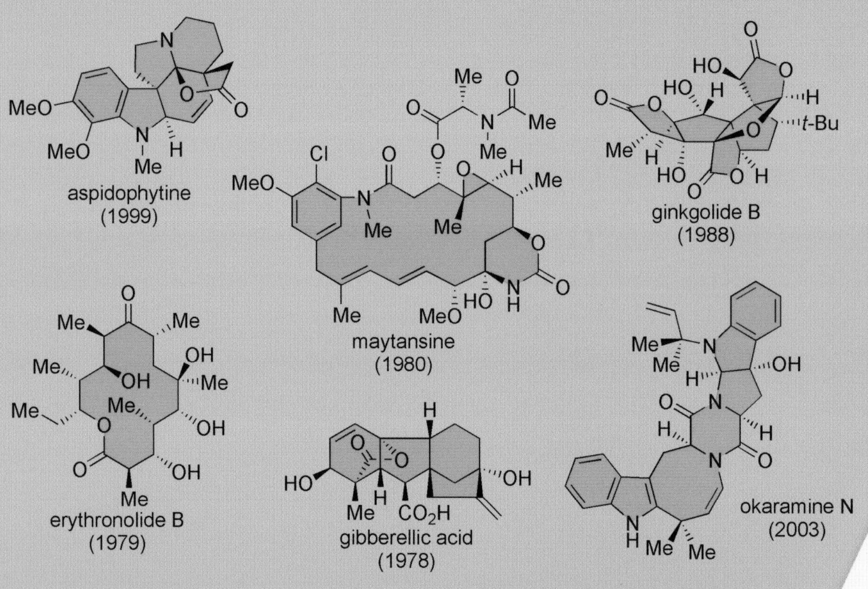

strategies for synthesizing complex molecules undisputed – so much so that the vast majority of total syntheses reported in the chemical literature today begin with a scheme depicting the retrosynthetic analysis of the target molecule. Furthermore, textbooks aimed at advanced students of chemical synthesis are based on retrosynthetic analysis as a means to convey the essence of the science. Thus, E. J. Corey, through his conquest of longifolene and his theory of retrosynthetic analysis, pioneered a true innovation in the history of the art and science of total synthesis.

E.J. Corey

Pinus roxburghii

Further Reading

E. J. Corey, M. Ohno, P. A. Vatakencherry, R. B. Mitra, Total Synthesis of *d,l*-Longifolene, *J. Am. Chem. Soc.* **1961**, *83*, 1251–1253.

E. J. Corey, M. Ohno, R. B. Mitra, P. A. Vatakencherry, Total Synthesis of Longifolene, *J. Am. Chem. Soc.* **1964**, *86*, 478–485.

E. J. Corey, W. T. Wipke, Computer-Assisted Design of Complex Organic Syntheses, *Science* **1969**, *166*, 178–192.

E. J. Corey, X.-M. Cheng, *The Logic of Chemical Synthesis*, Wiley-Interscience, New York, **1989**.

K. C. Nicolaou, E. J. Sorensen, *Classics in Total Synthesis*, Wiley-VCH, Weinheim, **1996**.

K. C. Nicolaou, S. A. Snyder, *Classics in Total Synthesis II*, Wiley-VCH, Weinheim, **2003**.

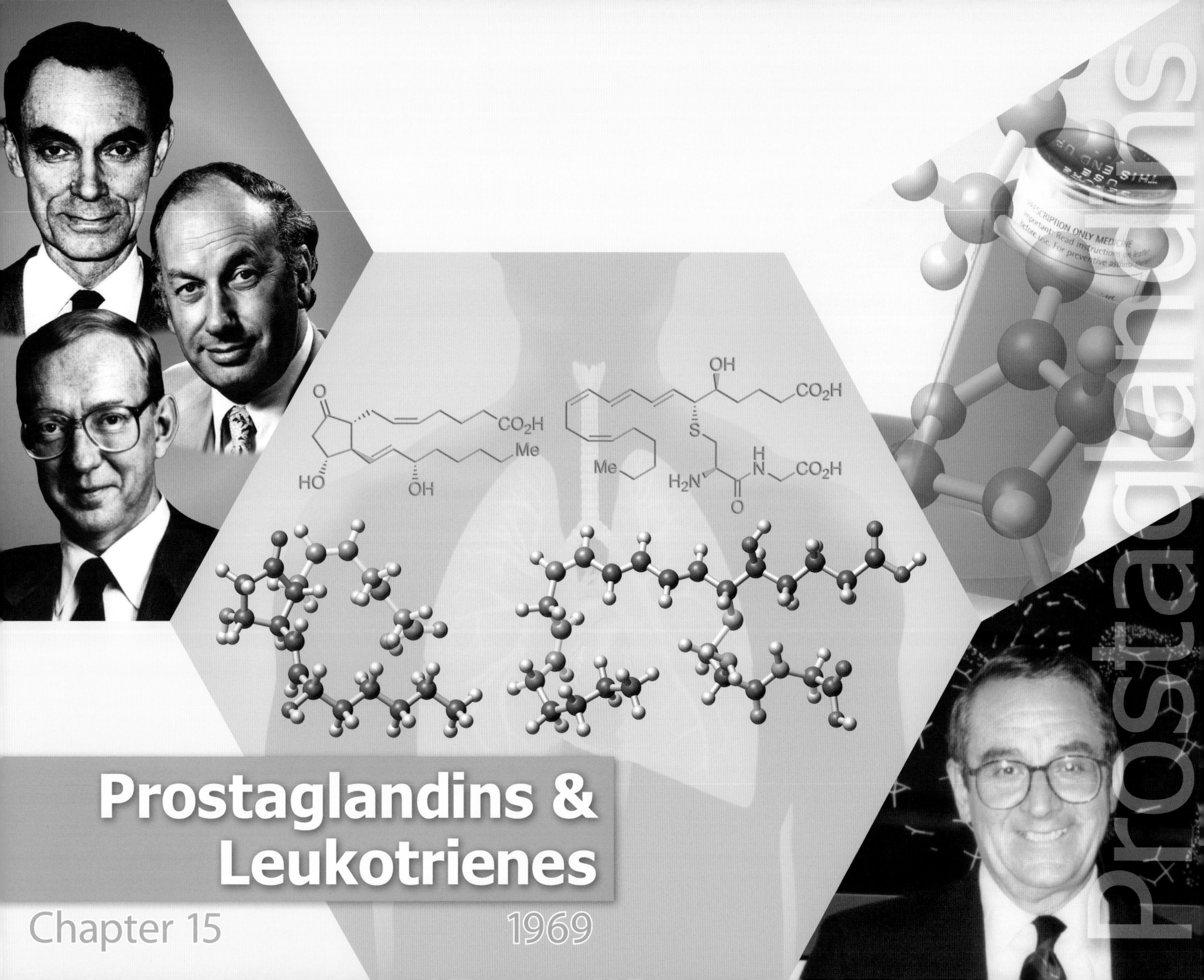
Prostaglandins &
Leukotrienes
Chapter 15
1969

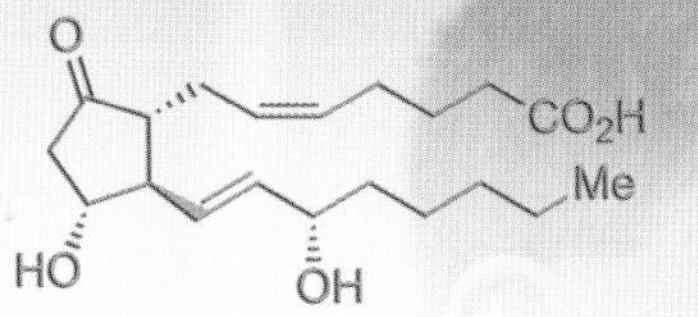

prostaglandin E_2 (PGE_2)

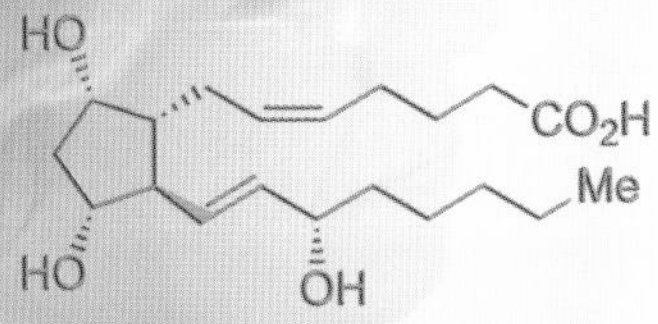

prostaglandin $F_{2\alpha}$ ($PGF_{2\alpha}$)

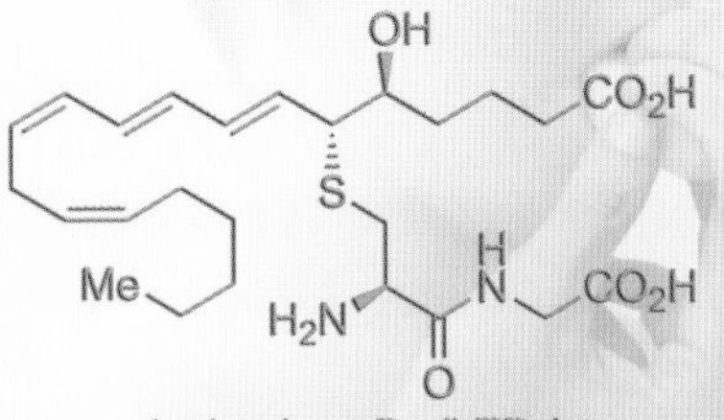

leukotriene D_4 (LTD_4)

Box 1 Common drugs used in asthma therapy

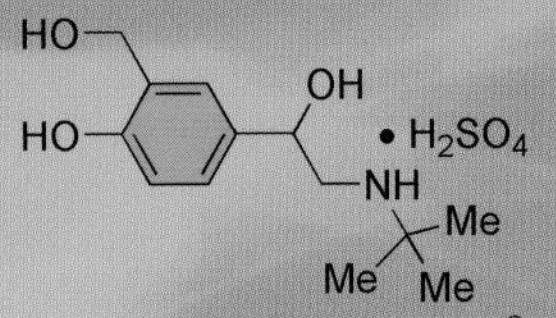

salbutamol sulfate [Ventolin®] (β-2 agonist)

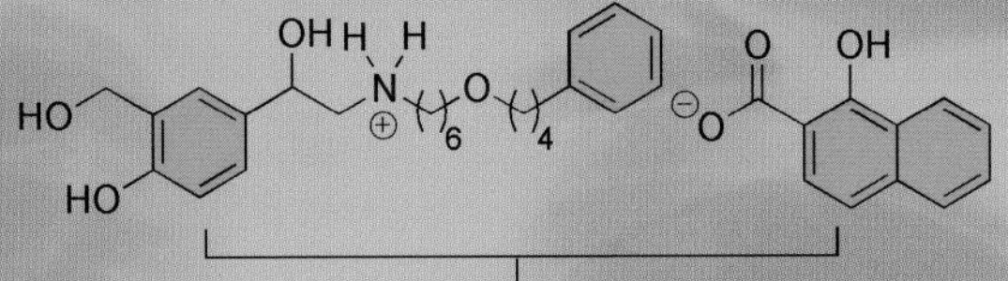

salmeterol xinafoate [Serevent®] (β-2 agonist)

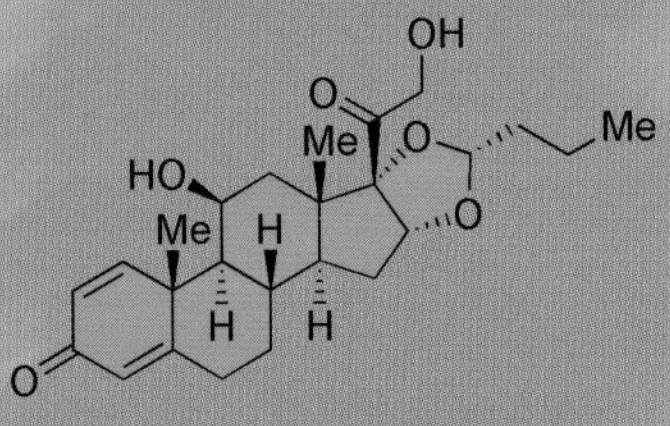

budesonide [Pulmicort®] (inhaled steroid)

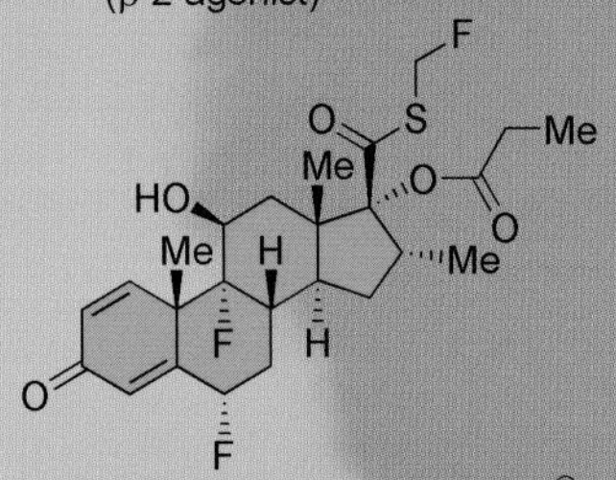

fluticasone propionate [Flovent®] (inhaled steroid)

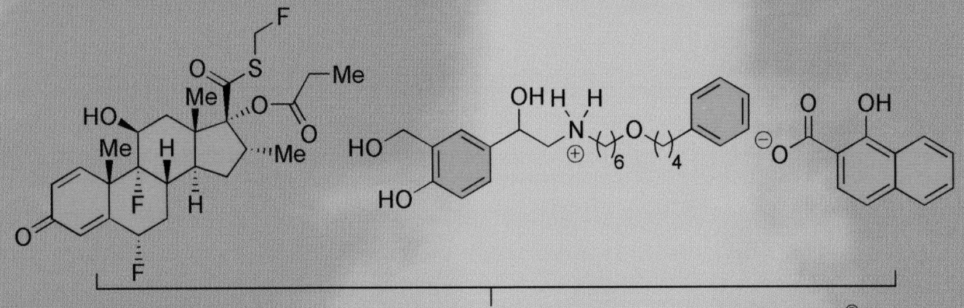

combination of fluticasone propionate and salmeterol xinafoate [Advair®] (inhaled steroid and β-2 agonist)

Asthma is a chronic inflammation of the lung airways marked by attacks of wheezing and a painful shortness of breath. It is a serious and indiscriminate disease, attacking adults and children with equal ferocity. It is also the most common chronic illness in children and the leading cause of absenteeism from school in the industrialized world. The World Health Organization (WHO) estimated that 100–150 million people worldwide (equivalent to about half the population of the USA) were suffering from asthma in 2000, with the global death toll exceeding 180,000 per year. Thus, the scale of the problem is huge, but perhaps more alarming is the recent rise in the occurrence of asthma. A rare condition a century ago, asthma has developed into an epidemic, particularly during the latter half of the twentieth century, with cases concentrated in industrialized nations. These features strongly suggest that human development and activity is in some way responsible for this scourge. Air pollution and the low levels of physical activity amongst children have been found to be significant contributors to the growing impact of asthma in our societies. This general point is aptly exemplified by the alarming rise in the number of sufferers in the USA between 1980 and 2000. Clearly, if this rise were to continue unabated, a huge proportion of the population would be affected over the coming decades. As scientists continue in their endeavors to reverse this frightening trend by seeking a fuller understanding of the disease, the search for new drugs also continues apace. In recent years, both environmental and genetic factors have been implicated in the development of asthma, and considerable progress has been made in deciphering some of the finer details relating to the trigger mechanisms of attacks.

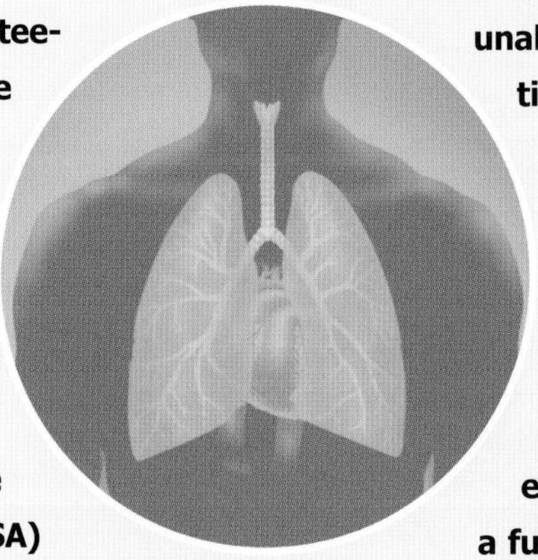

One of the biggest challenges in the field of anti-asthma drug development is the requirement that the drug should be particularly benign, so that it is acceptable for long-term use in both adults and young children. Treatments have traditionally targeted the relief of symptoms; inhaled steroids (like those found in Advair®, Flovent®, and Pulmicort®, Box 1) are used to heal damaged airways, and β_2-adrenergic receptor stimulants (such as those in Advair® and Ventolin®, Box 1) are employed to reduce constriction by relaxing the smooth muscle of the inflamed bronchial tract. Fortunately, the last few years have seen a dramatic breakthrough with the discovery and development of an innovative class of prophylactic agents called the leukotriene modifiers. This third major class of anti-asthma drugs has a more direct mode of action and fewer unwanted side effects than its predecessors. Certain candidates from this family are already approved by the regulatory authorities for general use and have quickly become exceedingly popular medicines. The most prominent representatives of this class are the formulations called Singulair® and Accolate®, developed by the pharmaceutical companies Merck and AstraZeneca, respectively (see Box 2).

Singulair® has also become a high profile prescription drug for treating seasonal allergies, thus underlining the biochemical link between allergies and asthma. Both of these conditions occur when the body's immune-based inflammation mechanisms overreact to the presence of an allergen (e.g. animal dander, dust mite residues, or pollen). In asthma, the harmful sequence of events begins with specialized immune cells, called mast cells, that line the bronchial tract. These cells are filled with a variety of potent inflammation mediating chemicals. When mast cells encounter allergens, or other stimuli, they become activated, releasing their payload of molecules and triggering a cascade of chemical reactions leading to the production of leukotrienes. These transient eicosanoid compounds are short-lived, but have powerful physiological properties. The localized release of leukotriene molecules initiates an intense inflammatory response that includes contraction of bronchial smooth muscle, secretion of mucus, and recruitment of eosinophils. These events culminate in airway edema, thus precipitating a full-scale asthma crisis. Eosinophils belong to the leukocyte family, which collectively make up the white blood cells. The newly recruited eosinophils rapidly migrate to areas of inflammation in the vascular endothelium where they excrete granules containing toxic proteins. Under ordinary circumstances, the purpose of these proteins is to kill invading germs; however, in an asthma attack, the intended target is misconstrued and the body's own endothelium

Mast cell

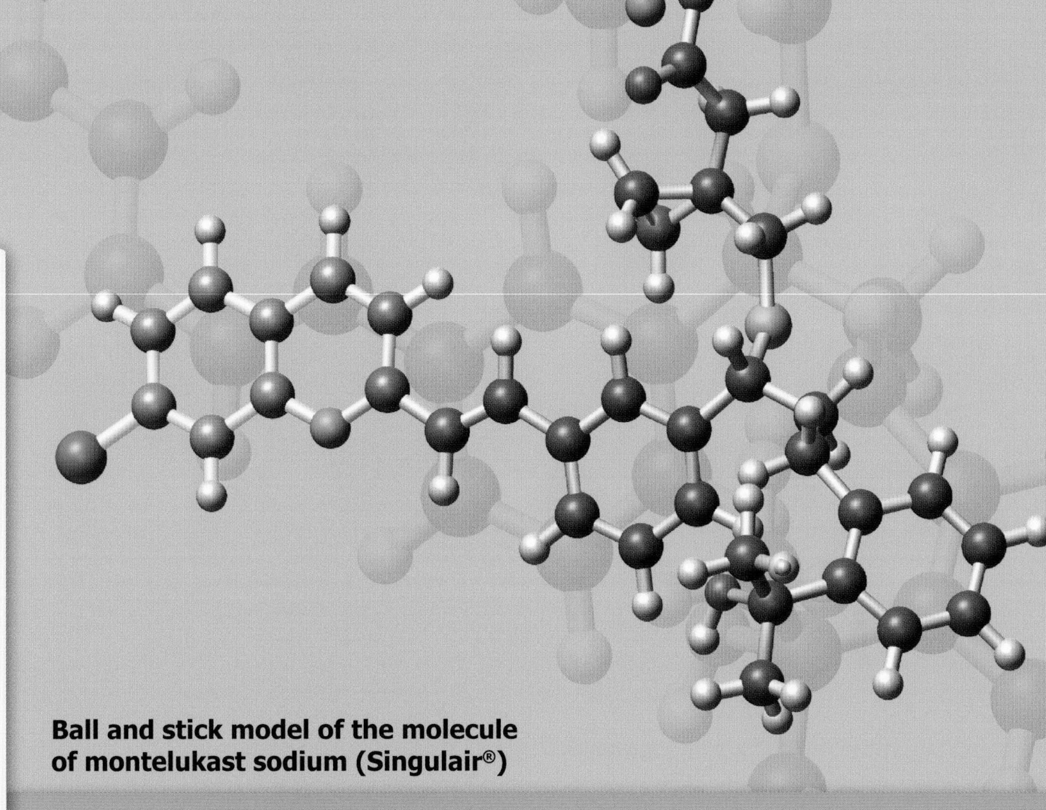

Ball and stick model of the molecule of montelukast sodium (Singulair®)

Box 2 Selected next-generation asthma drugs

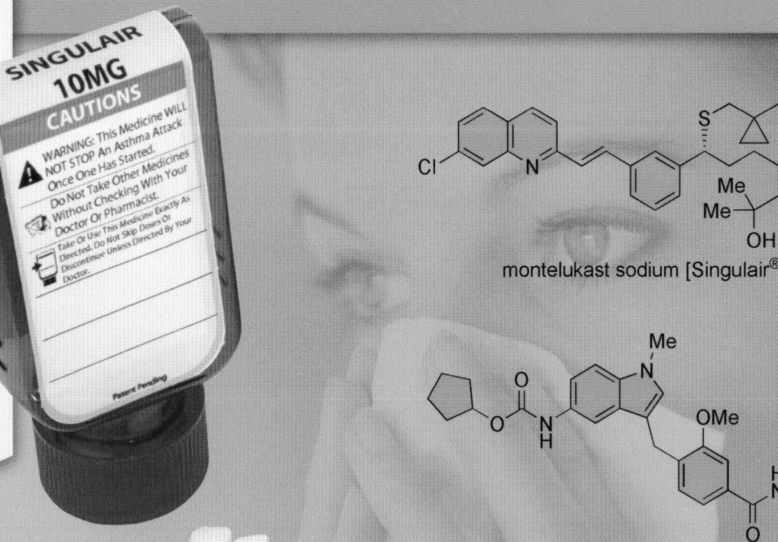

montelukast sodium [Singulair®]

zafirlukast [Accolate®]

Box 3 **Pathways of the arachidonic acid cascade**

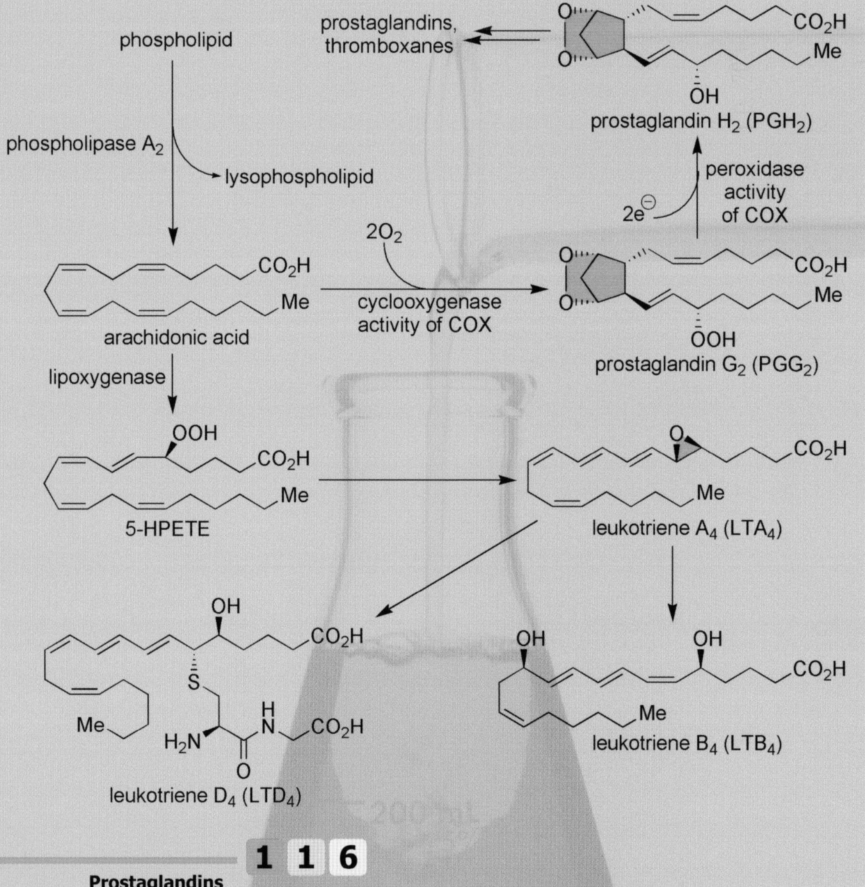

cells are destroyed, causing further damage to the airways of the unfortunate asthma sufferer. Thus, a spiraling cycle of airway inflammation followed by endothelium deterioration reverberates, and can potentially prove fatal unless treated effectively with an emergency symptom-alleviating medication. Singulair® and related drugs from the latest generation of anti-asthma and anti-allergy medications work quite differently. Rather than treating the symptoms, they act prophylactically to prevent the attack from occurring in the first place. These drugs are antagonists for the receptors that are normally triggered upon the release of leukotrienes nearby. They block the leukotriene chemical messengers from communicating with neighboring cells, thus inhibiting the cascade leading to inflammation of the patient's airways.

Leukotrienes are chemical relay agents involved in mediating many more, mostly inflammation-based, responses in the body. They are members of the class of transient, but vital, lipid-derived biomolecules introduced earlier in this account as the eicosanoids. These molecules all have very short lifetimes, but their effects are felt profoundly even at very low concentrations. In the early 1960s, Sune K. Bergstrøm and his team at the Karolinska Institute in Sweden determined the first eicosanoid molecular structures. In addition to the leukotrienes, the eicosanoid family includes the prostaglandins, thromboxanes, and prostacyclins, with the prostaglandins being the most common. Indeed, the prostaglandins are not made solely in the prostate, as thought in 1934 when they were discovered in sheep prostate gland secretions by the Swedish physiologist Ulf Svante von Euler-Chelpin, but in every nucleated cell; as such, they control a host of bodily functions.

Ulf Svante von Euler-Chelpin

In humans, all the eicosanoid molecules have the same primary source, arachidonic acid, which is an unsaturated fatty acid consisting of a linear chain of twenty carbon atoms. Arachidonic acid is stored as a conjugate in the phospholipid membranes of all cells and is released as required by the hydrolytic action of a phospholipase enzyme. In response to signals regarding a localized emergency, or reporting physical damage to an adjacent tissue, this fatty acid is freed, and a biochemical cascade begins through which it is converted into the requisite eicosanoid within fractions of a second (Box 3). Details of this process were unraveled in the 1970s primarily by Bengt I. Samuelsson (a former student of Sune Bergstrøm) and his colleagues, again working at the Karolinska Institute in Sweden. The first two steps of the main branch of the arachidonic acid pathway, yielding all the eicosanoids except the leukotrienes (which are produced through a separate offshoot), are cata-

CO_2H
Me
OH
thromboxane A_2 (TXA_2)

lyzed by a single 'super enzyme' called the cyclooxygenase (COX) enzyme. In 1971, Sir John Vane and his colleagues at the Wellcome Research Laboratories in the UK published a series of landmark papers, revealing that Aspirin® exerts its many beneficial effects (e.g. analgesia, fever moderation, protection from heart disease) by irreversibly disabling the COX enzyme. The inhibition of this pivotal biological catalyst is brought about by the transfer of an acetyl group from Aspirin® to the active site of the enzyme. The result is the prevention of the biosynthesis of all the prostaglandins, thromboxanes, and prostacyclins. Indeed, all the common painkillers belonging to the so-called non-steroidal anti-inflammatory drug family (NSAIDs, Box 4) act by inhibiting this enzyme. Furthermore, the later discovery of different forms of this enzyme (COX-1, COX-2, and recently, COX-3) afforded chemists the opportunity to develop a novel class of drugs, called COX-2 inhibitors (e.g. Celebrex® from Pharmacia, now part of Pfizer, and Vioxx® from Merck), although recent developments have cast doubt over the future of some members of this novel class of drugs. The story of Aspirin® and the other COX inhibitors is discussed in greater detail in Chapter 4, which also describes more fully the pioneering works of Sune Bergstrøm, Bengt Samuelsson and Sir John Vane, who shared the 1982 Nobel Prize in Physiology or Medicine "for their discoveries concerning prostaglandins and related biologically active substances."

Since most of us have, at some point, taken an over the counter analgesic (such as Aspirin®, Advil®, Aleve®, Nurofen®, or Tylenol®, Box 4), we can readily appreciate a second significant role played by the eicosanoids. Prostaglandins are intimately involved in the amplification and transduction of pain, as well as in fever modulation. However unpleasant, pain and fever have evolved as part of our complex systems for self-preservation. Fever is an integral part of our sophisticated immune system; raising the body's temperature (an action mediated by prostaglandins) results in a much less favorable environment for the replication of viruses, or the reproduction of bacteria, with the result that their proliferation is severely retarded. This restriction allows the cellular components of the immune system to tackle a significantly weakened enemy.

Similarly, pain protects us from hurting ourselves and warns us of illnesses or malfunctions within our body. Humans possess an enormously complicated network of neurons that transmit electrical pain signals to the brain, where they are received and processed and an appropriate response is organized. The combined length of all these nerves in a single person is estimated to be just under a billion kilometers, and the impulses they carry can travel at a speed of five hundred

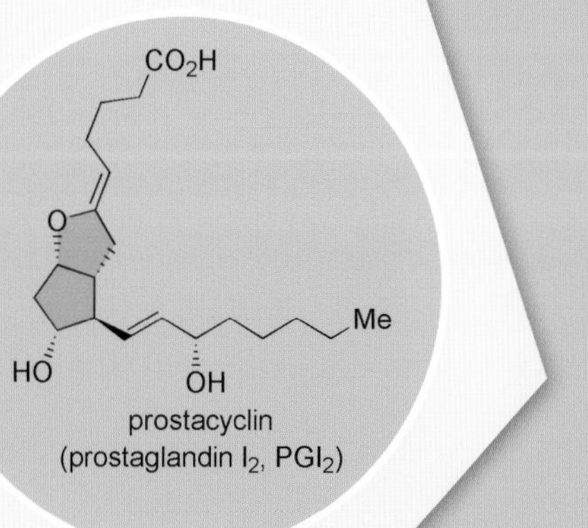

prostacyclin
(prostaglandin I_2, PGI_2)

Box 4 Selected non-steroidal anti-inflammatory drugs

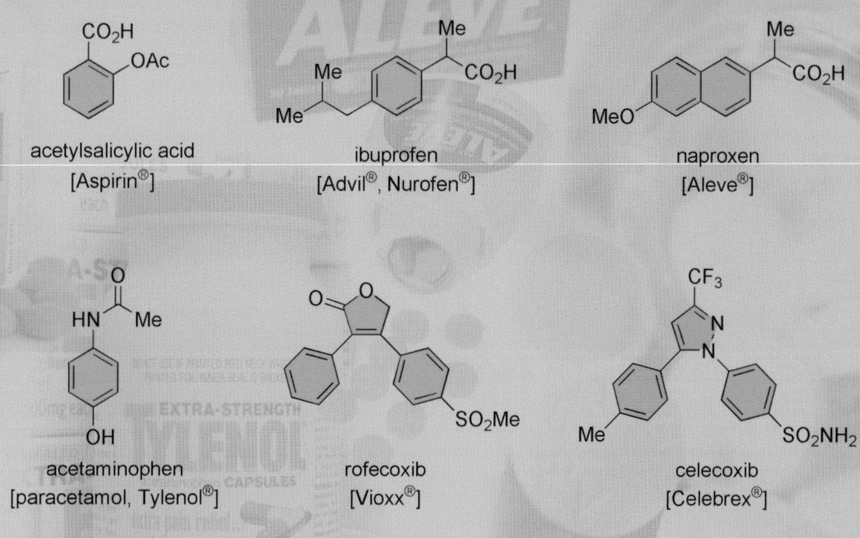

Box 5 Eicosanoid analogues used as drugs

latanoprost [Xalatan®]
(glaucoma treatment)

iloprost [Ilomedin®, Ventavis®]
(arterial occulsion, pulmonary hypertension)

misoprostol [Cytotec®]
(anti-ulcer)

E. J. Corey

Box 6 **E. J. Corey's retrosynthetic analysis of PGE_2 (the bicycloheptane strategy)**

Ball and stick model of the molecule of prostaglandin E_2

kilometers per hour. This impressive network is, however, merely one component of the system involved in the sensation of pain. An elaborate biochemical cascade, involving the release of chemicals from tissue damaged through illness or injury, always precedes the information transfer. The prostaglandins play a key role in this process and are also responsible for marshalling the components in the complex response to pain.

Many other tasks fall into the wide-ranging portfolio governed by the prostaglandin family, including the release of gastric fluids (i.e. hydrochloric acid and pepsin) in the stomach, control of renal water and electrolyte balance, induction of labor, and regulation of the sleep–wake cycle. The important roles held by members of the prostaglandin family have led to the development of a plethora of small molecule mimics of these chemical mediators by pharmaceutical companies. These drugs are employed over a range of therapeutic indications (Box 5). For example, Xalatan®, an optical solution produced by Pfizer, is a prostaglandin $F_{2\alpha}$ analog used to treat glaucoma, and Cytotec®, a prostaglandin E_1 analog formulation made by Searle, is used as an anti-ulcer drug. Cytotec® also has a controversial unregulated use for the induction of labor. This brief survey seeks to illustrate the paramount importance of the prostaglandins to the normal function of our bodies. It reveals how, despite their very fleeting existence, not a moment passes when these chemicals are not overseeing at least one critical operation.

Thus far, we have mentioned very little about the thromboxanes and prostacyclins, but, once again, we find that these small molecules wield immense power over the correct functioning of human physiology. The primary function under their control is the proper maintenance of blood circulation and pressure. Thromboxanes are produced by platelets (thrombocytes) and they play an active role in the formation of blood clots and in vasoconstriction. Prostacyclins are their antithesis and are made in vascular endothelial cells, from where they are released to promote vasodilation and inhibit platelet aggregation. The thromboxanes have been the target of intensive research because, when they engage in the blood clotting cascade under the wrong circumstances, they act as promoters of heart disease and stroke. In these two dreaded conditions, blood clots form

Blood clot in the brain

around areas of vascular tissue damage, such as atherosclerotic plaques, then break away, migrate to, and subsequently block small blood vessels in the heart or brain. Folklore and, more recently, health authorities have encouraged the consumption of garlic to reduce the risk of heart attacks and strokes. In the late 1970s, researchers discovered that chemicals found in garlic could inhibit thromboxane biosynthesis, lending biochemical backing to the traditional knowledge regarding the health benefits of this celebrated ingredient.

Inhibition of the COX-enzyme that mediates the arachidonic acid cascade also limits thromboxane production, and, indeed, it is by this latter mode of action that Aspirin® provides its well-known protection against heart disease and stroke. Looking at the issue from an alternate perspective, the Schering AG Pharmaceutical Company, for example, enlisted the positive effects of the counterbalancing prostacyclins by developing and formulating a stable prostaglandin I_2 analog, named Ilomedin® (Box 5). This drug is administered by inhalation or intravenous infusion to inhibit platelet aggregation and cause vasodilation in conditions where dangerous vascular dysfunction is occurring, such as occlusive arterial disease or pulmonary hypertension.

Laboratory syntheses of various eicosanoids were especially valuable in the 1960s and 1970s because even simple biological studies were being hampered by the instability and scarcity of many of these molecules. Structural elucidation had turned out to be a much slower and more painful process than originally anticipated, despite the relatively small size of these molecules and the timely advent of new mass spectrometry techniques. In the end, chemical synthesis proved to be an invaluable tool, successfully employed to establish (or at least confirm) many of the molecular structures proposed for numerous eicosanoids. In addition, chemical synthesis allowed the production of much larger quantities of these delicate compounds than were available at the time from natural sources, thus greatly facilitating the investigation of their biochemistry and physiology.

The establishment of this multidisciplinary approach to the eicosanoid problem was brought about by a meeting between the biochemists Bergstrøm and Samuelsson, and the illustrious organic chemist E. J. Corey. In 1957, Bergstrøm invited Corey to the Karolinska Institute, a visit that was to lead to a fertile collaboration between Corey and Samuelsson. The trip to Lund in Sweden came in the early days of Corey's career, while he was still at the University of Illinois (Urbana–Champaign, USA). Corey was so beguiled by the prostaglandins to which he was introduced during this trip that he focused a great deal of energy and attention on them. In the three decades that followed, he came to dominate the burgeoning field of eicosanoid chemistry. Almost in parallel, his insight and ingenuity led him to reign

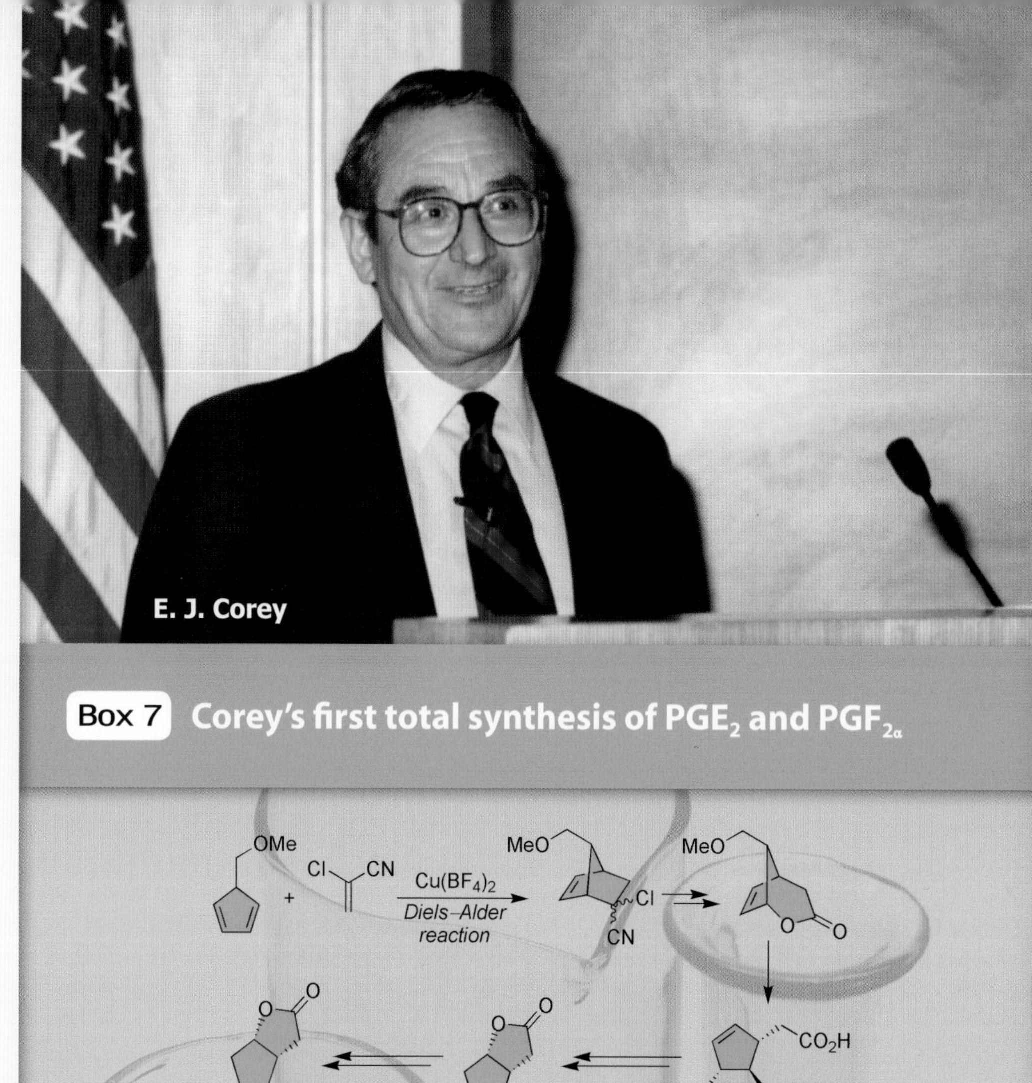
E. J. Corey

Box 7 Corey's first total synthesis of PGE_2 and $PGF_{2\alpha}$

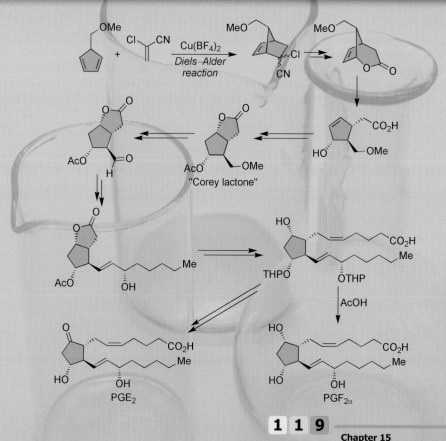

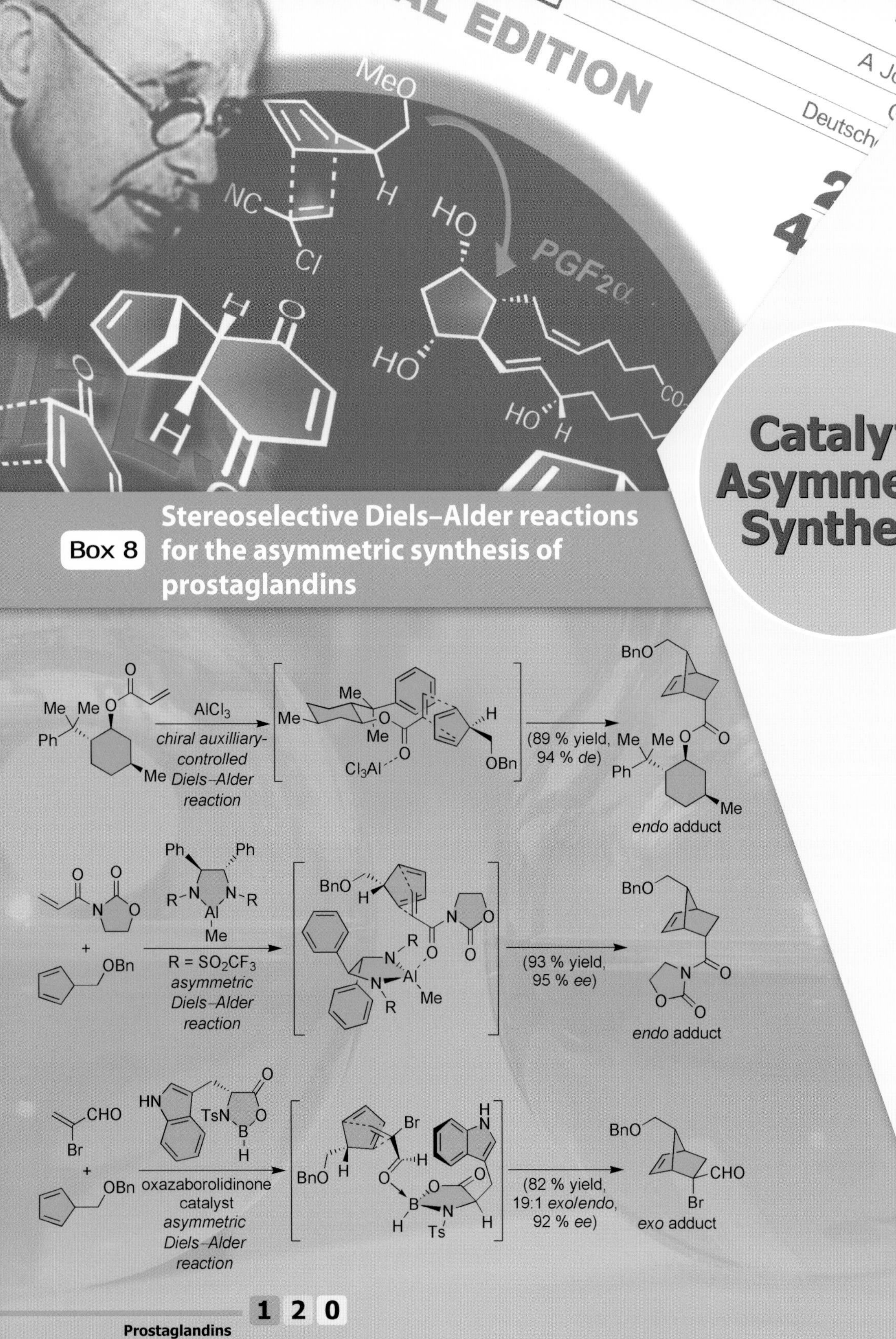

Box 8 Stereoselective Diels–Alder reactions for the asymmetric synthesis of prostaglandins

Catalytic Asymmetric Synthesis

supreme over synthetic organic chemistry in general during this era.

The evolution of eicosanoid synthesis is discussed in depth in *The Logic of Chemical Synthesis*, written by Corey and his student, Xue-Min Cheng, shortly before Corey was awarded the 1990 Nobel Prize in Chemistry for his contributions to organic synthesis. From this text, a detailed insight can be gained as to the challenges that dominated the thoughts of the chemists involved and how solutions were devised and improved as the knowledge surrounding these molecules grew. This process also exemplifies the theory of retrosynthetic analysis (see Chapter 14), which has characterized Corey's work. Because vast annals have been compiled describing the fine research undertaken in this area, from both the Corey laboratories and other groups, the interested reader is encouraged to delve deeper into this literature. Included here are just a few examples from this fascinating field.

The prostaglandins are characterized by a five-membered ring bearing two adjacent lipophilic carbon tails, attached to opposite faces of the core ring. Each molecule has up to five asymmetric centers, including one that is remote from all the others. The distal nature of this lone asymmetric center poses a significant hurdle to its selective introduction during the synthesis of these molecules. The sensitivity of the prevalent E-series of prostaglandins to both acidic and basic conditions motivated the Corey group's search for a synthesis employing mild conditions for all transformations, in order to preserve their delicate β-ketol motif. The results of this initial foray were the first chemical syntheses of not only selected E-series members, but also several other prostaglandins, along with the discovery of certain novel synthetic methods. From this point forward the successes multiplied, and new and refined syntheses were developed with improved access to a greater number of prostaglandins through efficiently constructed common intermediates. All these advances were made possible by ingenious designs and a growing knowledge and understanding of the intricacies and sensibilities of these targets. One approach employed to great effect by the Corey group is shown in Boxes 6 and 7. This bicycloheptane strategy is named for the structure of an early common intermediate, formed by a Diels–Alder reaction. Elaboration of this framework allowed access to numerous prostaglandins. Corey and his colleagues were at first satisfied with the accomplishment of these syntheses, which afforded the final products as racemic mixtures. However, Corey soon overturned even this feat, with the first series of syntheses to provide enantiomerically pure products by developing catalytic asymmetric Diels–Alder reactions which provided the early bicycloheptane intermediates *en route* to the prostaglandins as single enantiomers (Box 8). His inventions have since been applied in numerous other syntheses. The knowledge gained from his work on the prostaglandins probably provided the inspiration for the design of a new class of asymmetric catalysts, the oxazaborolidines, which ultimately brought Corey further fame. Such a catalytic system is widely used today for the asymmetric reduction of ketones. Corey's interest in the synthesis of the prostaglandins was revealed in 1968, with the publication of the first total synthesis, and did not end until well into the 1980s, when he reported modernized and elegant new approaches to these important biomolecules. During this time, Corey also accomplished the total syntheses of thromboxane B_2, prostaglandin I_2 (prostacyclin), and leukotrienes A_4, C_4, and D_4, to name but a few of his eicosanoid triumphs (Box 3). In the leukotriene field, his synthetic work was particularly valuable since it was intrinsic to gaining the final proof of the link between leukotriene B_4 and its biosynthetic precursor, 5-HPETE (Box 3).

Plexaura homomalla

Corey has also synthesized a number of eicosanoid congeners isolated from marine sources. These molecules are distinguished from their terrestrial counterparts by the fact that they are not derived from arachidonic acid, but from another, structurally similar, fatty acid. This observation provides evidence of the evolutionary divergence of these influential molecules. It is rather curious to note that the most prolific producer of prostaglandins discovered to date is a coral, *Plexaura homomalla*, underscoring the fact that the distribution of the eicosanoids in the living world is truly extensive and not at all limited to terrestrial mammals.

We hope that this summary of the eicosanoids and their chemistry has not only highlighted some of their chemical secrets, but has also revealed their power to direct our physiology. As sensitive molecules whose syntheses dictated a need for significant advances in chemical synthesis, they afforded worthy subjects for the endeavors of a chemist as respected as E. J. Corey. Just as investigating the eicosanoids has taught us much about how to develop new medications to treat conditions as diverse as pain, asthma, and heart conditions, so the study of Corey's eicosanoid syntheses can teach modern students of chemical synthesis a great deal about their subject. Indeed, his ingenious synthetic designs are found throughout this book, serving as both didactic examples and inspirations.

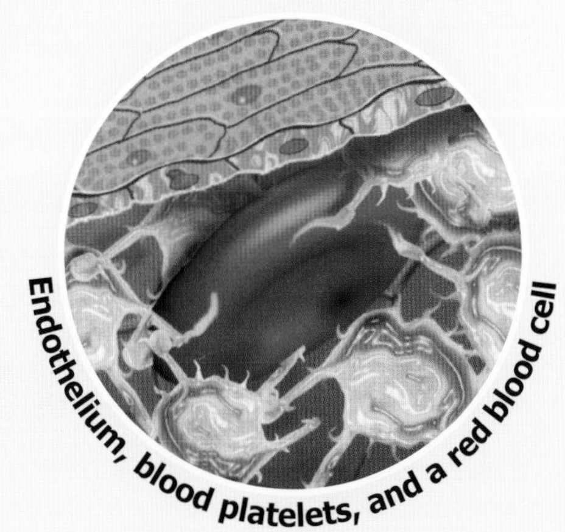
Endothelium, blood platelets, and a red blood cell

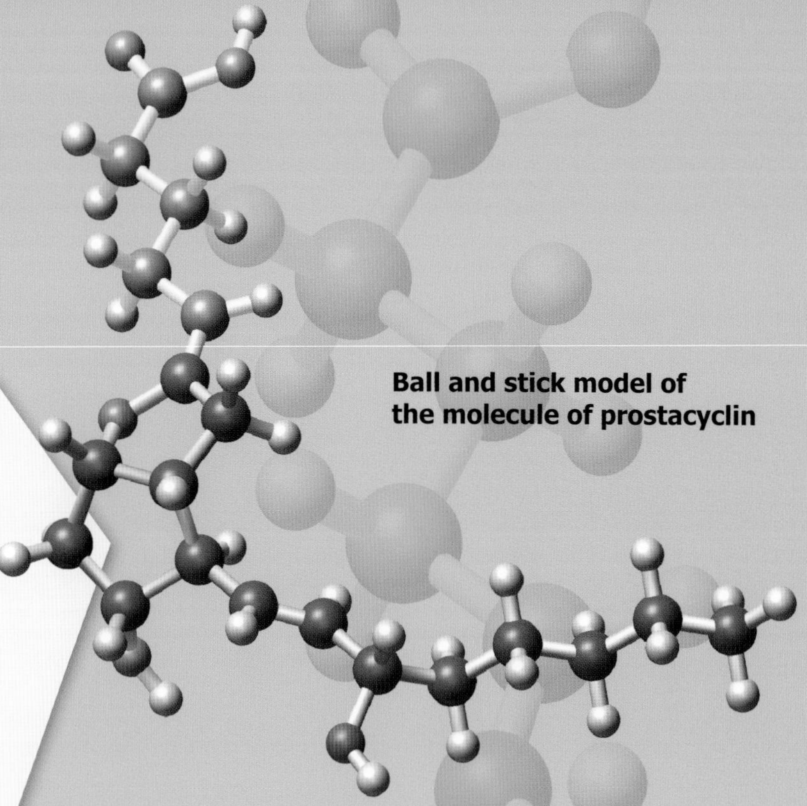

Ball and stick model of the molecule of prostacyclin

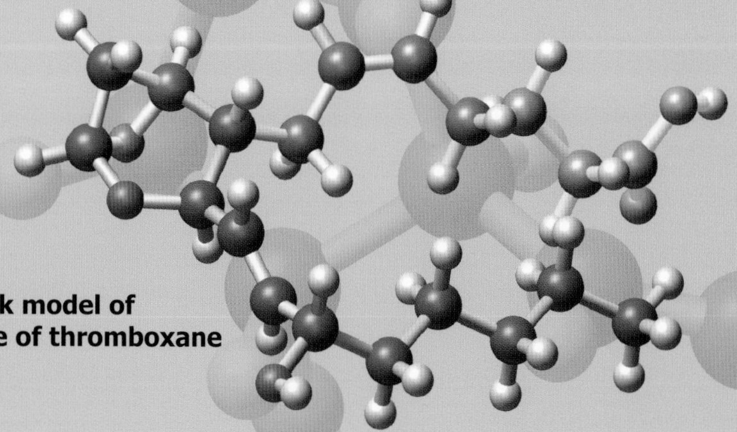

Ball and stick model of the molecule of thromboxane

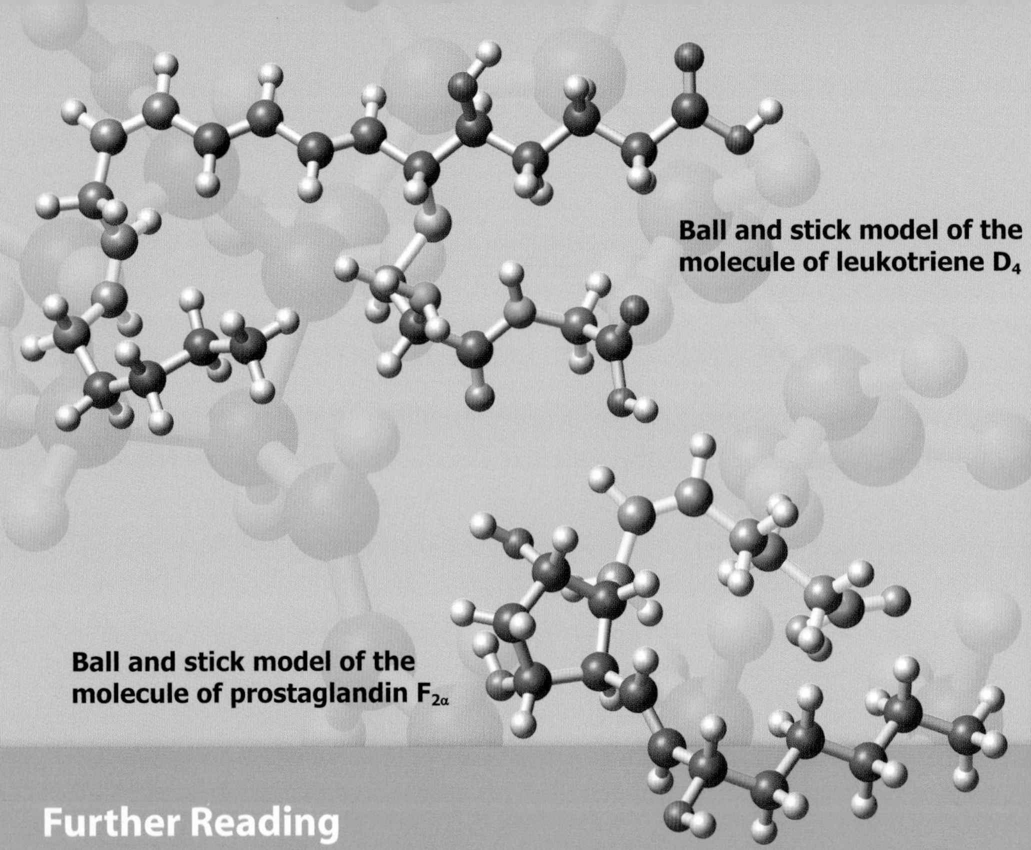

Ball and stick model of the molecule of leukotriene D_4

Ball and stick model of the molecule of prostaglandin $F_{2\alpha}$

Harvard University

Further Reading

E. J. Corey, N. H. Andersen, R. M. Carlson, J. Paust, E. Vedejs, I. Vlattas, R. E. K. Winter, Total Synthesis of Prostaglandins. Synthesis of the Pure *dl*-E_1, -$F_{1\alpha}$, -$F_{1\beta}$, -A_1, and -B_1 Hormones, *J. Am. Chem. Soc.* **1968**, *90*, 3245–3247.

E. J. Corey, I. Vlattas, N. H. Andersen, K. Harding, A New Total Synthesis of Prostaglandins of the E_1 and F_1 Series Including 11-Epiprostaglandins, *J. Am. Chem. Soc.* **1968**, *90*, 3247–3248.

E. J. Corey, N. M. Weinshenker, T. K. Schaaf, W. Huber, Stereocontrolled Synthesis of Prostaglandins $F_{2\alpha}$ and E_2 (*dl*), *J. Am. Chem. Soc.* **1969**, *91*, 5675–5677.

E. J. Corey, X.-M. Cheng, *The Logic of Chemical Synthesis*, Wiley-Interscience, New York, **1989**.

K. C. Nicolaou, E. J. Sorensen, *Classics in Total Synthesis*, Wiley-VCH, Weinheim, **1996**, pp. 65–82 and 137–151.

C. D. Funk, Prostaglandins and Leukotrienes: Advances in Eicosanoid Biology, *Science* **2001**, *294*, 1871–1875.

E. J. Corey, Catalytic Enantioselective Diels–Alder Reactions: Methods, Mechanistic Fundamentals, Pathways, and Applications, *Angew. Chem. Int. Ed.* **2002**, *41*, 1668–1698.

Karolinska Institute

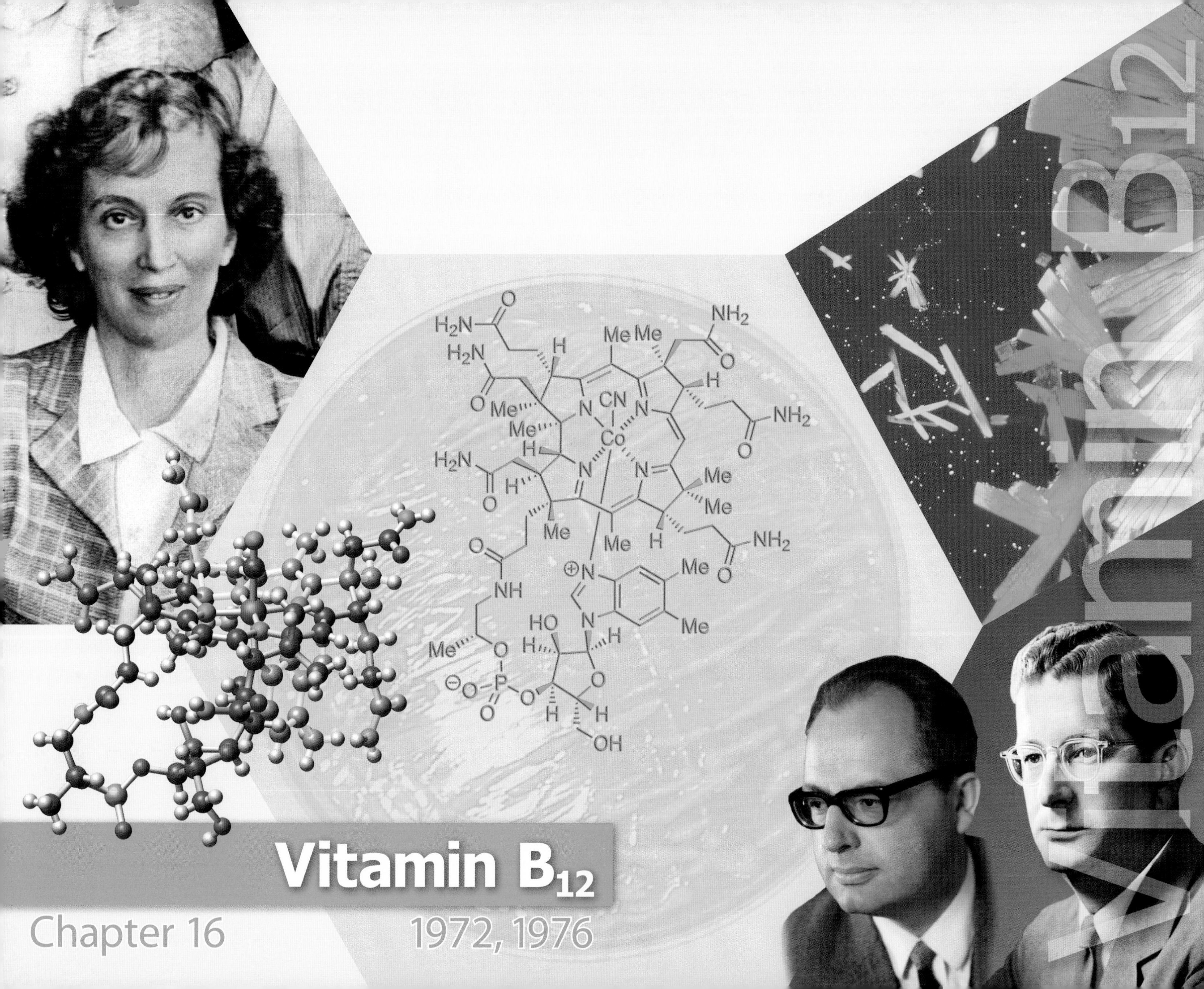

Vitamin B_{12}

Chapter 16

1972, 1976

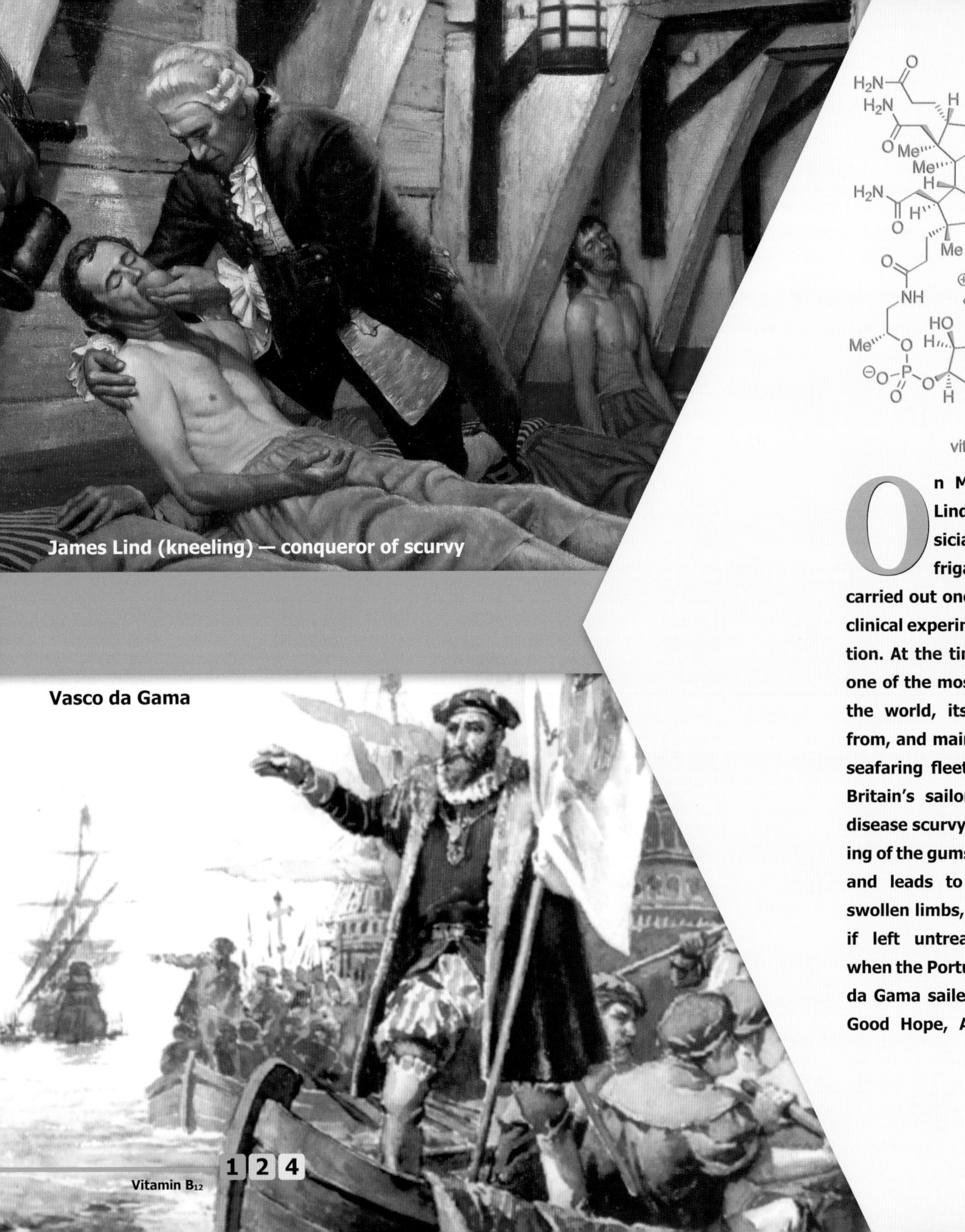

vitamin B_{12}

On May 20, 1747, James Lind, a British naval physician serving on the frigate *HMS Salisbury*, carried out one of the first recorded clinical experiments regarding nutrition. At the time, Great Britain was one of the most powerful nations in the world, its domination derived from, and maintained by, its mighty seafaring fleets. However, many of Britain's sailors suffered from the disease scurvy. Scurvy causes swelling of the gums and the loss of teeth, and leads to debilitating fatigue, swollen limbs, prolific bruising, and, if left untreated, death. Indeed, when the Portuguese explorer Vasco da Gama sailed around the Cape of Good Hope, Africa's southernmost tip, in 1497, one hundred of his one hundred sixty-strong crew died from scurvy. Lind was determined to identify the cause of this ailment, which he proposed was associated with lack of "acid" in the seamen's diet. On his ship, he divided sick sailors into groups and fed them one of the following prescriptions: a quart of cider; diluted elixir of vitriol (sulfuric acid!); two spoons of vinegar; a half pint of seawater; barley water; a concoction of nutmeg, garlic, mustard seed, gum myrrh and cream of tartar; or a combination of oranges and lemons. The sailors in the last group showed remarkable signs of healing in less than a week. Lind and his colleagues continued to ascribe their dramatic recovery to the acidity of the citrus fruits; therefore, because limes were thought to be more acidic, lime juice replaced oranges and lemons and was added to every sailor's daily rations. This beguiling tale constitutes the first discovery of the essential nature of ascorbic acid, more commonly known as vitamin C, and reveals the origins of the name 'Limeys' used for British sailors.

Little more was learned regarding the crucial cofactors required in our diet to facilitate the smooth operation of the human body until a Dutch doctor,

Christiaan Eijkman, studied beriberi in the Dutch colonies some one hundred fifty years later. At this time, the severe ravages of the disease, first noted one and a half millennia earlier in the writings of the ancient Chinese, led the Dutch government to appoint a special commission to study it in the field. Eijkman noticed a peculiar sickness amongst the hens belonging to the laboratory, which were suddenly being attacked by a form of paralysis. At first the birds would walk unsteadily, later they would not be able to perch, and finally they were driven to lie down on their sides unable to function normally. The talented scientist traced this phenomenon back to the feeding of these hens with 'polished rice' from the kitchen. Polished rice has had the fibrous husk removed from the kernel. When the hens were once again fed raw whole rice, they showed a rapid recovery. Eijkman was struck by the similarities in symptoms and pathology of the disease he had discovered in his birds and beriberi in humans. Although he was able to prove that the protective component in the rice husk was soluble in both water and alcohol, Eijkman believed that this chemical was required to defend against a poison or germ, rather than being an essential nutrient. However, his work inspired scores of investigators, and the concept of vitamins would not remain shrouded in mystery for much longer.

Christiaan Eijkman

Another investigator in this early quest for understanding of nutrition was

Box 1 Selected vitamins and their functions

Structure	Food sources	Selected functions	Deficiency disease
vitamin A family 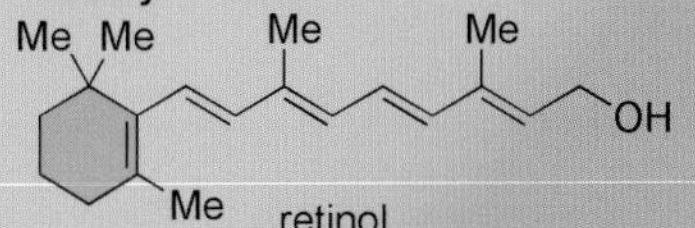 retinol	eggs, milk, liver, fruit, vegetables, fish oils	• vision • bone growth • immune system regulation • reproduction	night blindness
vitamin B family 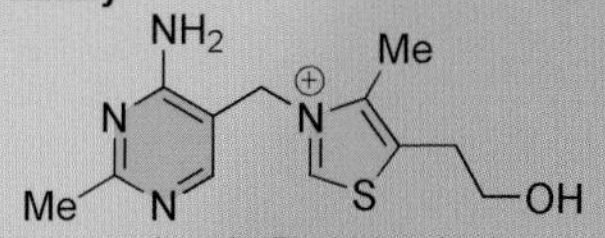 vitamin B_1 (thiamine)	liver, legumes, potatoes, wheatgerm, whole grains, produced by intestinal bacteria	• circulation • metabolism • nervous system • production of acetylcholine	beriberi
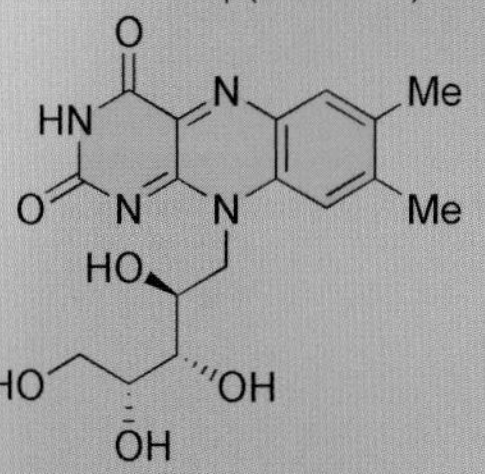 vitamin B_2 (riboflavin)	liver, kidney, soybeans, mushrooms, nuts	• needed to activate vitamin B_6 • adrenal gland function • growth • metabolism • red blood cell production	no specific reports
vitamin C 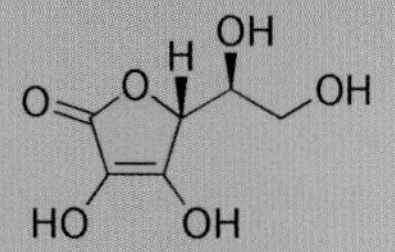 ascorbic acid	citrus fruits, tomatoes, broccoli, bell peppers	• protective antioxidant • collagen synthesis • reducing agent vital to many biochemical processes	scurvy
vitamin D family 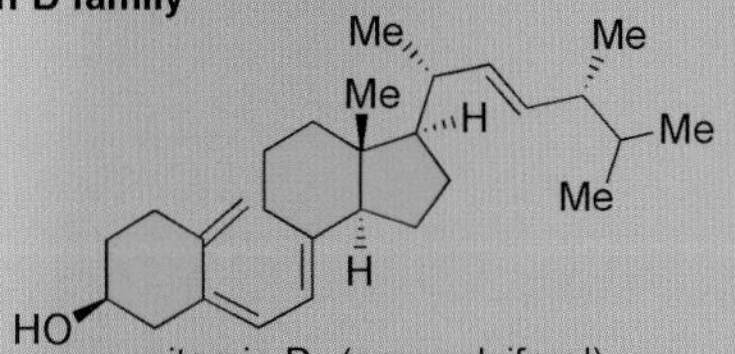 vitamin D_2 (ergocalciferol)	fish, fish oils, produced by the body after exposure to sunlight	• healthy bones • maintains normal blood levels of calcium and phosphorus	rickets, osteomalacia
vitamin E family 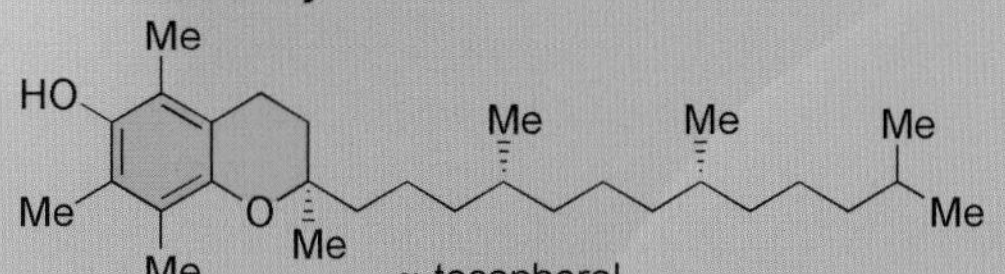 α-tocopherol	vegetable oil, nuts, leafy greens	• powerful antioxidant - protects cells from the effects of damaging free radicals formed by cellular metabolism	inhibition of sperm production
vitamin K family 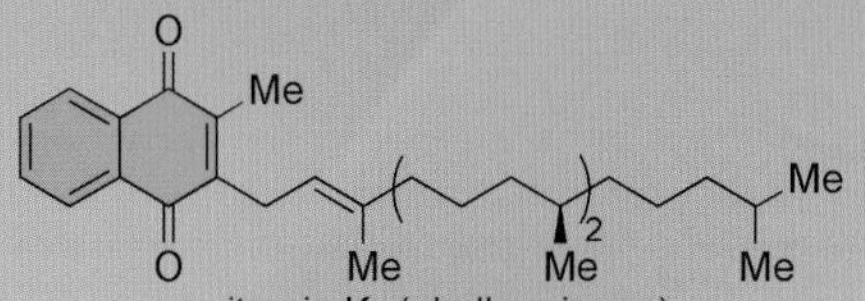 vitamin K_1 (phylloquinone)	oil (especially soybean), dark green vegetables	• blood clotting • bone integrity	subdermal hemorrhaging

Dorothy Crowfoot Hodgkin

Albert Szent-Györgyi

Walter N. Haworth

Sir Frederick Gowland Hopkins

Tadeus Reichstein

Sir Frederick Gowland Hopkins at the University of Cambridge (UK). Hopkins experimented with rats by carefully controlling their dietetic intake, a regime accompanied by fastidious weighing designed to measure precisely their growth requirements and rates. He found that purified lard (fat), starch (carbohydrate), and casein (protein) were not enough to sustain his rodent subjects; it was necessary to add 2–3 milliliters of milk. The milk represented only 1–2 % of the energy content of the rats' diet, but it led to sustainable growth. He described the milk as having a growth-promoting influence, which curiously did not originate from any of the known constituents of milk. Thus, he had pinpointed a new nutritional requirement. Hopkins' research had finally led to the acknowledgment that "accessory factors" exist which have no value in terms of energy, but which are absolutely essential to life. Eijkman and Hopkins went on to share the 1929 Nobel Prize in Physiology or Medicine, as a tribute to their respective investigations in nutrition.

In 1912, Polish biochemist Casimir Funk succeeded in isolating the growth factor from rice husks that had been identified by Eijkman thirty years earlier. The so-called thiamine molecule contained an amine group, prompting Funk to name all vital growth-factors "vitamines". It was later discovered that not all vitamines contain amine groups; however, the name stuck. Today, we have dropped the terminal "e" to give vitamin, and added a letter plus the occasional number in order to classify these compounds without having to resort to using their complicated full chemical names, e.g. vitamins A, B_1, B_2, D, E, and K (Box 1). Vitamins are substances that (almost without exception) the body cannot make, and thus must be obtained by embracing a broad diet that is rich in fish, nuts, fruits, and vegetables. Vitamins not only assist growth, but maintain health and vision and lessen the risk of many serious diseases ranging from heart attacks and Alzheimer's disease to various cancers. Vitamins in general can be subdivided into two classes depending on their solubility properties. The fat-soluble lipophilic vitamins (A, D, E, and K) can be stored in our bodies for some time, while the water-soluble vitamins (all the B's and vitamin C) are readily eliminated from the body, and therefore need to be ingested more frequently.

In the first half of the twentieth century, vitamins became a preoccupation with physicians and scientists alike. Elucidation of food sources containing specific vitamins was unremittingly pursued, the active constituents hunted down and isolated, the complex structures determined, and, in some cases, the vitamin actually synthesized in the laboratory. Of course, the speed of the progress within these investigations varied greatly according to the size and chemical nature of the particular vitamin in question. Vitamin C is a small and relatively simple molecule in terms of the number of constituent atoms and their connectivity. However, even with this most famous of examples, years passed in trying to track down this elusive, heat-sensitive and water-soluble vitamin. In 1921, Sylvester Zilva, working at the Lister Institute in London, made crude preparations of ascorbic acid by concentrating lemon juice. At the same time, Albert Szent-Györgyi, a Hungarian scientist working in Hopkins' laboratory in Cambridge, isolated a compound he called hexuronic acid from bovine adrenal cortex. After much squabbling, it was finally agreed that both men had isolated the same chemical. Later, Szent-Györgyi returned to Hungary where he discovered that peppers grown in the locality for paprika production were an exceedingly rich source of vitamin C. Very quickly he obtained large quantities of pure crystalline vitamin C from these peppers. His success in extracting vitamin C meant he was able to donate samples to scientists all over the world who were interested in solving the structure of this fascinating little molecule. The puzzle was finally solved with the chemical synthesis of the vitamin, which was independently accomplished by Walter N. Haworth at the University of Birmingham (UK) and Tadeus Reichstein at the Eidgenössische Technische Hochschule (ETH), the Swiss Federal Institute of Technology, in Zurich, Switzerland in 1933. A commercial synthesis from glucose followed soon afterwards conferring on vitamin C the honor of being the first pure vitamin available to the general public through large-scale industrial production. Haworth went on to receive the 1937 Nobel Prize in Chemistry for this discovery and for his work on carbohydrates. He shared the prize with Paul Karrer, a Swiss chemist and fellow key player in vitamin research at this time, who was recognized for his work on the carotenoids, flavins, and vitamins A and B_2. The very same year Szent-Györgyi received the Nobel Prize in Physiology or Medicine, at least in part, as recognition for his work on vitamin C. The following year, 1938, the Nobel Prize in Chemistry was again awarded for work on vitamins (and carotenoids), this time to the German chemist Richard Kuhn, showing just how the science of vitamins dominated this era.

Another early success in the realm of vitamin synthesis was achieved by Karl Folkers and his colleagues at the Merck Company (Rahway, New Jersey, USA), who accomplished the structural elucidation (as did Richard Kuhn in the same year) and concurrent synthesis of vitamin B_6 in

Vitamin B_6

Vitamin C

Crystals of vitamin B_{12}

1939. Prior to this event, vitamin B_6 (also known as pyridoxine) had had as colorful, and as multinational, a past as any other player on the vitamin stage. The first isolation of pyridoxine has been attributed to a Japanese scientist called Sator Ohdake, whose investigations took place as early as 1931. It would seem, however, that he had neither the resources, nor the experience to prevent his discovery from lapsing into obscurity by his failure to publish in an international journal. Therefore, scientists elsewhere continued to investigate the so-called "rat pellagra" factor, oblivious of Ohdake's work. Akin to vitamin C, the simple structure of pyridoxine belies the difficulty these workers had in determining its physiological role and molecular structure, and in devising a laboratory synthesis. In the tangled vitamin literature of the time, it went by various names, including antidermatitis factor for rats, vitamin H, factor Y, and factor 1. Finally, after years of comparative studies, Szent-Györgyi concluded that each of these factors were identical and in the future this vitamin should be named vitamin B_6, to indicate its relationship with other anti-pellagra vitamins (i.e. riboflavin, also called vitamin B_2). Subsequent work cleared the murky waters further when it became obvious that vitamin B_6 does not cure human pellagra because rat pellagra is not the same deficiency disease. In this context, the concisely described structural determination and synthesis of vitamin B_6 emanating from the Merck laboratories in 1939 would seem to signal a break in the pattern of confusion surrounding this particular vitamin. Merck's synthesis of vitamin B_6 involved just nine steps, including the complete assembly of the aromatic ring. Folkers continued to undertake esteemed work exploring the B vitamins, and from there moved on to garner further fame for his work on coenzyme Q10 and the first hypothalamic hormone, thyrotropin.

Dried apricots

Within the hothouse of vitamin research that existed around the 1920s, the story of the most architecturally demonic of these molecules was just beginning to gather steam. The tale of this complex compound, vitamin B_{12}, which brims with exciting anecdotes, stands as perhaps the greatest victory of collaborative chemical studies ever recorded. The story begins at the University of California Medical School (USA) in 1918, where George Hoyt Whipple was investigating anemia (lack of red blood cells). His desire was to see which foods could restore the red blood cell count to full strength after an animal had been purposefully bled. He identified liver and, surprisingly, apricots as being particularly effective at promoting the desired regeneration. Next, we cross the United States to Harvard on the East Coast where, in 1926, two physicians, George Richards Minot and William P. Murphy, took up the baton. Inspired by Whipple's work, they started to treat patients suffering from pernicious anemia by feeding them liver. Pernicious anemia, originally known as idiopathic anemia, was a hitherto fatal disease marked by irreversible neurological disturbance, gastrointestinal and cardiovascular problems, and anemia, leading to an oxygen deficit with associated symptoms such as fatigue and shortness of breath. The patients treated with liver exhibited remarkable recoveries, prompting feverish studies to begin all around the world aimed at fractionating liver extracts (separating the different components). The success of the liver therapy earned the three physicians, Whipple, Minot, and Murphy, the 1934 Nobel Prize in Physiology or Medicine.

By the end of World War II, despite various successes from, amongst others, the pharmaceutical company Eli Lilly, who had been selling 'Liver Extract 343' since 1928, and two Norwegian scientists, Per Laland and Aage Klem, who had prepared a high concentration, intensely orange, liver extract, the pure anti-pernicious anemia factor was still eluding scientists. However, the arrival of new assays and partition chromatography techniques greatly improved the situation such that, on December 11, 1947, Ed Rickes, a scientist at Merck, became the first person to isolate the deep red crystals of pure vitamin B_{12}. This event was closely followed by a similar announcement of success from a rival team working at the British pharmaceutical firm, Glaxo. Merck, Glaxo, and a number of academic laboratories around the world then focused all their efforts on discovering the exact make-up of the vitamin B_{12} molecule. Certain characteristics and segments of information regarding the vitamin's structure were determined, mostly in the laboratory of Lord Alexander R. Todd (University of Cambridge, UK), but the complete solution was still elusive. This group supplied Dorothy Crowfoot Hodgkin at Oxford University (UK) with a precious crystalline sample of a vitamin B_{12} derivative, and, in 1956, she finally conquered the structure of vitamin B_{12} using x-ray crystallography. Never before had the structure of such a complex natural product been derived using this technique, although x-ray crystallography and its talented pioneer, Dorothy Hodgkin, had both earned much fame from their contribution to the debate over the molecular structure of penicillin (see Chapter 13). The monumental accomplishment of deducing the true structure of vitamin B_{12} paved the way for further triumphs from the laboratory of this unassuming but brilliant scientist. She was awarded the 1964 Nobel Prize in Chemistry "for her determinations by x-ray techniques of the structures of important biochemical substances."

Vitamin B_{12} is an extraordinary molecule not only because of its size and function, but because it is packed full of

Box 2 Two approaches to the vitamin B_{12} corrin ring core structure: A→B and A→D macrocyclizations

Leopold Ružička and Vladimir Prelog

Box 3 The A→D approach to the corrin ring of vitamin B_{12} (Eschenmoser)

(1) hν (visible light)
(2) $CoCl_2$
(3) KCN

same advanced intermediate in vitamin B_{12} synthesis as targeted by Woodward (see Box 4)

Chlorophyll a

unique and esoteric chemical features. It contains an atom of cobalt at its heart, an exchangeable ligand (shown as cyanide, CN, in our diagram because this is how it is isolated, but which is adenosine in the biologically active form), formal charges, and, last but by no means least, a beautiful heterocyclic corrin ring reminiscent of, yet chemically quite different from, the porphyrin rings of haemin (Chapter 8) and chlorophyll. Choosing vitamin B_{12} as the target for research was quite clearly the major challenge of the time for synthetic and natural products chemists; however, just being in possession of Hodgkin's architectural details left them with sparse clues about how to achieve their goal. Fortunately, as early as 1960, an important step forward was made by Wilhelm Friedrich and Konrad Bernhauer, at the Technische Hochschule in Stuttgart, Germany, when they converted naturally occurring cobyric acid (the corrinoid core of vitamin B_{12}, see Box 5) into vitamin B_{12}. Thus, the problem of synthesizing vitamin B_{12} had been reduced in complexity to the problem of synthesizing the simpler precursor, cobyric acid. Vitamin B_{12} was no longer a remote chemical island whose chemical nature was known only from x-ray crystallography and not chemical degradation studies. Despite this improved scenario, cobyric acid still displayed a complexity that was without precedent in natural product synthesis at the time.

Synthetic chemists now yearned for a design that would allow the conquest of vitamin B_{12} by chemical synthesis. For Robert B. Woodward, the great artist of total synthesis of whose huge influence on the field we are already aware from the chapters discussing the steroids (Chapter 11) and strychnine (Chapter 12), accepting the challenge of attempting to synthesize vitamin B_{12} was a must. He had just successfully completed the synthesis of the related green plant pigment chlorophyll and his enthusiasm therefore abounded for this new trial of his talents.

Woodward had developed a strong friendship with the older grandmasters of organic chemistry, Leopold Ružička and Vladimir Prelog, at the ETH. This friendship led to a regular lecture series given by Woodward in Zurich, and culminated in the foundation of the Woodward Research Institute at the pharmaceutical company CIBA, in Basel (Switzerland). Woodward's close connection to ETH also laid the foundations for his subsequent extraordinary collaboration with the brilliant young chemist Albert Eschenmoser, the ultimate product of which was the joint conquest of vitamin B_{12} via a merging of two independent routes simultaneously developed to attain a common intermediate just steps away from cobyric acid.

Eschenmoser began his investigations towards a synthesis of vitamin B_{12} in 1960, immediately after finishing the first synthesis of the alkaloid colchicine.

He attacked the problem by first testing a basic strategy for the construction of the corrin ligand in a model system. By 1964 he had published a slick synthesis of a model corrin complex in which he pioneered the use of iminoester-enamine condensations to unite the two halves of the molecule. When applied, this reaction formed the necessary bridging motif that linked the various rings making up the corrin core. The last macrocyclization event of this sequence took place when the A and B rings were joined, so it became known as the A→B strategy for corrin synthesis. The structure of the synthetic corrin complex was proven by x-ray crystallographic analysis, undertaken by Jack D. Dunitz, also at ETH. This first corrin synthesis became the conceptual model for the synthetic strategy to be followed in one of the two eventual syntheses of cobyric acid. In the meantime, the Harvard group had already made significant inroads towards a synthesis of the stereochemically most demanding part of the vitamin B_{12} molecule, namely its left-hand half containing the A–D ring junction. Back at ETH, besides working on the model corrin synthesis, the Eschenmoser group was concomitantly exploring the problem of how to synthesize the right-hand side of the vitamin B_{12} molecule containing the B and C rings. They discovered that the iminoester-enamine condensation, which had served them so well in joining the rings together in their corrin model study, failed when attempts were made to use this reaction to join the real B and C rings (containing the sterically encumbering peripheral substituents of the vitamin B_{12} molecule). However, this turned out to be a fortunate failure, because it opened the door to some even more beautiful chemistry that would step in and take the place of the iminoester-enamine condensation in the final successful venture. The ETH group rationalized that the reluctance of the iminoester-enamine condensations to proceed in the real system could be countered by substituting the intermolecular iminoester condensation by an intramolecular thioiminoester-enamine condensation in which the two reacting centers are transiently attached to each other by a sulfur bridge. This tether is readily expelled during the critical condensation process (this reaction became known as the "sulfide contraction"). The advantage of this strategy lay in making the condensation an intramolecular reaction, facilitating the formation of the hindered bridging bond.

From this point forward the idea of a collaboration gradually matured and was finally formalized in 1965 because, apart from anything else, their two approaches were so eminently complementary and the enormity of the problems still facing each of them individually dictated such prudence in sharing burdens. The gift both men had for synthesis, protracted phone discussions, and many transatlantic airmail packages drove them forward through the seven-year odyssey that was to follow (see Boxes 2–5). At its height, when joining the various components together was taking its full toll on their patience, Eschenmoser recalls that, "They [the two groups] hung, so to speak, for years in the rocks, roped together, over an abyss of roughly fifty chemical steps, trying to reach the cobyric acid peak. During this time it could take an

Jack D. Dunitz

Albert Eschenmoser and R. B. Woodward

Box 4 The A→B approach to the corrin ring of vitamin B_{12} (Woodward–Eschenmoser)

cyanocorrigenolide

S-methyldithiocyanocorrigenolide

base, heat

advanced intermediate in vitamin B_{12} synthesis

Robert B. Woodward upon receipt of the Nobel Prize in Stockholm, 1965

Box 5 The final steps of the Woodward–Eschenmoser synthesis of vitamin B_{12}

advanced intermediate common to both strategies (see Boxes 3 and 4)

known method of Bernhauer and Friedrich

cobyric acid

vitamin B_{12}

entire year to move forward a single synthetic step, to have it initiated, secured, and optimized so that one could solidly build on it." Their patience and perseverance would pay dividends as the victory was not too far around the corner.

The accomplishment of another milestone in the vitamin B_{12} adventure was announced in 1969, again by Eschenmoser and his group at ETH. They had developed, inspired by the new thinking on chemical reactivity that had been brought to organic chemists with the advent of the theoretical development summarized in the Woodward–Hoffmann rules (see later), an ingenious photochemically-induced coupling reaction that could seal up the macrocycle by joining the A and D rings (an A→D variant). Whilst this radically new strategy was again tested first on a model corrin ring, it could soon be applied successfully to the synthesis of the real cobyric acid precursor (Box 3), and as such its core elements comprised one of the two strategies that were soon to meet with success. In its final application, this remarkable cycloisomerization reaction, which was initiated by light from the visible part of the spectrum, proceeded smoothly, furnishing the natural configuration at the A–D ring junction with high selectivity.

The route pursued at Harvard continued to follow the originally envisioned A→B macrocyclization strategy, the A→B variant (Box 4). This strategy was to share the winner's medal with Eschenmoser's photochemical A→D masterpiece. Both routes converged on the same cobyric acid precursor that was to be a key stepping stone on the path carved out towards the ultimate treasure, vitamin B_{12}. At Harvard, Woodward's group had conquered the construction of the more stereochemically complex and challenging half of the molecule, the A–D fragment, through a creative synthetic strategy. Using the sulfur bridging technique described earlier, the A–D and B–C fragments could be married to complete this variant.

The final triumph was secured in 1972, when the Woodward–Eschenmoser collaborative team was in a position to announce to the chemical community that cobyric acid had been successfully made in the laboratory by a relay chemical synthesis which borrowed and merged the best from both of the finely honed A→B and A→D strategies (Box 5). Later, at Harvard, a synthetic sample of cobyric acid was advanced through the Bernhauer and Friedrich transformations into vitamin B_{12}, and thus a small amount of wholly synthetic vitamin B_{12} was produced. A photograph was taken of this sample and invitations extended to a vitamin B_{12} party in Boston, not as famous as its "tea party" predecessor that had sparked the war for American independence, but to organic chemists almost as pivotal a moment, which ushered in a golden age for the powerful art and science of total synthesis.

In 1994, A. Ian Scott and colleagues at Texas A&M University (USA) completed the synthesis of hydrogenobyrinic acid (a metal-free analogue of cobyric acid) via a beautifully orchestrated enzymatic sequence exemplifying a synergy between chemical and biological synthesis. Each of the twelve enzymes involved in the biosynthesis of this vitamin B_{12} precursor had been meticulously identified and expressed in genetically modified bacteria. The enzymes were then mixed, along with the necessary cofactors, in a single flask, and the simple biosynthetic precursor 5-aminolevulinic acid added. Incubation of this mixture led to the isolation of hydrogenobyrinic acid in 20 % overall yield, which is truly remarkable considering that seventeen discreet operations had been successfully carried out in an enantioselective manner (Box 6). The conversion of this product to vitamin B_{12} can be completed via a short and efficient chemical synthesis, involving the completely regioselective incorporation of the nucleotide loop, in a self-assembly-mediated process developed by Eschenmoser. The enzyme-based synthesis of cobyric acid was only possible after the biosynthetic pathway to vitamin B_{12} had been carefully elucidated. This pioneering work was completed independently by the groups of A. Ian Scott in Texas and Sir Alan R. Battersby at the University of Cambridge, in collaboration with biochemistry and genetics groups led by Francis Blanche and Joel Crouzet, respectively, at Rhone-Poulenc Rorer in France.

Sir Alan R. Battersby

The vitamin B_{12} campaign also spawned an important chemical theory that is still familiar to, and employed by, all students of organic chemistry – the Woodward–Hoffmann Rules. The story of the emergence of this theory into the public domain is, however, beset with near misses and disputed claims about the origins of some of the pivotal ideas. The tale begins back in the late 1950s and early 1960s when a number of studies had begun to reveal to organic chemists certain patterns in the stereoselectivities of pericyclic reactions, both those that were thermally-initiated and those that were photochemically-induced. For example, Emanuel Vogel, a brilliant and highly accomplished professor of chemistry at the University of Cologne (Germany), made some stunning observations in this regard (see top of Box 7). Chemists rapidly became fascinated by these phenomena and began to search for a rationale that would make sense of the very distinct, alternating trends that were being observed. In 1961, E. Havinga and J. L. M. A. Schlatmann (University of Leiden, Holland) noted in a publication a further example of alternating stereospecificity for the thermally- and photochemically-induced cyclizations of a particular vitamin D related triene that they had been studying. What makes their report stand out from others, however, is a tantalizing paragraph referring to a personal conversation regarding possible explanations for such stereospecificity that occurred between the authors and Luitzen J. Oosterhoff, an out-

***Propionibacterium shermanii*, a producer of vitamin B_{12}**

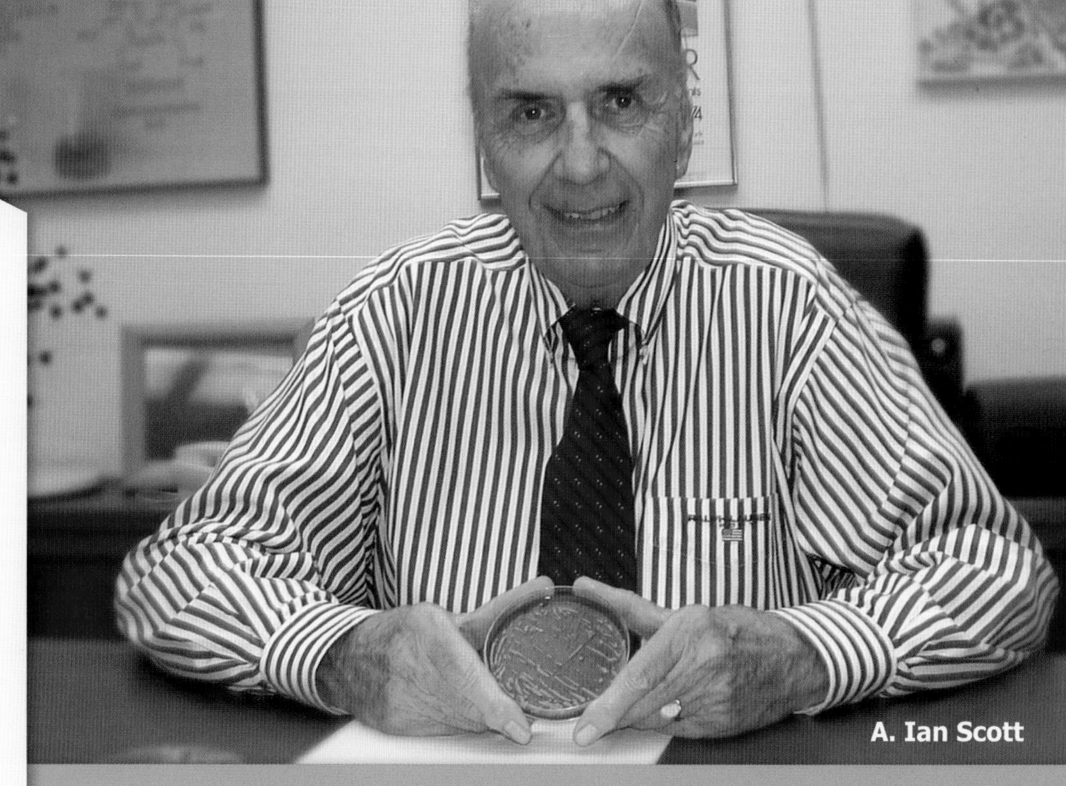

A. Ian Scott

Box 6 Enzymatic synthesis of a vitamin B_{12} precursor

5-aminolevulinic acid → (12 enzymes) → hydrogenobyrinic acid → (chemical steps) → vitamin B_{12}

Box 7 The Woodward–Hoffmann Rules

R. B. Woodward

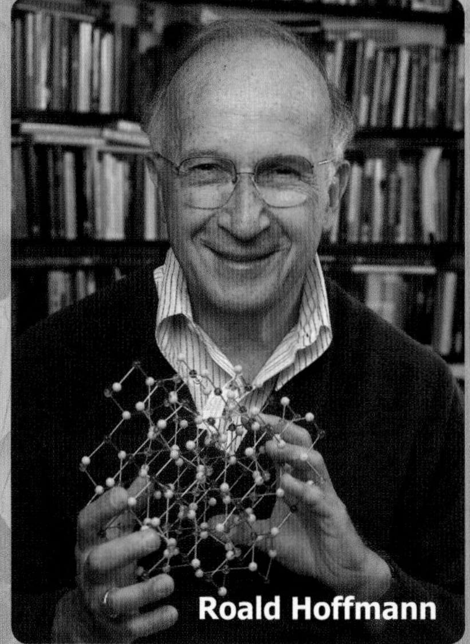

Roald Hoffmann

From Vogel's early cyclobutene/butadiene work:

E. Vogel, Small carbon rings. II. Thermal stability of the 4-membered carbon ring, *Ann.* **1958**, *615*, 14–21.

From Woodward's vitamin B_{12} work:

vitamin B_{12} intermediate

"I REMEMBER very clearly – and it still surprises me somewhat – that the crucial flash of enlightenment came to me in algebraic, rather than in pictorial or geometric form. Out of the blue, it occurred to me that the coefficients of the terminal terms in the mathematical expression representing the highest occupied molecular orbital of butadiene were of opposite sign, while those of the corresponding expression for hexatriene possessed the same sign. From here it was but a short step to geometric, and more obviously chemically relevant, view that in the internal cyclisation of a diene, the top face of one terminal atom should attack the bottom face of the other, while in the triene case, the formation of a new bond should involve the top (or pari passu, the bottom) faces of both terminal atoms."

R. B. Woodward, *The Arthur C. Cope Award Lecture*, Chicago, August 28, 1973. (Published in the Benfey-Morris book referenced in the Further Reading section.)

"On May 4, 1964, I suggested to my colleague R. B. Woodward a simple explanation involving the symmetry of the perturbed (HOMO) molecular orbitals for the stereoselective cyclobutene/1,3-butadiene and 1,3,5-hexatriene/cyclohexadiene conversions that provided the basis for the further development of these ideas into what became known as the Woodward-Hoffmann rules."

E. J. Corey, *Chem. Eng. News* **2004**, *82*, 42–44; E. J. Corey, *J. Org. Chem.* **2004**, *69*, 2917–2919.
See also: R. Hoffmann, *Angew. Chem. Int. Ed.* **2004**, *43*, 2–6.

standing theoretical chemist who also worked at the University of Leiden. They quote Oosterhoff as having suggested that the source of the switch in stereospecificity between the thermally-controlled reaction and its corresponding photochemically-induced reaction may reside in the symmetry characteristics of the highest occupied π-orbital of the conjugated system. Without getting too involved in chemical intricacies, we can say that we now know Oosterhoff was correct in his conjecture. Woodward was confronted by the puzzle in an example from his own work on vitamin B_{12} (Box 7). Joined by Roald Hoffmann, then a junior fellow at Harvard University, Woodward's milestone analysis of his problem in relation to other empirical examples drawn from the chemical literature would give rise to a set of predictive rules describing the stereochemical outcome of all concerted reactions of this type. The formalized result was published in a series of seminal papers authored by the two Harvard chemists in 1965.

Kenichi Fukui

In the years that followed, Woodward and Hoffmann received all the credit for this major contribution to chemical theory, but lately some have questioned whether this discovery can really be attributed so clearly as a bolt from the blue visited upon an individual, or whether the story of the emergence of the Woodward–Hoffmann rules includes more players than originally thought. Prominent among those is E. J. Corey who, in summarizing his contribution to organic chemistry on the occasion of receiving the 2004 Priestley Medal (the highest honor of the American Chemical Society), made a startling revelation (Box 7). We urge the interested reader to immerse themselves in the references we give, both in Box 7, and at the chapter's end, because they serve to highlight this particular tale as an example of just how discoveries in science are frequently attributed not to the individual who has some indescribable spark of inspiration, but to he who recognizes the importance of a research result. In other words, the discoverer whom history credits, rightly or wrongly, is often the individual who draws general conclusions and succeeds in bringing information to the attention of a broad audience. Prior to the much touted discovery date, many other scientists may have supplied vital clues and ideas, or relevant examples, to the discoverer that were essential in the final analysis.

E. J. Corey

Returning to the subject of the Woodward–Hoffmann theory, Kenichi Fukui (Kyoto University, Japan) developed the concept from an alternative perspective. He derived mathematical models explaining the reactivity of molecules by examination of the symmetry of the most accessible electronic molecular orbitals. The result of Fukui's work has been called

Emanuel Vogel

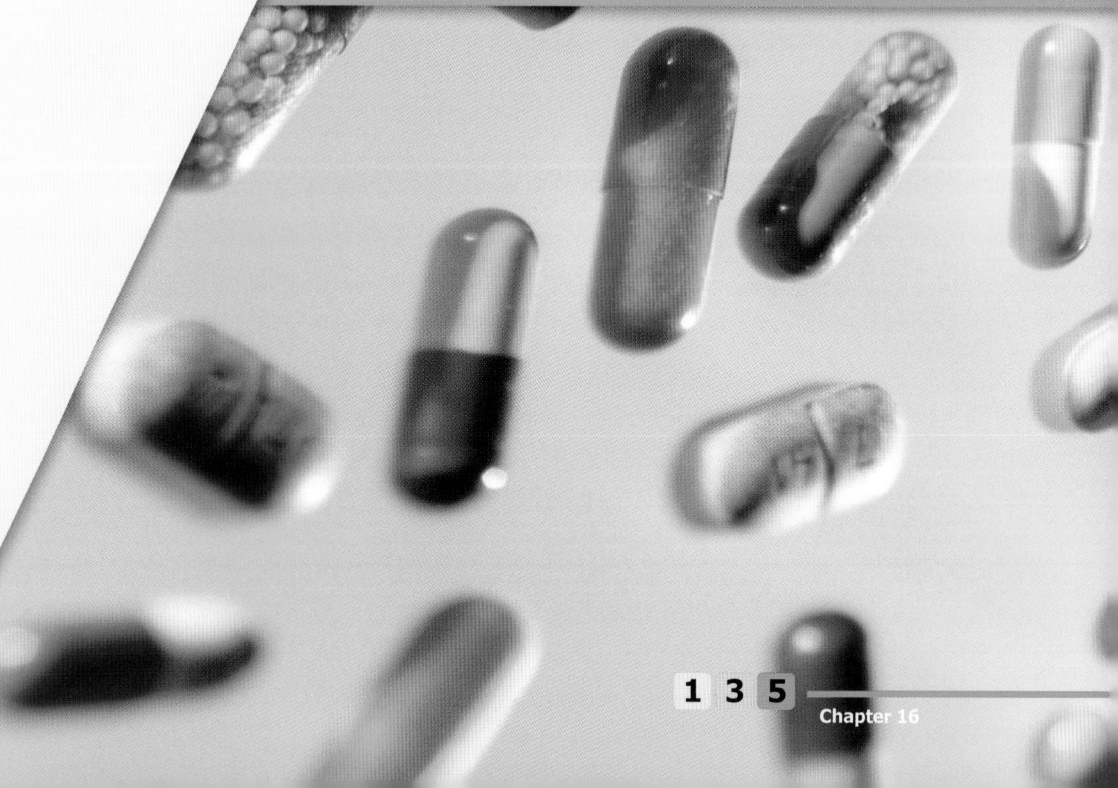

Frontier Molecular Orbital Theory, commonly shortened to FMO Theory. Although these theories may seem esoteric to the reader they decisively impacted the understanding and, therefore, the practice of synthetic organic chemistry. After Woodward's death, the independently developed, but mutually compatible theories, earned Hoffmann and Fukui the 1981 Nobel Prize in Chemistry.

We chose to begin our vitamin narrative in 1747, although deficiency diseases have been written about ever since writing was invented and across the full span of cultures, from the Ebers papyrus (Egyptian) and ancient Chinese manuscripts to the logs of various explorers. Given this history and the scourge that diseases of vitamin deficiency have perpetrated against humanity, it is striking that it was not until the dawning of the twentieth century that the concept of vitamins was truly born and that we finally understood that we need traces of important chemicals in our diet to facilitate the normal function of our body's machinery. That the advances in the vitamin field parallel those in organic synthesis is no coincidence. Indeed, with this new subclass of natural products, synthetic chemistry once again rose to the challenge, and its diligent and talented practitioners conquered even the most architecturally intransigent member of the class, vitamin B_{12}.

Further Reading

The vitamin B_{12} synthesis:
S. A. Harris, K. Folkers, Synthesis of Vitamin B_6, *J. Am. Chem. Soc.* **1939**, *61*, 1245–1247.

R. B. Woodward, The Total Synthesis of Vitamin B_{12}, *Pure Appl. Chem.* **1973**, *33*, 145–177.

A. Eschenmoser, C. E. Wintner, Natural Product Synthesis and Vitamin B_{12}, *Science* **1977**, *196*, 1410–1420.

K. C. Nicolaou, E. J. Sorensen, *Classics in Total Synthesis*, Wiley-VCH, Weinheim, **1996**, pp. 99–136.

Robert Burns Woodward: Architect and Artist in the World of Molecules, O. T. Benfey, P. J. T. Morris, Eds., Chemical Heritage Foundation, Philadelphia, **2001**.

A. I. Scott, Discovering Nature's Diverse Pathways to Vitamin B_{12}: A 35-Year Odyssey, *J. Org. Chem.* **2003**, *68*, 2529–2539.

The Woodward–Hoffmann rules:
R. Hoffmann, R. B. Woodward, Conservation of Orbital Symmetry, *Acc. Chem. Res.* **1968**, *1*, 17–22.

J. A. Berson, *Chemical Creativity: Ideas from the Work of Woodward, Hückel, Meerwein and Others*, Wiley-VCH, Weinheim, **1999**.

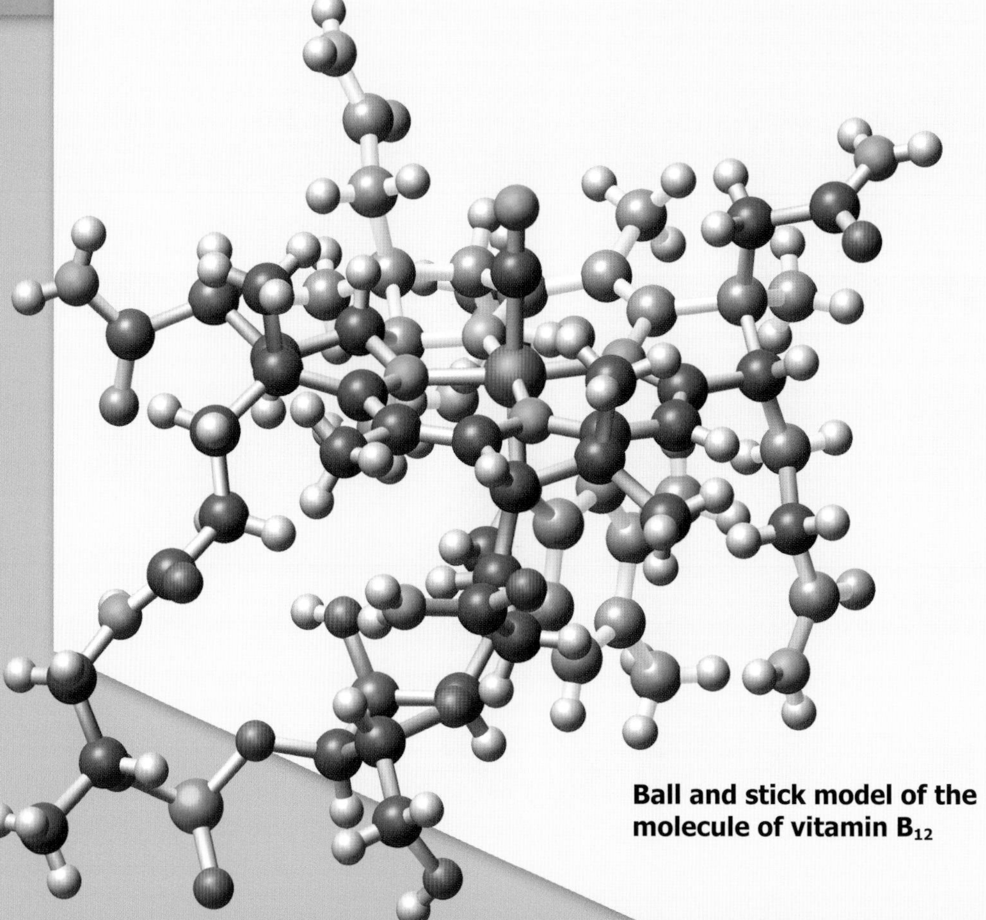

Ball and stick model of the molecule of vitamin B_{12}

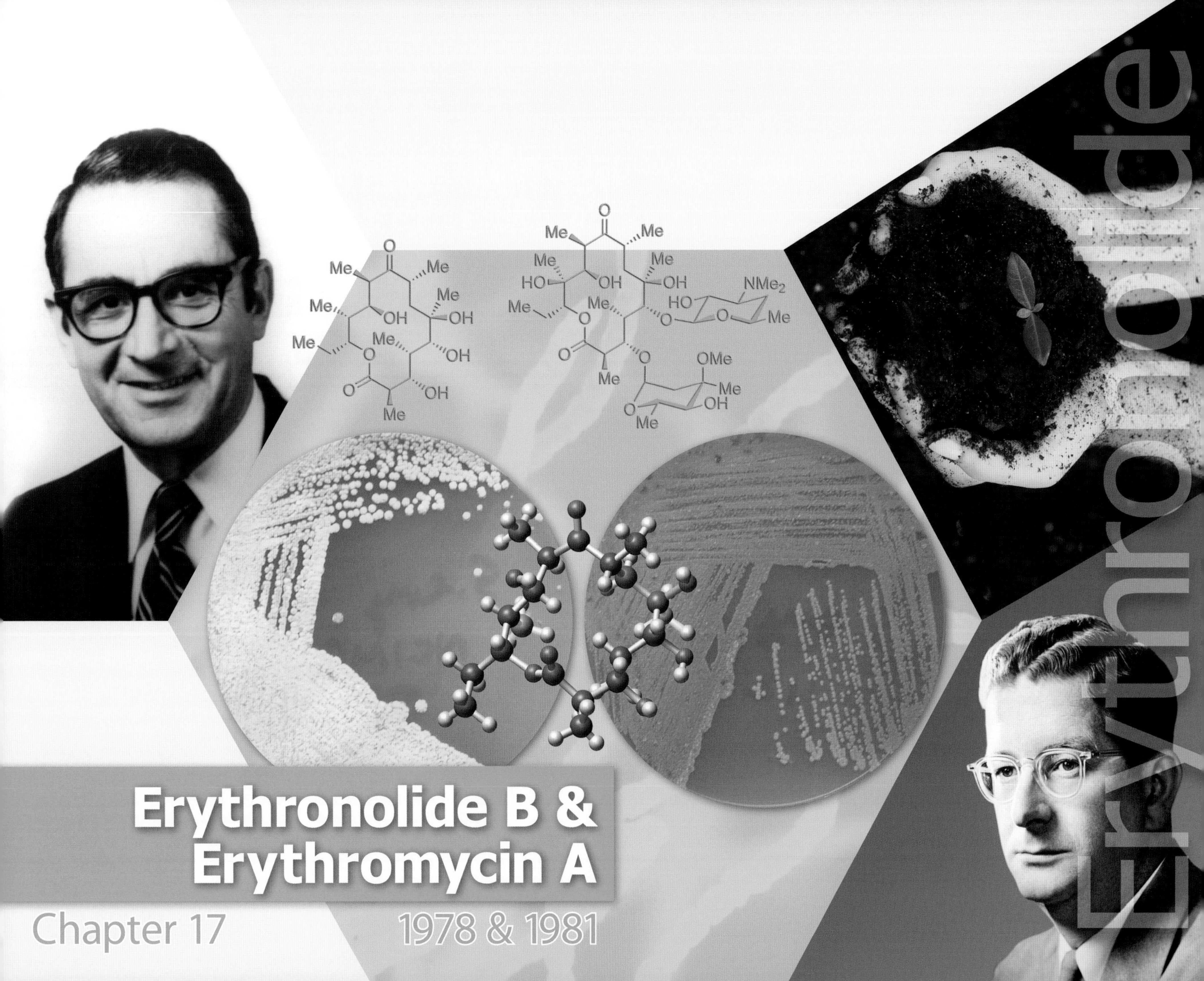

Erythronolide B & Erythromycin A

Chapter 17

1978 & 1981

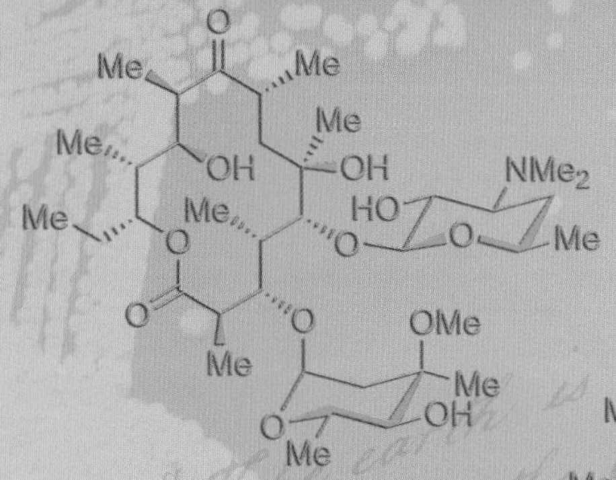

erythromycin B

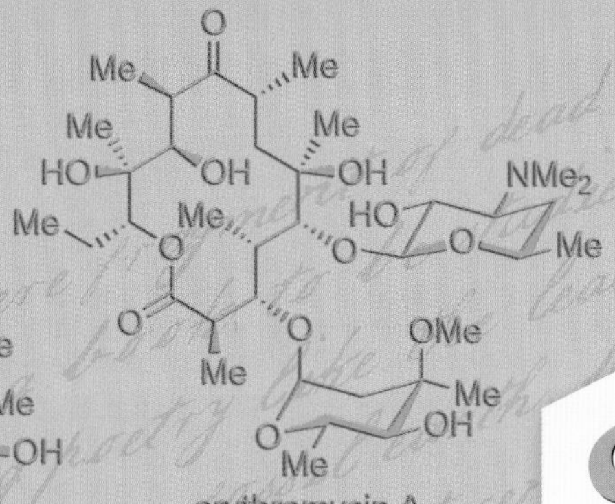

erythronolide B

erythromycin A

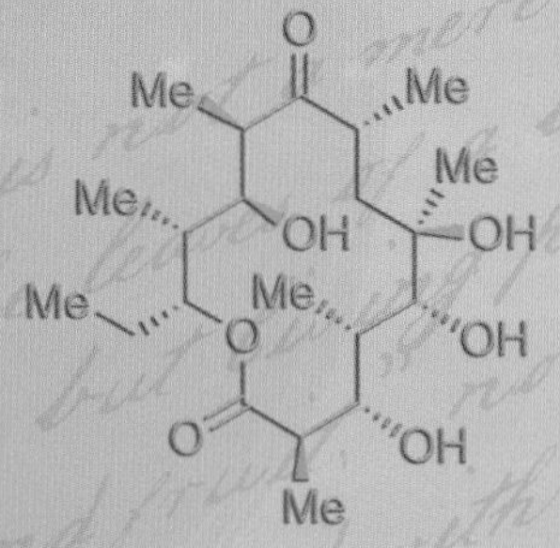

Box 1 Examples of natural products obtained from soil micro-organisms and adapted for human use

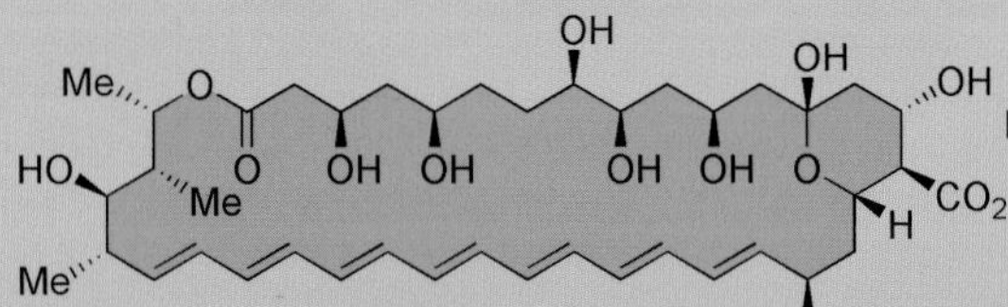

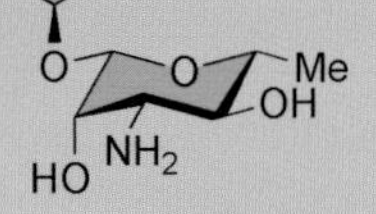

amphotericin B
(Amphocil®, AmBisome®, Abelcet®, Fungilin®, Fungizone® - antifungal drug)
(*Streptomyces nodosus*)

avermectin B_{1A}
(Affirm®, Agri-Mek®, Avid®, Vertimec®, Zephyr®, Abamectin® - agricultural insecticide)
(*Streptomyces avermitilis*)

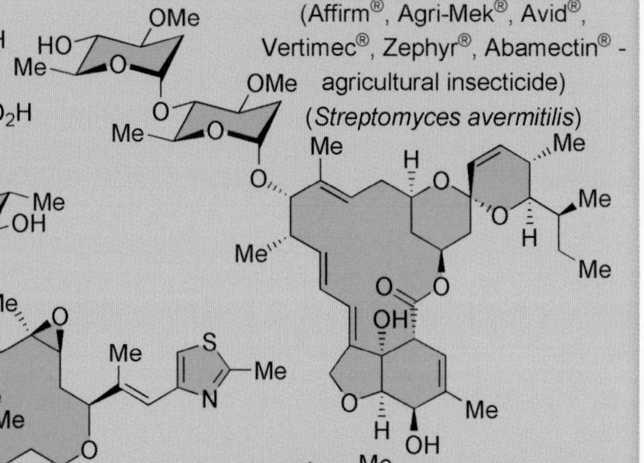

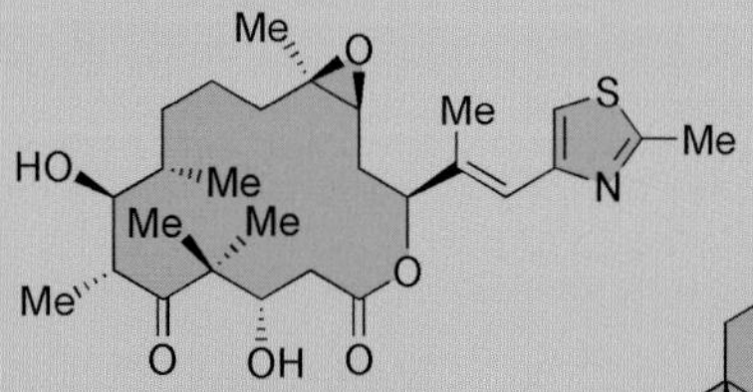

epothilone B
(anticancer agent in clinical trials)
(*Sorangium cellulosum*)

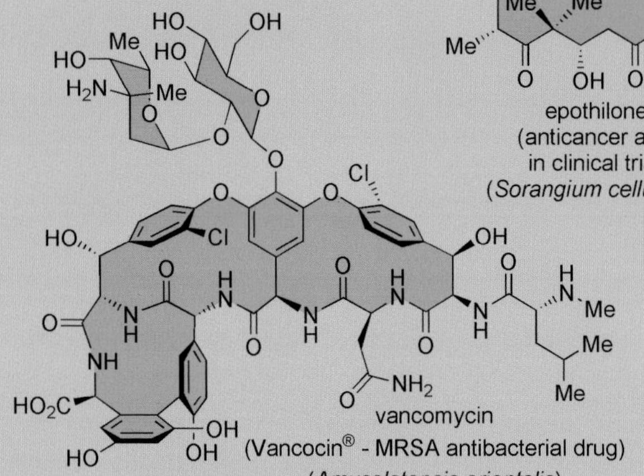

vancomycin
(Vancocin® - MRSA antibacterial drug)
(*Amycolatopsis orientalis*)

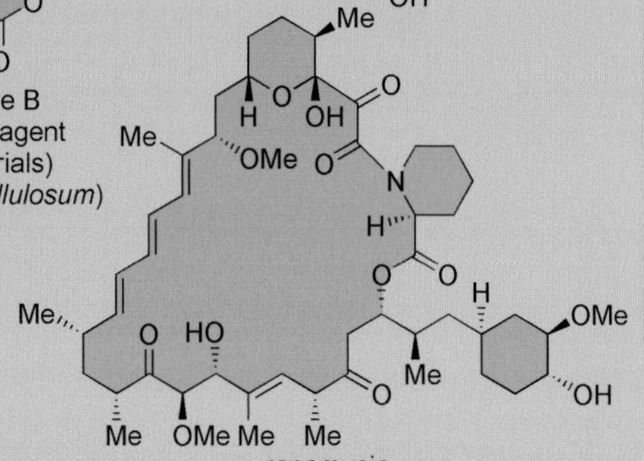

rapamycin
(Rapamune® - immunosuppressant drug)
(*Streptomyces hygroscopicus*)

"The earth is not a mere fragment of dead history, stratum upon stratum like the leaves of a book, to be studied by geologists and antiquaries chiefly, but living poetry like the leaves of a tree, which precede flowers and fruit, – not a fossil earth, but a living earth; compared with whose great central life all animal and vegetable life is merely parasitic."

Henry David Thoreau (US philosopher, writer, and naturalist, 1854)

Henry David Thoreau

While we may not feel quite as strongly as Henry David Thoreau, whose vision of poetry and vitality in the earth and its detritus is so emphatically expressed in the passage above, we can agree that soil is disparaged too readily and undeservedly bears an ill repute. These aspersions can be seen in our language – dirty, soiled, stick-in-the-mud, and mud-raking. This negativity is unjust, since, quite apart from the pivotal role of soil in nourishing plants, including our crops, it contains a secret treasure trove of natural products with enormous medicinal value. These medicinal molecules (Box 1) have a surprising diversity and have had a pronounced impact on mankind. In the account that follows, we chronicle two such molecules, erythronolide B and its relative, the widely used antibacterial agent erythromycin A. Both of these eye-catching macrocyclic natural products are secondary metabolites of *Streptomyces* bacteria, a genus of mud-dwelling microbes that is a particularly prolific producer of bioactive compounds.

Since this story is about antibiotics, we should clarify what we mean when we use this commonplace word. The term antibiotic is casually employed today, most often meaning an antibacterial drug obtained from any source. This usage may be functional, but it is not very accurate. The designation antibiotic was first coined in 1889 by the French biologist Paul Vuillemin, who, in studying pathogenic organisms, was following in the footsteps of the father of microbiology, his compatriot Louis Pasteur. Pasteur had revolutionized medicine several decades earlier by identifying microbes and then developing the germ theory for infectious disease. Before this monumental advance, these illnesses were attributed to fearful metaphysical origins (see Penicillin, Chapter 13). Vuillemin referred to antibiosis as "the destruction of one creature by another to sustain its own life." The term was adjusted a decade later by the British biologist H. M. Ward, now becoming centered on "the antagonistic relations between microorganisms." The meaning remained largely unchanged until an intense period of research that followed the landmark discoveries and the initial applications of the microbial natural products penicillin and streptomycin. At this point, Selman A. Waksman, an idiosyncratic but incisive Ukrainian émigré scientist working at Rutgers University (New Jersey, USA), was asked, in his capacity as the discoverer of numerous germicidal natural products, to assign a formal definition to the term antibiotic for *Biological Abstracts*. This series of reference books was a key repository for important scientific data in the pre-computer era. Waksman defined antibiotics as, "...chemical substances that are produced by microorganisms and that have the capacity, in dilute solution, to selectively inhibit the growth of and even to destroy other microorganisms." This definition was quickly adopted by scientists and laymen alike, although, much to the chagrin of its author, it also evolved, losing clarity and becoming increasingly vague with time. The modern definition of antibiotic, to which we shall adhere in our writing, is "a secondary metabolite produced by a microorganism that inhibits the function or has lethal activity against a defined cell type, be it cancerous, bacterial, or fungal." The associated term, antibacterial, is in one sense more general as it encompasses other sources, including wholly synthetic agents, and in another sense more specific in that therapeutically the sole targets are bacteria.

Waksman was awarded the Nobel Prize in Physiology or Medicine in 1952 for "his discovery of streptomycin, the first antibiotic effective against tuberculosis." Streptomycin was the antecedent of all the natural products issuing from soil microorganisms that developed into life-saving drugs. Working from the long-held knowledge that the lethal tubercle bacillus was rapidly destroyed in the soil, Waksman became convinced that the demise of these

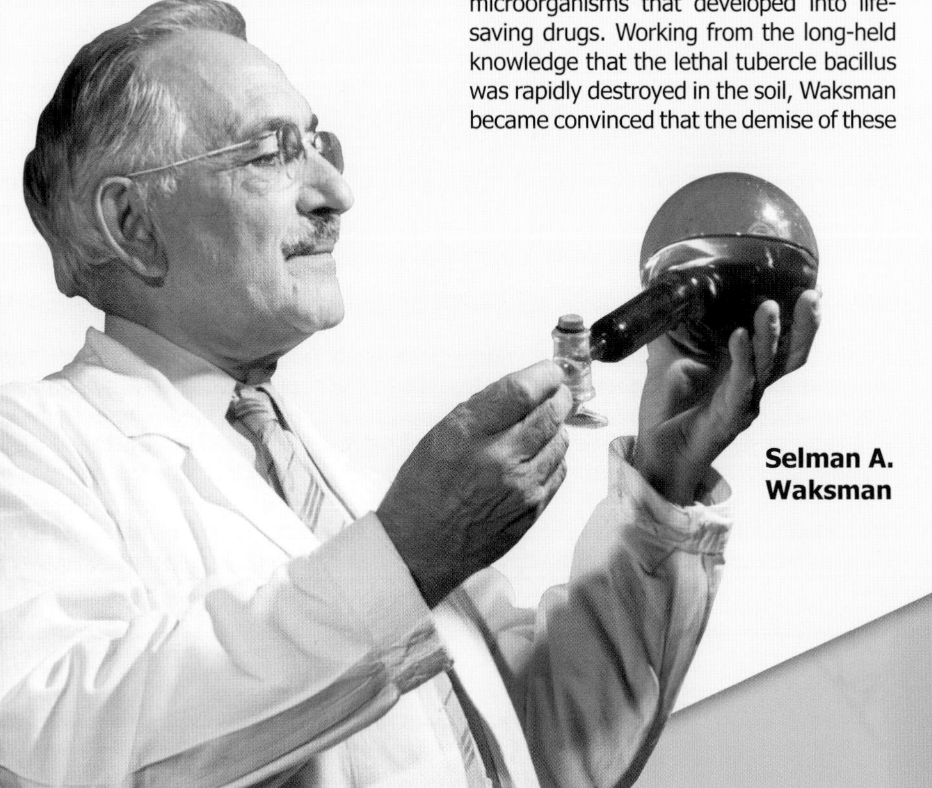

Selman A. Waksman

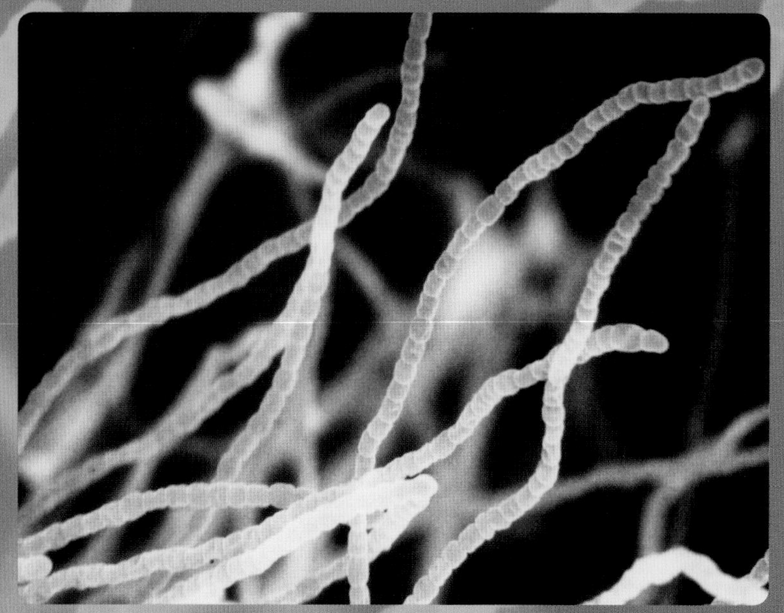

Streptomyces griseus

Box 2 **Waksman's famous early isolates from Streptomyces bacteria**

streptothricin F
(toxic compound)

streptomycin
(tuberculosis drug)

actinomycin D [Cosmegen®]
(anticancer drug)

Box 3 Semi-synthetic derivatives of erythromycin A used as antibacterial drugs

azithromycin [Zithromax®]

clarithromycin [Biaxin®]

Ball and stick model of the molecule of erythronolide B

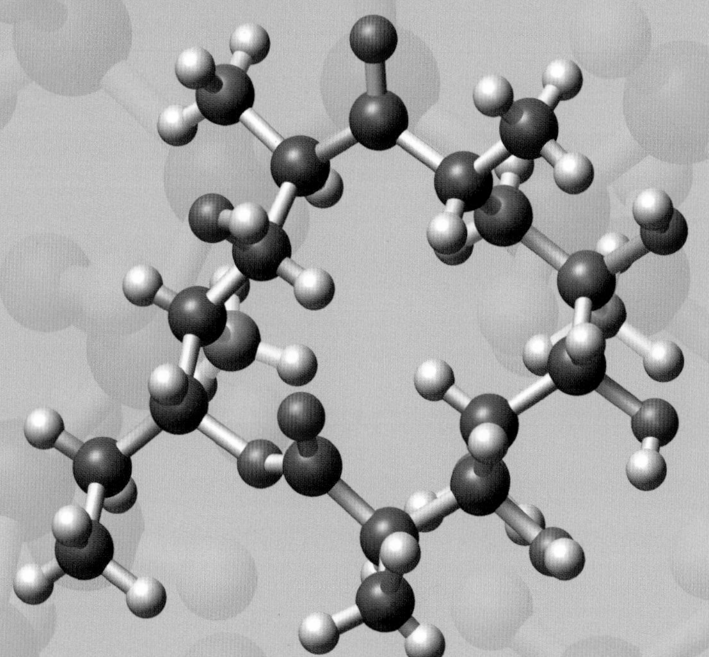

Tuberculosis bacteria

deadly bacteria was caused by a chemical antibiotic produced by a competing microorganism. He began a systematic campaign to identify the microorganisms producing such antibiotics, for these compounds had the potential to halt the epidemic of tuberculosis. Early in his career, he had isolated a strain of filamented and fungi-like actinomycete bacteria (first named *Actinomyces griseus*, but later reclassified as *Streptomyces griseus*) that survived under many poor soil conditions. He felt sure that this ability to persevere suggested that bacteria belonging to this class were elite warriors who had developed sophisticated chemical weapons to ensure their victory in the intense contest for nutrients. Despite this visionary assessment, his triumph did not come easily. He would test in excess of 10,000 soil microbes over an eleven-year period, dismissing several candidates (including actinomycin D, which later became an anticancer agent, and streptothricin) that were too toxic to be administered to humans, before he discovered streptomycin in 1943 (Box 2). In a strange twist of fate, streptomycin was found when Waksman's young assistant, Albert Schatz, re-isolated and examined the very same streptomyces strain (*Streptomyces griseus*) that Waksman had originally identified back in 1915, and which had been responsible for setting this train of investigation into motion in the first place.

It was quickly found that the aminoglycoside streptomycin was remarkably effective as a treatment for tuberculosis, even in its most vicious manifestations, tubercle meningitis and miliary tuberculosis, both of which had previously been regarded as being rapidly and inevitably fatal. Tuberculosis has been found in ancient Egyptian mummies, demonstrating that the scourge of this disease existed even in some of our earliest civilizations. From then on, and almost without respite, it has plagued mankind, becoming one of the biggest killers to ever afflict humans. In this context, streptomycin was a true Godsend, saving countless lives. Tuberculosis, also known as consumption, which had previously spread like a wild fire through affected communities, suddenly became rare, and retreated to the point where it was a problem restricted to societies within poorer nations where streptomycin could not be accessed or easily afforded. Unfortunately, today, tuberculosis is fast re-emerging. It now kills a quarter of the eight million people

Egyptian mummy of Amonred

infected each year; especially frightening is a drug-resistant form that accounts for approximately 4 % of the current total infections. We shall, however, leave discussion of the growing problem of generalized bacterial resistance for the forthcoming chapter about vancomycin, a magnificent natural product that has become the so-called drug of last resort in such instances (Chapter 31).

Saccharopolyspora erythraea

The phenomenal success of streptomycin not only in combating tuberculosis, but also in treating other nasty Gram-negative bacterial infections, including some unresponsive to penicillin, triggered a trend for searching the soil for antibiotic natural products. This pervasive fashion has not really faded with time, although its focus has broadened to include a broader variety of biological tests. Practitioners have collected soil samples from every geographical outpost. For example, the immunosuppressant rapamycin (Chapter 22) came from Easter Island samples, and the anticancer drug candidates the epothilones (Chapter 29) originated in soil taken from the edge of the Zambezi River. Each specimen gathered is meticulously scrutinized in search of novel organisms, particularly strains of streptomyces, which might produce natural products that can be commandeered into becoming drugs (Box 1). After streptomycin, the macrolide antibiotics, a class of natural products that includes the erythromycins, were the next category of mud-sourced antibiotics to be discovered. Overall they are the most prolific structural group belonging to this proud lineage. This feature is illustrated by the examples shown in Box 1, where only vancomycin is not a macrolide (a macrocyclic lactone). More generally, and outside the bounds of the antibiotic classification, the macrolide structural spectrum is perhaps one of the richest in natural product chemistry, and you will find many other examples of this important class throughout this book.

The erythromycins themselves were discovered and isolated in the first antibacterial 'gold rush' of the 1950s, which yielded a host of such macrolides from soil organisms. They are a large family of compounds with antibacterial activity, bearing only minor differences between members, and all made by *Saccharopolyspora erythraea*. The pharmaceutical company Eli Lilly (Indianapolis, USA) was responsible for their initial isolation and structural elucidation. This fervent research was part of the drive within the blossoming pharmaceutical industry to seek out new wonder drugs with the potential to outshine the existing combination of streptomycin and penicillin. Huge vats of fermentation broth were cultured and the bioactive materials subsequently isolated using laborious and relatively inefficient paper chromatography methods. This process serves as a reminder that the technology available in those days was still relatively unsophisticated, despite the innovations that had resulted from the dramatic wartime Penicillin Project a decade earlier

Box 4 The Corey–Nicolaou double activation macrolactonization method

E. J. Corey and K. C. Nicolaou

Box 5 Corey's retrosynthetic analysis of erythronolide B

P = protecting group

erythronolide B

Box 6 Highlights of Corey's total synthesis of erythronolide B

(resolution)

erythronolide B

(Chapter 13). Within ten years, erythromycin A was approved as a broad spectrum antibacterial agent used to control both Gram-positive and Gram-negative bacterial infections, as well as certain other common infections that fall outside the bounds of this simple classification. Erythromycin exerts its effects by binding to bacterial ribosomes, disrupting microsomal protein synthesis. It has proved to be an enduring drug, despite the advent of a plethora of newer antibacterial drugs. Added to this phenomenal success is the fact that a number of its derivatives (made semi-synthetically) now belong on the list of top-selling patented drugs. For example, Zithromax® (azithromycin, Pfizer) and Biaxin® (clarithromycin, Abbott Laboratories) had sales of \$1.5 and \$1.2 billion, respectively, in 2001 (Box 3).

Examination of the gross architecture of the erythromycins reveals three distinct structural units, the macrolide ring and the two different appended sugar units. The naked macrolides, without the appended sugars, are themselves natural products (the erythronolides). These structures serve as obvious synthetic precursors to the parent erythromycins. The erythronolides, with their stereochemically rich macrocyclic structures, include the majority of the challenges associated with the chemical synthesis of the erythromycins. Indeed, the great synthetic chemist Robert B. Woodward remarked in 1956, "Erythromycin, with all our advantages, looks at present quite hopelessly complex, particularly in view of its plethora of asymmetric centers."

Macrolides are defined as being large cyclic lactones, usually made from a linear precursor chain by uniting a terminal carboxylic acid with a suitably disposed hydroxyl group. Of course, as the length of the tether between these two functional groups increases, the rate of ring formation slows and lower concentrations are required to prevent competitive dimerization. Due to the proliferation of these compounds in nature, methods to overcome these hurdles and effect the aforementioned cyclization reaction have taken on great importance over the years. E. J. Corey and his group at Harvard University (USA) accomplished the first erythronolide synthesis in 1978. Their synthesis showcases one of the earliest successful macrolactonization procedures (Boxes 4–6). This rationally designed protocol, known as the Corey–Nicolaou double activation procedure, permits macrolactonization under mild conditions and has been used not only in the total synthesis of a number of complex natural products, but also provided the inspiration for many of the more powerful methods that followed (Box 4).

E. J. Corey

Box 7 Woodward's retrosynthetic analysis of erythromycin A

Before the Corey group could even contemplate investigating the macrolactonization, the obstacle of forming the array of stereocenters present in the linear precursor needed to be tackled. The targeted fourteen-membered erythronolide B macrocycle possesses a daunting array of ten stereocenters, including five arranged in a contiguous series, which needed to be installed selectively. The Corey group's retrosynthetic analysis of erythronolide B (Box 5) reveals a strategy using a six-membered ring as a scaffold upon which the stereochemistry of the target could be assembled efficiently. In the event, this strategy proved highly effective (Box 6). The densely functionalized cyclohexane ring was expanded using the Baeyer–Villiger reaction. Further manipulation, including a Mukaiyama coupling, provided the complete carbon chain of the target, setting the stage for the successful Corey–Nicolaou lactonization. A few routine transformations then yielded erythronolide B, completing this landmark total synthesis. For the sake of practicality, the Corey group actually prepared erythronolide B in racemic form, although optical resolution was shown to be possible at an early stage, allowing access to enantiomerically pure material. The completion of this synthesis, and the utility of the Corey–Nicolaou macrolactonization reaction, spurred many others to attempt and complete total syntheses of other macrolides, such that today these natural products no longer present the daunting obstacle they once did.

The challenges offered by the erythromycins and erythronolides made them popular target molecules and in the 1980s many other groups finished total syntheses of various members of this class. The first synthesis of an erythromycin, including the attached sugars, was accomplished by R. B. Woodward's group (Harvard University) and was published in 1981, two years after his death. The Woodward group used degradation studies to find a pathway through the later synthetic steps and benefited from the Corey–Nicolaou macrolactonization method when it came to cyclizing the delicate linear molecule they had assembled. Highlights of their synthesis of enantiomerically pure erythromycin A, in which the sugar attachment was reserved until the last steps and effected on the closed and fully functionalized macrolide, are shown in Boxes 7 and 8. As in Corey's synthesis, the Woodward approach made use of temporary smaller rings in order to control the stereochemistry of the precursor, in this case a bicyclic thioacetal. Such cyclic template strategies were invaluable in meeting the challenge of synthesizing extended carbon chains bearing multiple stereocenters, until the development of reliable methods for acyclic stereocontrol in subsequent years.

Robert B. Woodward

Box 8 Highlights of Woodward's total synthesis of erythromycin A

The erythromycins were the first macrolide antibiotics isolated from soil to enter the clinic, and as such they ushered in a new era of medicines to be harvested from the land for the benefit of mankind. Chemists would later go on to isolate and develop more antibacterial agents, anticancer drugs, and immunosuppressants, all made by soil microorganisms and all containing a large macrocyclic lactone. Interestingly, the general lesson of this story appears in a biblical quotation, "the Lord hath created medicines out of the earth and he that is wise will not abhor them" (*Ecclesiasticus, 38:4*). As this verse suggests, we should appreciate the soil, its inhabitant organisms, and their natural products, which furnish us with such an amazing array of life saving drugs. To be sure, there are many more to be discovered.

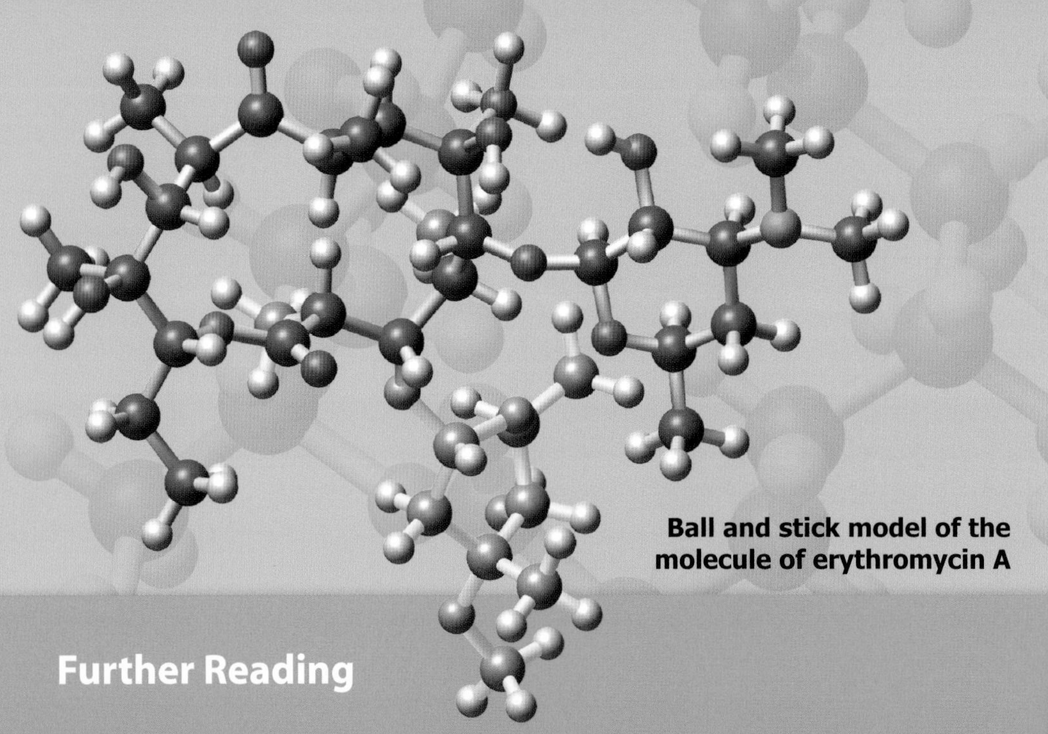

Ball and stick model of the molecule of erythromycin A

Harvard seal

Further Reading

E. J. Corey, K. C. Nicolaou, An Efficient and Mild Lactonization Method for the Synthesis of Macrolides, *J. Am. Chem. Soc.* **1974**, *96*, 5614–5616.

E. J. Corey, S. Kim, S. Yoo, K. C. Nicolaou, L. S. Melvin, Jr., D. J. Brunelle, J. R. Falck, E. J. Trybulski, R. Lett, P. W. Sheldrake, Total Synthesis of Erythromycins. 4. Total Synthesis of Erythronolide B, *J. Am. Chem. Soc.* **1978**, *100*, 4620–4622.

R. B. Woodward, E. Logusch, K. P. Nambiar, K. Sakan, D. E. Ward, B.-W. Au-Yeung, P. Balaram, L. J. Browne, P. J. Card, C. H. Chen, R. B. Chênevert, A. Fliri, K. Frobel, H.-J. Gais, D. G. Garratt, K. Hayakawa, W. Heggie, D. P. Hesson, D. Hoppe, I. Hoppe, J. A. Hyatt, D. Ikeda, P. A. Jacobi, K. S. Kim, Y. Kobuke, K. Kojima, K. Krowicki, V. J. Lee, T. Leutert, S. Malchenko, J. Martens, R. S. Matthews, B. S. Ong, J. B. Press, T. V. Rajan Babu, G. Rousseau, H. M. Sauter, M. Suzuki, K. Tatsuta, L. M. Tolbert, E. A. Truesdale, I. Uchida, Y. Ueda, T. Uyehara, A. T. Vasella, W. C. Vladuchick, P. A. Wade, R. M. Williams, H. N.-C. Wong, Asymmetric Total Synthesis of Erythromycin. 2. Synthesis of an Erythronolide A Lactone System, *J. Am. Chem. Soc.* **1981**, *103*, 3213–3215.

R. B. Woodward, E. Logusch, K. P. Nambiar, K. Sakan, D. E. Ward, B.-W. Au-Yeung, P. Balaram, L. J. Browne, P. J. Card, C. H. Chen, R. B. Chênevert, A. Fliri, K. Frobel, H.-J. Gais, D. G. Garratt, K. Hayakawa, W. Heggie, D. P. Hesson, D. Hoppe, I. Hoppe, J. A. Hyatt, D. Ikeda, P. A. Jacobi, K. S. Kim, Y. Kobuke, K. Kojima, K. Krowicki, V. J. Lee, T. Leutert, S. Malchenko, J. Martens, R. S. Matthews, B. S. Ong, J. B. Press, T. V. Rajan Babu, G. Rousseau, H. M. Sauter, M. Suzuki, K. Tatsuta, L. M. Tolbert, E. A. Truesdale, I. Uchida, Y. Ueda, T. Uyehara, A. T. Vasella, W. C. Vladuchick, P. A. Wade, R. M. Williams, H. N.-C. Wong, Asymmetric Total Synthesis of Erythromycin. 3. Total Synthesis of Erythromycin, *J. Am. Chem. Soc.* **1981**, *103*, 3215–3217.

K. C. Nicolaou, E. J. Sorensen, *Classics in Total Synthesis*, Wiley-VCH, Weinheim, **1996**, pp. 167–184.

Monensin
Chapter 18
1979
Rumensin
TROUGH TREATMENT
ELANCO
FOR ORAL USE IN CATTLE ONLY

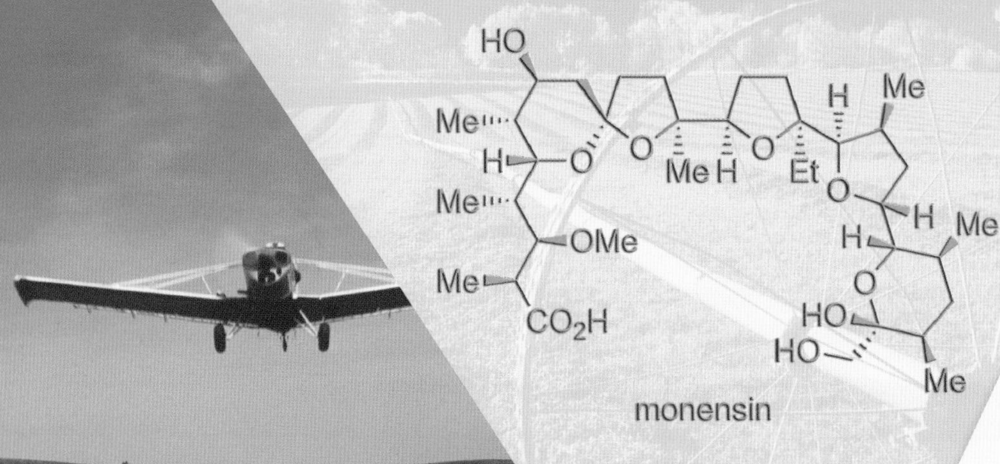

In 1930 the population of the world was around 2 billion; today the figure is approximately 6.5 billion and, according to the United Nations, it is projected to rise still further to reach 9 billion by 2050. The 1990s saw the fastest worldwide population growth ever and the elder members of our societies have witnessed a three-fold increase in the human population. These figures are quite astounding, and are the result of very complicated patterns of cause and effect.

Amongst the many contributing factors, it is undeniable that broader access to better healthcare has led to a significant reduction in death rates, improved infant mortality figures, and an extension in life expectancy for most citizens of the world. Alongside this healthcare revolution, agriculture has played a pivotal role and has been subject to its own radical changes. Productivity and crop yields have grown at phenomenal rates across the board, augmented by the development of improved farming technologies and the emergence of efficiency-boosting agrochemicals. This expansion has allowed us to feed a population once thought to be unsustainable. We have seen many instances where synthetic organic chemistry has helped to secure breakthroughs in medicine, but in this chapter we shall describe one of the many ways in which chemistry has also revolutionized food production. Though often pilloried by environmentalists, sometimes with just cause, pesticides, herbicides, fertilizers, and animal feed additives have all played an essential part in underpinning the ability of farmers to feed the world's booming societies. From this vast area we have chosen the natural product monensin as an example. Monensin is not only of immense agricultural significance, but also of great chemical importance due to its intricate and oxygen-rich structure. As a synthetic target it was first successfully tackled in 1979. Monensin is also remembered for its role in the early investigations into an important class of molecules, the ionophores, particularly the medicinally important sub-group the polyether antibiotics.

Ionophores, a large and structurally diverse set of compounds, have the ability to complex cations (positively charged ions) and assist in their translocation through lipophilic environments that are normally hostile to these charged species. The last fifty years have seen an enduring interest in ionophores amongst chemists. Synthetic ionophores have been designed and prepared bearing precise specifications, giving them an affinity for just one particular type and size of cation. These synthetic molecules have been put to many uses, from the sequestering of contaminating ions, to employment as analytical tools. The 1987 Nobel Prize in Chemistry was awarded to Charles J. Pedersen (DuPont, USA), Donald J. Cram (University of California, Los Angeles, USA), and Jean-Marie Lehn (Université Louis Pasteur, Strasbourg, France) for "their development and use of molecules with structure-specific inter-

Charles J. Pedersen

Jean-Marie Lehn

Donald J. Cram

actions of high selectivity." Pedersen described methods to synthesize perhaps the simplest of ionophores, cyclic polyethers, which he named crown ethers (see Box 1). Jean-Marie Lehn built on this work by developing bicyclic crown ether-type molecules that he called cryptands. These molecules bound the guest cation more tightly from a greater number of directions than their crown ether predecessors. Lehn and Cram later went on to synthesize a series of more complex organic compounds containing fissures and cracks where low molecular weight compounds having exactly the correct dimensions could dock snugly. These studies heralded the inception of new and powerful fields of chemistry: molecular recognition and supramolecular chemistry.

Of course, there are also a variety of naturally occurring ionophores, and these are involved in a vast array of physiological roles. The protein haemoglobin, found in red blood cells (erythrocytes), consists of four polypeptide subunits, each carrying an identical haemin ionophore (see Chapter 8). This cyclic ligand is coordinated to an iron cation in its center, which absorbs and releases oxygen at the appropriate moment depending on whether it is in an environment that is oxygen-rich (the lungs) or oxygen-poor (any part of the body requiring fresh oxygen). The discovery and understanding of naturally occurring ionophores continues to be a vibrant area of research. The 2003 Nobel Prize in Chemistry was awarded to two scientists, Peter Agre (Johns Hopkins University, Baltimore, USA) and Roderick MacKinnon (Rockefeller University, New York, USA), who made fundamental discoveries concerning how water and ions move through cell membranes using biological macromolecular ionophores.

In this chapter we shall focus on another subset of natural ionophores, namely polyether antibiotics produced by microscopic organisms (usually *Streptomyces*) as defense chemicals. These compounds protect the microorganisms that produce them by destroying competing organisms found within their bustling ecosystems. In 1967, a group led by Amelia Agtarap (Lilly Research Laboratories) isolated monensic acid, now known as monensin, from *Streptomyces cinnamonensis*, and elucidated its intricate structure. This event was a milestone in polyether antibiotic research because, although monensin was the fifth such compound to be discovered, it was the first to have its structure fully unraveled and unveiled to the waiting scientific community. Polyether antibiotics all share a number of structural features. They are built from an array of tetrahydrofuran and/or tetrahydropyran rings (saturated 5- or 6-membered rings, respectively, containing a single ethereal oxygen atom); they frequently contain one, or more, spiroketal junction; and

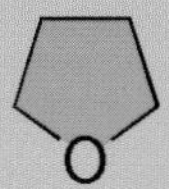

Tetrahydrofuran ring

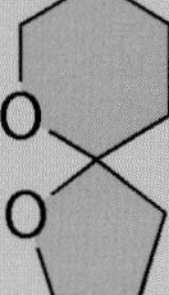

Spiroketal ring system

Tetrahydropyran ring

Peter Agre

Roderick MacKinnon

Box 1 Selected examples of synthetic ionophores

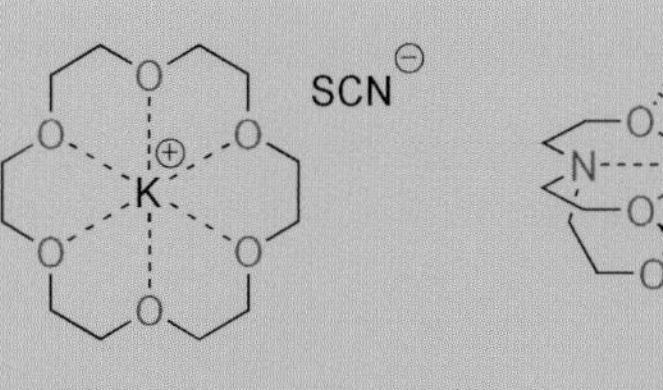

a crown ether (18-crown-6) (C. J. Pedersen)

a cryptand (J.-M. Lehn)

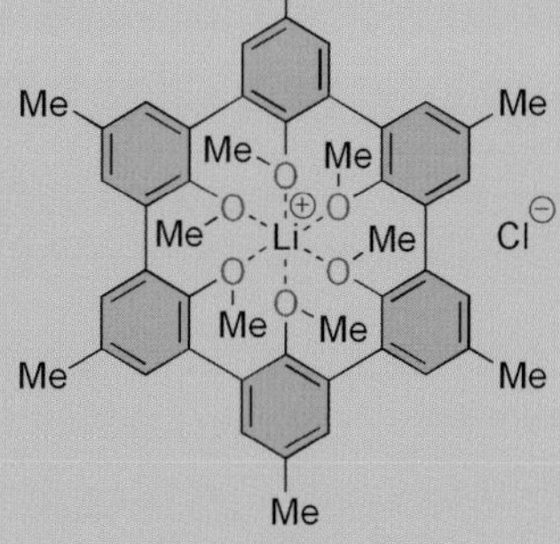

a host-guest complex (D. J. Cram)

O, N = Lewis basic interior of ionophore
$K^{\oplus}$, $Li^{\oplus}$ = cationic guest

their structures are terminated at one end by a carboxylic acid group (Box 2). In addition to the molecular structure, the original isolation report noted that monensin was a potent inhibitor of alkali metal cation transport into rat liver mitochondria and, more importantly, had broad-spectrum anticoccidial activity. This latter activity led to the launch of monensin onto the market as early as 1971 as an additive for poultry and cattle feed.

Coccidial parasites are single-celled protozoa that infect the intestines of young farm animals, causing severe damage to the epithelial lining of the gut and leading to diarrhea, poor nutrient absorption, and imbalances of both iron and water. These factors culminate in poor growth and low meat quality in the affected animals. Mature animals are generally immune to these germs, which are related to the infectious agents that cause toxoplasma and cryptosporidium in humans, but they are still regularly dosed with monensin for other reasons. Monensin, marketed under the brand name Rumensin®, has been used for the last thirty-five years all over the world as a food additive to prevent coccidial infection in susceptible farm animals. As its trade name suggests, monensin plays a second important role in improving digestion and, therefore, growth potential in ruminant animals, especially cattle. The antibiotic effects of monensin extend to the beneficial altering of the natural balance of microbial populations in the digestive tracts of cattle by selectively controlling the numbers of Gram-positive bacteria. Gram-negative bacteria tend to be resistant to monensin and related ionophores (e.g. lasalocid A, salinomycin, and narasin, Box 2) due to their less permeable cell exterior. The ionophore antibiotics work by disrupting the movement of ions across the bacterial cell membrane. The bacteria must then work harder, expending a great deal of additional energy, in order to maintain intracellular pH and ion balance. Individual bacteria are weakened and fail to grow and reproduce at levels sufficient to sustain the colony; as a result, it eventually dies out altogether. The change in the ratio of Gram-negative to Gram-positive bacteria favors the production of propionate and succinate (from ingested food) and suppresses acetate, butyrate, hydrogen, ammonia, and lactic acid production. In simple terms, the increase in propionate production increases energy efficiency by raising the metabolic value of the feed. Ionophore food additives also improve protein metabolism and prevent ketosis in lactating animals.

Monensin and related antibiotics are employed as general food additives, rather than to tackle any specific disease, and many people question the wisdom of such a practice. The problem of multiple drug resistance evolving in bacterial strains, as discussed in the chapters describing penicillin and vancomycin (Chapters 13 and 31), leads to concerns about the indiscriminant dosing of livestock antibiotics. The practice is, however, justified by the following arguments, although it is for the individual to decide the merits of the case. Proponents assert that the ionophore antibiotics are not related to any antibacterial agent used to treat humans,

Box 2 Selected polyether antibiotics used as food additives for farm animals

lasalocid A

salinomycin

narasin

and that any resistance developed toward them will be the result of a mechanism that is highly specific for the ionophores. Furthermore, the bacteria targeted by the general treatment of mature livestock are benign members of the intestinal fauna, and are not responsible for causing illness. However, cautious observers may counter that, whilst every effort must be made to meet the huge food requirements of our burgeoning population, our past experiences with other antibiotics suggest that the low and continual dosing of animals with mild antibiotics is almost certain to end in tears, even if, at present, the precise knowledge of how this might happen does not exist.

Arguments about the use of monensin do not detract from its chemical beauty. It is a typical polyether antibiotic, and as such bears the distinguishing features of this class as described earlier. However, it also possesses a stereochemical intricacy that made it a particularly challenging synthetic target for its time. Of the twenty-six carbon atoms which comprise the backbone of this molecule, seventeen are asymmetrically substituted, including one especially tough section where six such carbons are arranged contiguously. Each of these centers needs to be formed as selectively as possible if an efficient synthesis is to result. A further structural curiosity of monensin is that, in both its crystalline form and in solution, it has a cyclic structure maintained by two intramolecular hydrogen bonds between the terminal carboxylic acid residue at one end and the two hydroxyl groups at the other. This organization produces an arrangement whereby the external surface of the molecule is a largely hydrocarbon, lipophilic region, while the many Lewis basic functional groups (all oxygen based) are directed towards the interior. This provides an ideal environment for the coordination of a metal cation, making monensin a perfect ionophore.

Polyether antibiotics in general, and monensin in particular, spearheaded a dramatic expansion in the power and scope of the methods used to control stereochemistry in acyclic systems. During this time, the crucial importance of being able to derive stereochemical purity in synthetic compounds was beginning to be appreciated throughout the field of organic chemistry. The origins of our current ability to understand and predict the stereochemical course of such processes as hydroboration, epoxidation, and carbonyl addition reactions (including asymmetric aldol reactions) can be traced back to this period. This knowledge is of fundamental importance to the job of a modern organic chemist. The man who led the first successful conquest of monensin, Yoshito Kishi (Harvard University, USA), is not only an extremely successful synthesizer of many of the most challenging natural products, but also a pioneer in the development of asymmetric methods and their application to chemical synthesis.

Kishi's highly convergent synthesis of (+)-monensin, in 1979, came just one year after his group had completed the first synthesis of the polyether antibiotic lasalocid A (Box 2). Kishi and his group developed a sophisticated plan (Box 3) to reach their difficult new target in which they effectively utilized experience they had gained in the synthesis of lasalocid A.

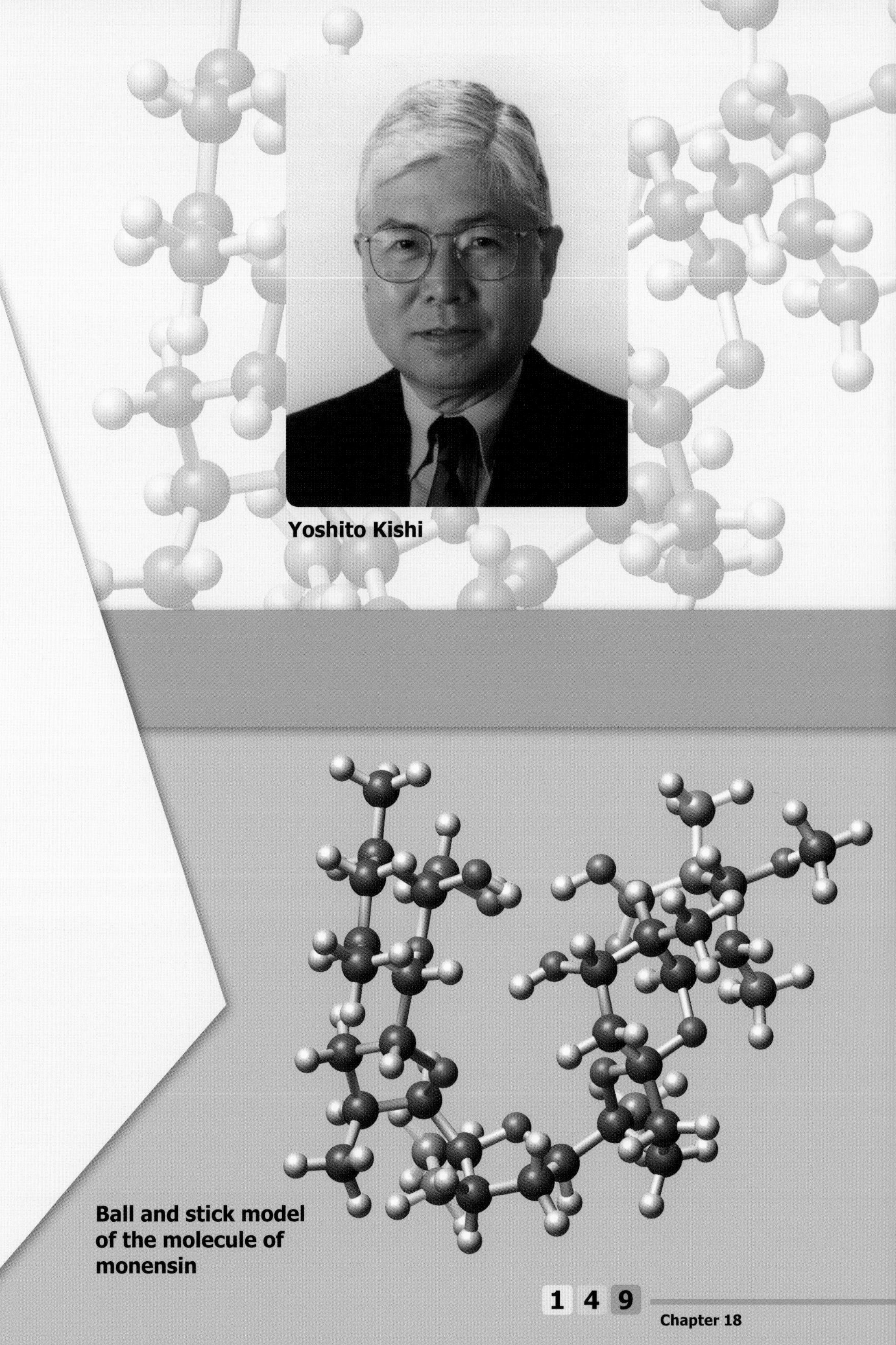

Yoshito Kishi

Ball and stick model of the molecule of monensin

Box 3 Kishi's retrosynthetic analysis of monensin

aldol reaction
spiroketalization
monensin
hydroboration
hydroboration
bromoetherification
hydroxy epoxide cyclization
ring closure

'left hand wing' (furan)

'left hand wing' (aldehyde)

'right hand wing' (ketone)

Box 4 The final stages of Kishi's total synthesis of monensin

aldol reaction
spiroketal formation
aq. NaOH, MeOH

sodium salt of monensin

A crucial decision was to leave the spiroketal formation until almost the end of the synthesis, because it was rationalized that this would afford the best opportunity to form this ring juncture with the correct stereochemistry. In almost all cases natural spiroketals exist in the most stable form (the thermodynamic isomer), where steric interactions and anomeric effects are optimally balanced, because spiroketals can open and close under rather mild conditions, thereby allowing them to equilibrate to the most stable isomer. One therefore stands the best chance of getting the stereochemistry of a spiroketal correct, without the necessity of reverting to equilibration, if the linear precursor is as close an analogue of the natural product as possible. After this key disconnection, the molecule was retrosynthetically severed into two portions, the smaller 'left hand wing' and the larger 'right hand wing'. The lasalocid A synthesis had given the team confidence that the stereochemistry of the aldol reaction needed to join these two fragments could be predicted by a theoretical system, called the Cram–Felkin–Anh model after its discoverers. It was then envisioned that the smaller left hand segment of the molecule, bearing five contiguous asymmetric centers, could be constructed in a stepwise manner, using internal induction to provide the requisite stereochemical control (in this context, induction is where the stereochemical formation of each new center is ordained by a pre-existing stereocenter and so stereochemical bias is translated along a growing chain of stereocenters). Key to this analysis was the use of two substrate-controlled hydroboration reactions, each setting two new stereocenters. The olefins required as precursors for these hydroboration reactions were assembled using Wittig and Horner–Wadsworth–Emmons olefination reactions (see Chapter 20). Within this fragment of the molecule, the carboxylic acid was replaced by a surrogate furan. This swap was done because furans can be readily transformed into carboxylic acids and are much easier to work with and handle than easily ionizable carboxylic acids. The synthetic plan demanded by the right hand section was more complex, as this portion possessed nineteen of the twenty-six carbon atoms of the backbone of the molecule, and had twice the number of stereocenters to be considered. The right hand side also bears a section consisting of two adjacent tetrahydrofuran rings followed by a tetrahydropyran ring. The assembly of the right hand domain of the target included several imaginative tactics, such as a hydroxyl-epoxide cyclization and a Johnson *ortho*-ester Claisen rearrangement, used to reach the required molecular complexity rapidly.

With the detailed blueprint in hand, Kishi and his team set forth on their synthetic adventure, and after two early resolutions to separate racemic mixtures, the visionary stereochemical control strategy was applied with a high degree of success. Box 4 highlights the last climactic steps used to complete the challenge of synthesizing (+)-monensin. At the time, Kishi's achievement was trumpeted as

Box 5 Hydroboration

On his graduation from the University of Chicago (USA) in 1936, Herbert C. Brown was presented with a copy of Alfred Stock's book, *The Hydrides of Boron and Silicon*, by his future wife, Sarah. Inspired, Brown began what was to become an eminent career studying the chemistry of organoboranes by joining the group of Hermann I. Schlesinger (University of Chicago) to study for his Ph.D. In 1947, he was appointed as a professor at Purdue University (Indiana, USA) and, in the decades that followed, Brown came to dominate the field of organoborane chemistry, which he noted is strikingly appropriate, given his initials (H for hydrogen, C for carbon and B for boron). He developed a catalog of tools essential to modern organic synthesis. This feat was recognized when the 1979 Nobel Prize in Chemistry was awarded to H. C. Brown and Georg Wittig (see Chapter 20) "for their development of the use of boron and phosphorous containing compounds, respectively, into important reagents in organic synthesis."

Herbert C. Brown

Hydroboration, discovered by Brown in 1956, is an important reaction in which a boron hydride is added across a carbon–carbon double or triple bond, usually in a regioselective fashion, with the boron adding to the least substituted carbon of the multiple bond. Subsequent oxidation of the intermediate organoborane introduces the versatile hydroxyl group into the molecule, as exemplified by Kishi in his synthesis of monensin. The use of chiral borohydrides, often derived from terpenes, allows asymmetric additions to carbon–carbon multiple bonds. Brown also extended this principle to the asymmetric formation of carbon–carbon bonds using allylic boranes.

The scope of boranes in organic chemistry extends to many other areas of chemical synthesis. Examples include simple reductants, Corey's efficient asymmetric reduction catalysts, boronic acids used in Suzuki palladium-catalyzed cross-couplings, and boron enolates employed in aldol and alkylation reactions.

Nobel Prize medal

(a) The hydroboration/oxidation reaction

R–CH=CH₂ + R'_2BH → (*anti-Markovnikov addition*, *hydroboration*) → [R–CH(H)–CH₂–BR'_2] (*organoborane intermediate*) → (H_2O_2, NaOH; *oxidation*) → R–CH(H)–CH₂–OH

(b) Some commonly employed R'_2BH reagents

borane•THF (H_3B←O(THF)); thexylborane; catecholborane; 9-BBN; $(Ipc)_2BH$

(c) Applications of the hydroboration/oxidation reaction in Kishi's total synthesis of monensin

(1) borane; (2) KOH, H_2O_2 (85 %)

(1) borane; (2) KOH, H_2O_2 (80 %)

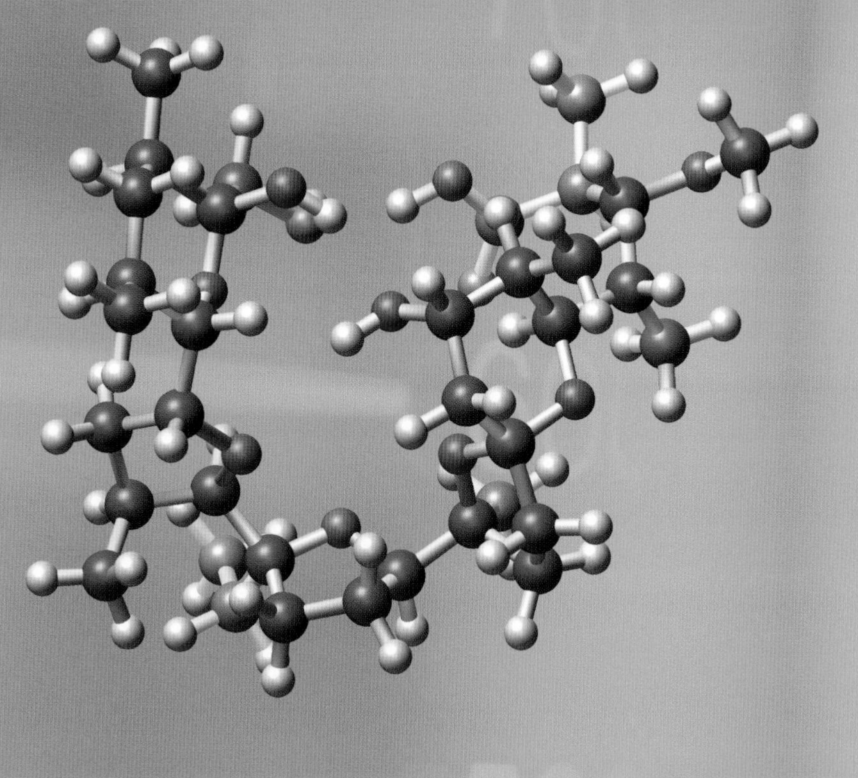

Further Reading

A. Agtarap, J. W. Chamberlin, M. Pinkerton, L. Steinrauf, The Structure of Monensic Acid, a New Biologically Active Compound, *J. Am. Chem. Soc.* **1967**, *89*, 5737–5739.

G. Schmid, T. Fukuyama, K. Akasaka, Y. Kishi, Total Synthesis of Monensin. 1. Stereocontrolled Synthesis of the Left Half of Monensin, *J. Am. Chem. Soc.* **1979**, *101*, 259–260.

T. Fukuyama, C.-L. J. Wang, Y. Kishi, Total Synthesis of Monensin. 2. Stereocontrolled Synthesis of the Right Half of Monensin, *J. Am. Chem. Soc.* **1979**, *101*, 260–262.

T. Fukuyama, K. Akasaka, D. S. Karanewsky, C.-L. J. Wang, G. Schmid, Y. Kishi, Total Synthesis of Monensin. 3. Stereocontrolled Total Synthesis of Monensin, *J. Am. Chem. Soc.* **1979**, *101*, 262–263.

D. B. Collum, J. H. McDonald, W. C. Still, Synthesis of the Polyether Antibiotic Monensin. 1. Strategy and Degradations, *J. Am. Chem. Soc.* **1980**, *102*, 2117.

K. C. Nicolaou, E. J. Sorensen, *Classics in Total Synthesis*, Wiley-VCH, Weinheim, **1996**, pp. 185–209 and 227–248.

a groundbreaking victory. Fortunately for synthetic organic chemistry, it was by no means the last time that a Kishi synthesis was hailed as a milestone. The Japanese born and trained chemist (he obtained both his B.Sc. and Ph.D. from the University of Nagoya) has had a distinguished career full of firsts. In 1994, Kishi's group synthesized palytoxin, the largest secondary metabolite yet to be synthesized in a chemist's laboratory. You can read more about this particularly difficult synthetic expedition in Chapter 24.

Monensin was synthesized for a second time just a year later (1980) by W. Clark Still (Columbia University, USA), thereby consolidating its inspirational role in the development of stereoselective synthesis. For today's synthetic chemists, a week rarely passes when knowledge garnered from this feverish period in the development of synthetic organic chemistry is not applied in one form or another. If one adds this critical status of monensin in chemistry to its importance in agriculture, it can be seen that this relatively small molecule has played a major role in all of our lives.

W. Clark Still

Avermectin
Chapter 19
1986, 1987

The Kitasato Medal

The preceding chapter described the role of the natural product monensin as an agricultural antibiotic, used to treat a wide range of bacterial infections in livestock, and to improve the yield and quality of animal products. The increase in agricultural productivity over the last century has been driven by the demands of the world's growing population. In turn, the greater access to higher quality food has helped increase life expectancy and, thereby, further fueled the population increase. Organic chemistry has played a pivotal role in revolutionizing farming through the development of agrochemicals, allowing communities to meet the ever-increasing need for food. Many natural products have been exploited in agriculture, including monensin (Chapter 18) and the molecules discussed in this chapter, the avermectins. In addition, synthetic chemists have added a number of modified natural products as well as wholly synthetically designed molecules to the ever-growing list of agrochemicals. As a result, many pests that once ravaged crops or damaged herds can now be controlled effectively. Derivatives of the avermectin natural products are now used to control two major groups of pests, insects and parasitic nematode worms.

Streptomyces avermitilis MA-4680

The discovery of the avermectins can be traced to a scientist playing golf. He spotted a particularly healthy patch of grass on the golf course and took a soil sample back to his laboratory for analysis. Although the initial investigation did not reveal the reason for the lushness of the grass, Satoshi Ômura and his group at the Kitasato Institute (Japan) sent their samples to colleagues at the Merck Research Laboratories (New Jersey, USA) as part of an ongoing collaboration. The soil sample taken from the golf course was found to contain a bacterial species identified as *Streptomyces avermitilis* MA-4680. When grown in culture, this organism produced a set of eight similar compounds that all showed potent nematicidal activity. The chemical structures of these molecules were quickly elucidated by the same group at Merck through the use of x-ray crystallographic analysis and other analytical techniques. The new compounds were named the avermectins, and each was given a specific code according to their substitution patterns (Box 1).

The structures and activity of these compounds were first reported in 1979, and, as early as the late 1980s, the avermectins and their offspring had become great successes in the agrochemical industry. They were being used extensively to control insect pests in greenhouses and to treat a range of internal and external parasites in livestock, as well as in commercial cockroach and ant baits. In addition to their originally discovered anti-nematode actions, the avermectins are also efficient general insecticides, with effective acaricidal (anti-mite) activity. Crucially, the avermectins show almost no toxicity to humans. They can cause minor eye and skin irritations on exposure to high concentrations, but have no major effects associated with their ingestion. Thus, regulatory authorities require only a short time interval between the final application of the pesticide and the harvesting of the crop, a valuable feature for farmers. Today, avermectin-based agrochemicals can be found in a range of formulations, for various uses, carrying colorful trade names such as Affirm®, Agri-Mek®, Avid®, Vertimec®, and Zephyr®.

The avermectins exert their pesticidal activity by interfering with the neurotransmission of the invertebrate targets. They block receptors for the neurotransmitter GABA (γ-amino butyric acid) in the nerve synapses, leading to neuron malfunction and paralysis of the organism. This mode of action is particularly useful, as it remains effective against pests that have developed resistance to other common insecticides, such as the organophosphates and pyrethroids. Other advantages of the avermectins include their very low dosage requirement and long-lasting effects, negating the need for repetitive dosing. In plants, this endurance arises from the absorption of the avermectins through the leaves and into the veins. The pesticide then persists within the plant, providing long-term pest protection; just one low-dose application can provide protection against numerous pests. One disadvantage of these compounds is their toxicity to bees and fish. Their use must, therefore, be contained within appropriate environments.

Following the discovery of the avermectins, the effect of their various structural features on their biological activity was closely examined. For example, a rationale was sought as to why avermectin B_{1a} was more active than avermectin B_{2a} when administered orally to sheep that had been deliberately infected with helminthes (intestinal worms), but less active when administered by injection. The data from these studies indicated several possible structural modifications that might enhance the pesticidal activity of the parent compounds. Chemists then set about realizing these changes

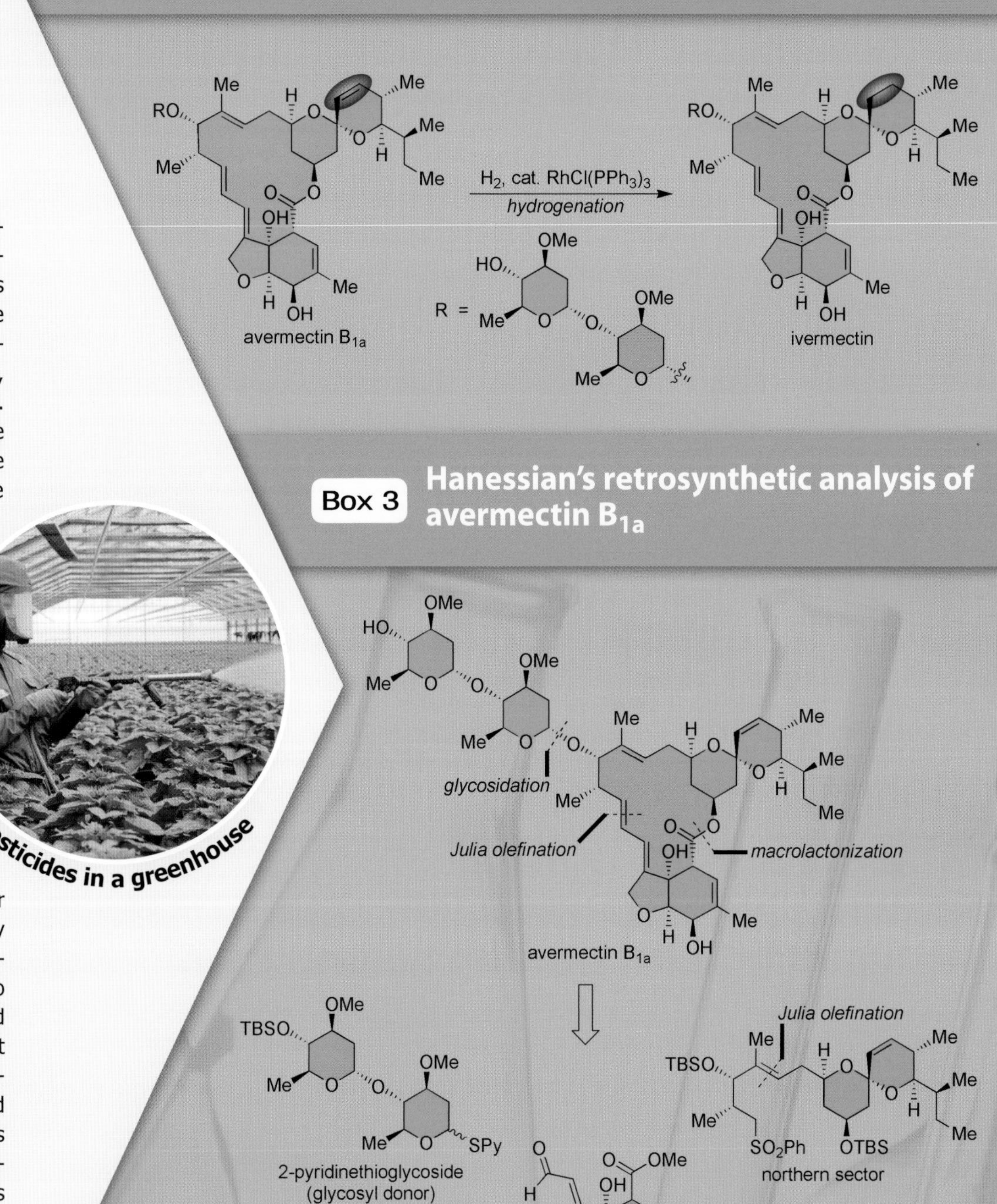

Pesticides in a greenhouse

Box 4 Hanessian's synthesis of the northern sector of avermectin B_{1a}

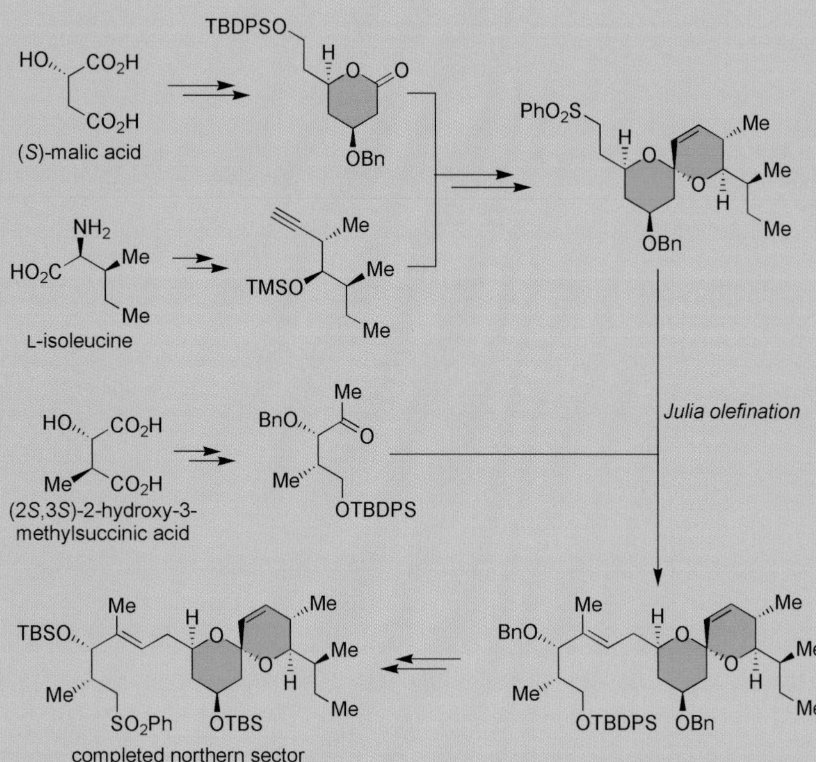

in the laboratory, using short sequences, or even single chemical reactions. Thus, a new series of compounds, named the ivermectins, was made available for testing. Some of these semi-synthetic molecules have now become very successful antiparasitic agents for treating crops and animals.

From a synthetic viewpoint, the avermectins excited chemists not only because structural modification by chemical synthesis held so much promise, but also because the 16-membered macrolide system, encompassing many functional groups, posed significant synthetic challenges. Stephen Hanessian and his group at the University of Montreal (Canada) published pioneering work on the avermectins, reporting a synthesis of avermectin B_{1a} in 1986. The Hanessian team had to rely on the degradation of naturally occurring avermectin B_{1a} to supply two pivotal building blocks for their synthesis (Box 3). In 1989, a group led by Samuel J. Danishefsky (then at Yale University, now at the Sloan–Kettering Institute, and Columbia University, New York, USA) accomplished the first total synthesis of avermectin A_{1a}, from scratch, so to speak, since they used small chemical building blocks as starting materials to assemble the molecule. Many other groups have since published work directed towards the synthesis of various avermectins. Here we shall confine ourselves to a brief discussion of the two pioneering adventures.

Streptomyces avermitilis

Hanessian's retrosynthetic analysis of avermectin B_{1a} severed the molecule into three key fragments, the spiroketal northern segment, the bicyclic southern section, and the disaccharide domain (Box 3). Both the southern sector and the disaccharide domain were obtained by degradation of the natural product, whilst the stereochemically complex northern fragment was synthesized in the laboratory beginning from several enantiomerically pure building blocks. Such starting materials may include hydroxy carboxylic acids and amino acids, such as malic acid and isoleucine, which were employed in this synthesis (Box 4), or naturally abundant carbohydrates and small terpenes. Collectively, these components, from which chemists can choose their favorite starting materials, are said to constitute the chiral pool. Hanessian has been a true pioneer in developing this approach to the total chemical synthesis of enantiomerically pure substances. This approach still complements the more modern strategy of employing asymmetric reactions to construct enantiomerically pure compounds. The process of constructing a molecule in the laboratory using large portions obtained from natural sources is known as semi-synthesis and a number of important drugs are produced in this way. For example, in the synthesis of β-lactam antibiotics, such as the penicillins and cephalosporins (Chapter 13), the central β-lactam portion is produced by the fermentation of molds, and the desired side chains are appended using chemical synthesis.

Having assembled the spiroketal segment, starting with their chiral pool building blocks, the Hanessian group proceeded to join it, through a Julia coupling reaction, to the bicyclic aldehyde system derived from degradation of the natural product (Box 5). This process led, upon appropriate elaboration, to the desired precursor for the anticipated macrocyclization step. This was accomplished through the use of DCC (*N,N'*-dicyclohexylcarbodiimide) and DMAP (4-dimethylaminopyridine), leading to the avermectin aglycon (the non-sugar part of the molecule). With the complete aglycon now in hand, the Hanessian group appended the disaccharide domain, which was also derived from natural material, and isomerized a double bond from a conjugated to a non-conjugated position. Finally, removal of the protecting groups (see Box 8) completed the first laboratory synthesis of avermectin B_{1a}. At this juncture, we should also note the early and influential studies of Hanessian in the field of oligosaccharide synthesis. His pioneering work on the use of 2-pyridinethioglycosides allowed him and others to construct glycosidic bonds effectively and with considerable stereochemical control.

Danishefsky's retrosynthetic analysis led him to a spiroketal moiety (the northern segment), a five-membered cyclic ketone representing the southern section, and a glycal precursor to the disaccharide domain. It is here that some of Danishefsky's most useful contributions to synthetic methodology were so elegantly applied. Dienes of the type used to prepare the sugar groups of the avermectins (Box 6) were developed within the Danishefsky group, and have come to be known as Danishefsky dienes. Many other groups have also used these useful Diels–Alder substrates in the synthesis of natural products. Their utility arises from their special reactivity, and the ease with which they undergo enantioselective cycloadditions under the influence of chiral auxiliaries and catalysts. Following the Diels–Alder reaction, the resulting enol derivative can be transformed easily into the synthetically versatile enone group. In the case of the avermectins, Danishefsky employed a chiral diene to prepare the enantiomerically pure sugar units, and a similar achiral diene to construct a portion of the aglycon.

Following the construction of the individual aglycon fragments, these intermediates were joined through the venerable aldol reaction (Box 7). In a daring strategy, the bicyclic unit was completed via a late-stage cyclization. It should be noted that the construction of this compact fused ring system had been causing serious problems for other groups

Box 5 Hanessian's synthesis of avermectin B_{1a}

DCC, DMAP
macrolactonization

avermectin B_{1a}

Box 6 Danishefsky's synthesis of the avermectin sugar unit

Danishefsky diene
(Aux* = 8-phenylmenthol
chiral auxiliary)

Lewis acid
catalyst
*hetero-
Diels–Alder
reaction*

CF_3CO_2H

Samuel J. Danishefsky

Box 7 Highlights of Danishefsky's total synthesis of avermectin A_{1a}

avermectin A_{1a}

attempting the synthesis of the avermectins. As in Hanessian's synthesis, closure of the macrocyclic ring and attachment of the sugar, along with some other necessary manipulations, completed the total synthesis of avermectin A_{1a}. The total synthesis of avermectin A_{1a} was a great accomplishment for synthetic organic chemistry, but as with several other synthetic targets that are available in large quantities from natural sources one may ask of what benefit it is, and is it worth the effort. The answer is a resounding yes, for the true value of such endeavors lies in the new chemical knowledge generated en route to the target molecules. For example, as a result of the quest for the avermectins, chemists developed a number of new ways to prepare carbohydrates, which are of broad importance to chemistry and biology. Additionally, the perfection of techniques for joining sugars to the avermectin aglycon added new synthetic technologies to the repertoire of glycoside-forming reactions and allowed investigators to carry out structure-activity relationship (SAR) studies within this class of molecules in search of more desirable biological properties.

The avermectin story does not end here; it takes one more twist in the form of a gratifying example of how effectively and profoundly human suffering can be alleviated through insightful scientific observation. In certain parts of tropical Latin America and in much larger areas of equatorial Africa, human populations are ravaged by a disease known as river blindness, or onchocerciasis. This terrible affliction is spread by the common blackfly, and can lead to a life of virtually indescribable pain and misery. The blackflies carry a microscopic parasitic worm called *Onchocerca volvulus* which wreaks havoc as it grows and proliferates within the victim's body. Avoiding the blackfly bite is just not possible; in some villages it has been recorded that one person may be bitten up to 10,000 times in a single day. Once the worm has entered the body, it replicates at an astonishing rate, producing millions of tiny offspring each year. The constant migration of these worms, whose numbers can reach 200 million in a single victim, through a person's skin causes intense irritation, itching, and disfigurement. When these invaders reach the eyes, lesions are formed that lead to the characteristic blindness that marks this dreadful disease. Adult worms do not cause any major debilitating effects directly, but instead they group together in large nodules that protrude through the victim's skin. Each adult female worm can grow up to two feet long and produces microscopic offspring throughout its fifteen-year lifespan. Until 1987 the primary treatment for this scourge was a medicine called DEC (diethylcarbamazine). Unfortunately, the awful side effects of this drug were almost as nightmarish as the infection itself, since its use frequently led to blindness or even the death of patients. Ivermectin was destined to revolutionize treatment of river blindness, thanks to the determination of the Merck team involved in its development.

Onchocerca volvulus

In 1978, a veterinary researcher at Merck tested ivermectin against gastrointestinal worms in horses. In this context, ivermectin proved to be very efficient in dealing with a class of worms called *Onchocerca cervicalis*. The same scientist realized that these worms were very similar to the terrible *Onchocerca volvulus*. With encouragement from an enlightened management, including the then CEO of Merck, P. Roy Vagelos, the group involved in this work began to explore whether ivermectin could be used in humans as a treatment for river blindness. After much detailed study and the long process of regulation, both of which are required before a new drug can be approved as being fit for use in humans, an ivermectin formulation called Mectizan®, was ready for use in the field. Mectizan® is an ideal drug in that just a single dose immobilizes the adult worms for a full year and prevents their reproduction whilst killing almost all of the microscopic offspring. Patients have to take the drug for fifteen years to cover the full lifecycle of the worm because the drug does not actually kill the adult worms, but with just one tablet per year, this regime is far from difficult to follow. In 1987, Merck undertook a huge humanitarian program under which they donated Mectizan®, free of charge, to the needy regions of Africa. Their generous gift added to a World Bank sponsored spraying scheme that began in 1972 (The Onchocerciasis Control Program) aimed at eliminating the black-

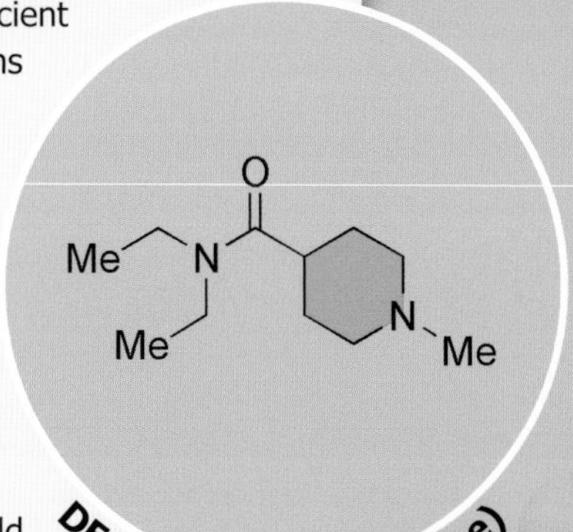

DEC (diethylcarbamazine)

P. Roy Vagelos

Box 8 Protecting groups in chemical synthesis

In the early days of natural product synthesis, most of the targets were fairly small molecules with only a few functional groups. As synthesis and chemistry as a whole developed, larger and more complex natural products were discovered and synthesized. With the increasing complexity and size of the targets, the challenge of reacting one domain of the growing molecule selectively, without damaging other parts of the structure, becomes a serious issue. Sometimes chemists can take advantage of inherent differences in the reactivity of various functional groups; however, when the differences are small, or when the less reactive group must be transformed, it becomes necessary to mask temporarily the reactivity of another group within the molecule. This can be achieved through the use of so-called 'protecting groups'. This term covers a wide range of different structural moieties used in various ways to protect different compound types. Some of the most common types are shown below. Many different groups have been developed to allow the selective manipulation of a number of similar functional groups within a single compound. If two or more similar structural motifs are protected with complementary groups, each can be manipulated at will, at least in theory. The concept of using various groups with complementary introduction and removal conditions is known as orthogonality, and protecting groups lie within orthogonal sets. There is a degree of overlap between sets, which can be problematic or, in certain cases, useful.

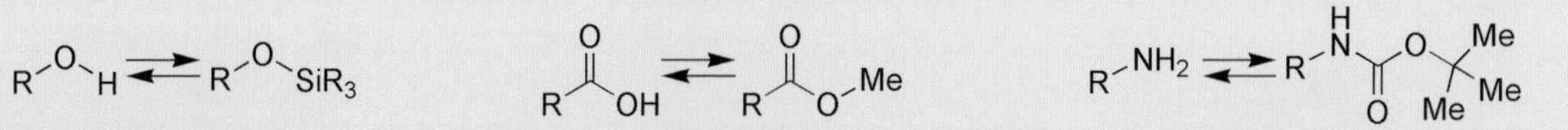

silyl ether (alcohol protection)

ester (carboxylic acid protection)

carbamate (amine protection)

The choice of protecting group used depends on many factors. The group should be easily and efficiently introduced, and it must be stable to all the reaction conditions used prior to its removal. The removal of the protecting group should ideally be accomplished using mild and highly selective conditions in order to avoid damaging the carefully constructed molecule. If met, these criteria minimize the impact of the two extra synthetic steps required whenever a protecting group is used. The correct choice of protecting group strategy can have a marked influence on the efficiency, elegance, and even success of a synthesis.

fly larvae. With the employment of this dual strategy, life soon began to return to normal in the previously devastated villages. In 2002, The Onchocerciasis Control Program was terminated, leaving Mectizan® as the only defense against river blindness. Thankfully, this miracle drug continues to bring hope and progress to communities where none had previously existed.

Further Reading

T. W. Miller, L. Chaiet, D. J. Cole, L. J. Cole, J. E. Flor, R. T. Goegelman, V. P. Gullo, H. Joshua, A. J. Kempf, W. R. Krellwitz, R. L. Monaghan, R. E. Ormond, K. E. Wilson, G. Albers-Schönberg, I. Putter, Avermectins, New Family of Potent Anthelmintic Agents: Isolation and Chromatographic Properties, *Antimicrob. Agents Chemother.* **1979**, *15*, 368–371.

J. C. Chabala, H. Mrozik, R. L. Tolman, P. Eskola, A. Lusi, L. H. Peterson, M. F. Woods, M. H. Fisher, W. C. Campbell, J. R. Egerton, D. A. Ostlind, Ivermectin, a New Broad-Spectrum Antiparasitic Agent, *J. Med. Chem.* **1980**, *23*, 1134–1136.

S. Danishefsky, Siloxy Dienes in Total Synthesis, *Acc. Chem. Res.* **1981**, *14*, 400–406.

S. Hanessian, A. Ugolini, P. J. Hodges, P. Beaulieu, D. Dubé, C. André, Progress in Natural Product Chemistry by the Chiron and Related Approaches – Synthesis of Avermectin B_{1a}, *Pure Appl. Chem.* **1987**, *59*, 299–316.

S. J. Danishefsky, D. M. Armistead, F. E. Wincott, H. G. Selnick, R. Hungate, The Total Synthesis of Avermectin A_{1a}, *J. Am. Chem. Soc.* **1989**, *111*, 2967–2980.

Mectizan® website (access date 2/07): http://www.mectizan.org.

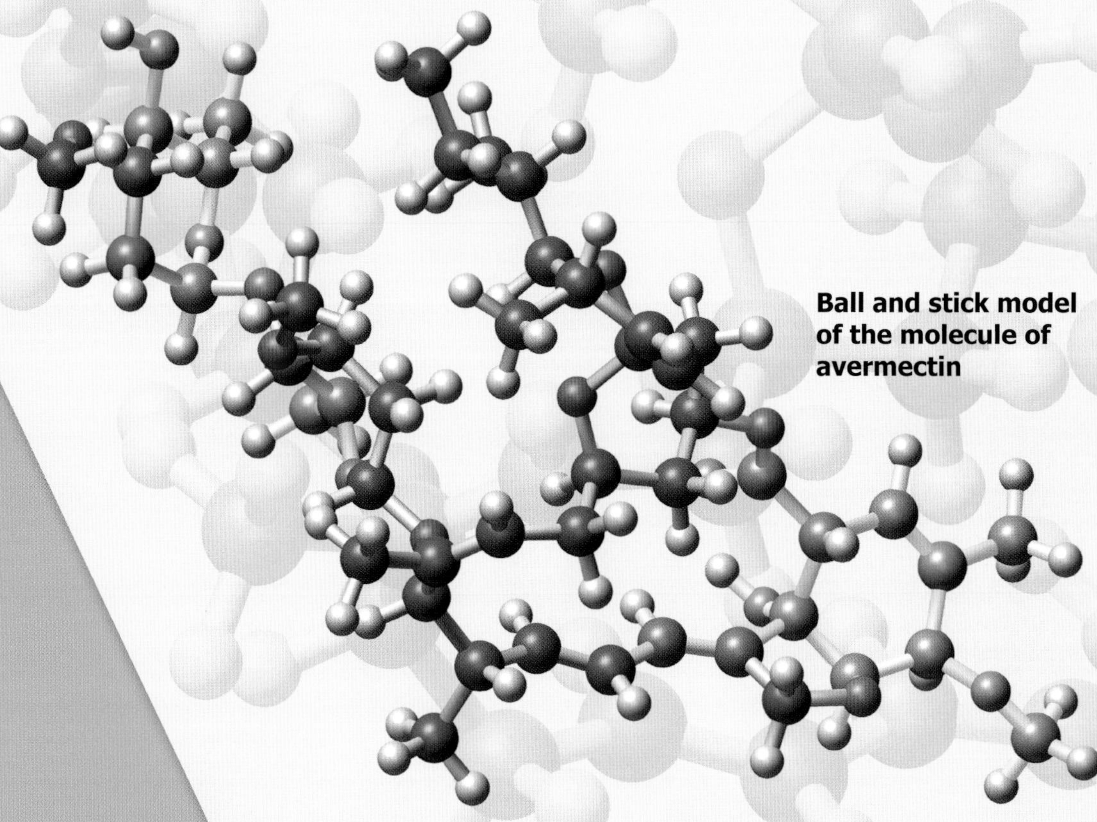

Ball and stick model of the molecule of avermectin

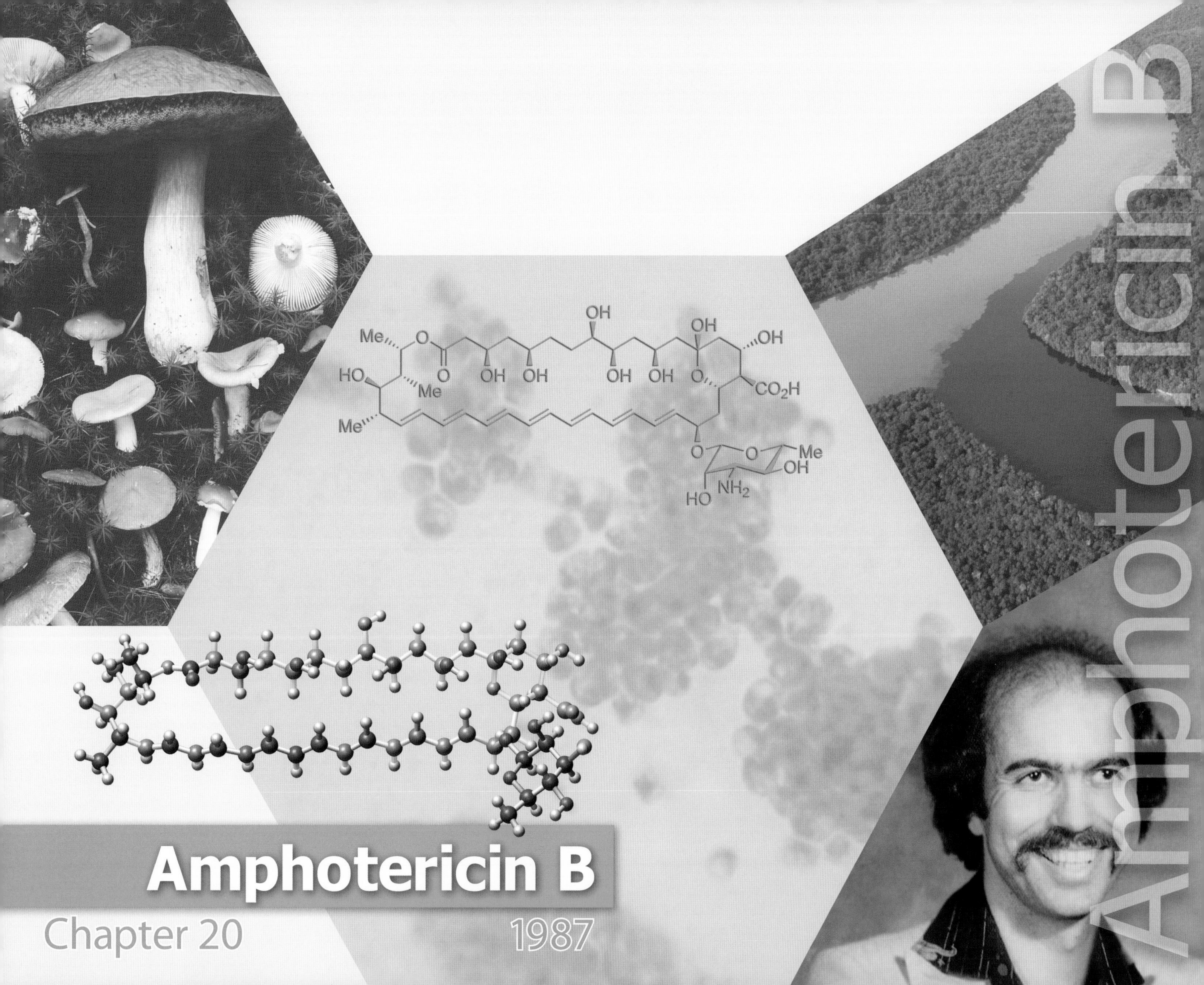
Amphotericin B
Chapter 20
1987
Amphotericin B

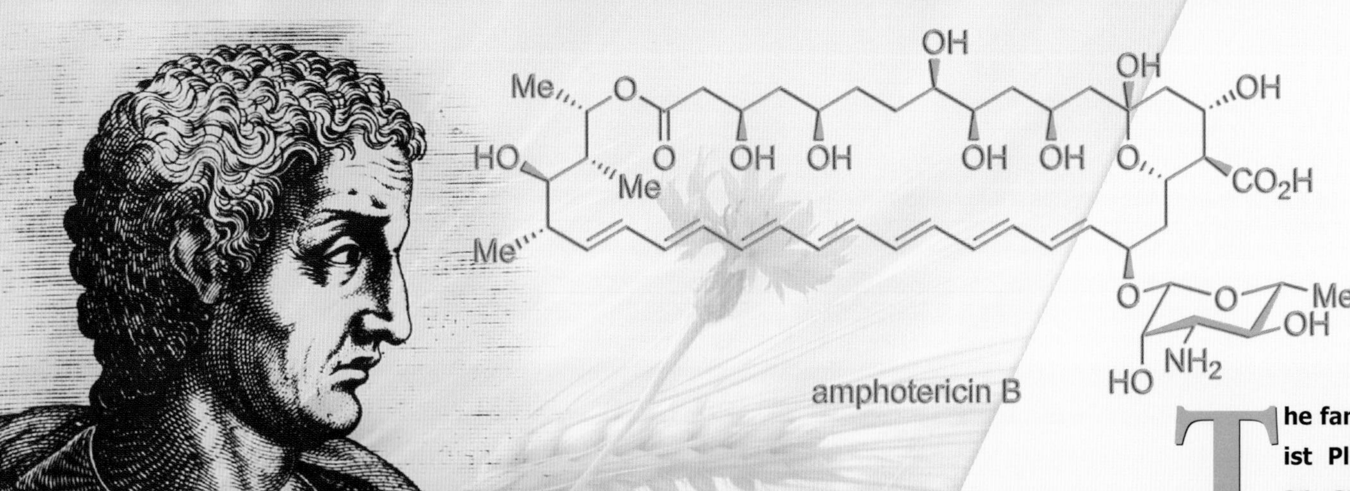

amphotericin B

Pliny the Elder

The famous Roman naturalist Pliny the Elder wrote an authoritative account of his observations of the world around him, within which he described the mushroom as having mystical intrigue. To his mind, it was the greatest miracle of nature, because he could not understand how it survived when it bore no discernible network of roots. Fungi, which include mushrooms, molds, yeasts, and many other organisms, constitute a rather curious group of life forms. They are neither plants, nor animals, but are classified in their own separate kingdom, defined by their metabolic features and reproductive mechanisms. Despite their strangeness, fungi have been an integral part of our lives since the dawn of civilization, particularly in food production. For example, the rising of bread is caused by the carbon dioxide released as the yeast ferments the sugars present in the dough. Similarly, the ethanol present in alcoholic beverages is the metabolic by-product of another type of yeast. The distinctive flavors of many cheeses are the result of fungal metabolism. The French Camembert and Roquefort and the English Stilton cheeses develop their flavors by hosting colonies of *Penicillium camembertii* and *Penicillium roquefortii*, respectively, with the latter species being associated with both of the blue cheeses.

Of course, by no means are all fungi beneficial to us, and many exhibit detrimental properties. Some produce powerful toxins, such as the aflatoxins and trichothecenes, or hallucinogenic compounds (Box 1). *Amanita muscaria*, the classic red and white mushroom featured in children's stories, has been used (or perhaps abused) as an intoxicant in Siberia and other parts of Asia for thousands of years. However, hallucinations brought about by fungi are not always deliberately sought, and many historical accounts of accidental poisonings can be found. The madness that led to the infamous persecution and witch trials

Box 1 Unpleasant and dangerous compounds made by fungi

aflatoxin B_1
(made by *Aspergillus* molds found in peanuts)

trichothecolone
(made by *Fusarium* molds found in bad beer)

lysergic acid - precursor of LSD
(made by *Claviceps purpurea* found on grain)

Camembert and Roquefort cheeses

of Salem (Massachusetts, USA) in 1692 have been attributed to a mold growing on rye grain, a staple of the diet at the time. Modern-day investigators have found evidence that the weather during that year would have provided ideal conditions for the growth of *Claviceps purpura* on the rye crop. This organism produces ergot alkaloids, such as lysergic acid (Box 1), which are related to the modern illicit hallucinogen LSD (lysergic acid diethylamide). When ingested, these chemicals cause delirium, convulsions, and trances accompanied by strange and incomprehensible speech emanating from the victim of the ergot poisoning. Young women are particularly susceptible to these symptoms, which may have led the people of Salem to believe the poisoned girls were possessed with evil spirits.

In addition to causing a number of diseases, fungi can also help in curing them. Many fungi live in crowded environments with tough competition for resources, and often produce chemical weapons to give them a competitive edge over their neighbors. Thus, fungi produce a huge range of natural products, some of which have been harnessed as life-saving drugs, including penicillin (Chapter 13) and Mevacor® (Chapter 26). Likewise, bacteria produce their own chemical weapons, which they employ to attack other bacteria or invading fungal colonies. Some of these compounds can also be used as drugs. An antifungal metabolite produced by bacteria, amphotericin B, is one such compound, which is now used as a medicine to treat fungal infections.

Fungal infections of our bodies, known as mycoses, fall into one of two categories; they can be either systemic, or superficial. Systemic infections penetrate deep into organs and tissues while superficial infections attack membranes (e.g. lungs) and surfaces (e.g. skin, hair, nails). Both of these types of infection were once rare. Hippocrates (460–377 B.C.) recorded a case of oral *Candidiasis* infection (thrush) but, after him, fungal infections received little attention in medical writings. An edition of the rather morbid journal *Disease and Casualties of the Week*, published in London in September 1665, contains the earliest records of deaths arising from fungal infection. Six deaths were attributed to the fungal disease thrush; however, the bacterial infections that cause consumption (tuberculosis) and the plague killed 6,673 people in the same week.

Recent decades have witnessed a continuous rise in fungal infections, for a number of reasons. One factor is the widespread use of broad-spectrum antibacterial agents in human or veterinary health and in agriculture. This practice has severely decreased populations of non-pathogenic bacteria that, under normal conditions, compete with and, there-

Hippocrates

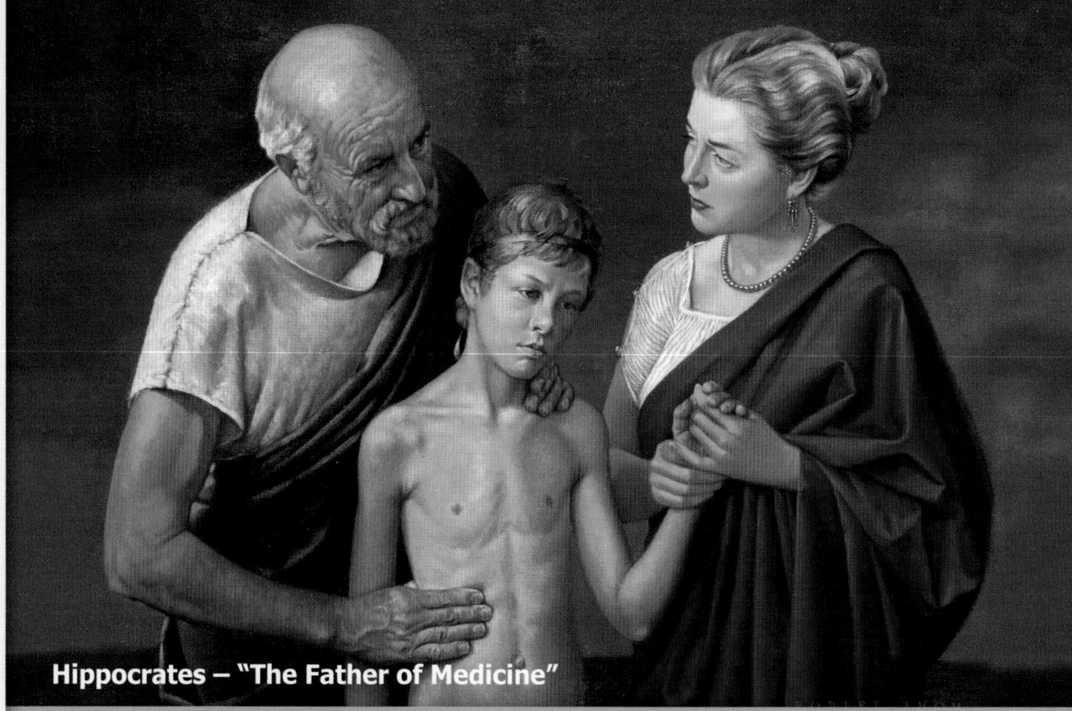

Hippocrates – "The Father of Medicine"

Box 2 Selected antifungal agents

griseofulvin [Grifulvin®, Fulvicin®] (natural product - treats tinea (ringworm) infections)

fluconazole [Diflucan®] (synthetic compound - treats *Candidiasis* infections (thrush))

terbinafine hydrochloride [Lamisil®] (synthetic compound - treats onychomycosis (nail bed) infections)

flucytosine [Ancobon®] (synthetic compound - treats systemic cryptococcosis (meningitis), mucormycosis (sinus/brain/lung infections), and *Candidiasis* infections)

itraconazole [Sporanox®] (synthetic compound - treats onychomycosis infections)

Box 3 The chemical structure of nystatin A_1

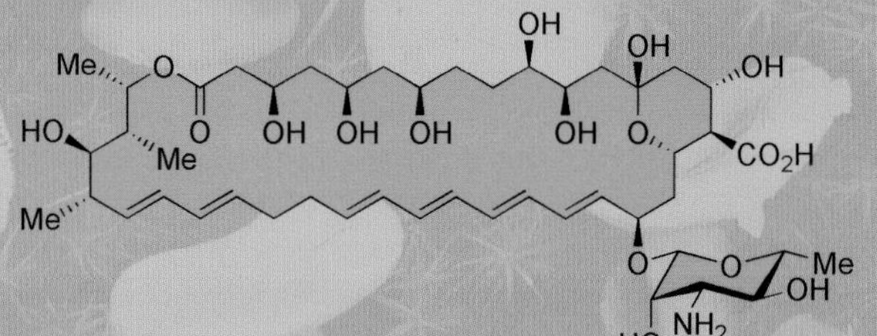

nystatin A_1

Box 4 A summary of the Nicolaou team's retrosynthetic analysis of amphotericin B

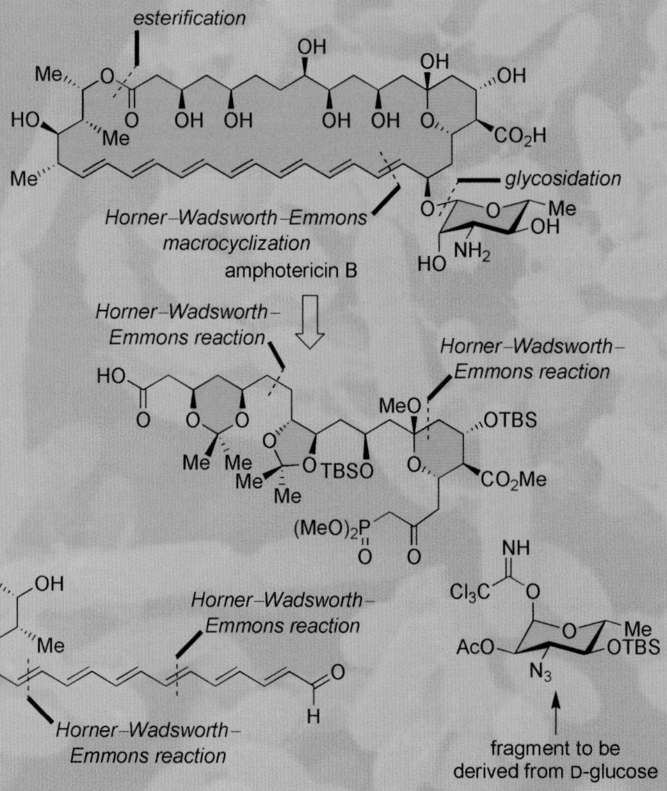

esterification

glycosidation

Horner–Wadsworth–Emmons macrocyclization

amphotericin B

Horner–Wadsworth–Emmons reaction

Horner–Wadsworth–Emmons reaction

Horner–Wadsworth–Emmons reaction

Horner–Wadsworth–Emmons reaction

fragment to be derived from D-glucose

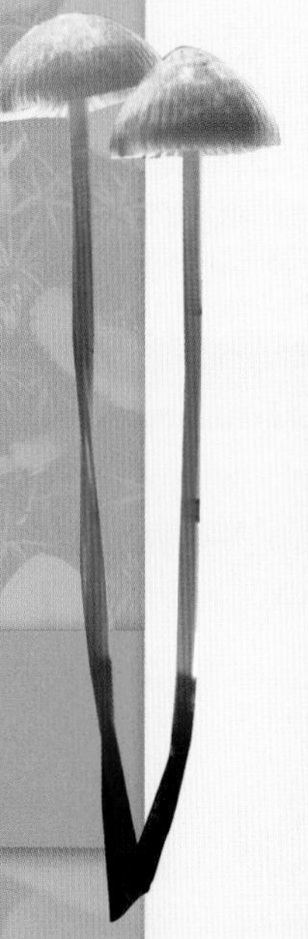

fore, control fungi. A second factor is the increase in the number of people living with compromised immune systems. These include transplant patients, whose immune systems must be suppressed with drugs to prevent rejection of the transplanted organ, patients undergoing aggressive cancer treatments, such as chemotherapy and radiotherapy, and the large number of individuals carrying the human immunodeficiency virus (HIV). Weakened immune systems leave people vulnerable to attack by opportunistic fungi that would otherwise be fought off without causing disease. A diverse group of antifungal compounds, both fungistatic agents (which prevent the growth of fungi) and fungicidal agents (which kill fungi), have been developed in order to combat these infections (Box 2). The inspiration for these drugs has come from a variety of sources. However, when measured against the broad range and comparative safety of antibiotics, the collection of antifungal medications remains rather small, and is compromised by greater risks and a higher incidence of side effects. Fungal cells are more advanced than bacteria, at least in evolutionary terms. They are defined as being eukaryotic, meaning that they have a nucleus that contains their genetic material. As such, they have more similarities to mammalian cells and their metabolism is more complex than the more primitive prokaryotic bacterial cells. This has made the design and discovery of potent and selective drugs to tackle detrimental fungi in many ways a much more challenging task than the development of antibacterial agents.

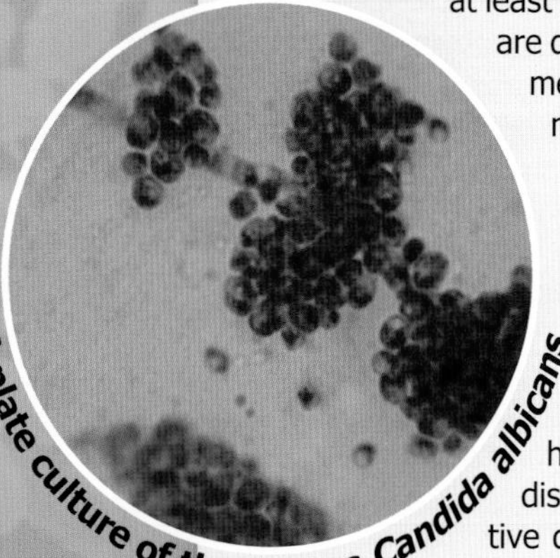

A plate culture of the fungus *Candida albicans*

Today, a number of designed antifungal agents strengthen our armament and provide comfort in our fight against fungi. These include a wide variety of compounds, and are used to treat various types of fungal infection, both systemic and superficial. Examples include Diflucan®, Lamisil®, Ancobon® and Sporanax® (Box 2).

In the 1930s, scientists began to search for natural products that might be employed as antifungal agents. They began with microbial broths, which were also providing leads in other therapeutic areas. The first success was griseofulvin (Box 2), initially isolated in 1939 from the mold *Penicillium griseofulvum*, a close relative of the organism that first gave us penicillin (*Penicillium notatum*). The structure was not known at that time, but the compound was re-isolated and more fully investigated in 1947. Griseofulvin is a microtubule-binding fungistatic agent that is still in use as a drug in special cases today (marketed under the trade names Grifulvin® and Fulvicin®), more than sixty years after it was first discovered. Unfortunately, it has a very narrow spectrum of activity, meaning that it is effective against only a few select fungal infections. In 1951, at the laboratories of the New York State Institute for Health (USA), scientists isolated another molecule, containing a giant macrolactone ring and a long chain of conjugated double bonds, from a *Streptomyces* broth. The researchers involved discovered that this molecule exhibited fungicidal activity, hence its first name, fungicidin. The compound was subsequently developed as a drug and renamed nystatin, after the institute where it was originally discovered (Box 3). Nystatin, like griseofulvin, is still in use today, but due to exceptionally poor absorption over mucous membranes its use is limited to treating topical skin or gastrointestinal tract infections. In 1956, a second polyene antibiotic was isolated from a *Streptomyces nodosus* broth cultured using sediments collected from a Venezuelan riverbed. Bearing structural similarities to nystatin, the new compound was named amphotericin B because of its amphoteric nature. A molecule is amphoteric when it can behave either as an acid or as a base; amphotericin contains both a basic amine and an acidic carboxylic acid group. Amphotericin B, like nystatin, is not well absorbed from the lining of the gut and must therefore be injected in order to treat systemic infections effectively. Once in the blood it becomes protein-bound and takes some time to be excreted from the body. It slowly accumulates in the kidneys, causing renal toxicity, its most common unwanted side effect.

These polyene macrolide antibiotics work by binding to cell membranes and then interfering with their permeability and transport functions. Their lipophilic polyene section associates with the ergosterol clusters present in fungal cell membranes such that the hydrophilic hydroxyl-bearing domain creates a pore through which an unregulated and catastrophic loss of potassium ions occurs. Amphotericin B, the most prominent member of the ever expanding, polyene antibiotic class, exhibits some selectivity in its action, being most active against fungi and protozoa. It has a lower affinity for mammalian cells and virtually none for bacterial cells. This selectivity arises from the different sterol present in each of the corresponding cell membranes; all the polyene antibiotics preferentially bind the fungal sterol ergosterol, but can only form a weaker complex with the mammalian sterol cholesterol. This lower affinity for cholesterol is not sufficient to prevent nystatin and, to a marginally lesser extent, amphotericin B having a relatively high toxicity, which is only tolerated in their medical applications because they are amongst the most effective antifungal agents known. Sophisticated drug formulations have recently succeeded in reducing these problems. Formulations of amphotericin B with lipids, such as Amphocil® and Abecelet®, increase the solubility, stability, and absorption of the drug, while a new innovation that encapsulates amphotericin B in liposomes (AmBiosome®) has been shown to reduce its toxicity quite significantly. The intensity of work directed towards developing these adaptations reflects the limited clinical choices available to treat fungal infections and indicates that amphotericin B will be clinically important for some time to come.

The chemical structure of amphotericin B presents a daunting prospect to any synthetic chemist charged with synthesizing this relatively large 36-membered macrocylic lactone. Undeterred by this challenge, K. C. Nicolaou and his team, then at the University of Pennsylvania, (Philadelphia, USA), undertook the synthesis of this important compound in the early 1980s. By 1987, they were able to announce the first total synthesis of amphotericin B. To date,

Box 5 Preparation of key intermediates using asymmetric epoxidation and chiral pool approaches

Sharpless asymmetric epoxidation

D-xylose

L-xylose

K. C. Nicolaou in laboratory at the University of Pennsylvania

Box 6 The completion of the total synthesis of amphotericin B

DBU, LiCl

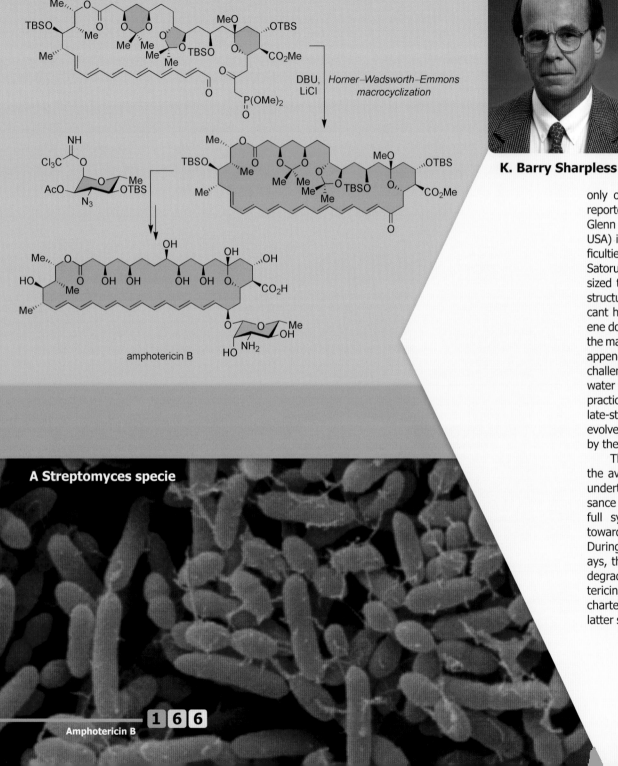

Horner–Wadsworth–Emmons macrocyclization

amphotericin B

A Streptomyces specie

K. Barry Sharpless

Tsutomu Katsuki

only one other total synthesis has been reported, emanating from the group of Glenn J. McGarvey (University of Virginia, USA) in 1996, illustrating the inherent difficulties of a project of this nature. In 1988, Satoru Masamune and his group synthesized the aglycon of amphotericin B. The structure amphotericin B presents significant hurdles, including the sensitive polyene domain, the fourteen stereocenters on the main lactone ring, and the amino-sugar appendage. Added to the purely synthetic challenges, the compound is insoluble in water and unstable at 37 °C, presenting practical problems for the manipulation of late-stage intermediates. The final, much evolved retrosynthetic strategy employed by the Nicolaou team is shown in Box 4.

The Nicolaou team took advantage of the availability of the target molecule by undertaking several detailed reconnaissance missions before embarking on the full synthetic adventure towards amphotericin B. During these initial forays, they derivatized and degraded natural amphotericin B, and, by doing so, charted strategies for the latter stages of the anticipated total synthesis. Due to space limitations, only a few highlights are discussed here. The retrosynthetic analysis reveals several key building blocks and a hidden symmetry within the target structure. This allowed much of the northern portion of amphotericin B to be constructed from the two enantiomers of the sugar xylose (Box 5). This chiral pool approach in which the stereochemistry of the target is derived from a naturally occurring chiral starting material was complemented by an asymmetric synthesis approach, whereby the same building blocks were constructed from an achiral precursor through asymmetric catalysis. In this case, the Nicolaou team made use of the Sharpless asymmetric epoxidation reaction to prepare two of the required fragments. This process was developed by K. Barry Sharpless and his associate Tsutomu Katsuki, then at the Massachusetts Institute of Technology (MIT, Cambridge, USA), and employs a chiral catalyst derived from diethyl tartrate to epoxidize a prochiral allylic alcohol. The epoxy alcohol products from this reaction are formed essentially as single enantiomers, and the reaction has been used in countless total synthesis campaigns as well as several industrial processes for the production of drugs. Sharpless was awarded a share of the 2001 Nobel Prize in Chemistry for his contribution to asymmetric catalysis (see also Chapter 26). In the total synthesis of amphotericin B, diethyl tartrate served as a chiral ligand and also as a chiral pool

Glenn J. McGarvey

Satoru Masamune

Box 7 The Wittig and Horner–Wadsworth–Emmons olefination reactions

The formation of carbon–carbon double bonds (olefins or alkenes) is of great importance within organic chemistry, as exemplified by the total synthesis of amphotericin B. In the early days of chemical synthesis, methods for the construction of the olefinic bond relied heavily on elimination and dehydration reactions. However, the poor selectivity of these reactions often led to mixtures of products. The Wittig reaction, first reported in 1953 by Georg Wittig and his co-worker, Georg Geissler, allowed such linkages to be installed with complete regioselectivity and often with good control over the geometry of the newly formed double bond.

Georg Wittig was an accomplished chemist who developed many reactions over the span of his eminent career. He spent the major part of his scientific career at the University of Heidelberg, in the Federal Republic of Germany. In 1979, he was awarded the Nobel Prize in Chemistry (jointly with Herbert C. Brown, see Chapter 18) for his development of phosphorus-containing reagents for organic synthesis.

The Wittig reaction (he was also responsible for developing the unrelated Wittig rearrangement) makes use of organo-phosphorus ylides. These chemical species can be represented with paired charges or with a carbon–phosphorus double bond. Phosphorus ylides react with aldehydes or ketones to form an alkene and a phosphine oxide. The strength of the phosphorus–oxygen double bond provides a strong driving force for this reaction. By the 1960s, the Wittig process was already being used industrially by the German company BASF in the production of vitamin A. Since then, this reaction has become a key transformation in organic synthesis.

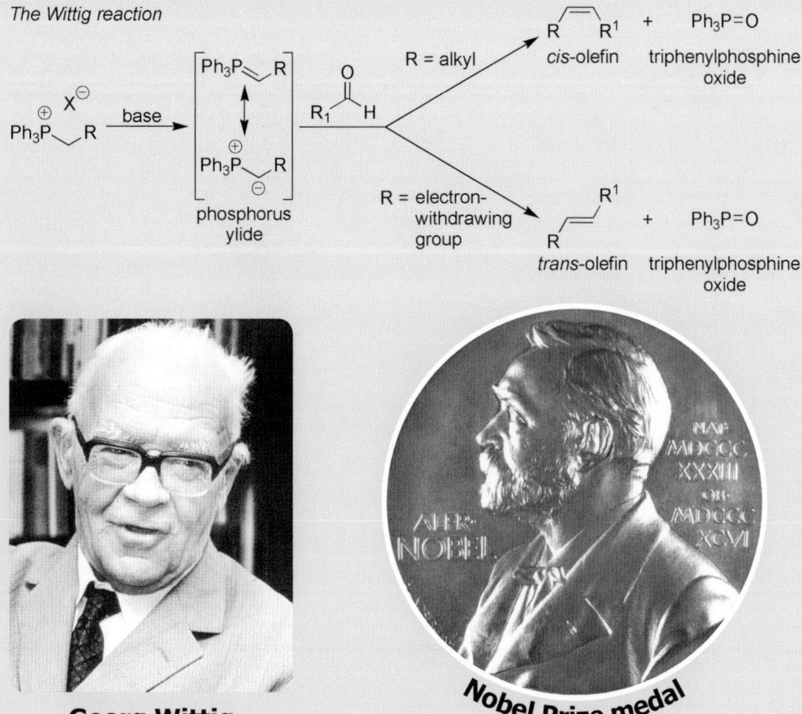

Georg Wittig

Nobel Prize medal

Leopold Horner

In 1958, Leopold Horner reported the first olefination reactions using alkyl phosphonates. Three years later, two American chemists, William S. Wadsworth, Jr., and William D. Emmons, developed this process further, providing a generally applicable variant of the Wittig reaction. The Horner–Wadsworth–Emmons (HWE) reaction, as it is now widely known, has the advantage that the by-product is not a phosphine oxide, but a water-soluble phosphate, which is easily removed from the alkene product. The two processes are complementary in that simple alkyl phosphonates give *trans*-alkenes selectively, while the analogous Wittig reactions give predominantly *cis*-products. Furthermore, modifications to the phosphorus substituents and the reaction conditions permit almost complete control of the product geometry in both reaction variants.

The Horner–Wadsworth–Emmons reaction

base

trans-olefin + water-soluble phosphate by-product

EWG = electron withdrawing group (e.g. CO_2Et, CN, C(=O)R)

Sculpture of Benjamin Franklin at the University of Pennsylvania

starting material for the construction of two other fragments.

With the various fragments in hand, the Nicolaou team began to stitch the molecule together, making use of another important synthetic process, the Horner–Wadsworth–Emmons (HWE) olefination reaction (Box 6). This reaction, like the related Wittig reaction (see Box 7), provides a mild means to create carbon–carbon double bonds. In the total synthesis of amphotericin B, this reaction was employed to unite the two fragments of the northern domain of the target and to construct its sensitive polyene segment. The two halves of the target structure, the polyene and the protected polyhydroxyl domains, were then joined by an esterification, setting the stage for the crucial ring-closing step. Once again, the Horner–Wadsworth–Emmons reaction was used, in this case in an intramolecular sense, providing the desired 36-membered ring without compromising the delicate structure of the molecule. Following this key reaction, reduction of the carbonyl group, coupling of the sugar to the generated core, and a few deprotection steps provided synthetic amphotericin B. The completion of the total synthesis of this complex antifungal agent marked a key event in the art and science of total synthesis, pioneering the use of asymmetric reactions and ketophosphonate–aldehyde coupling reactions in the assembly of complex molecular architectures, practices that were to be adopted and exploited in many syntheses to come.

Ball and stick model of the molecule of amphotericin B

Further Reading

R. S. Baldwin, *Fungus Fighters: Two Women Scientists and Their Discovery*, Cornell University Press, New York, **1981**.

K. C. Nicolaou, R. A. Daines, T. K. Chakraborty, Y. Ogawa, Total Synthesis of Amphotericin B, *J. Am. Chem. Soc.* **1987**, *109*, 2821–2822.

K. C. Nicolaou, E. J. Sorensen, *Classics in Total Synthesis*, Wiley-VCH, Weinheim, **1996**, pp. 421–450.

J. M. Fostel, P. A. Lartey, Emerging Novel Antifungal Agents, *Drug Discov. Today* **2000**, *5*, 25–32.

Ginkgolide B

Chapter 21

1988

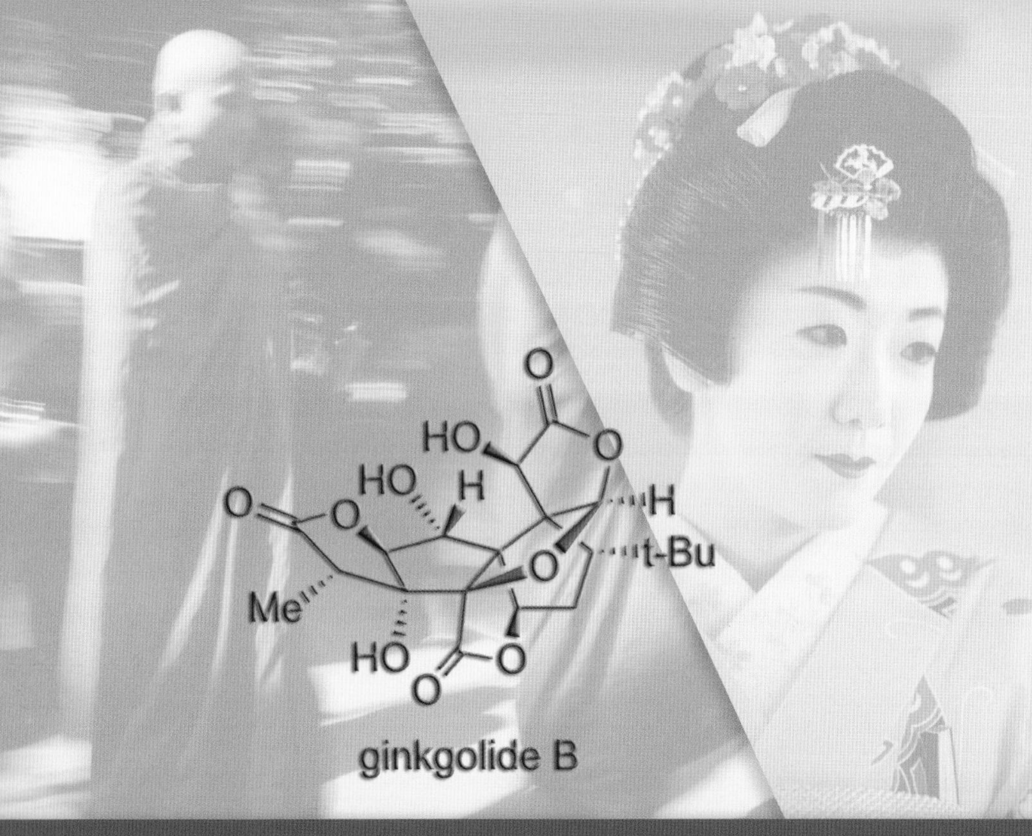

Box 1 ***"Ginkgo biloba"*, by the German philosopher-poet Johann Wolfgang von Goethe**

"This leaf from a tree in the East
Has been given to my garden.
It reveals a certain secret,
Which pleases me and thoughtful people.
Does it represent One living creature
Which has divided itself?
Or are these Two, which have decided
That they should be as one?
To reply to such a question,
I found the right answer:
Do you notice in my songs and verses
That I am One and Two"

Johann Wolfgang von Goethe

"It appeals to the historic soul: we see it as an emblem of changelessness, a heritage from worlds too remote for our intelligence to grasp, a tree which has in its keeping the secrets of the immeasurable past."

Sir Arthur Seward
(Paleobotanist, 1938)

Aside from the poetry and philosophy for which he is better known, Johann Wolfgang von Goethe also had a predilection for the sciences, especially the study of botany. Whilst pursuing this interest he encountered perhaps the world's most intriguing tree, the *ginkgo biloba*, and was inspired by its unique double-lobed leaves to write the metaphoric poem of the same name in 1815 (Box 1). The *ginkgo biloba* tree has captured the interest of philosophers, poets, and painters alike through the ages due to its resilience, longevity, and many varied applications, both past and present, across many cultures.

The ginkgo, or maidenhair, tree may well be the oldest living seed-plant species. It is the sole survivor from an order of plants called the Ginkgoales, whose ancestry can be traced back over 250 million years to a time when the first dinosaurs were just starting to roam our planet. In the moist and warm climate at the beginning of the Mesozoic era, the ginkgo tree flourished and specimens covered the entire surface of the earth's single super continent, Pangea. Later, geological cataclysms and an ice age eradicated it from the now separated landmasses of North America and Europe, so that it survived only in the remotest regions of Asia. The history of this group of species can be traced by studying and dating the fossilized leaves from ginkgo trees which have been found distributed across the world. On an individual level, this tree possesses an unparalleled endurance, as the ginkgo life span may easily encompass 1,000 years. It is this trait that lies behind the deep reverence paid to the tree by Buddhist and Taoist monks, who have been planting it in their temple grounds since the eleventh century. In so doing, they encouraged and facilitated the proliferation of the ginkgo out of China to Japan and Korea, where it became an intrinsic part of these cultures. A curious example of the pervasive influence of the ginkgo can be seen in the formal hairstyles worn in Japan by sumo wrestlers, samurai warriors, and brides, all of whom have a topknot beautifully arranged in the shape of a ginkgo leaf.

Charles Darwin

During the Renaissance period, European society pored over new objects brought back from around the globe by the many explorers of the

time. Engelbert Kaempfer, a German physician and botanist, was responsible for reintroducing the ginkgo to Europe after returning from an expedition to Japan (1690–1692) funded by the Dutch East-India Company. He planted seeds in gardens in Utrecht (Holland), and from there the tree rose to fame and became the darling of plant collectors across Europe, such that every respectable botanical institute vied to have a ginkgo tree as part of its collection. Charles Darwin, the great naturalist, is reputed to have been mesmerized by the ginkgo tree, aptly describing it as a "living fossil" in 1859 because the species had not changed over so many millennia.

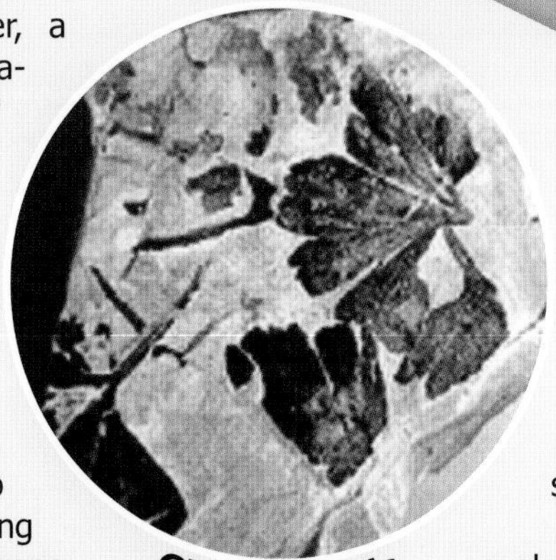

Ginkgo fossils

The hardy resilience evolved by the ginkgo tree has allowed it to evade destruction even in the most extreme situations. In 1923, a terrible earthquake shook Tokyo, causing a great fire to rage through the city, destroying everything in its path. However, one temple survived because it was surrounded by ginkgo trees which secreted a chemical from their bark and leaves that acted as a fire retardant, protecting the buildings inside from the inferno. Furthermore, from the darkest moment in Japan's history, we see once again the immutability of the ginkgo tree. One kilometer from the center of the atomic explosion in Hiroshima on August 6, 1945, a ginkgo tree stood as the only surviving structure in the days that followed this terrible event. The temple within which it had been planted was, for the most part, razed to the ground, yet soon thereafter the ginkgo tree was again in bud, sprouting and flowering as a sign of hope and regeneration in the devastated city. The ginkgo tree can also withstand temperature extremes and environmental toxins, and it appears not to be susceptible to attack from any known insect or microbe. Today, this demonstrated resistance makes the ginkgo tree a favored choice to line the streets in some of the busiest and most polluted urban environments.

Ginkgo tree (Hiroshima, 1945)

Ginkgo tree on the campus of Tohoku University, Sendai (Japan). The tiny figure to the right of the tree is Koji Nakanishi.

Koji Nakanishi

Box 2 Some of the compounds isolated from *ginkgo biloba* tree extracts

ginkgolide A : X = H, Y = H, Z = OH
ginkgolide C : X = OH, Y = OH, Z = OH
ginkgolide J : X = H, Y = OH, Z = OH

ginkgetin

quercetin derivative

bilobalide

In Asia, particularly China, the fruit of the ginkgo has been a delicacy consumed at banquets ever since the Sung Dynasty (960–1279 A.D.). Indeed, one of the Chinese names for the ginkgo tree is *Yin hsing*, which translates to silver apricot. The ginkgo tree is a dioecious species, meaning that it has separate male and female trees, the female trees being much less popular as ornamental plants because the grayish fruit they bear has a notoriously foul odor. The reason this foul-smelling item found its way into the most impressive feasts in China was the belief in its medicinal properties. Ginkgo fruit was believed to aid digestion, relieve diarrhea, and prevent drunkenness, thus tempering the effects of overindulgence. In Japan, the fruits are also used to soothe digestive upsets and moderate the effects of alcohol, frequently being served with saki. Documentation of these effects in Chinese writings is alleged to date back almost five thousand years (2,700 B.C.); although, the ancient texts no longer exist so confirmation is not possible. What is certain is that traditional Chinese medicine has long embraced the use of ginkgo constituents (roots, bark, leaves, and fruits) to treat a wide variety of ailments. Two thousand years ago, in the midst of the Han Dynasty, ginkgo leaves were promoted as a means to improve blood circulation and lung function. In the great herbal tome, *Pen ts'ao kang mu*, compiled by Li Shih-Chen in 1578, the seeds (*baigo*) are recommended for the treatment of a disparate collection of maladies, including asthma, coughs, bladder irritability, blennorrhea, and uterine cancer. Furthermore, the influence of *ginkgo biloba* in ancient Asiatic medicinal rites stretches as far as the Indian sub-continent, where Vedic writings inform us that it is also an ingredient of the Hindu rejuvenating tonic called *soma*.

Ginkgo fossil

Modern Western medicine, practiced from an entirely alternate perspective when compared to its elder Chinese

Ginkgo female leaves and nuts; ginkgolides are contained in leaves, nuts, and root bark.

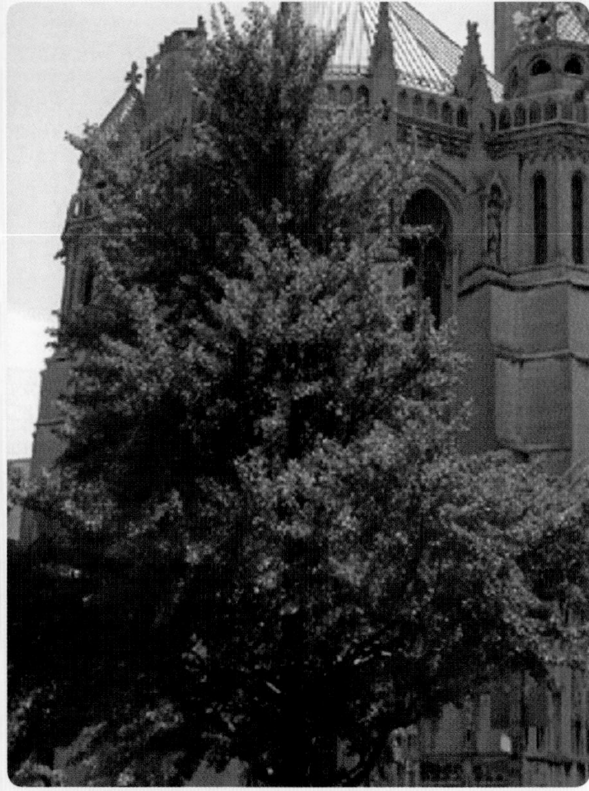

***Ginkgo biloba* in front of Riverside Church, Manhattan, New York City**

counterpart, learned of the stimulating properties ascribed to *ginkgo biloba* extracts, especially its supposed enhancement of memory and cognitive function, and envied these assets. The desire to learn more about these oriental secrets during the last half century has prompted an ever growing frenzy of research directed at unraveling the constituents and physiological effects of *ginkgo biloba* preparations. Organic chemistry has played a central role in the quest to achieve these goals. The story of the modern chemical investigation of the *ginkgo biloba* tree begins in Tokyo, Japan, where, in 1932, chemist Shu Furukawa succeeded in isolating "bitter principles" from the root-bark of a *ginkgo biloba* specimen. Barring a few limited exceptions, however, the molecular structures of the constituents of this complex mixture of extracted chemicals were not known until a fine spectroscopic investigation was published by Koji Nakanishi (then at Tohoku University, Japan, now at Columbia University, USA) in 1967 (Box 2). The striking molecular architectures of the ginkgolides not only captivated synthetic chemists, but also spurred research toward understanding the physiological effects of these natural products. Nakanishi and his team found the ginkgolide molecules to be remarkably robust, withstanding harsh physical and chemical treatment that would be expected to degrade most natural products. Seemingly,

Ginkgolide crystal (left)

Koji Nakanishi

Box 3 Highlights of Corey's retrosynthetic analysis of ginkgolide B

oxidation; epoxide opening; aldol reaction; ginkgolide B; epoxidation; oxidation; ring-closure; hydroxylation; Baeyer–Villiger oxidation; [2+2] cycloaddition; Mukaiyama aldol reaction; Sonogashira coupling

Box 4 Summary of Corey's total synthesis of ginkgolide B

(1) $(COCl)_2$; (2) n-Bu_3N, heat; intramolecular [2+2] ketene-olefin cycloaddition; Ph_3COOH, NaOH, Baeyer–Villiger oxidation; $H^{\oplus}$, transketalization-cyclization; ginkgolide B

even the compounds the *ginkgo biloba* produces have inherited its famed resilience.

The challenge of synthesizing the uniquely twisted and intricate hexacyclic ginkgolide molecules was solved in a creative and elegant way by E. J. Corey, some of whose other accomplishments can be found elsewhere in this book (Chapters 14, 15, 17 and 28). The ginkgolide skeleton is classified as a diterpene, revealing its general biosynthetic origins (see Chapters 6 and 25). It has six rings fused and knitted together in a complex and compact array. The task of synthesizing such an interwoven and rigid structure is far from trivial; however, Corey made an incisive observation that greatly simplified the task. A summary of the retrosynthetic analysis used to disentangle the problem is shown in Box 3. The crux of Corey's synthetic strategy was the realization that construction of the two central carbocyclic five-membered rings arranged in a spirocycle at an early stage in the synthesis could provide a scaffold for achieving all the further synthetic operations stereoselectively. Therefore, these rings were constructed first, and the remainder of the molecule built onto this core using a series of graceful transformations (Box 4). Thus, in 1988, Corey and his group completed the total syntheses of both ginkgolide A and ginkgolide B. Corey's inspired contributions to chemistry were recognized shortly afterwards by the award of the 1990 Nobel Prize in Chemistry. Incidentally, the Corey group also synthesized bilobalide (Box 2), a related terpene isolated from *ginkgo biloba* samples, through an entirely different synthetic strategy. Such is the complexity of the ginkgolide skeleton that, despite the great interest in these molecules, only one further total synthesis of a member of this family has been reported. This came in 1999, from the group of Michael T. Crimmins (University of North Carolina at Chapel Hill, USA).

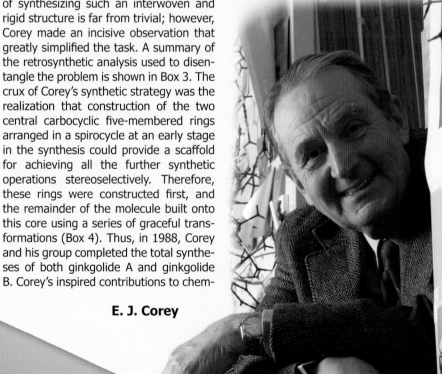

Despite the considerable efforts of modern science to unravel the physiological effects and medicinal potential of

Michael T. Crimmins

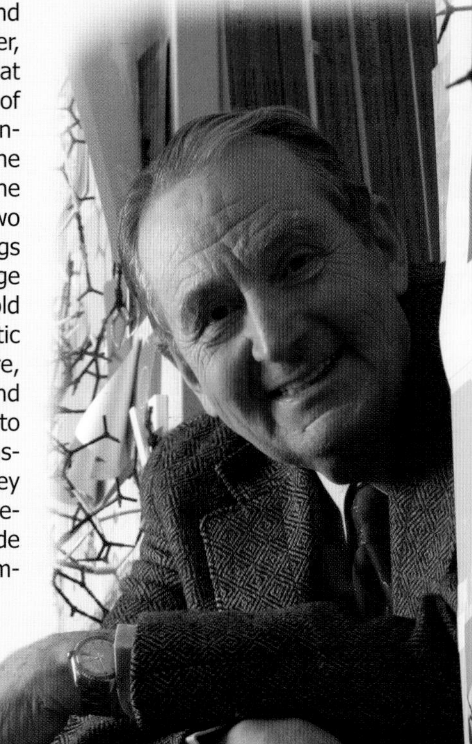

E. J. Corey

ginkgo extracts, the field remains clouded. At best a great deal more systematic study is required to evaluate and quantify the effects of *ginkgo biloba* extracts. These extracts contain many chemicals, and a standardized extract called EGb 761 is generally used for research studies. As yet, the evidence for its benefits is questionable and sometimes overplayed by interested parties. Today it is the most widely taken herbal supplement, with sales of *ginkgo biloba* in 1998 exceeding $300 million dollars in the US alone. Many people believe ginkgo extracts can augment their cognitive function, an effect that is becoming increasingly relevant to modern aging societies afraid of conditions such as Alzheimer's disease. Four million Americans are currently diagnosed with Alzheimer's disease; their plight was highlighted by the relatively recent death of the much loved former President Ronald Reagan. Alzheimer's disease destroys a person in the cruelest way by first attacking their memory, judgement, and language skills before any physical decline is seen, and it can be accompanied by violent outbursts and anger from patients, regardless of their prior temperament. The severity of the disease may be the reason we cling to believing in a herbal medicine cure-all that offers us hope. Furthermore, traditional Chinese medicine has provided many leads toward therapeutic agents for treatment of vicious illnesses (e.g. artemisin for treating malaria), and these successes are sometimes called upon as validation of the claims made for the potential of *ginkgo biloba* extracts.

Ginkgo leaf

It has been suggested in some quarters that the symptoms of Alzheimer's disease can be alleviated by *ginkgo biloba* extracts. However, for each study suggesting cognitive function improvements in Alzheimer's patients who were prescribed standardized *ginkgo biloba* extracts, there is another reporting a negligible benefit. It should be noted that the analysis of cognitive function in the mentally impaired is not easy to measure or quantify, and it may be the subjectivity of the models employed that leads to the variation in the conclusions drawn from these studies. Ginkgo extracts are not generally claimed to treat the actual causes of Alzheimer's disease, or to retard the progression of the illness. Rather, such preparations may bring some temporary measure of symptomatic relief to its sufferers. It is undisputed that the antioxidant properties of *ginkgo biloba* extracts can protect neurons from the damaging free radicals that cause age-related changes in the brain; but whether these extracts can protect healthy young brains from the danger of developing Alzheimer's disease in the future is a question that the combined forces of the medical, biological, and chemical sciences still need to probe further.

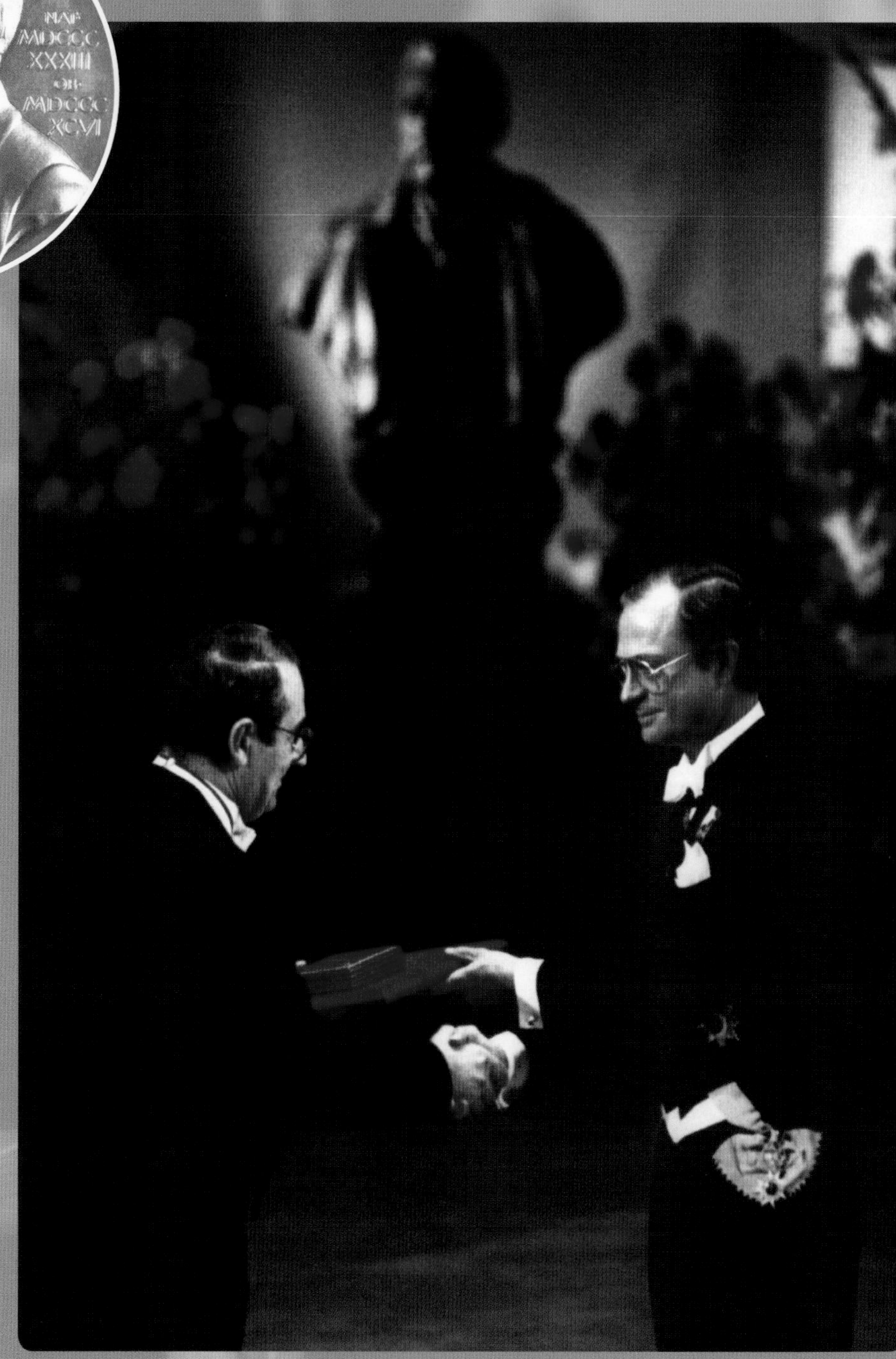

E. J. Corey receiving the Nobel Prize from the hands of his majesty King Carl Gustaf of Sweden

To Elias Corey
Congratulations and best wishes, Ronald Reagan

E. J. Corey receiving the prestigious National Medal of Science from President Ronald Reagan

Ginkgo biloba extracts have also been shown to widen blood vessels, reduce cholesterol, inhibit platelet aggregation (ginkgolide B is the most active component in this action), and enhance the absorption of glucose and choline by the brain (both are necessary for cognitive function), so maybe there are more truths remaining to be uncovered in the specific case of treating Alzheimer's disease with *ginkgo biloba* extracts. To be sure, there are many new discoveries to be made in the field of the brain biochemistry. The list of physiological effects described above suggests that *ginkgo biloba* may also have a number of more general applications outside the brain. It is promises of this type, with the potential to ameliorate the pain of terrible diseases and afflictions, which provide the impetus and motivation to the scientific community to continue, in unison, pushing the frontiers forward in so many disparate fields.

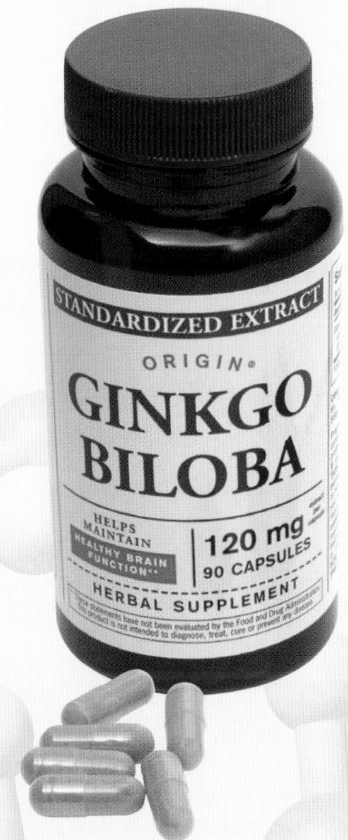

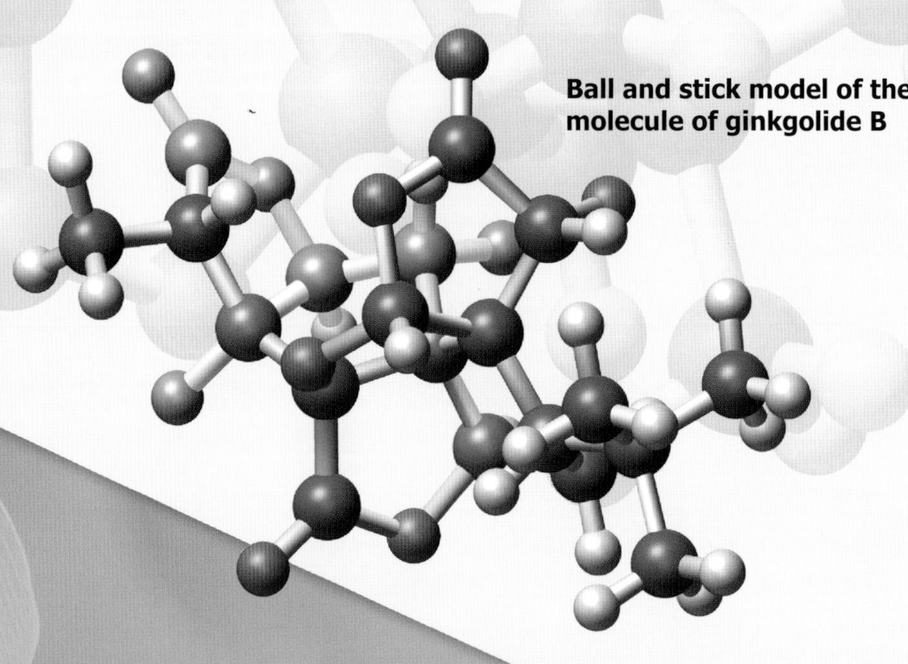

Ball and stick model of the molecule of ginkgolide B

Further Reading

K. Nakanishi, The Ginkgolides, *Pure Appl. Chem.* **1967**, *14*, 89–113.

K. Nakanishi, Terpene Trilactones from *Gingko biloba*: From Ancient Times to the 21st Century, *Bioorg. Med. Chem.* **2005**, *13*, 4987–5000.

E. J. Corey, M.-C. Kang, M. C. Desai, A. K. Ghosh, I. N. Houpis, Total Synthesis of (±)-Ginkgolide B, *J. Am. Chem. Soc.* **1988**, *110*, 649–651.

E. J. Corey, X.-M. Cheng, *The Logic of Chemical Synthesis*, Wiley-Interscience, New York, **1989**.

P. E. Gold, L. Cahill, G. L. Wenk, The *Lowdown* on *Ginkgo Biloba*, *Sci. Am.* **2003**, *288*, 86–91.

K. Stromgaard, K. Nakanishi, Chemistry and Biology of Terpene Trilactones from *Ginkgo Biloba*, *Angew. Chem. Int. Ed.* **2004**, *43*, 1640–1658.

Rapamycin
Cyclosporin, FK506,
& Rapamycin
Chapter 22
1989 & 1993

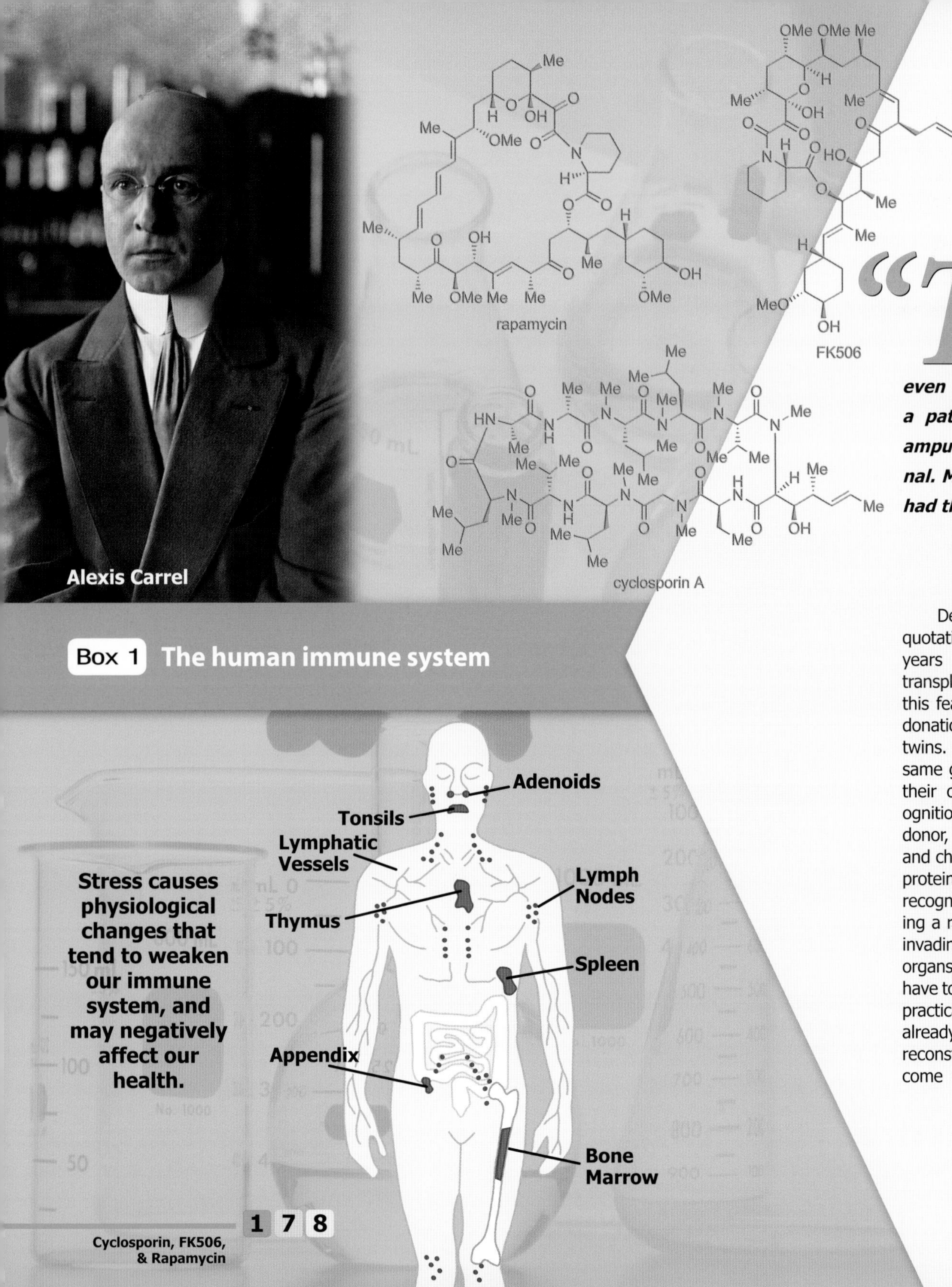

"The idea of replacing diseased organs by sound ones, of putting back an amputated limb or even of grafting a new limb on to a patient who has undergone an amputation, is far from being original. Many surgeons before me have had this idea...."

Alexis Carrel (Surgeon and Nobel Laureate, 1912)

Despite the optimism of the above quotation, it would be a further forty-two years before the first successful organ transplant took place. As significant as this feat was, this first success lay in the donation of a kidney between identical twins. Identical twins have exactly the same genetic makeup, so the surfaces of their cells express indistinguishable recognition proteins. The cells of any other donor, even relatives as close as mother and child, carry subtle differences in these proteins, which allow the alien organ to be recognized by the immune system, provoking a reaction against what it sees as the invading tissue. The free transplantation of organs between unrelated patients would have to wait many more years to become a practical reality, even though surgeons had already developed the necessary tissue reconstruction skills. Success would only come when organic chemists embraced the problem, and developed new drugs to counter tissue rejection effectively. In this chapter, we will explore a number of such molecules. Prominent amongst these so-called immunosuppressors is a natural product, FK506, with a most unremarkable name which belies its fascinating chemistry and biology. We will also discuss cyclosporin A and rapamycin, two other naturally occurring molecules that have had a major impact on medicine, particularly in the field of organ transplantation.

The 1912 Nobel Prize in Physiology or Medicine was awarded to the surgeon Alexis Carrel for his advancement of vascular surgery, and the quote introducing this chapter is an excerpt from his acceptance speech. His speech indicated that tissue transplantation had been occupying the minds of surgeons for some time. Carrel was accomplished in the practice of removing limbs from laboratory animals and reattaching them, either onto their original host or onto another individual. After describing all the procedures, precautions, and precision engineering that was required to undertake these delicate operations, he noted, "...vascular surgery can today be considered as completed from the standpoints both of technical and experimental results." While caution is generally prudent before announcing that an advance in any given field marks the summit, with no further progress possible, Carrel perhaps understood the limiting issues better than many of his successors. He knew his dogs, with their transplanted limbs, would die soon after surgery, not because of any mechanical dysfunction of the limb or its blood vessels, but because of some biological factor communicating between the dog's tissue and the new appendage. Thus, he was correct in saying that there was little to be gained from refining surgical techniques until, "...the biological relationships existing between living tissues…" could be understood and the problems highlighted by his investigations fully addressed.

Despite Carrel's awareness and his accurate assessment of the challenges, many people continued to attempt hopeless transplants. The donor species was even ignored; sheep, goats, and pigs were all having their organs fruitlessly interchanged. Then, with the onset of World War II, this research gained a new impetus as many casualties, both civilian and military, suffered terrible burns. There was no hope for these patients unless new skin could be grafted onto their ravaged bodies. Encouraged by the Medical Research Council in Great Britain, a talented professor of zoology at Oxford University, Peter B. Medawar, began to investigate the problem of achieving a successful permanent skin graft with skin donated by a healthy person to a burn victim. From this research, Medawar collected data and developed theories for achieving transplantation immunity, work for which he was awarded the 1960 Nobel Prize in Physiology or Medicine. He showed that if fetal or newborn mice were injected with cells from an incompatible mouse strain early enough, not only would the cells not be rejected immediately, but when the juvenile mouse grew to reach maturity it could accept grafts from the donor mouse whose cells it had received early in life. Medawar and his colleagues went on to show that a combination of lymphocytes and antibodies circulating in the blood destroyed foreign cells in test animals that had not been given the opportunity to develop this tolerance. Thus, Medawar had identified the two distinct components of the body's immune defense system (see Box 1), humoral (antigen-antibody responses) and cellular (lymphocytes), and began to untangle their roles in the rejection of grafts. Of course, in so doing, he had also finally discovered Carrel's mysterious biological factor.

Sir Peter B. Medawar

Following Medawar's work, early investigators in the field of human genetics had noticed that identical twins did not have this immunological barrier. For this reason, the first organ transplant was between the identical twins Richard and Ronald Herrick. In December 1954, Richard received one of Ronald's kidneys under the direction of surgeon Joseph E. Murray at the Brigham and Women's Hospital in

Richard and Ronald Herrick

Joseph E. Murray

George H. Hitchings **Gertrude B. Elion**

Box 2 Selected antimetabolite drugs developed by Hitchings and Elion

azathioprine [Imuran®]
(still used as part of a cocktail of drugs administered to prevent graft rejection, certain leukemias and Crohn's disease)

trimethoprim [Proloprim®, Monotrim®]
(originally an anti-malarial drug, now used to combat urinary and respiratory tract infections)

methotrexate [Trexall®, Rheumatrex®]
(still used in cancer chemotherapy and to treat rheumatoid arthritis and psoriasis)

6-mercaptopurine [Purinethol®]
(still used as part of a cocktail of drugs administered to prevent certain leukemias and Crohn's disease)

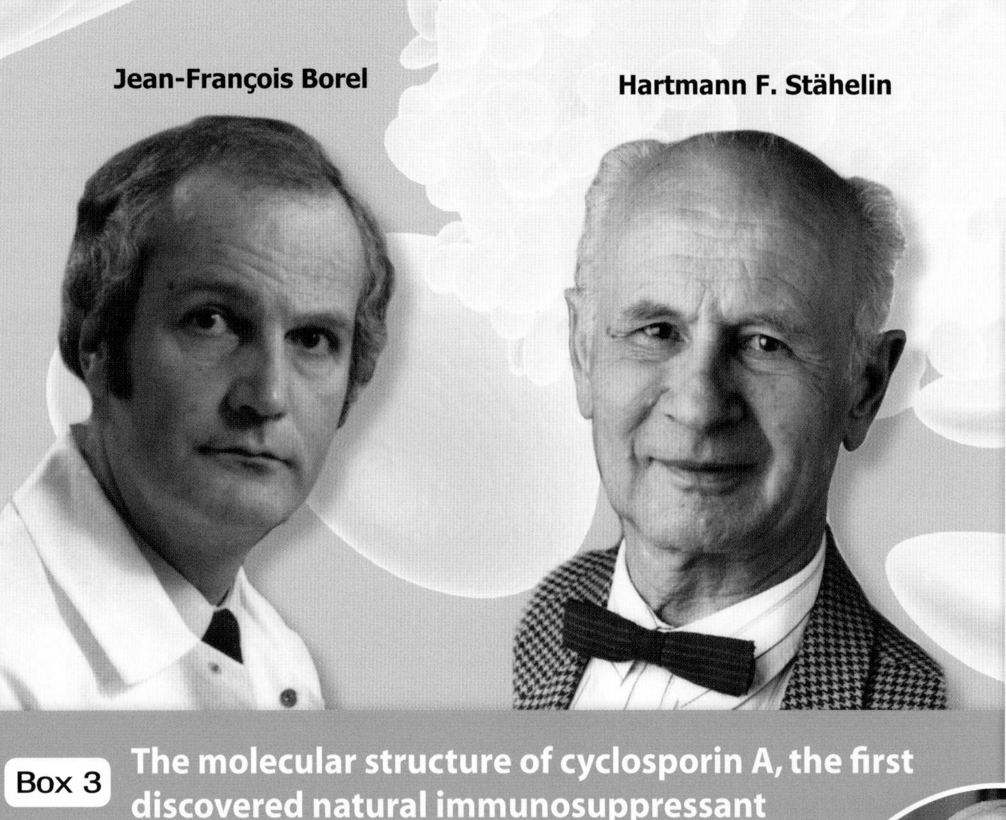

purine

pyrimidine

Boston (Massachusetts, USA). Murray had first taken Richard and Ronald to the police in order to analyze their fingerprints to ascertain whether they were indeed identical twins. The success of the transplant was sealed soon afterwards, when a thriving Richard Herrick married the recovery room nurse who had tended to him. They had two children before disaster struck; Richard Herrick died eight years later, when the original kidney disease recurred in his newly transplanted organ. Murray and another pioneering transplant clinician, E. Donnall Thomas, from the Fred Hutchinson Cancer Research Center in Seattle (Washington, USA), received the 1990 Nobel Prize in Physiology or Medicine for "their discoveries concerning organ and cell transplantation in the treatment of human disease." In their writings, preceding the award of the Nobel Prize, both of these doctors reiterated that the great turning point in the long term success of transplant surgery came with the discovery and use of immunosuppressant drugs, beginning with antimetabolites such as azathioprine (Box 2).

E. Donnall Thomas

In the early 1950s, medicinal chemists George H. Hitchings and Gertrude B. Elion, working at the Wellcome Research Laboratories in Tuckahoe (New York, USA), became interested in *The Antimetabolite Theory*. This concept had first been discussed in relation to the action of the sulfonamide antibiotics (see Chapter 13), which were proposed to act by mimicking a component (*para*-aminobenzoic acid, PABA) of the dihydrofolate metabolic pathway. This reaction sequence is responsible for the biosynthesis of nucleotide units for incorporation into DNA and the supply of cellular energy in the form of ATP (adenosine triphosphate). The sulfonamide substitute is incorporated up to a certain point, after which it blocks the dihydrofolate cycle. When the bacteria cannot make DNA they are unable to reproduce, and eventually die out. Hitchings theorized that, since the latter stages of this pathway are essentially the same in mammalian cells, small molecule inhibitors could also be used to attack other rapidly dividing cells, such as those that make up tumors. The naturally occurring purine and pyrimidines, which form the foundation of the genetic code and are components of nucleotides, were obvious templates on which to base such inhibitors. Elion undertook the synthesis of molecular mimics of these natural bases, analogs that would later be tested for potential biological activities. Elion and

Box 3 The molecular structure of cyclosporin A, the first discovered natural immunosuppressant

cyclosporin A

Ball and stick model of the molecule of cyclosporin A

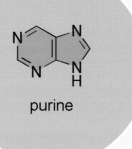

Hitchings made hundreds of purines and pyrimidines as part of their drug discovery program. Such programs typically involve the design, synthesis, and biological testing of many molecules in an iterative process that aims to fine-tune the structure of a compound in order to optimize its pharmacological properties. Early success came with the discovery of 6-mercaptopurine (6-MP), used to treat leukemia, and then azathioprine (Box 2). In 1958, two Boston doctors, Robert Schwartz and William Dameshek, rationalized that the body's immune response leads to an overproduction of lymphocytes in a manner not dissimilar to that involved in leukemia. Schwartz showed that Elion and Hitchings' 6-MP, when administered to rabbits for several days, prevented the production of antigens by the animal's immune system when blood from an alien species was injected into the test subject. A British surgeon, Sir Roy Y. Calne (Royal Free Hospital, London, UK), stimulated by the promise these results held for kidney transplantations, investigated such grafts in laboratory animals. He had significant success with 6-MP, but eventually the performance of this drug was overshadowed by that of azathioprine. By 1962, kidney transplants between unrelated human patients had become a therapeutic reality when accompanied by immunosuppression induced by azathioprine. E. Donnall Thomas subsequently managed to diminish organ rejection further with a treatment employing another of Hitchings and Elion's compounds, the cytotoxic anticancer compound methotrexate (Box 2). Over the ensuing fifteen years, some 25,000 kidney transplants were carried out, with the first liver (1963), lung (1963), and heart (1967) transplants also recorded in rapid succession. Later investigations showed that the success of this approach is predicated on the fact that T-cell (T-lymphocyte) proliferation, which is key to the immune system response, is critically dependent on *de novo* synthesis (synthesis from scratch) of purines because these cells have a far reduced capacity to salvage and reuse nucleotides when compared with other cell types. Hitchings and Elion were awarded the 1988 Nobel Prize in Physiology or Medicine "for their discoveries of important principles for drug treatment." This prize was also shared with Sir James W. Black, another pioneer of medicinal chemistry. More about these biomedical researchers can be found in Chapter 33 (Small Molecule Drugs).

Azathioprine, methotrexate, and certain other related compounds are still in use today in cancer chemotherapy, and to counter certain autoimmune diseases in which the body's own immune system malfunctions and attacks itself, such as rheumatoid arthritis, multiple sclerosis, and systemic lupus. However, their use in post-transplant therapy has, for the most part, been superseded by natural products discovered in the 1970s and 1980s, beginning with cyclosporin (Box 3).

Selman A. Waksman's discovery of the natural antibiotic streptomycin from a soil sample in 1943, and its subsequent huge success in combating the scourge of epidemic tuberculosis, sparked the scientific fashion of examining soil samples in order to find potential drugs. The large number of medicinal leads discovered by this method has been ascribed to the abundance of defense chemicals produced by

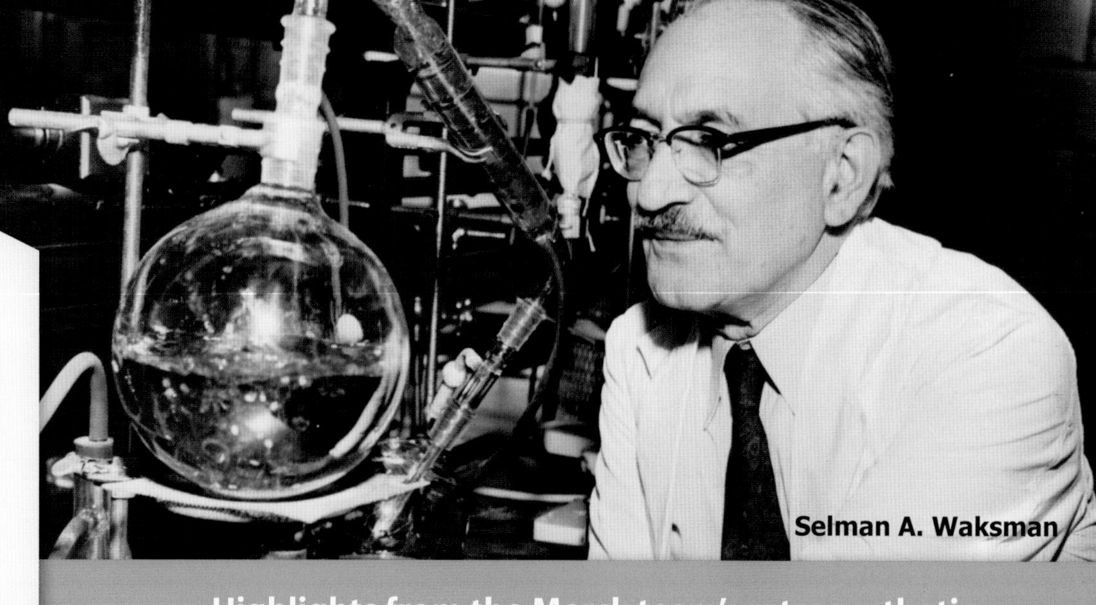
Selman A. Waksman

Box 4 Highlights from the Merck team's retrosynthetic analysis of FK506

cyclization

macrolactamization

FK506

esterification

Horner–Wittig reaction

chain extension

Box 5 **The use of the Evans auxiliary in the Merck total synthesis of FK506**

A, Et_3N, n-Bu_2BOTf
Evans aldol reaction

B, Et_3N, n-Bu_2BOTf
Evans aldol reaction

the millions of organisms living in the soil as they compete with each other for nutrients. This chemical warfare between germs has afforded us with one of the richest sources of bioactive molecules, and throughout this book you will encounter many examples, including erythromycin, avermectin, amphotericin, the epothilones, and thiostrepton (Chapters 17, 19, 20, 29, and 32, respectively). Cyclosporin was isolated in 1971 from a Norwegian soil fungus, *Tolypocladium inflatum Gams*, by a team under the leadership of two scientists from the Sandoz Pharmaceutical Company (now part of Novartis) in Switzerland. Antibacterial chemotherapy was their main investigative focus, and Jean-François Borel and Hartmann F. Stähelin initially found no practical therapeutic use for their new and complex discovery, since it did not exhibit useful levels of activity in any of their bioassays. Fortunately, they were sufficiently intrigued by its chemical properties to continue their research. Their persistence paid substantial dividends when it became evident from immune modulation studies that cyclosporin was a selective suppressor of T-cells. Sandoz then rushed to develop cyclosporin for immunosupression chemotherapy, thus rescuing the compound from obscurity. They started with full structural elucidation and synthesis of the peptidic fungal metabolite (comprising eleven amino acid units, including one previously unknown amino acid). It is interesting to note that the progress of cyclosporin to the clinic nearly met with disaster on one more occasion, because there was some debate within the Sandoz management group as to whether organ transplantation would ever become a practical operation with broad applicability. In 1978, Sir Roy Calne, the British surgeon who had pioneered the use of antimetabolites in kidney transplant surgery, tested cyclosporin, first in pigs (which are good animal models for many drugs as they have much in common with humans) and then in human subjects, with excellent results. In the human patients, rejection was inhibited in five out of the seven cases where a new kidney had been received from mismatched deceased donors. By 1983, cyclosporin A had been approved by the regulatory authorities and entered the market as Sandimmun®. Throughout the 1980s, Sandimmun® (also sold as Neoral®) provided a lifeline for people whose failing organs had led them to depend on transplant therapy. Though still used today in more extreme cases, it quickly emerged that cyclosporin had a serious drawback; not only is it highly toxic, thus requiring cautious administration and dos-

Tolypocladium inflatum

Streptomyces hygroscopicus

age control, but it also increases patients' risk of developing cancer. It directly affects tumor growth, encouraging cancerous cells to divide and spread by inducing the synthesis of transforming growth factor-β (TGF-β). Furthermore, patients must also receive a high dose of steroids along with cyclosporin, to ward off the damaging side effects of this powerful drug. Even under these circumstances, however, the side effects meant that some patients experienced only a marginal improvement in their health after their transplant operation. Fortunately, new immunosuppressive agents that would improve the prognosis of transplant patients considerably were soon to appear on the horizon.

In 1975, Suren Sehgal and co-workers from the pharmaceutical company Ayerst Laboratories (now part of Wyeth) reported the identification of a new antibiotic produced by a streptomyces culture (*Streptomyces hygroscopicus*) contained in soil samples collected from Easter Island. Named rapamycin, after the native name for Easter Island (Rapa Nui), it had weak antifungal properties and marginal activity against a limited number of Gram-positive bacteria. As with cyclosporin, rapamycin could easily have become lost in the vast annals of unutilized natural products. It had simply been tested in the 'wrong' bioassay when isolated, although this could not have been recognized at the time. Three years later, John A. Findlay and co-workers (University of New Brunswick, Canada) completed the structural elucidation of the rapamycin molecule, based on spectroscopic analysis and x-ray crystallography. Their painstaking work unveiled an ornate macrocycle incorporating a highly sensitive triene system. Despite the new information regarding its beautiful structure, the development of rapamycin still stalled because a medicinal niche had yet to be found. A change in its fortune would only occur when it was reconsidered in the light of the discovery of FK506, another macrocyclic natural product with exciting biological properties.

Above circle: Suren Sehgal (left), discoverer of rapamycin, and Dr. Joe Camardo (right), who was responsible for development of Rapamune® in the clinic and its current use in transplantation.

Thomas E. Starzl

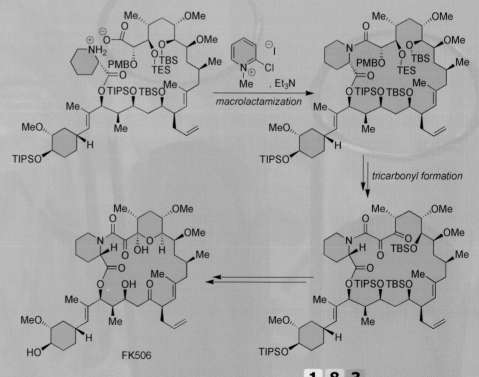

Box 6 The final stages of the Merck total synthesis of FK506

macrolactamization

tricarbonyl formation

FK506

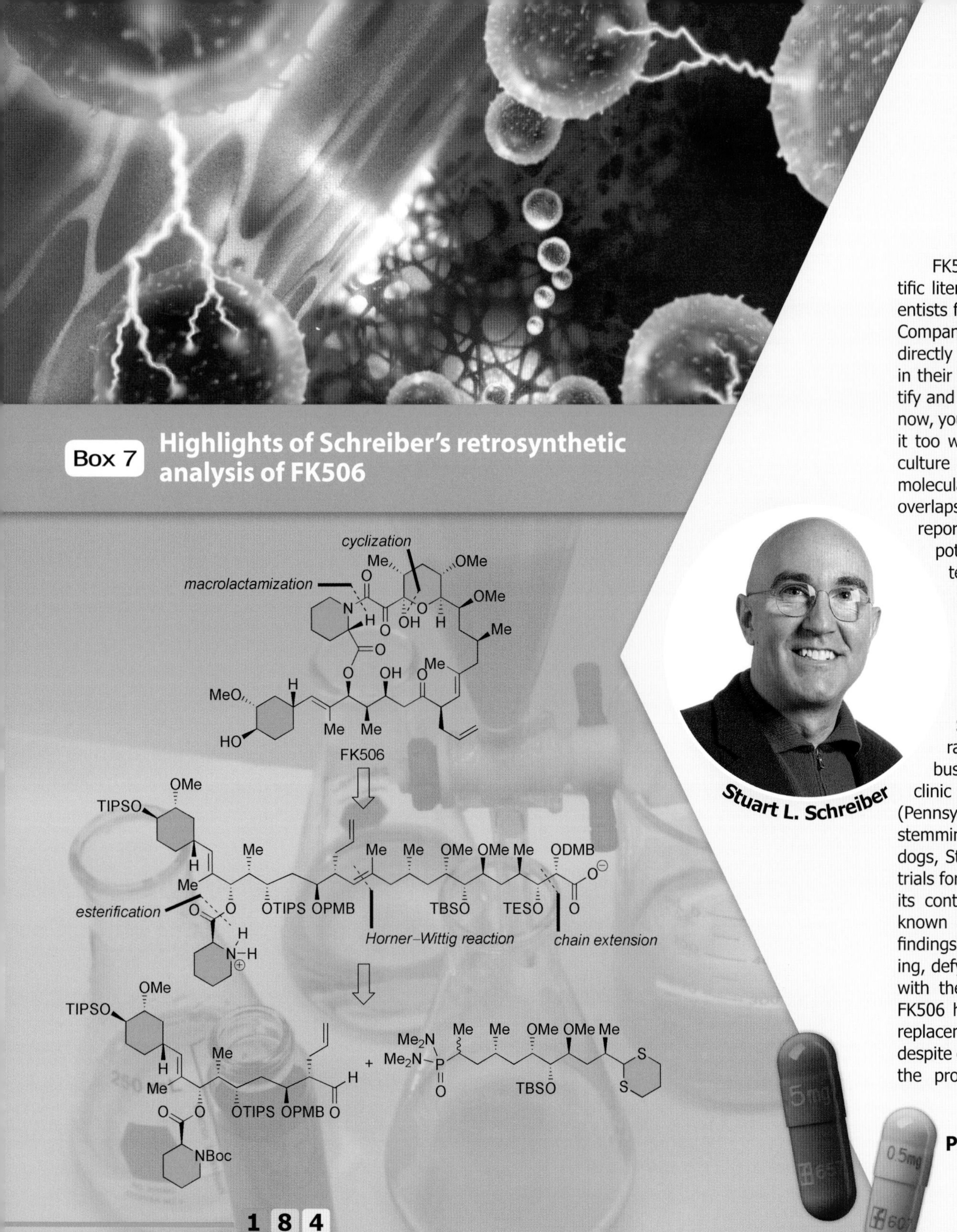

Stuart L. Schreiber

FK506 was first reported in the scientific literature in 1987 by a group of scientists from the Fujisawa Pharmaceutical Company (Japan), who had been looking directly for immunosuppressor molecules in their bioassay-driven program to identify and isolate novel natural products. By now, you will not be surprised to learn that it too was isolated from a streptomyces culture (*Streptomyces tsukubaensis*). Its molecular structure features remarkable overlaps with that of rapamycin. Initial reports revealed that FK506 was more potent *in vitro* and *in vivo* (in the test tube and in live test animals, respectively) than cyclosporin. FK506 (sometimes also called tacrolimus), a smaller molecule than cyclosporin, although perhaps more structurally complex, was championed by Thomas Earl Starzl, a charismatic surgeon who ran the world's most successful and busiest, at the time, transplantation clinic at the University of Pittsburgh (Pennsylvania, USA). Despite early fears stemming from fatalities with laboratory dogs, Starzl had directed the first clinical trials for this drug by 1989 and had urged its continuing development in his well-known assertive and driven style. The findings from his studies were breathtaking, defying belief amongst those familiar with the field. The initial application of FK506 had been in patients whose new replacement organs were being rejected despite cyclosporin therapy. FK506 turned the prognosis of these people upside down; some, who had been so ill that they were facing a second, and usually hopeless, transplant to replace failing and rejected organs, were rescued. In almost every case where FK506 was administered, dramatic improvements were seen, and in many of these patients the episodes of rejection petered out completely. FK506 was estimated to be between 50 and 100 times more powerful than cyclosporin on a gram per gram basis, a figure reflected in the dramatic increase in one-year survival rates seen once FK506 (marketed as Prograf®) became established as the preferred therapy to accompany transplant surgery. FK506 proved to be a resounding success, and it continues to be a wonderful drug today. In 2001, it recorded annual sales of $416 million, and this figure looks set to grow as the use of cyclosporin declines.

As with any molecule that exhibits important medicinal properties, investigations relating to FK506 hastily became an intense cauldron of activity, heated by anticipation, competition, and rivalry. The very best minds from a range of scientific disciplines focused their energies on determining its mechanism of action and biochemistry, and devising plausible strategies for its total synthesis. At the center of the efforts, originating from both industrial and academic settings, emerged a young and talented chemistry professor, Stuart L. Schreiber, from Harvard University (Massachusetts, USA). He not only exhibited a strong passion for chemical synthesis, but was also a major proponent of research at the interface of chemistry and biology. Simultaneously, the Schreiber group, in collaboration with Matthew Harding (a rising star in immunology from Yale Medical School, Connecticut, USA), and a team from the pharmaceutical giant Merck (Rahway,

Prograf® capsules

New Jersey, USA) independently identified and expressed quantities of the cellular target protein for FK506. The protein, named FKBP-12, was found to bind FK506, along with calcium and another cellular protein, calmodulin, inside the T-cell, thus disrupting the function of an enzyme called calcineurin. Details of the resultant cellular reverberations are beyond the scope of this text, but the effects are felt throughout a complex sequence of signal transduction events that govern the machinery of T-cells. Ultimately, the pathways that are blocked are those signaling and activating other T-cells (following the secretion of chemical messengers, such as interleukin-2). In addition, the FKBP-FK506 complex blocks the entry of resting immune cells into the reproductive section of a normal cell cycle (the transition from the G_0 to the G_1 phase of the cell cycle is prevented). Hindering T-cell activity in this manner prevents their proliferation with the net result being the shutdown of the immune system (Box 1). Studies by Robert E. Handschumacher and his colleagues at Yale University demonstrated that cyclosporin binds to a different protein, called cyclophilin. However, the resultant complex also inhibits the enzyme calcineurin, so, in essence, the same mechanism of action is in play for both of these agents. The proteins that bind these two immunosuppressors were given a common classification and collectively named the immunophilins. Alongside these intriguing biological investigations, Schreiber had also been working on the total synthesis of FK506. The task was completed within a year, along with the synthesis of a collection of analogues, which were to be used to probe the binding characteristics of the FKBP-FK506 complex. As is often the case in such circumstances, the synthesis had been a two horse race, and in the end the winners were a team of chemists from Merck led by Todd K. Jones, who published their FK506 total synthesis less than five weeks ahead of that of the Schreiber group (Boxes 4–9).

The FK506 molecule includes a rich array of stereochemistry, a striking 21-membered macrolactam ring, a pyran ring and a sensitive pipercolic acid motif. Of the numerous possible disconnections that could be envisaged for FK506, a number were selected by both the Merck team and the Schreiber group independently. Furthermore, in several key instances the two teams relied on the same powerful reactions to forge pivotal FK506 bonds in the synthetic direction. For example, an

Ball and stick model of the molecule of FK506

Box 8 Highlights from Schreiber's synthesis of FK506

$Ti(O\textit{i}\text{-}Pr)_4$ (7 mol %), L-(+)-diisopropyltartrate (9 mol %), *t*-BuOOH

Sharpless asymmetric epoxidation

L-arabitol

$Ru_2Cl_4[(S)\text{-BINAP}]_2Et_3N$ (1 mol %), H_2 (1500 psi)

Noyori asymmetric hydrogenation

(*S*)-BINAP =

Box 9 The completion of Schreiber's total synthesis of FK506

macrolactamization

Et_3N

FK506

The Nicolaou rapamycin team

Box 10 Highlights from the Nicolaou team's retrosynthetic analysis of rapamycin

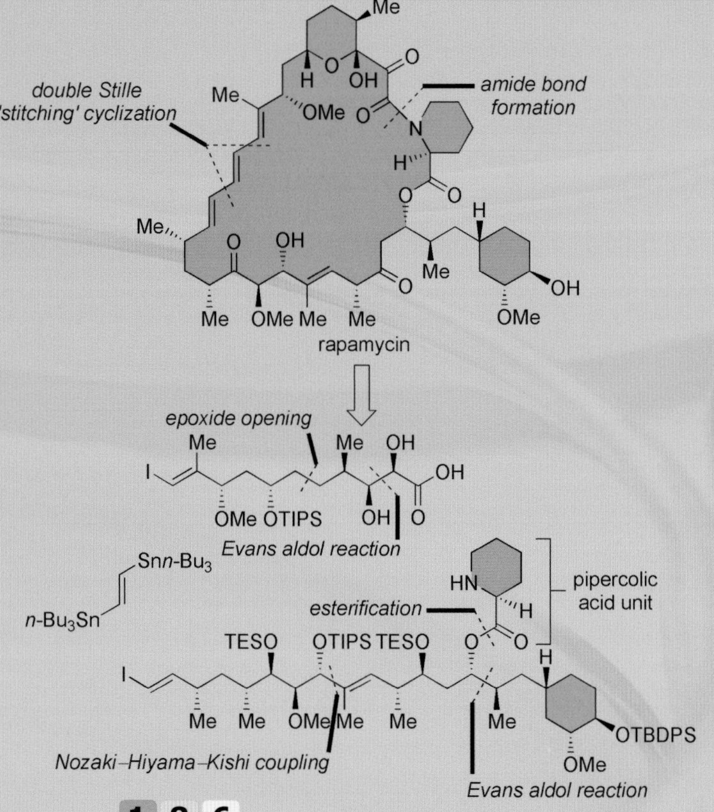

olefination was chosen by both groups to construct the endocyclic double bond, and, likewise, macrolactamization was used to join the pipercolic ring to an appropriate terminal carboxylic acid, thus accomplishing the final fusion of the macrocycle (Boxes 4 and 7). The two groups, however, employed different conditions in these pivotal reactions and used distinctly different strategies to build up the stereochemically complex linear precursors. Merck relied heavily on the stalwart chemistry developed by David A. Evans (Harvard University, USA) wherein oxazolidinone chiral auxiliaries are employed to direct aldol reactions (see Box 5 and Chapter 31). Schreiber's group relied on a more eclectic series of reactions to obtain the desired stereochemical arrays (Box 8). A Sharpless asymmetric epoxidation (see Chapter 26) forged two new stereocenters, an inexpensive sugar (arabitol) provided a series of asymmetric centers in another fragment, and finally the asymmetric reduction of a ketone provided the correct stereochemistry at yet another key site. The ruthenium catalyst used in this last reaction, which employs a chiral phosphine ligand known as BINAP, was developed by Ryoji Noyori (see Chapter 28). The final delicate orchestrations of these approaches are summarized in Boxes 6 and 9. Since the reports of these syntheses in 1989 and 1990, a number of other groups have made worthy contributions in the form of formal syntheses of FK506 (i.e. novel syntheses that converge on a late intermediate from a previously known route).

Amos B. Smith, III

Steven V. Ley

When we introduced FK506 we commented on its structural overlaps with rapamycin. These similarities did not escape the attention of scientists working in the frenzied world of FK506 development. Suddenly, rapamycin was drawn into the limelight as its own immunosuppressive activity was confirmed. The race then began in earnest to tap into the potential of this natural product, and the chemical literature came to be filled with synthetic and biological studies focused on rapamycin. In 1993, three groups accomplished total syntheses of this challenging molecular architecture; first was that of K. C. Nicolaou of The Scripps Research

Samuel J. Danishefsky

Institute and the University of California, San Diego (USA), followed in close succession by the groups of Schreiber and Samuel J. Danishefsky (then at Yale University, now at the Sloan–Kettering Institute and Columbia University, New York, USA). Later, Amos B. Smith, III (University of Pennsylvania, USA), Charles J. Sih (University of Wisconsin, USA), Steven V. Ley (University of Cambridge, UK), and Robert E. Ireland (University of Virginia, USA) added their own innovative variations. With these syntheses of the molecule accomplished, it quickly became apparent that rapamycin effects its immunosuppressive activity through a slightly different mechanism than FK506. This mechanistic dichotomy occurs despite the structural similarities between the two, a reminder of the functional intricacy and complexity of biological systems. Before we discuss its biological activity, we will highlight some details of how the first synthesis of rapamycin was accomplished by the Nicolaou group.

K. C. Nicolaou

Rapamycin possesses the same delicate pipercolic system as FK506, united to a similar pyran ring by the same 1,2-ketoamide unit. Rapamycin, however, contains the additional synthetic obstacle of a sensitive triene array as part of its intimidating 29-membered macrocyclic structure. The Nicolaou team chose this latter unstable motif as the first that should be excised in their retrosynthetic analysis. This proposal ensures that its inclusion would be left until a late stage of the synthesis, so that complications arising from its sensitivity might be avoided (Boxes 10 and 11). They daringly proposed that this unit could be introduced via a double inter- then intramolecular, Stille reaction in a cascade sequence, simultaneously forming the triene moiety and closing the macrocyclic ring. Other key retrosynthetic disconnections included operations to remove the pipercolic ring motif by cleavage of the amide and ester bonds, thus revealing two major fragments for synthesis. The macrocyclic system of rapamycin is even more stereochemically rich than that of FK506, and the Nicolaou team opted to employ a broad range of techniques to access the desired stereoisomers. Certain stereocenters were derived from readily available naturally occurring substances; examples include a mannitol derivative (a sugar), citronellene (a naturally occurring terpene), and a derivative of vitamin C (ascorbic acid). A number of powerful asymmetric reactions were also used to synthesize pivotal building blocks with a high degree of convergency, such as several Evans auxiliary-mediated aldol reactions (see Chapter 31) and an asymmetric boron-mediated allylation, developed by Herbert C. Brown (see Chapter 18). After the many hurdles were overcome, the stage was set at last for the audacious macrocyclization event.

Irving Weissman

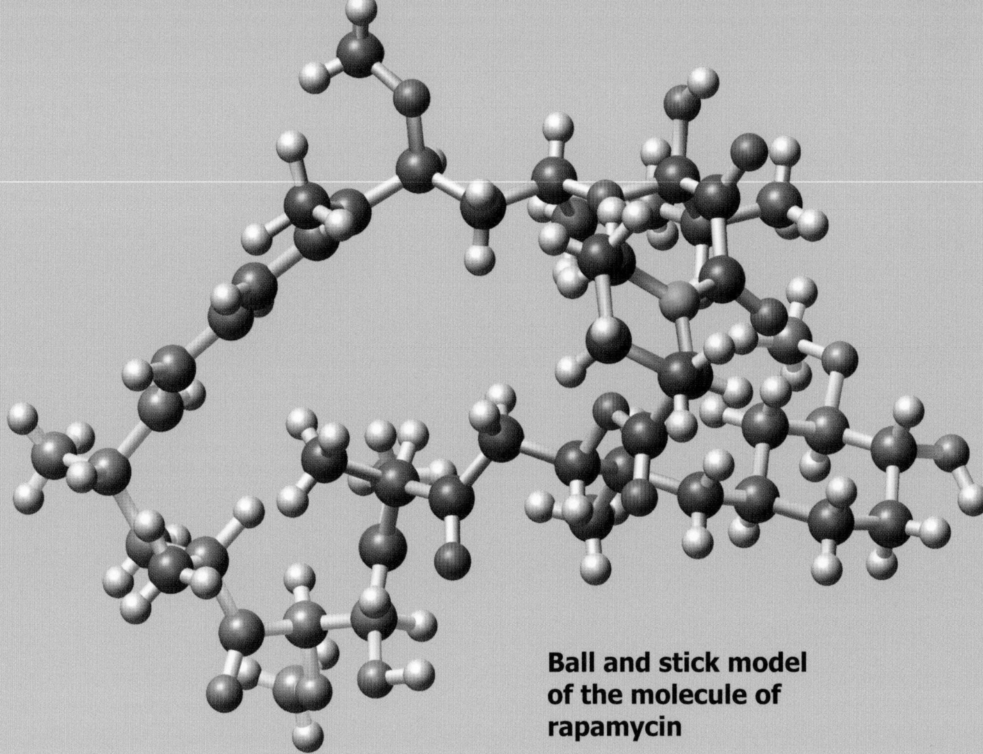

Ball and stick model of the molecule of rapamycin

Box 11 The double-Stille reaction to complete the Nicolaou team's total synthesis of rapamycin

n-Bu_3Sn–CH=CH–$Sn$$n$-$Bu_3$, cat. $PdCl_2(MeCN)_2$, i-Pr_2NEt

inter- then intramolecular Stille reaction cascade

rapamycin

Box 12 Palladium-Catalyzed Cross-Coupling Reactions

The striking double-Stille coupling used by the Nicolaou group to complete the synthesis of rapamycin (Box 11) belongs to a class of related reactions known collectively as palladium-catalyzed cross-couplings. These reactions constitute an important group of carbon–carbon bond-forming processes that has come of age only in the last thirty years. Although initially investigated using stoichiometric palladium (i.e. one part metal for each part substrate), these processes have now been refined to use catalytic quantities of this rather expensive metal.

The earliest example of this class involves the coupling of an alkenyl or aryl halide or triflate with an alkene, which has come to be known as the Heck reaction, in honor of one of its pioneers, Richard F. Heck. The first examples of this coupling were reported by Mizoroki and Heck in 1971 and 1972, respectively. Since then, a number of other coupling processes have been developed which differ primarily in the nature of the nucleophilic component. In each case, the reactions have been named after their primary developer, although it is important to remember that such developments are often influenced by the results of several groups. The Stille reaction, pioneered by John K. Stille and David Milstein, involves the coupling of an aryl or alkenyl halide with an alkyl tin reagent. This coupling reaction has found widespread use in total synthesis due to the mild conditions and convenient preparation of the organotin coupling partners. A related process involves the coupling of aryl and alkenyl boronic acids and esters, which has become known as the Suzuki reaction. The Sonogashira reaction replaces the organometallic component with a terminal alkyne. The Negishi coupling uses organozinc reagents, which can be prepared directly from organohalides, or from other organometallics. This reaction predates the Stille and Suzuki procedures, but has gained popularity in total synthesis only recently.

The Tsuji–Trost reaction, another palladium-catalyzed coupling process, involves nucleophilic substitution at an allylic position. Allylic esters and carbonates are the most commonly employed electrophiles, and they couple with a range of carbon nucleophiles. This process is particularly useful for the coupling of two alkyl units, and the asymmetric variant is commonly used in total synthesis. The scope of palladium-catalyzed coupling reactions continues to expand and, in addition to the reactions described here, a range of other processes to form carbon–carbon and carbon–heteroatom bonds are gaining popularity. The latter reactions have been developed recently primarily by Stephen L. Buchwald (Massachusetts Institute of Technology, USA) and John F. Hartwig (University of Illinois at Urbana-Champaign, USA).

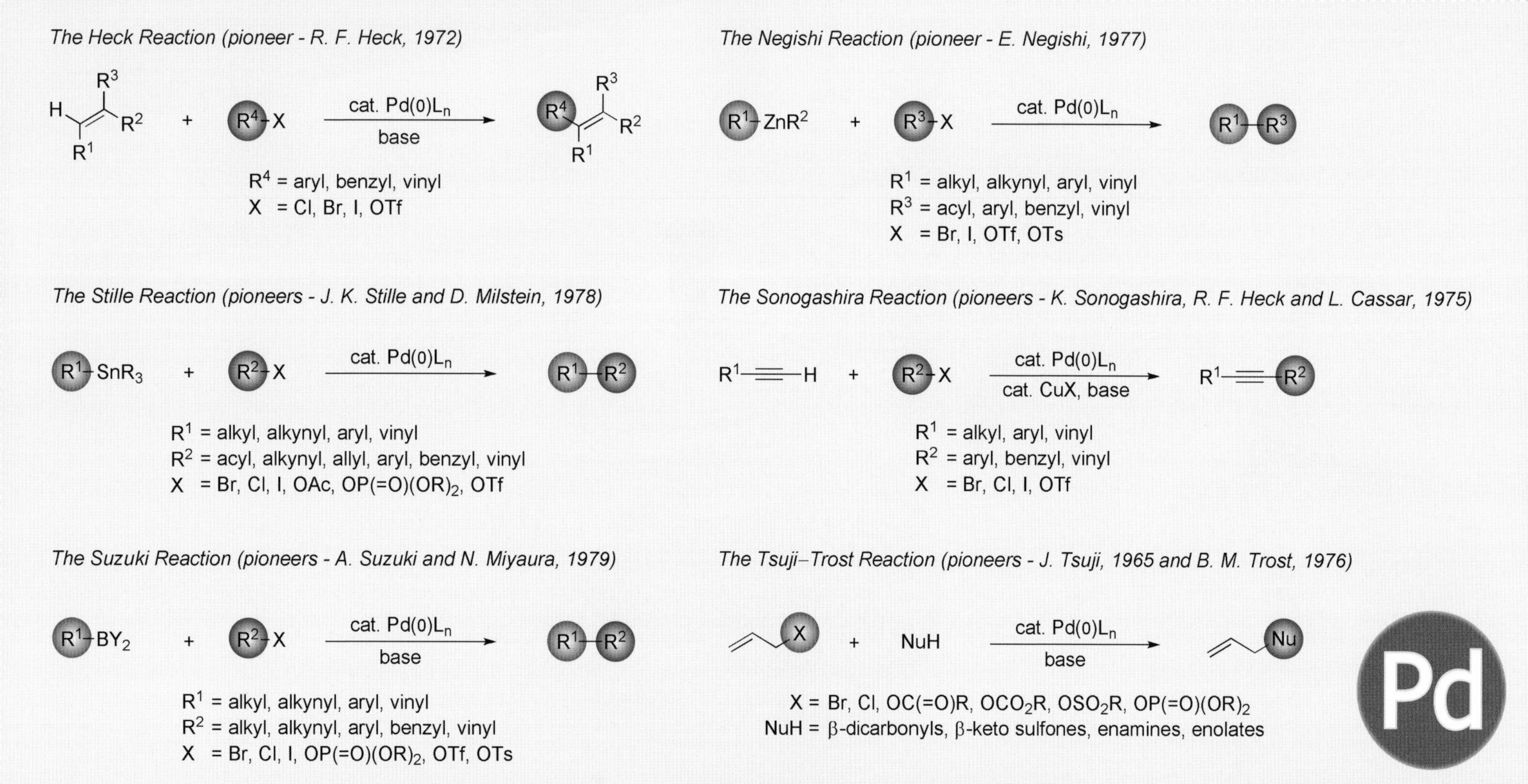

Barry M. Trost

Richard F. Heck

John F. Hartwig

Akira Suzuki

John K. Stille

Stephen L. Buchwald

Ei-ichi Negishi

David Milstein

Jiro Tsuji

Fortunately, their daring strategy paid off and rapamycin, triene intact, was isolated from this cascade Stille reaction (Box 12), thereby completing the total synthesis (Box 11).

The elucidation of the mechanism of action of rapamycin was once again spearheaded by Schreiber, but with several other scientists playing prominent roles, most notably Irving Weissman at Stanford University (California, USA). Their discoveries were both fascinating and highly relevant to the future practice of immunotherapy. In addition, the new knowledge that was obtained explained much about cytoplasmic signal transduction, an area of cellular biology that had long been shrouded in mystery. Early studies had shown that rapamycin binds to the same protein, FKBP-12, as FK506 in the cytoplasm of the cell. It came as no surprise to the investigators that the most important binding sites on the two natural products were located in their common structural domain. Both of the natural products were shown to inhibit a rotamase activity of the enzyme, but then their effects diverge. Rapamycin does not inhibit calcineurin, and therefore does not act on the same signaling pathway as FK506 and cyclosporin; rather, it blocks a regulatory kinase involved in a later pathway associated with T-cell activation. The FKBP-rapamycin complex blocks the progression of many cell types through an early phase of the cell cycle (at the G_1 to S phase transition), thereby causing cell cycle arrest. Ultimately, this difference has presented benefits to patients with respect to other proliferative disorders such as cancer. As mentioned earlier, cyclosporin, and to some extent FK506, has a propensity to promote tumor growth. With cancer as one of the main causes of death in post-operative transplant patients, rapamycin, fortunately, have been found to have the opposite effect in that it inhibits primary and metastatic tumor growth. Relatively small doses of rapamycin can cause regression in tumors, but only after angiogenesis. In simple terms, rapamycin stops, or at least slows down, the growth of the blood supply that the tumor develops to nourish itself after it has grown too big to rely on hijacking the blood supply of proximal tissue. Furthermore, this remarkable molecule also moderates antibody production, another major constituent of the immune response. The study of the biochemistry of rapamycin is still relatively youthful, and as it is further unraveled it promises to teach us much more in coming years. It is quite possible that a host of other medical benefits will emerge from this intrepid journey of discovery.

Rapamycin, also sometimes known as sirolimus, was approved by regulatory authorities (Food and Drug Administration, USA) in September 1999 and is marketed by Wyeth as Rapamune®. It looks increasingly likely that Rapamune® will become the immunosuppressant of choice for transplant patients, given its advantages over FK506 and cyclosporin. As we herald the dawning of these miracle drugs, we should keep in mind that transplants have only become as routine as they are today due to the heroic efforts of the chemists, biologists, and clinicians who discovered these natural products and developed them into the useful medications.

Further Reading

M. W. Harding, A. Galat, D. E. Uehling, S. L. Schreiber, A Receptor for the Immunosuppressant FK506 is a cis-trans Peptidyl-Prolyl Isomerase, *Nature* **1989**, *341*, 758–760.

T. K. Jones, S. G. Mills, R. A. Reamer, A. Askin, R. Desmond, R. P. Volante, I. Shinkai, Total Synthesis of the Immunosuppressant (–)-FK506, *J. Am. Chem. Soc.* **1989**, *111*, 1157–1159.

J. J. Siekierka, S. H. Y. Hung, M. Poe, C. S. Lin, N. H. Sigal, A Cytostolic Binding Protein for the Immunosuppressant FK506 has Peptidyl-Prolyl Isomerase Activity but is Distinct from Cyclophilin, *Nature* **1989**, *341*, 755–757.

T. K. Jones, R. A. Reamer, R. Desmond, S. G. Mills, Chemistry of Tricarbonyl Hemiketals and Application of Evans' Technology to the Total Synthesis of the Immunosuppressant (–)-FK506, *J. Am. Chem. Soc.* **1990**, *112*, 2998–3017.

M. Nakatsuka, J. A. Ragan, T. Sammakia, D. B. Smith, D. E. Uehling, S. L. Schreiber, Total Synthesis of FK506 and an FKBP Probe Reagent, $(C_8,C_9\text{-}^{13}C_2)$-FK506, *J. Am. Chem. Soc.* **1990**, *112*, 5583–5601.

S. L. Schreiber, G. R. Crabtree, The Mechanism of Action of Cyclosporin A and FK506, *Immunology Today* **1992**, *13*, 136–142.

M. K. Rosen, S. L. Schreiber, Natural Products as Probes of Cellular Function: Studies of Immunophilins, *Angew. Chem. Int. Ed. Engl.* **1992**, *31*, 384–400.

K. C. Nicolaou, T. K. Chakraborty, A. D. Piscopio, N. Minowa, P. Bertinato, Total Synthesis of Rapamycin, *J. Am. Chem. Soc.* **1993**, *115*, 4419–4420.

B. Werth, *Billion-Dollar Molecule: The Quest for the Perfect Drug*, Simon and Schuster, New York, **1994**.

K. C. Nicolaou, E. J. Sorensen, *Classics in Total Synthesis*, Wiley-VCH, Weinheim, **1996**, pp. 565–631.

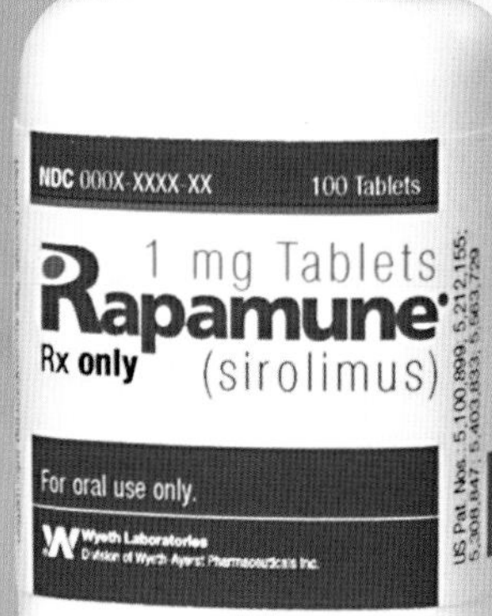

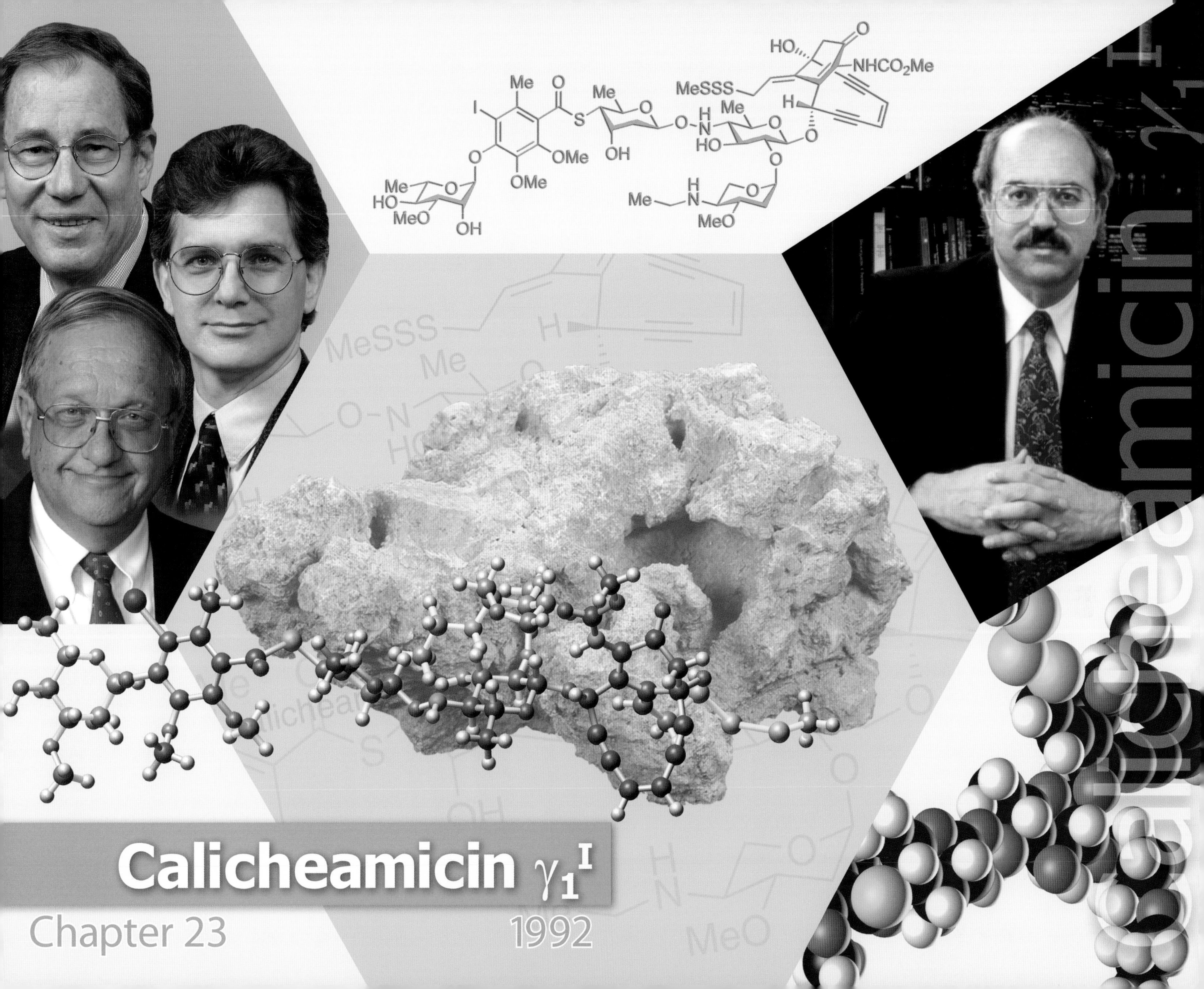

Calicheamicin γ_1^I

Chapter 23

1992

Box 1 The Bergman cycloaromatization reaction

The pebble (*caliche*) from which the calicheamicin γ_1^I producing bacteria were isolated

Sometime in the 1980s, beside a Texas highway, a touring scientist from Lederle Laboratories picked up a pebble. Living inside this chalky rock were bacteria, subsequently identified as *Micromonospora echinospora*, subspecies *calichensis*, the name having been derived from the Greek word for a small lime stone pebble, *chaliche*. These bacteria were cultured in the laboratory and their metabolites screened for useful biological activity. This led to the discovery of a phenomenally cytotoxic agent, called calicheamicin γ_1^I, which was destined to become the most prominent member of the then newly emerging enediyne family of antitumor antibiotics. Apparently, these bacteria had evolved over the eons with a super-weapon for waging chemical warfare against their competitors and potential predators.

Harnessing calicheamicin γ_1^I and using the valuable knowledge derived from this beautiful molecule, scientists have been able to develop a new weapon for use against cancer, in the form of Mylotarg®. This drug represents a significant new advance in the treatment of cancer wherein the cytotoxic molecule is delivered, like a 'guided missile', directly to cancer cells. To achieve this feat, calicheamicin γ_1^I was chemically linked to a monoclonal antibody, forming a unique molecular assembly that constitutes an exciting new approach to cancer chemotherapy (for a more detailed discussion of cancer and its treatment, see Chapter 25). To fully appreciate the development and efficacy of this approach, we must first consider the chemistry of calicheamicin γ_1^I and, in particular, the properties of the enediyne structural motif.

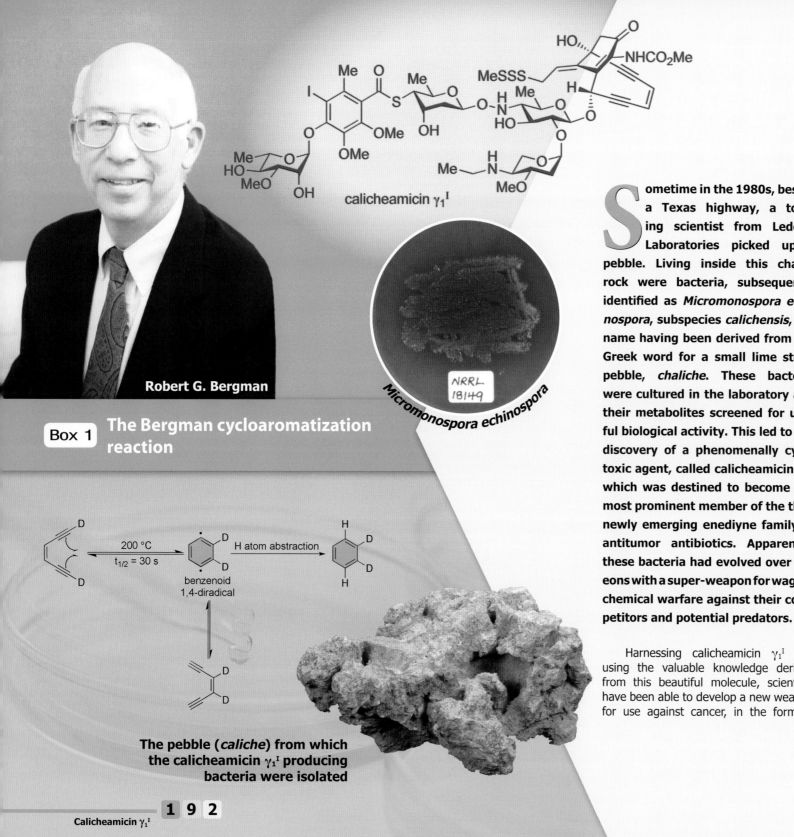

Annulenes and aromaticity make seemingly esoteric subjects for study within organic chemistry. In spite of this, these topics captivated a generation of pioneering organic chemists during the second half of the twentieth century. In 1971, Robert G. Bergman, then at the California Institute of Technology (CalTech, USA), made an important observation. Following the works of Franz Sondheimer (University College London, UK) and Satoru Masamune (University of Alberta, Canada), he designed experiments that demonstrated the cycloaromatization reaction of conjugated enediyne compounds through a hypothetical benzenoid diradical intermediate (Box 1). This process came to be known as the Bergman cyclization reaction. Bergman's work would later become central to the study of the calicheamicins and related naturally occurring enediynes. A Bergman cyclization is the key event in the mechanism by which these compounds exert their cytotoxic activity. Cell death is caused when the calicheamicin γ_1^I molecule undergoes a series of transformations within the target cell that alter its conformation, triggering a Bergman cycloaromatization to give a benzenoid diradical that severs both strands of any nearby DNA molecule.

An early indication of the small victory over cancer that was to follow came in 1985 when the antitumor antibiotic neocarzinostatin was discovered in Japan (Box 2). This material consists of a highly unstable chromophore molecule, which includes an epoxy-diyne moiety shielded within a protective protein molecule. Further investigations into this seminal finding were, however, overshadowed by a spate of papers published in 1987 by scientists from the Lederle Laboratories and Bristol-Myers companies in which the structures of the calicheamicins and their cousins, the esperamicins, were disclosed. The biological assay-driven identification of these extraordinarily potent anticancer antibiotics had provided the impetus for their isolation and structural elucidation, which revealed that this new class of molecules boasted magnificent molecular architectures, including a particularly intriguing and unprecedented, conjugated, enediyne structural motif. Since these early reports, a number of other enediyne natural products have been discovered. Their mesmerizing chemical structures, combined with their spectacular biological properties, generated great interest and excitement and prompted many chemists to initiate programs directed towards their total synthesis. In order to meet this formidable synthetic challenge, chemists first had to develop an understanding of how these beautiful molecules exert their lethal action in nature.

Even more than the other naturally occurring enediynes, calicheamicin γ_1^{I} is a masterpiece of molecular design. The carefully engineered central core incorporates the deadly enediyne motif along with a number of finely balanced architectural features, ensuring that the cytotoxic effects of the molecule are manifested only in the right circumstances. Thus, the producing bacteria can manufacture this powerful cytotoxin and yet avoid falling victim to it themselves. The first key element of the design is the incorporation of the enediyne group within a ten-membered ring. The acyclic enediynes originally investigated by Bergman (Box 1) require excessively high temperatures to effect the cyclization. Chemists have since constructed enediynes within a range of ring sizes and studied their stability. Those within ten-membered rings were found to undergo Bergman cyclization at body temperature. Within calicheamicin γ_1^{I}, the ends of the enediyne motif are held at a distance by the rigid shape of the bridged bicyclic skeleton, preventing cyclization. The first step in the cascade that brings about the Bergman cyclization is the attack of an external nucleophile (such as glutathione) on the trisulfide unit of the natural product (Box 3). This generates a thiolate anion, which is perfectly positioned to attack the enone group in an intramolecular conjugate addition. The conversion of the bridgehead enone into a tricyclic ketone causes a structural change, bringing the ends of the enediyne system closer together and triggering the Bergman cyclization reaction. The resulting diradical abstracts hydrogen atoms from the sugar-phosphate backbone of the DNA of the target cell, initiating a further sequence leading

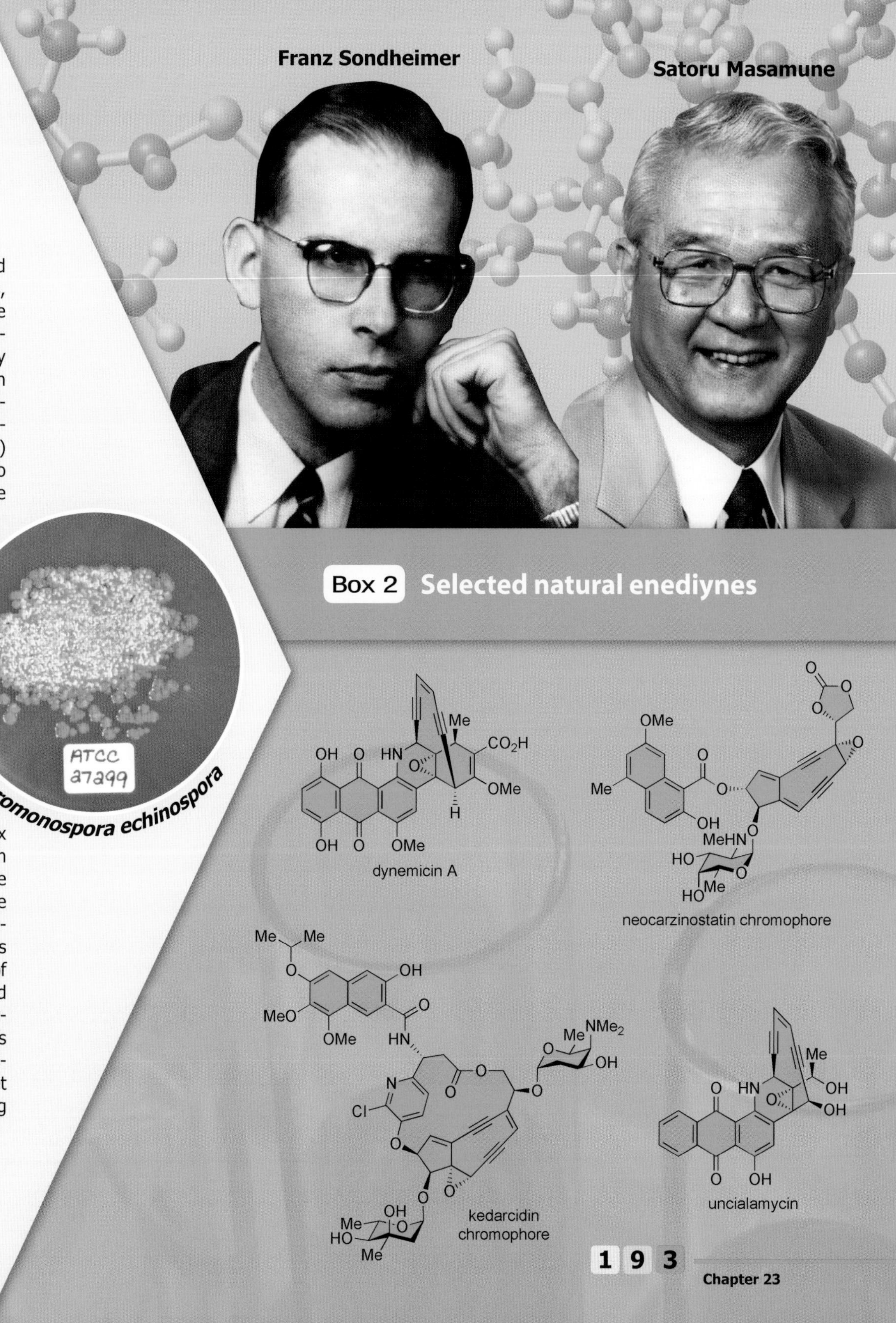

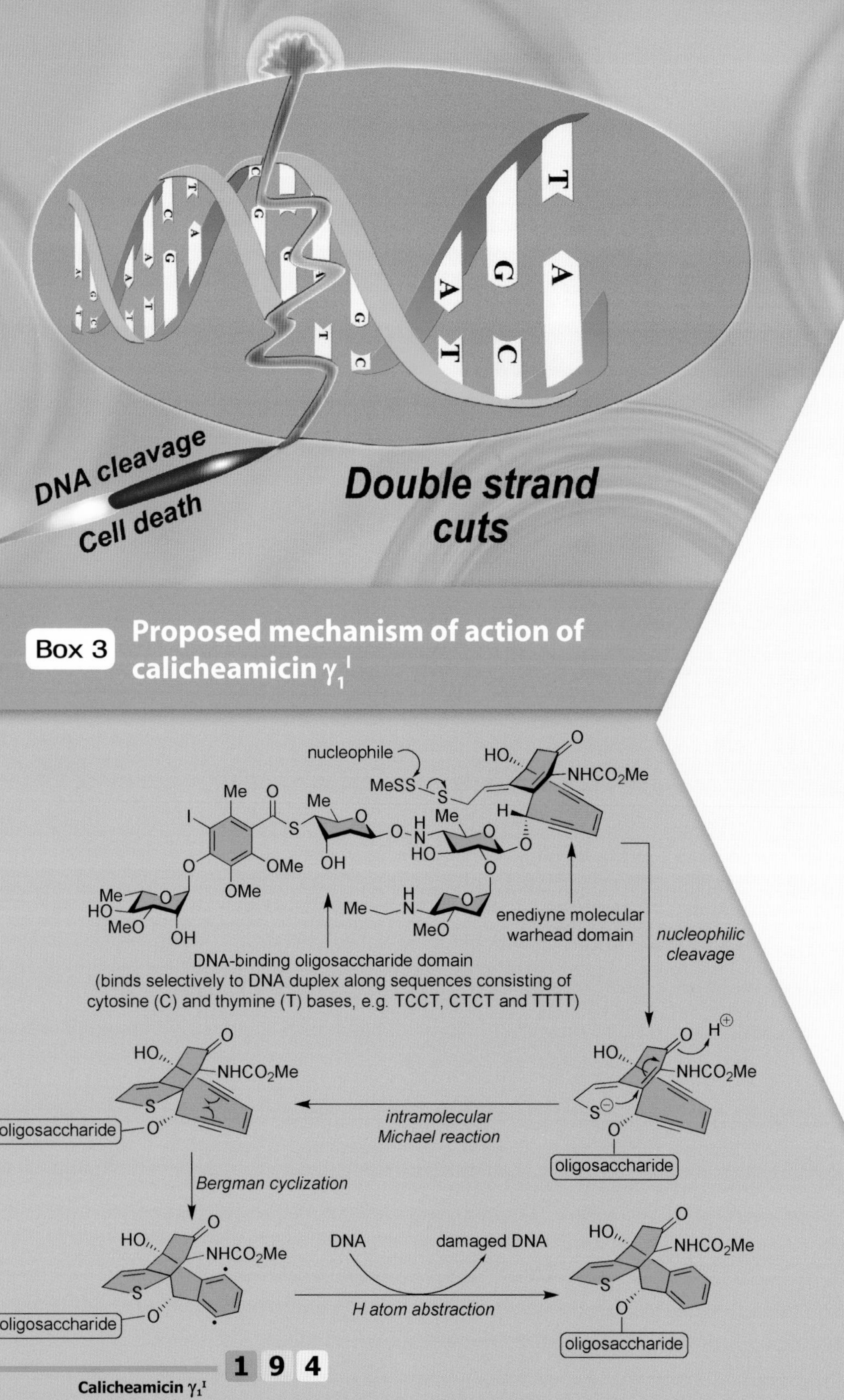

to double strand cuts in the DNA and eventual cell death. Furthermore, the oligosaccharide 'tail' of calicheamicin γ_1^I acts as a targeting device, selectively delivering the 'warhead' to the target DNA by recognizing specific base-pair sequences in the minor groove of the double helix. The sophistication and elegance of this beautiful design proved irresistible to chemists, who hoped to adapt this highly potent compound for use in the treatment of cancer and to mimic its action by designing their own enediyne molecules.

Inspired by the stunning architecture of calicheamicin γ_1^I, chemists began to consider how to achieve its total synthesis. Despite the daunting challenge posed by such a target, some dared to embark on a journey that was sure to be full of adventure and risk, reasoning that the opportunities it held were too valuable to ignore. The first to complete this task were K. C. Nicolaou and his group, who published their total synthesis of calicheamicin γ_1^I in 1992, after a five-year campaign. Their work on the enediynes was multifaceted from the start and involved concurrent investigations on several fronts, including molecular design, chemical synthesis, and chemical biology studies. The crowning achievement was the total synthesis of calicheamicin γ_1^I, one of the most complex naturally occurring substances ever to be synthesized in the laboratory. The retrosynthetic analysis shown in Box 4 indicates the final strategy employed, although, as is common in such endeavors, the original plan had gone through many revisions over the course of the campaign. The latter stages of the synthesis are shown in Box 5. These schemes cannot, however, convey the frustrating lows and exhilarating highs this synthetic adventure held for the chemists involved, and the reader is encouraged to delve into the more detailed narratives listed in the Further Reading box for a full account.

Calicheamicin γ_1^I bound to DNA

Calicheamicin γ_1^I is composed of two distinct structural domains. The first, a highly complex oligosaccharide domain, confers the sequence specificity to the binding of this molecule within the minor groove of the DNA duplex, while the second domain, the aglycon, consists of the enediyne warhead. The oligosaccharide fragment is far from being just a simple carbohydrate anchor for the impressive enediyne moiety. Its own intimidating structure includes a fragile N–O bond that bridges two of the carbohydrate units through a challenging β-glycoside bond, and a thioester linkage connecting a rare hexasubstituted aromatic ring to one of the sugar units. Each of the building blocks identified in the retrosynthetic analysis (Box 4) of this part of the molecule would require its own careful assembly. For the aglycon, the retrosynthetic analysis suggested the strategy shown in which

Samuel J. Danishefsky

Andrew G. Myers

Masahiro Hirama

key bonds would be formed by an intramolecular acetylide addition to an aldehyde, an intramolecular nitrile oxide/olefin cycloaddition, and a Sonogashira coupling. Having decided upon a synthetic strategy, the Nicolaou group proceeded to construct the key building blocks required for the final stages of the synthesis. Needless to say, the roads to these goals were both arduous and demanding, for each of these intermediates presented unique synthetic challenges, but the endeavor also proved rewarding and, in a relatively short period of time, the two main fragments of calicheamicin γ_1^I had been synthesized and were ready for unification. A successful glycosidation reaction, facilitated by a Lewis acid at low temperature, locked the entire carbon skeleton of the target molecule in place (Box 5). Following the introduction of the trisulfide unit (the molecule's triggering device) and the unmasking of several protected functional groups, the total synthesis of calicheamicin γ_1^I was completed on the September 29, 1992, at the Scripps laboratories of the Nicolaou team. With this extraordinary structure reached by total synthesis, the field was entering a golden age of unprecedented accomplishments, as the subsequent decade was to witness.

Samuel J. Danishefsky and his group, then at Yale University (USA) accomplished a second total synthesis of calicheamicin γ_1^I, in 1994, adding a host of further discoveries and inventions to the collection of those made during the Nicolaou synthesis. The challenges posed by the enediyne class led many groups to attempt syntheses of the various members, however, only a few such adventures have ended in success. Amongst these successes are two independent syntheses of dynemicin A (Box 2), published in 1995, by the groups of Andrew G. Myers (then at CalTech) and Danishefsky (now

K. C. Nicolaou

Box 4 The Nicolaou group's retrosynthetic analysis of calicheamicin γ_1^I

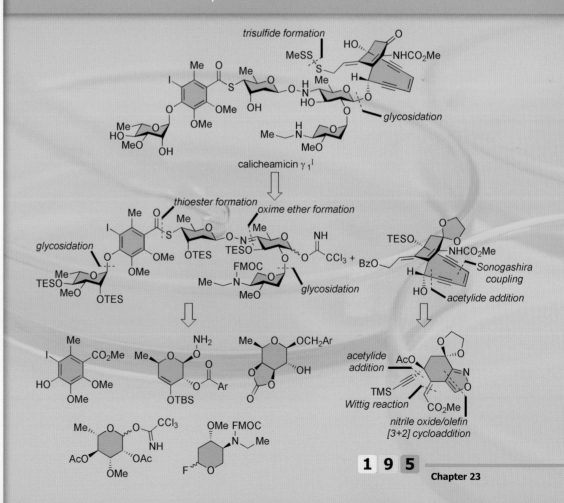

trisulfide formation

glycosidation

calicheamicin γ_1^I

thioester formation

oxime ether formation

glycosidation

glycosidation

Sonogashira coupling

acetylide addition

acetylide addition

Wittig reaction

nitrile oxide/olefin [3+2] cycloaddition

The Nicolaou calicheamicin γ_1^I team

Box 5 The final stages of the Nicolaou group's total synthesis of calicheamicin γ_1^I

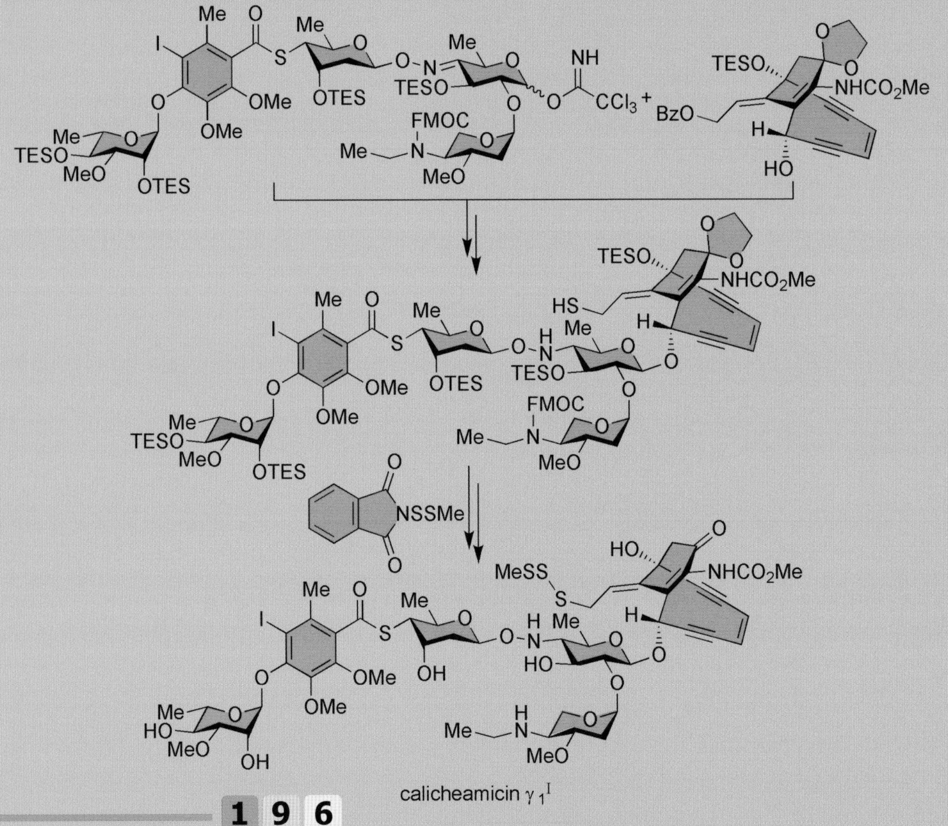

calicheamicin γ_1^I

at the Sloan–Kettering Institute and Columbia University, USA), and two of the neocarzinostatin chromophore, by the Myers group (now at Harvard University, USA) in 1998, and the Masahiro Hirama group (Tohoku University, Japan) in 2006.

César Milstein

Despite falling to the advances of total synthesis in the early 1990s, it would be a further eight years before calicheamicin γ_1^I could be tamed to afford an approved drug for the treatment of cancer. The development of Mylotarg® was made possible not only by advancements in chemistry, but also by the increased understanding of our immune system, which had developed over the preceding thirty years. In 1984, the Nobel Prize in Physiology or Medicine was awarded to Niels K. Jerne, Georges J. F. Köhler, and César Milstein, in part for their "discovery of the principle for production of monoclonal antibodies." To generate a monoclonal antibody (which has been chosen because it will attack disease ridden cells), the antibody-synthesizing cells (B cells) are fused to specially treated tumor cells, which proliferate rapidly. The result is a hybrid cell line that can produce large numbers of monoclonal antibodies with absolute integrity (i.e. clones). Early attempts to use this approach to treat disease in humans were complicated by the fact that the patient's own immune system often attacked the foreign antibodies, causing a worsening of the illness. The problem of rejection has been overcome in recent years by transplanting the engineered antigen recognition domain of the cloned antibody onto human antibodies to furnish chimeric (~66 % human) or humanized (~90 % human) antibodies. This new technique has now been applied to the production of antibodies designed to attack specific tumor cells with admirable success. The anticancer drugs Rituxan® (developed by Idec, now Biogen Idec) and Herceptin® (developed by Genentech) have recently been introduced to treat non-Hodgkin's lymphoma and breast cancer, respectively, and are already significantly enhancing the prospects for patients. The monoclonal

Georges J. F. Köhler and Niels K. Jerne

antibodies are frequently administered in combination with a chemotherapeutic agent, such as Taxol® (see Chapter 25), in order to double the force of the attack on the tumor. The application of monoclonal antibodies has expanded considerably as a result of phage display. This technique was pioneered by George P. Smith (University of Missouri, Columbia, USA), and further developed by the groups of Richard A. Lerner (The Scripps Research Institute, La Jolla, USA) and Sir Gregory Winter (Medical Research Council Laboratories, University of Cambridge, UK), for the production and screening of very large combinatorial libraries of antibodies. This application of phage display was exploited by scientists at Abbott Laboratories to discover Humira®, a monoclonal antibody for the treatment of rheumatoid and psoriatic arthritis. Further discussion on antibody therapies can be found in Chapter 34.

Mylotarg® represents the natural evolution of the antibody/chemotherapy combination strategy, as it consists of a chemotherapeutic agent, calicheamicin γ_1^I, chemically bound to a delivery system targeting certain leukemia cells, a monoclonal antibody. The antibody domain was designed to bind specifically to an adhesion protein found on the surface of leukemia cells. The pioneering research that culminated in the approval of Mylotarg® for the treatment of relapsed acute myeloid leukemia was led by George Ellestad, Philip Hamann, and Janis Upeslacis, working at Lederle Laboratories (later American Home Products and now Wyeth). Their team modified calicheamicin γ_1^I and designed a linker that maintained the function of the cytotoxic agent as well as that of the antibody. The overall result was a powerful molecular weapon capable of delivering a very effective and carefully directed two-fold blow against leukemia cells, which utilized the very best of what the biological and chemical sciences had to offer at the end of the twentieth century. This advance, which comes closer than any previous attempts to reaching the magic bullet ideal of chemotherapy, is continuing to impress clinicians, who are now able to treat certain types of cancer with unprecedented rates of success. During the twenty-first century, we can expect the development of more targeted chemotherapeutic agents in order to help in the fight against cancer.

The design of calicheamicin γ_1^I has been honed by nature to form a molecular machine that is ideally suited to perform its given function. The evolutionary refinement of this molecule has now been combined with the ingenuity of scientists to furnish a formidable weapon against

Leukemia cell

Herceptin®

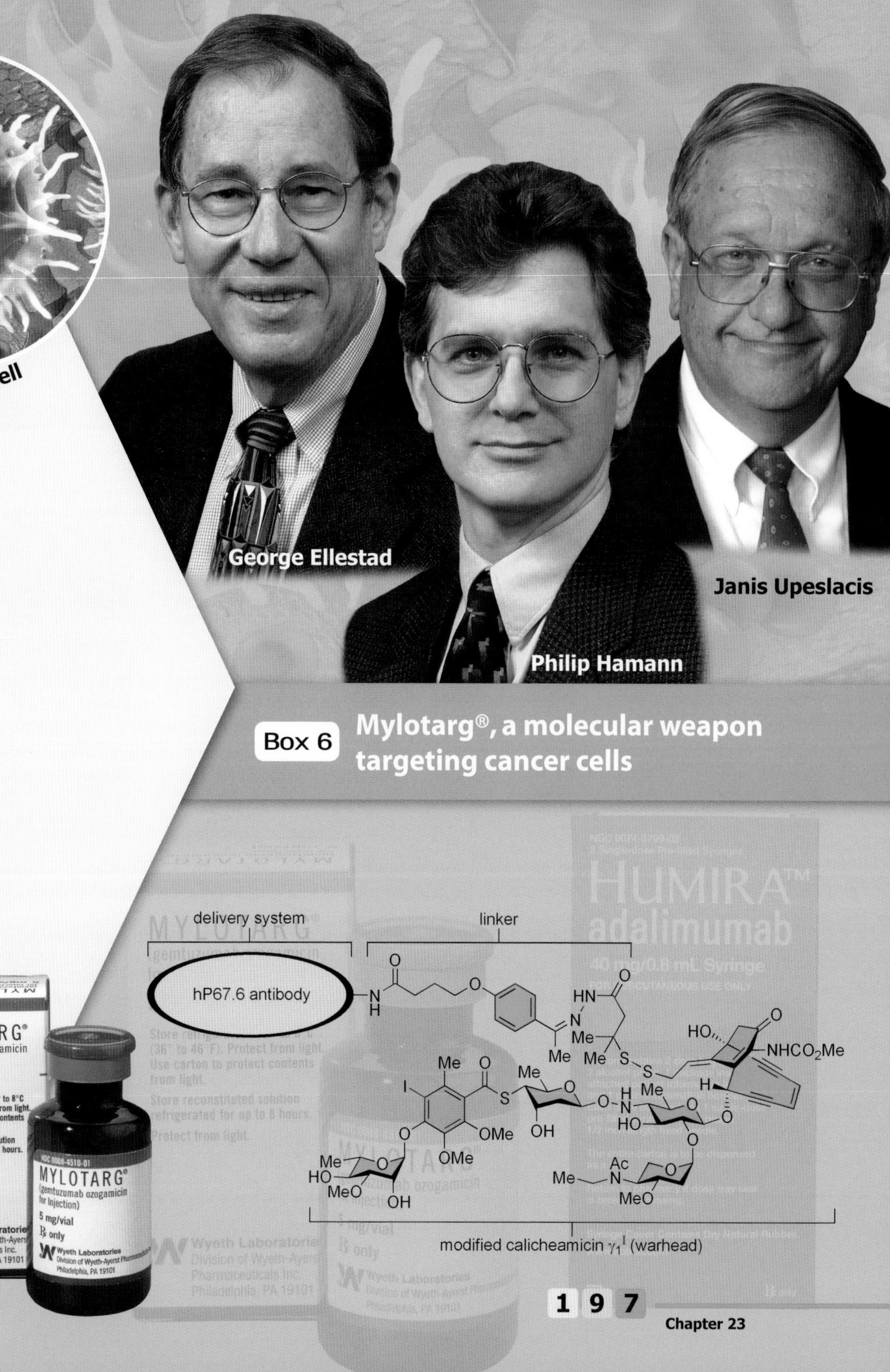

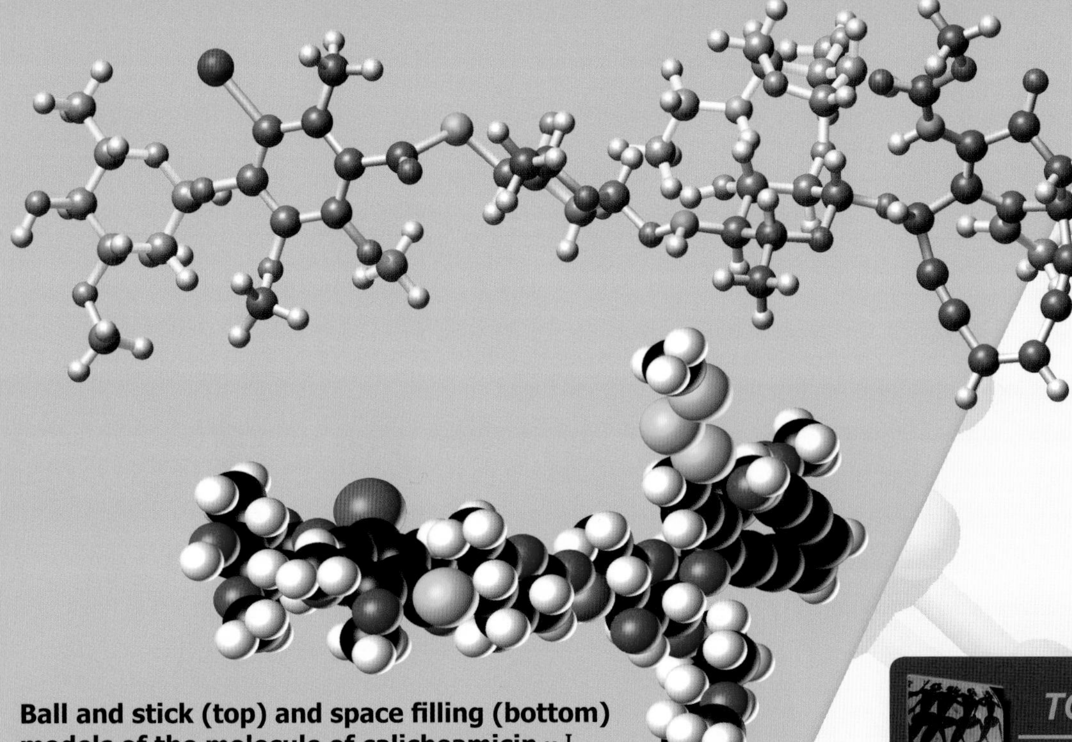

Ball and stick (top) and space filling (bottom) models of the molecule of calicheamicin γ_1^I

cancer. The conquest of calicheamicin γ_1^I by chemical synthesis facilitated the design and synthesis of numerous enediyne molecules that mimic the biological action of this natural product (some examples are shown below). Synthetic enediynes have allowed chemists to probe multiple chemical and biological issues relevant to biomedical research. Synthetic organic chemistry has, once again, played a pivotal role in organizing and driving forward the frontiers of the combined biological and chemical sciences for the betterment of mankind.

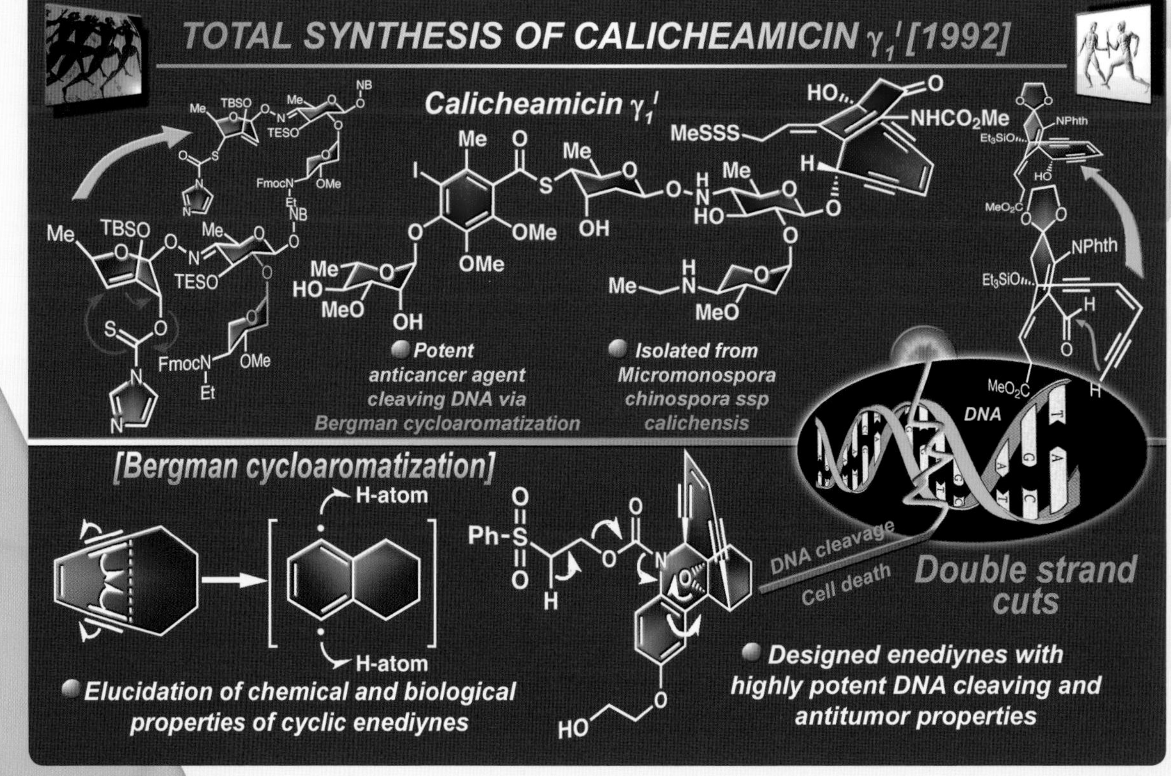

TOTAL SYNTHESIS OF CALICHEAMICIN γ_1^I [1992]
Calicheamicin γ_1^I
Potent anticancer agent cleaving DNA via Bergman cycloaromatization
Isolated from Micromonospora chinospora ssp calichensis
DNA
[Bergman cycloaromatization]
H-atom
H-atom
Elucidation of chemical and biological properties of cyclic enediynes
DNA cleavage
Cell death
Double strand cuts
Designed enediynes with highly potent DNA cleaving and antitumor properties

Further Reading

K. C. Nicolaou, W.-M. Dai, Chemistry and Biology of the Enediyne Antibiotics, *Angew. Chem. Int. Ed. Engl.* **1991**, *30*, 1387–1416.

K. C. Nicolaou, The Battle of Calicheamicin, *Angew. Chem. Int. Ed. Engl.* **1993**, *32*, 1377–1385.

K. C. Nicolaou, E. J. Sorensen, *Classics in Total Synthesis*, Wiley-VCH, Weinheim, **1996**, pp. 523–564.

K. C. Nicolaou, The Magic of Enediyne Chemistry, *Chem. Br.* **1994**, *30*, 33–37.

P. Carter, Improving the Efficacy of Antibody-Based Cancer Therapies, *Nature Rev. Cancer* **2001**, *1*, 118–129.

E. L. Sievers, M. Linenberger, Mylotarg: Antibody-Targeted Chemotherapy Comes of Age, *Curr. Opinion Oncol.* **2001**, *13*, 522–527.

C. Ezzell, Magic Bullets Fly Again, *Sci. Am.* **2001**, *285*, 34–41.

Palytoxin
Chapter 24
1994
palytoxin

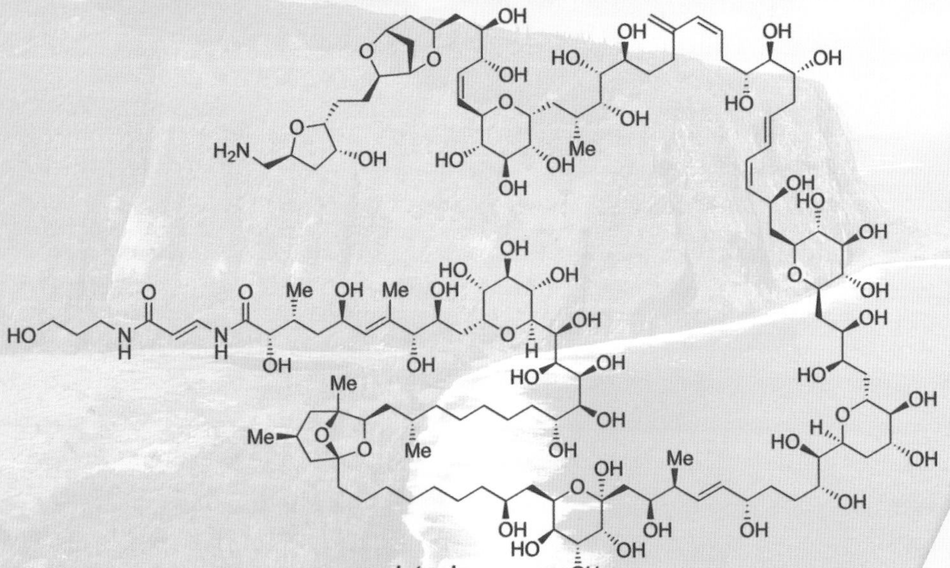

palytoxin

Box 1 Molecular structures of selected polyether marine toxins

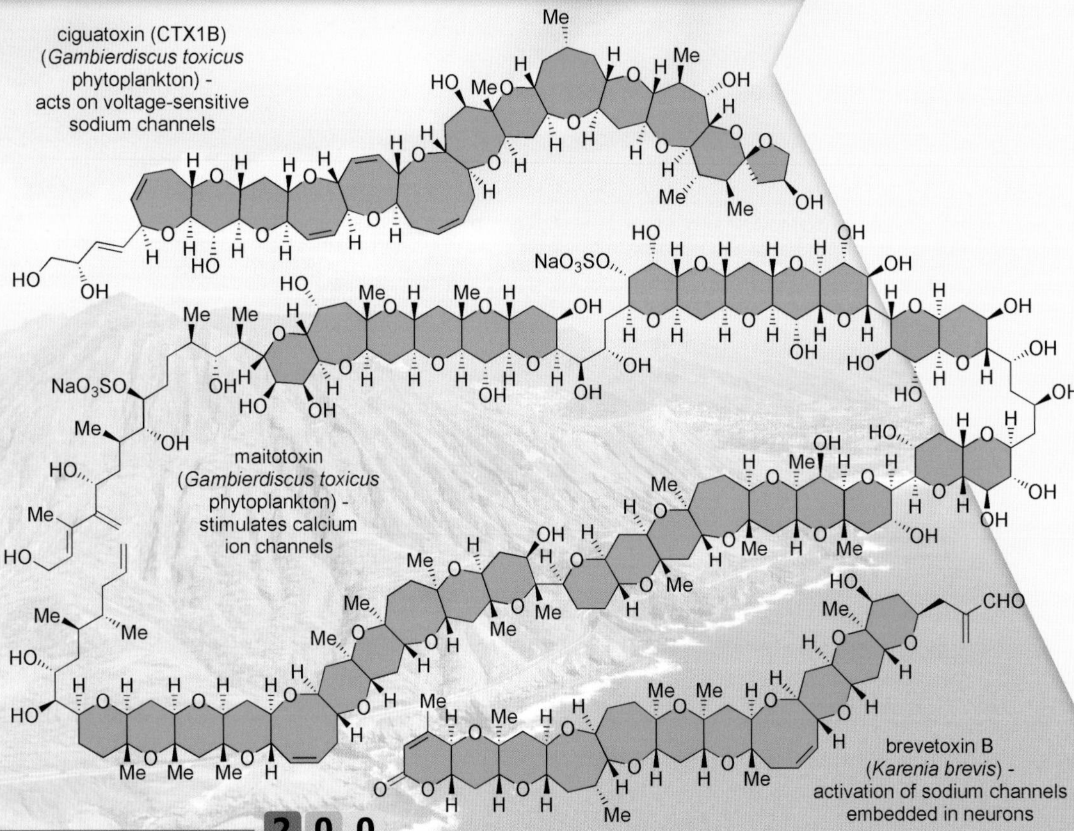

ciguatoxin (CTX1B) (*Gambierdiscus toxicus* phytoplankton) - acts on voltage-sensitive sodium channels

maitotoxin (*Gambierdiscus toxicus* phytoplankton) - stimulates calcium ion channels

brevetoxin B (*Karenia brevis*) - activation of sodium channels embedded in neurons

According to a legend from the oral story-telling traditions of the Hawaiian archipelago, there was once a man living in the tropical paradise of Hana (a region on the southeastern coast of the island of Maui) who attracted people's attention because of his feverish planting and harvesting habits. He appeared to tend his taro patch from dawn until dusk every single day, and within the local community he was regarded with deep suspicion for this compulsive devotion to his crops. Tragedy beset his neighborhood when, after each fishing trip, the group of fishermen returning from a night's industrious casting of the nets would arrive one companion short. For a time, the people chattered busily about the cause of this fearful mystery, searching for and expounding on many disparate explanations. Finally, the consensus of opinion settled on the taro patch farmer as somehow being responsible, provoking a band of the village's menfolk to confront the strange man. When they ripped the clothes off his broad muscular frame they discovered the fearsome jaws of a shark on his back. Their terrified response was to kill the man and burn his body. His ashes were then tossed into the sea in the hope that they would be washed far away, never to plague the village again, but instead they landed in a nearby tide pool. Soon afterwards, the seaweed (*limu*) in the spot where his remains had settled became poisonous, and the tide pool was declared taboo (*kapu*). Locals covered the pool with stones and its location was shrouded in secrecy, never discussed openly since it was understood that disaster would befall anyone who attempted to gather the deadly seaweed of Hana (*limu make o Hana*).

In 1961 a group of researchers from the University of Hawaii in Honolulu (USA), led by Paul J. Scheuer and Richard E. Moore, intrigued but undeterred by the legend of the toxic *limu*, set out to find the pool and gather the seaweed if, indeed, it did exist. These particular chemists had a long-standing fascination with

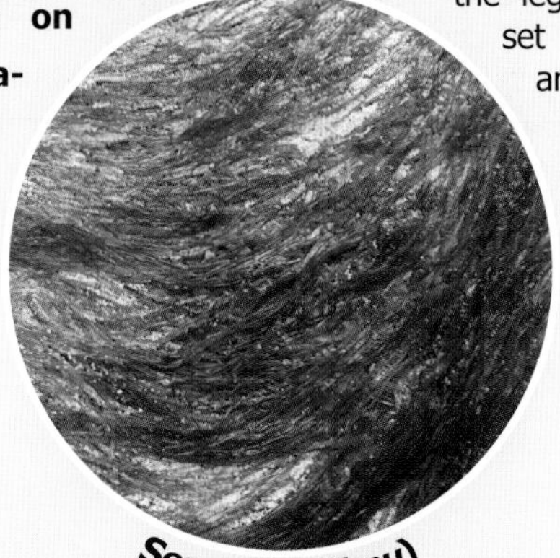

Seaweed (*limu*)

the study of marine toxins, having already investigated the biological origins of the ciguatoxins (Box 1), which accumulate in certain fish and cause many incidences of severe food poisoning in humans. When these researchers arrived in Hana they were met with some antagonism and were reminded frequently of the impending misfortune dictated by the *kapu* should they insist on harvesting the *limu*. Eventually, however, a chain of informants led them to a tide pool located at the end of a lava flow at Muolea, and the *limu* was sampled and taken back to the laboratory. No misfortune appeared to accompany their raid on the *limu*, that is unless a fire of unknown origin that destroyed the main buildings of the Hawaii Marine Laboratories the very same afternoon can be counted!

Palythoa coral

Extracts from the sampled organism were shown to be highly toxic, such that even after significant dilution the solutions were powerful enough to kill mice injected with very small doses. Alongside these investigations into the lethality of the extracts obtained from the tide pool material, the scientists were also focusing on the taxonomy of the toxin producer. It was found that the organism was not a seaweed at all, but an animal belonging to the *Coelenterate* phylum, order *Zoantharia* and genus *Palythoa*, hence the toxin was named palytoxin. These colonial creatures are polyps, similar to the more familiar sea anemones. Many zoanthids have since been shown to produce palytoxin, or similar compounds, which are found in the mucous they excrete as a defense mechanism to protect their soft immobile bodies from predators. It should be noted that the manufacturer of the poison may still turn out to be a microorganism (e.g. a dinoflagellate) existing symbiotically within the body of the zoanthid, as has been the case with many other marine toxins (Box 1).

With regard to the structure of palytoxin, the intrigued researchers could do little more at the time than estimate its molecular weight as being around 3,300. This figure was extraordinary within the context of the toxins familiar to these experts, such as tetrodotoxin, saxitoxin, or batrachotoxinin A, all of which register molecular weights of under 500 (Box 2). This feature was made all the more remarkable when the scientists determined that palytoxin had no bio-polymeric fragments, such as repeating amino acid chains or sugar units that are easily assembled in biological systems to yield compounds with high molecular weight. Palytoxin was instead a fantastically large and complex secondary metabolite. To this day only one larger secondary metabolite has been identified, a gigantic molecule named maitotoxin, another marine toxin (Box 1).

Puffer fish (fugu)

Palytoxin was first collected in December 1961, and its isolation published in the journal *Science* in 1971. After a marathon investigation by Richard Moore at the University of Hawaii, its

Box 3 Retrosynthetic analysis of palytoxin

Nozaki–Hiyama–Kishi coupling

Wittig reaction

Suzuki coupling

Nozaki–Hiyama–Kishi coupling

amide bond formation

Wittig reaction; then hydrogenation

Horner–Wadsworth–Emmons olefination

palytoxin

gross structure was finally elucidated in 1981. A second group, led by Yoshimasa Hirata (Nagoya University, Japan), also published the completed gross structure of palytoxin in 1981. The composition of this water-soluble poison defied belief, as it consisted of a huge linear chain, interspersed with pyran rings, which could only be easily drawn in diagrams on an ordinary page if the molecule was curled back around on itself several times. Palytoxin is one of the most toxic natural products known. Only a few proteins, such as ricin (from plants), tetanus toxins (from bacteria), the infamous botulinum toxin (now used in Botox® therapy), and the marine toxins ciguatoxin and maitotoxin, induce deadly effects at lower concentrations than this powerful marine toxin. Nanomolar concentrations of palytoxin (one nanomole = one thousand millionth of a mole or 10^{-9} moles) are enough to increase membrane permeability of vascular muscle cells for sodium and potassium cations. At higher concentrations the same cells become permeable to larger molecules such as adenosine triphosphate (ATP). The increased permeability leads to depolarization and muscle contraction/spasm in every susceptible organ. Consequently, palytoxin has the highest coronary vasoconstricting activity of any known substance, and it can also affect nerve cells. Perhaps naively, palytoxin was initially tested as a potential anticancer therapy and as a local anesthetic, but despite possessing the desired activity for both these conditions it is rather unlikely that it could ever be used because of its extreme toxicity.

Yoshimasa Hirata

One eminent synthetic chemist, Yoshito Kishi (first at Nagoya University, Japan, and now at Harvard University, USA), when casting around for targets to challenge his synthetic skills, has given a priority to architecturally striking natural toxins. His fascination with these intricate and dangerous molecules began with the puffer fish toxin, tetrodotoxin (Box 2), which he synthesized to much acclaim in 1972. Puffer fish (fugu) is a fashionable oriental delicacy, part of its attraction being the tingling and euphoric sensations brought on by the consumption of tetrodotoxin. The line between enjoying one's dinner and dying, however, is rather fine, so many places allow the preparation of puffer fish only by strictly licensed chefs. Despite this caution, every year a number of diners become victims of accidental death due to consumption of too much fugu, or samples where the organs in which tetrodotoxin concentrates have not been carefully removed. Kishi later went on to complete the synthesis of more toxins from such diverse sources as shellfish to those exuded from the skin of Colombian poison arrow frogs, such as batrachotoxinin A and histrionicotoxin (Box 2). The climactic achievement in this research field was the monumental accomplishment of the total synthesis of palytoxin, completed by the Kishi group in 1994.

Kishi had been introduced to the proposed gross structure of palytoxin in Nagoya by his Ph.D. supervisor, Yoshimasa Hirata (who contributed one of the two simultaneous structural elucidations), after the death of a friend and mentor had left Kishi dismayed. The person mourned by Kishi was his postdoctoral advisor, the legendary chemist Robert B. Woodward, of whom you have heard much throughout this book. Hirata knew that more powerful than kind words of consolation was the provision of inspiration for Kishi to continue his work in synthetic organic chemistry as a fitting memorial to the genius of his teacher. However, before a program directed towards the synthesis of this colossal target could begin, the stereochemical ambiguities had first to be addressed. Palytoxin possesses seventy-one stereochemical elements, comprising sixty-four asymmetric carbon atoms and seven geometric double bond isomers. When the structure of palytoxin was disclosed in the two literature reports of 1981, it was what is referred to as a gross structure only, meaning the atoms and their connectivities were known, but the discrete spatial arrangement (stereochemistry) was not established. In effect, the reported structure encompassed 2^{71} (2,361,183,241,434,822,606,848) possible stereoisomeric combinations, only one of which represented the true structure of palytoxin. While certain inferences could be made, and some parts of this huge molecule were better appreciated than others, Kishi and his group faced a monumental obstacle; the absolute arrangement remained unclear for at least twenty-seven of the sixty-four asymmetric carbons. As a result, the Kishi group spent a laborious two-year period just validating the structure of their chosen target. By using a combination of degradation studies, synthetic investigations, and spectroscopic analysis they finally identified the *bona fide* composition of palytoxin in its entirety. Now the real chase could begin in earnest, a campaign of enormous magnitude to synthesize a molecule that would require extreme levels of dedication, persistence, and insight.

Palythoa coral

Normally, convergent syntheses are spoken of as being desirable as they increase the efficiency and elegance of the work. However, in this instance, the massive size of palytoxin dictated such an approach to ensure the task could be completed at all. Highlights of the retrosynthetic analysis employed by the Kishi group, which broke the molecule up into several more manageable fragments, are shown in Box 3. During the synthetic odyssey that followed, these researchers pioneered a number of valuable new reactions and methods, foremost of which was

Box 4 The Nozaki–Hiyama–Kishi coupling reaction in the total synthesis of palytoxin

$CrCl_2$, $NiCl_2$ cat. (90 % yield)

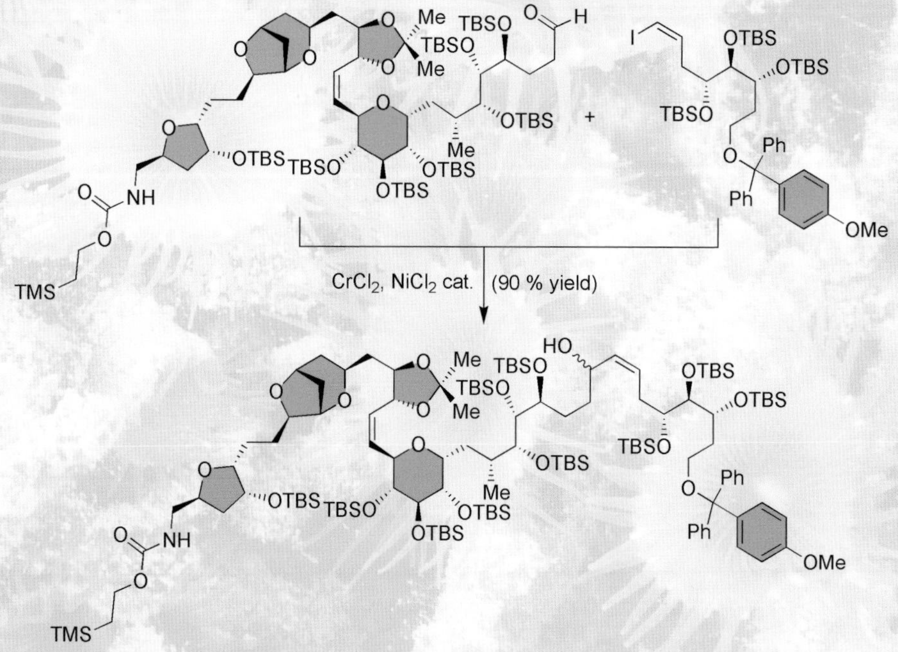

Box 5 Final steps of Kishi's total synthesis of palytoxin

palytoxin carboxylic acid

palytoxin

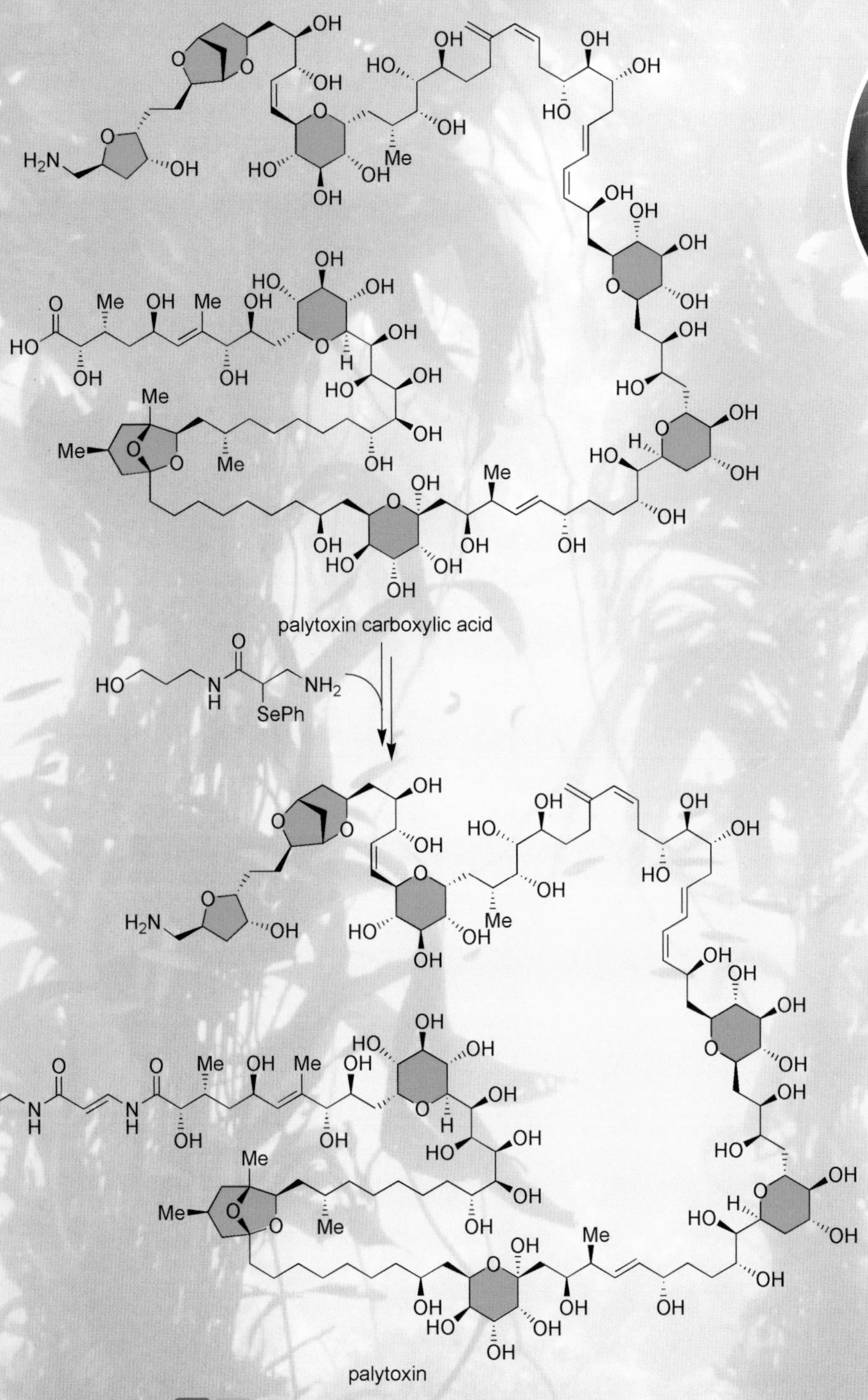

Seaweed (*limu*)

a selective addition of vinyl halide to an aldehyde under mild conditions to afford the corresponding allylic alcohol. The reaction, first discovered in a primitive form by Hitosi Nozaki (Kyoto University, Japan) and co-workers in 1983, involves an organochromium species with nickel catalysis playing a pivotal role. One early example of the use of this synthetic method in complex situations can be found in the synthesis of palytoxin (Box 4), and this reaction has since found widespread application in the synthesis of complex molecules. Interestingly, the advantageous role of the nickel was discovered by pure serendipity, for this metal was a contaminant in commercially available chromium(II) chloride. Kishi's synthesis of palytoxin also showcased a palladium-catalyzed, thallium hydroxide-assisted construction of a *cis-trans* conjugated diene from a vinyl iodide and a vinyl boronic acid, as well as a new means to construct *N*-acyl vinylogous ureas.

The Kishi group had completed the construction of the bulk of the molecule by 1989, reaching an advanced staging post called palytoxin carboxylic acid (PCA); however, the ascent to the ultimate summit was not to be achieved for a further five years. PCA, itself a natural product isolated from Okinawan *Palythoa*, is extremely sensitive to both acidic and basic conditions, meaning that the formation of the required amide bond to transform it into palytoxin was a formidable task. Furthermore, the *N*-acyl vinylogous urea precursors initially employed yielded exclusively the undesired *cis*-geometry about the newly installed double bond. Fortunately, a solution was found in the form of a phenylselenide variant accompanied by photochemical isomerization of the

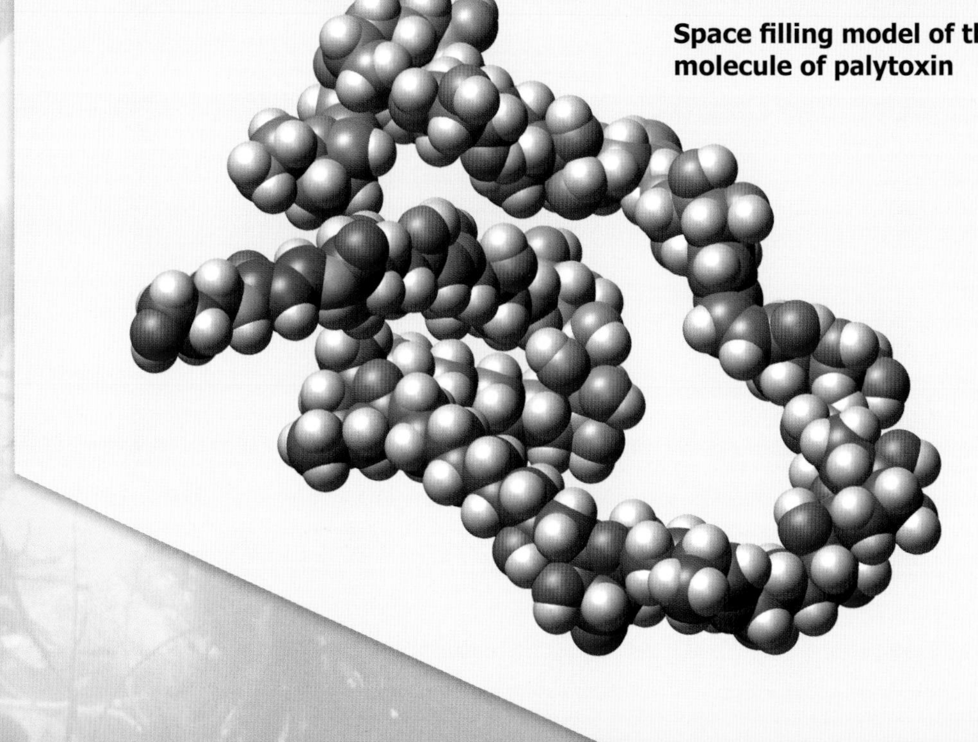

Space filling model of the molecule of palytoxin

mixed-geometry intermediate in order to favor the desired *trans*-isomer (Box 5). By these means, palytoxin, the largest natural product yet conquered by chemical synthesis, was constructed in 1994, fifteen years after Kishi had first been tantalized by its mesmerizing architecture.

The inspirational synthesis of palytoxin represents one of the prime triumphs of chemical synthesis, a science the reach of which seems to have no limit. Its contemporary feats are viewed by many with awe, however, this admiration should not distract us from considering its future with even more optimism. There are many more intricate natural products to be discovered and to be made in more efficient and expedient ways using new technologies and strategies. Let us not rest on our laurels, but use such accomplishments as the spur to continue the march forward, seeking out new discoveries and inventions that are sure to benefit mankind in a myriad of diverse ways, some as yet unimaginable.

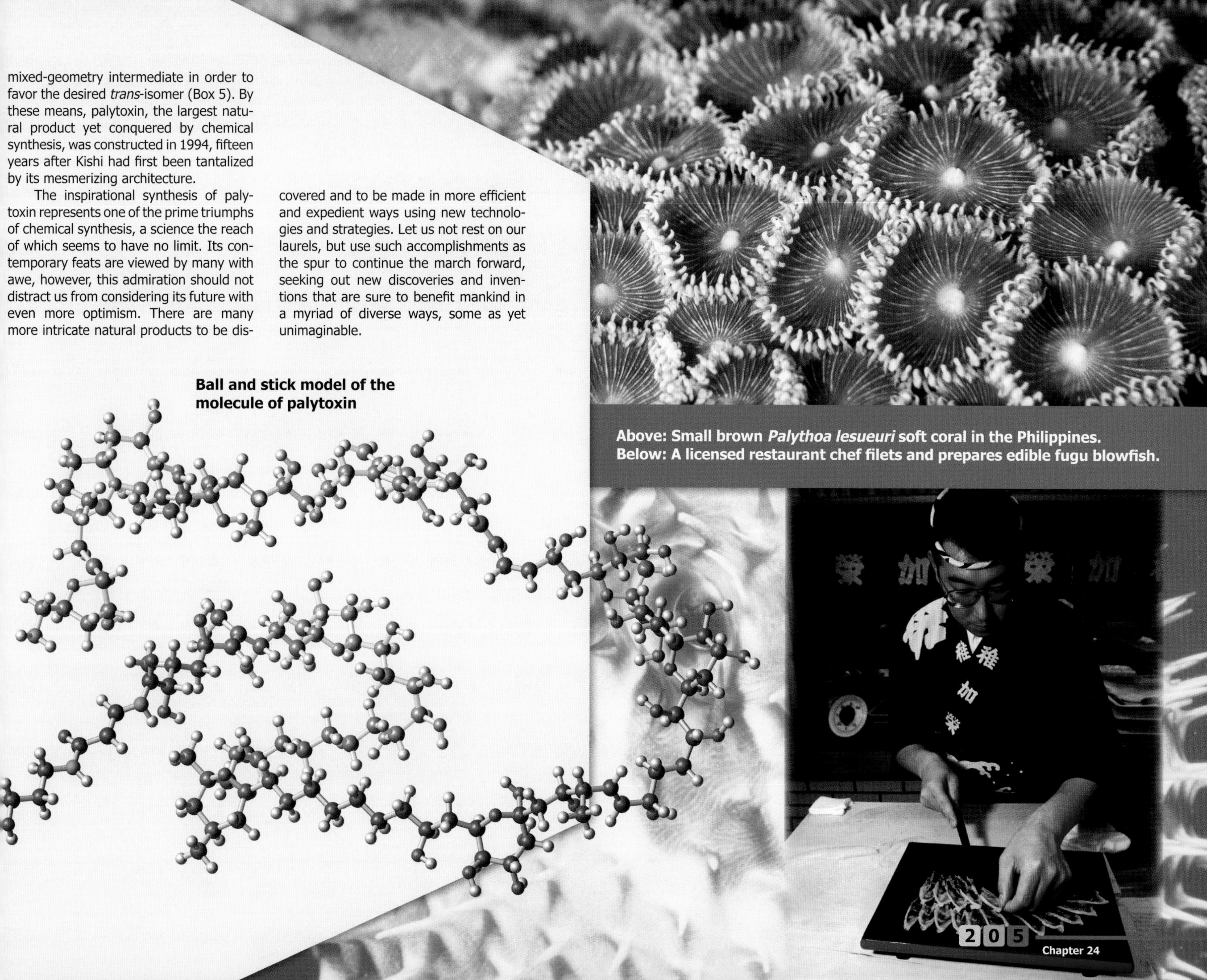

Ball and stick model of the molecule of palytoxin

Above: Small brown *Palythoa lesueuri* soft coral in the Philippines.
Below: A licensed restaurant chef filets and prepares edible fugu blowfish.

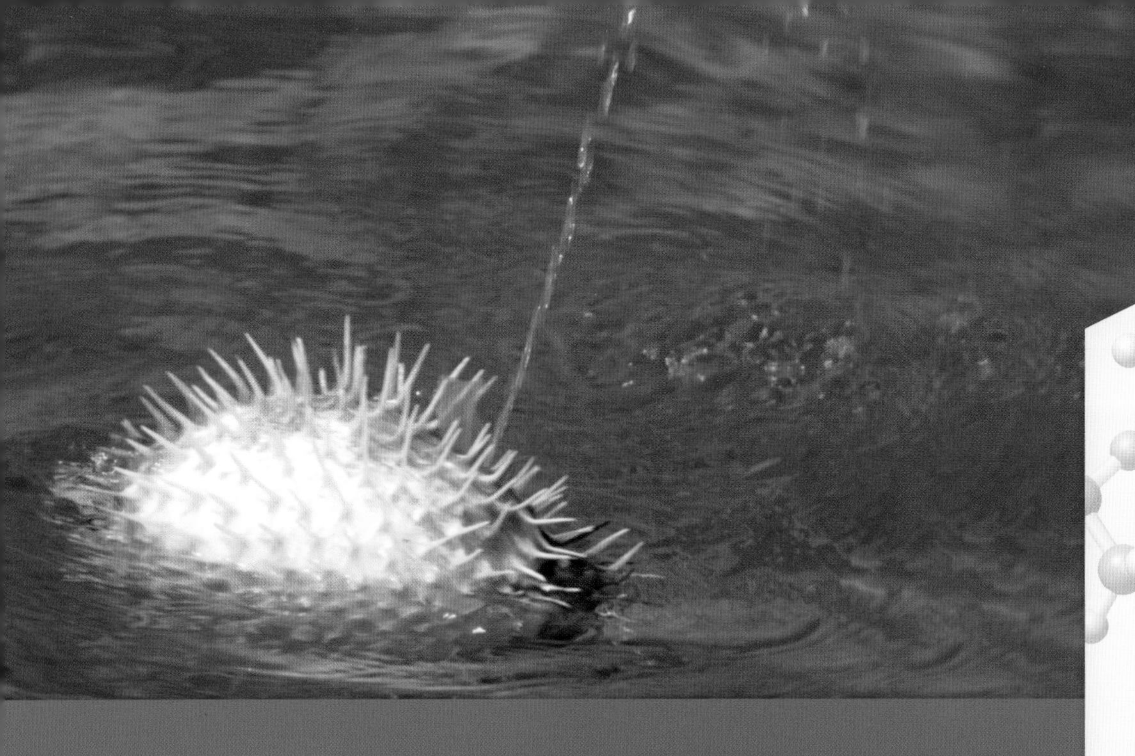

Further Reading

R. E. Moore, P. J. Scheuer, Palytoxin: A New Marine Toxin from a Coelenterate, *Science* **1971**, *172*, 495–498.

R. E. Moore, G. Bartolini, Structure of Palytoxin, *J. Am. Chem. Soc.* **1981**, *103*, 2491–2494.

D. Uemura, K. Ueda, Y. Hirata, H. Naoki, T. Iwashita, Further Studies on Palytoxin. II. Structure of Palytoxin, *Tetrahedron Lett.* **1981**, *22*, 2781–2784.

R. W. Armstrong, J.-M. Beau, S. H. Cheon, W. J. Christ, H. Fujioka, W.-H. Ham, L. D. Hawkins, H. Jin, S. H. Kang, Y. Kishi, M. J. Martinelli, W. W. McWhorter, Jr., M. Mizuno, M. Nakata, A. E. Stutz, F. X. Talamas, M. Taniguchi, J. A. Tino, K. Ueda, J. Uenishi, J. B. White, M. Yonaga, Total Synthesis of a Fully Protected Palytoxin Carboxylic Acid, *J. Am. Chem. Soc.* **1989**, *111*, 7525–7530.

Y. Kishi, Natural Products Synthesis: Palytoxin, *Pure Appl. Chem.* **1989**, *61*, 313–324.

P. J. Scheuer, Some Marine Ecological Phenomena: Chemical Basis and Biomedical Potential, *Science* **1990**, *248*, 173–177.

E. M. Suh, Y. Kishi, Synthesis of Palytoxin from Palytoxin Carboxylic Acid, *J. Am. Chem. Soc.* **1994**, *116*, 11205–11206.

K. C. Nicolaou, E. J. Sorensen, *Classics in Total Synthesis*, Wiley-VCH, Weinheim, **1996**, pp. 711–730.

Taxol®
Chapter 25
1994
AcO
O
OH
Me
Ph
O
O
Me
Me
Ph
N
H
O
A
Me
B
C
OH
D
O
HO
H
OBz
OAc
Taxol®

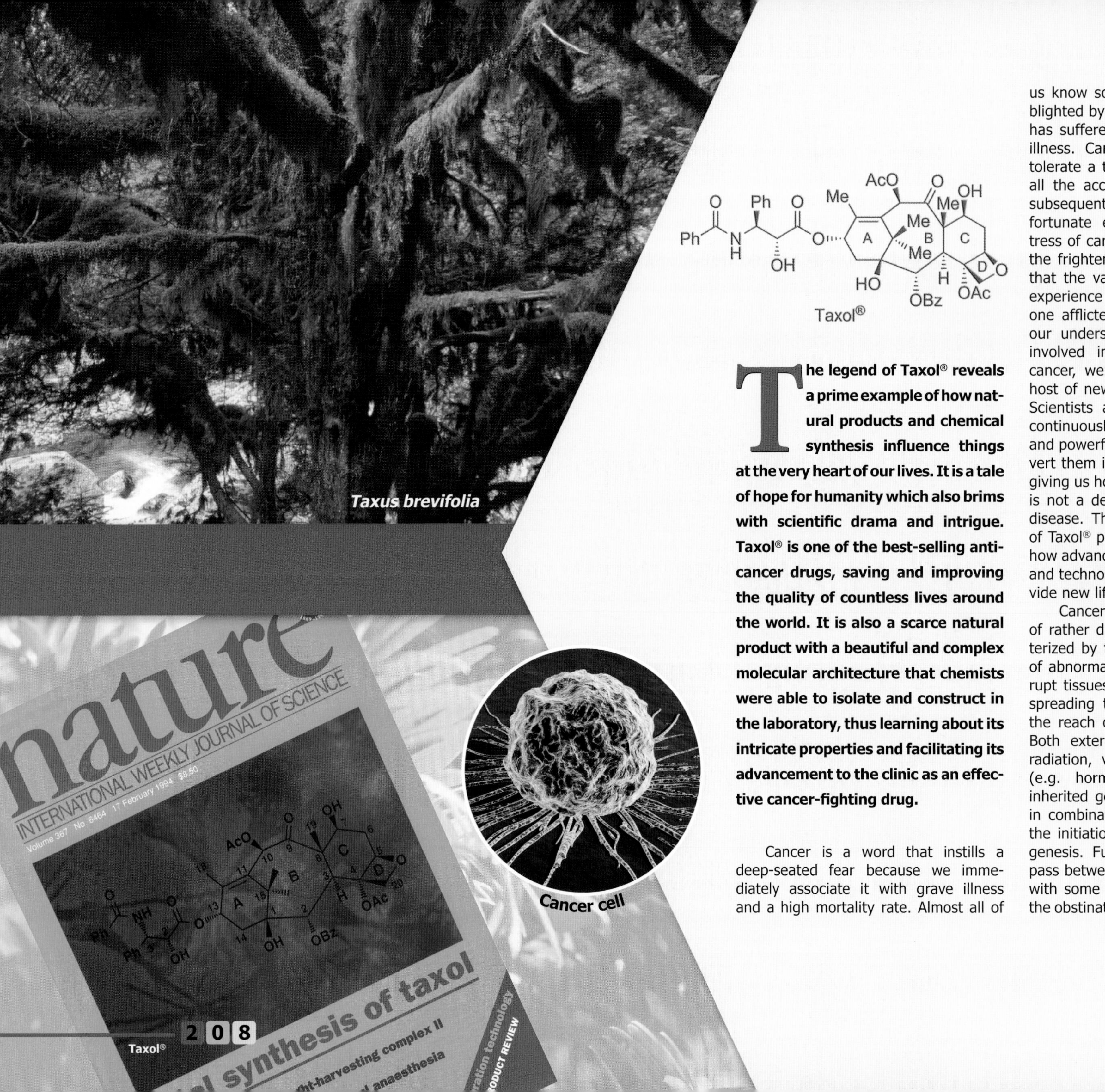

Taxus brevifolia

Cancer cell

The legend of Taxol® reveals a prime example of how natural products and chemical synthesis influence things at the very heart of our lives. It is a tale of hope for humanity which also brims with scientific drama and intrigue. Taxol® is one of the best-selling anti-cancer drugs, saving and improving the quality of countless lives around the world. It is also a scarce natural product with a beautiful and complex molecular architecture that chemists were able to isolate and construct in the laboratory, thus learning about its intricate properties and facilitating its advancement to the clinic as an effective cancer-fighting drug.

Cancer is a word that instills a deep-seated fear because we immediately associate it with grave illness and a high mortality rate. Almost all of us know someone whose life has been blighted by a cancer diagnosis, and who has suffered the prolonged pain of the illness. Cancer patients are forced to tolerate a tough treatment regime with all the accompanying side effects and subsequent problems. Few people are fortunate enough to escape the distress of cancer over their lifetime, since the frightening statistics would suggest that the vast majority of us will either experience it first hand, or have a loved one afflicted. However, as we advance our understanding of the mechanisms involved in causing and propagating cancer, we are gradually uncovering a host of new leads and hopes for cures. Scientists and clinicians diligently and continuously harness such intelligence and powerful resources, laboring to convert them into practical strides forward, giving us hope for a future where cancer is not a death sentence, but a curable disease. The story of the development of Taxol® provides a striking example of how advances in several fields of science and technology can be combined to provide new lifesaving therapies.

Cancer is a collection of a number of rather disparate diseases, all characterized by the uncontrolled proliferation of abnormal cells which invade and disrupt tissues, beginning locally and then spreading through the body to extend the reach of their destructive behavior. Both external causes (e.g. chemicals, radiation, viruses) and internal factors (e.g. hormones, immune conditions, inherited genes), acting either alone or in combination, may be responsible for the initiation and promotion of carcinogenesis. Furthermore, many years can pass between cause and detection, and, with some types of cancer, there exists the obstinate problem of detecting malignant growths early enough for intervention to stand any chance of success.

The global impact of cancer cannot be overstated. It constitutes a major public health problem with an ever growing worldwide occurrence. In 2003, approximately 1.34 million people were diagnosed with cancer (not including basal and squamous cell skin cancers) and more than half a million patients died from the disease in the United States alone. The only condition attributed with more deaths per annum is cardiovascular disease, which is responsible for one in every four deaths in the industrialized world. The American Cancer Society estimates that men have slightly less than a one in two lifetime risk of developing cancer; for women the risk is slightly more than one in three. Quite apart from the personal suffering caused by this high incidence, there is an immense financial cost to both the individual and society as a whole in the form of direct medical expenses and lost productivity. As a consequence, each small step scientists make in advancing our ability to treat this heinous group of diseases translates into a major unburdening for society, not to mention the lives saved.

Bark of *Taxus brevifolia*

The seriousness of the problem was recognized in the United States in the 1950s, prompting the rapid passage of new legislation over the ensuing two decades as part of a campaign against the disease. This legislation included the 1971 National Cancer Act which aimed to invigorate cancer research through a directed and substantial increase in funding. As a result of this act proclaiming "the conquest of cancer a national crusade," several new cancer centers were established around the country, and research in the field was accelerated considerably. However, while the long and arduous journey of Taxol® to the clinic would certainly benefit from this infusion of funds, the story begins much earlier. A specific project directed by the National Cancer Institute (NCI), already underway during the 1960s, aimed to conduct a widespread screening of substances and extracts obtained from a variety of natural sources for antineoplastic activity. Thus, United States Department of Agriculture (USDA) botanist Arthur S. Barclay collected samples of bark from the relatively rare Pacific yew, *Taxus brevifolia*, while on a field trip to an Oregon forest in the summer of 1962. Barclay's choice of the yew was an astute one, since these trees had a long and proud history of use by mankind, and buried within this ancient knowledge were the clues that they might contain cytotoxic compounds. For example, Julius Caesar recorded that, after defeat at the hands of the Roman

Needles of *Taxus baccata*

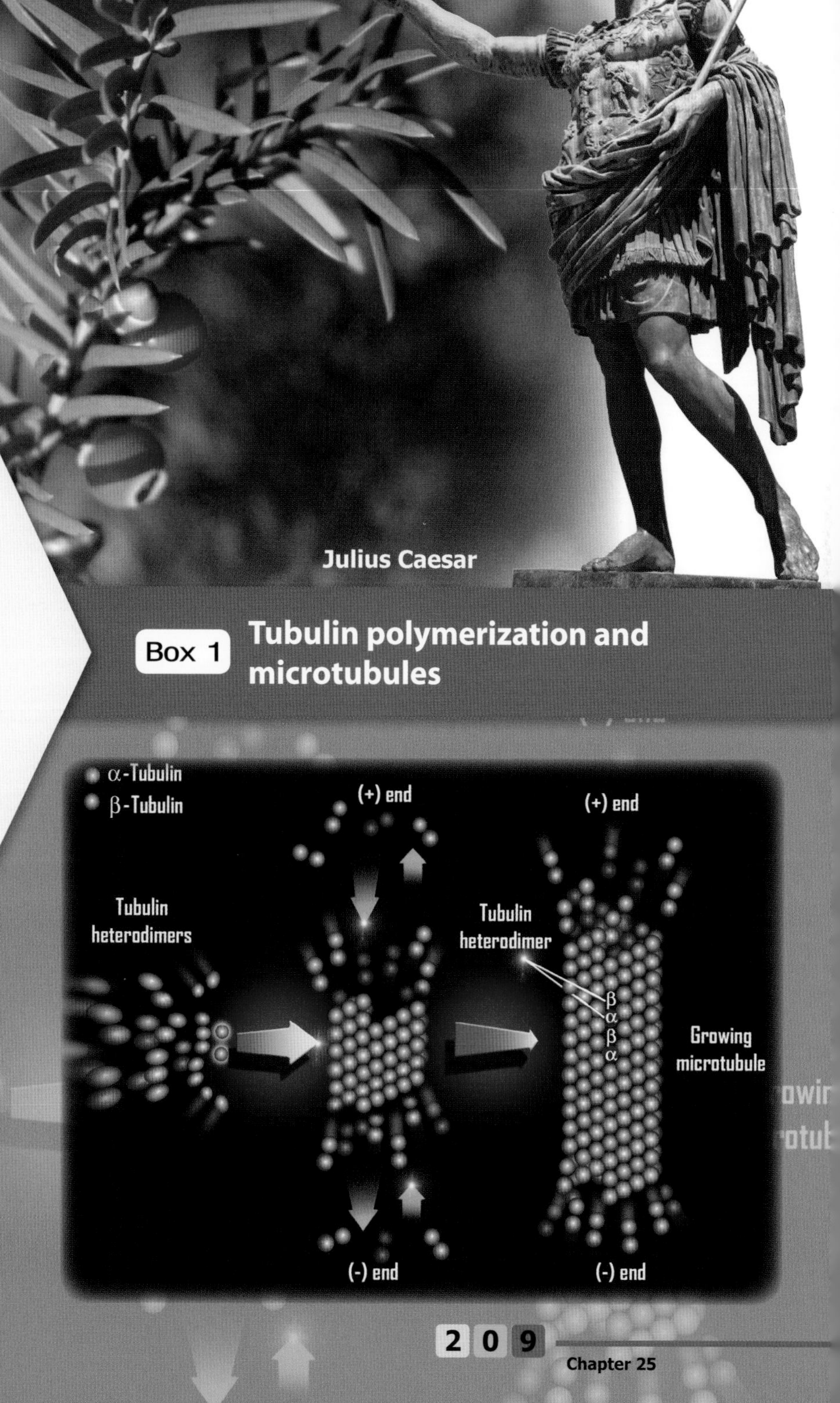

Julius Caesar

Box 1 Tubulin polymerization and microtubules

Monroe E. Wall and Mansukh C. Wani

Susan B. Horwitz

Box 2 The cell cycle and the mechanism of action of Taxol®

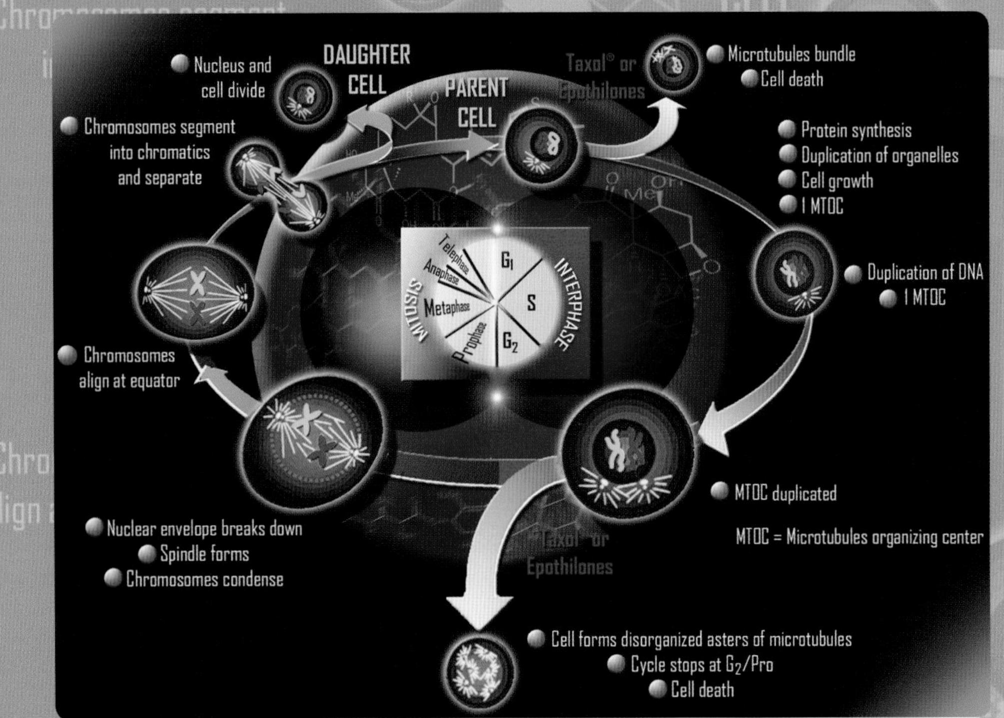

legions, the Gallic chieftain Cativolcus committed suicide by drinking tea made from yew bark. Barclay's samples were eventually packed up and sent to the Research Triangle Institute (North Carolina, USA) in 1964, where two chemists, Monroe E. Wall and Mansukh C. Wani, were charged with their investigation. The crude extracts of the yew bark were shown to exhibit an unusually broad spectrum of cytotoxic activity against leukemia cells, as well as against a variety of other cancer cells. This activity spurred on Wall and his colleagues to isolate the active constituent of the extracts, the complete structure of which was definitively established, in collaboration with researchers at Duke University (North Carolina, USA), by x-ray crystallographic analysis in 1971. Of the more than 110,000 compounds from 35,000 plant species tested by the NCI between 1960 and 1981, this active ingredient, named Taxol®, proved to be the most promising. Despite its promise, however, Taxol® was set to languish on laboratory shelves for almost another full decade before the next advance was made. The reluctance to pursue investigations into the potential therapeutic benefits of Taxol® was largely due to difficulties associated with its isolation and low solubility, as well as the belief that it was simply another microtubule-destabilizing agent akin to the other available natural products, colchicine and the vinca alkaloids, already in use as anticancer agents. However, in 1979, Susan B. Horwitz (Yeshiva University, New York, USA) and coworkers reported in the journal *Nature* that Taxol® actually possessed the unique characteristics of stabilizing and promoting the formation of microtubules. This exciting disclosure led to new research momentum, since this mode of action constituted an entirely new mechanism of intervention against the aberrant replication of cancerous cells (Boxes 1 and 2).

Microtubules are involved in many aspects of cellular biology; they give shape and structure to the cell, assist in reorganizing organelles, and, most importantly for cancer, play an essential role in mitosis, the process of cell division. Their pivotal role in the growth, function, and division of cells led to their description in the 1990s as "the most strategic subcellular targets of anticancer chemotherapeutics." Microtubules are predominantly composed of two similar protein subunits, α- and β-tubulin, which combine to form a heterodimer (Box 1). The tubulin dimer binds two molecules of guanosine 5′-triphosphate (GTP), and in the presence of magnesium these dumbbell-shaped dimers begin to unite in a head-to-tail fashion forming protofilaments. These, in turn, assemble in a staggered manner, leading to the left-handed helices that make up the microtubule. Typically, exchanges are set up at both ends of the microtubule with regular loss and gain of tubulin subunits at relative rates that are often different, thus endowing the tubule with a growing directionality. In general, microtubules are not static structures and, after a certain period of growth, the microtubule and free tubulin will reach a constant equilibrium concentration. At this point, termed the critical concentration, disassembly and growth are finely balanced in a situation regulated by the bound GTP molecules. What Horwitz's group showed was that Taxol® affects the tubulin-microtubule equilibrium. Taxol® decreases both the concentration of free tubulin (to almost zero) and the induction time for polymerization, with the result that microtubules forming in the presence of Taxol® have a distinctly different morphology from the control variants. They have a shorter average length and are resistant to change under conditions that would depolymerize normal microtubules. In addition, the Taxol® promoted polymerization of tubulin does not require GTP.

K. C. Nicolaou

From the brief discussion above, we can now understand the interactions of Taxol® with tubulin and the subsequent effects on the gross architecture of microtubules, but how does this translate into the powerful cytotoxicity of the drug? Most cells, excluding a few that cannot replicate, spend the majority of their time in a nonproliferative state called quiescence, yet they are able to switch rapidly into a reproductive cycle (mitosis) when their populations are low and in need of a boost. The mitotic cycle itself has a number of defined stages (Box 2). Cancer cells, unlike their normal counterparts, do not enter the quiescent state when the population density is high, and furthermore, during the mitotic cycle, cancer cells suffer from a deficiency in the checkpoints which regulate the rate of division so that the problem of excessive expression of these cells is grossly exacerbated. Horwitz showed that tumor cells treated *in vitro* with Taxol® were arrested at the transition from metaphase to anaphase. Morphologically these cells exhibited unnatural bundles of microtubules and no mitotic spindle. Other research groups quickly extended these findings, proving that Taxol® causes the general and irreversible formation of bundles of microtubules at several cell lifecycle stages, visually reminiscent of a log jam. The net effect of this phenomenon is the sequestering of tubulin in the form of stable structures, ultimately preventing the formation of a mitotic spindle, halting cell division, and leading to rapid cell death by apoptosis (suicidal death). Furthermore, the disturbance to microtubule mobility was also shown to have dramatic effects on other processes within the cell, such as inhibiting the secretion of certain proteins, leading to the general conclusion that the full nature of the attack of Taxol® on cancerous cells is likely to be a complex conglomeration of cell function shutdowns.

The Nicolaou Taxol® team

Box 3 The Nicolaou retrosynthetic analysis of Taxol®

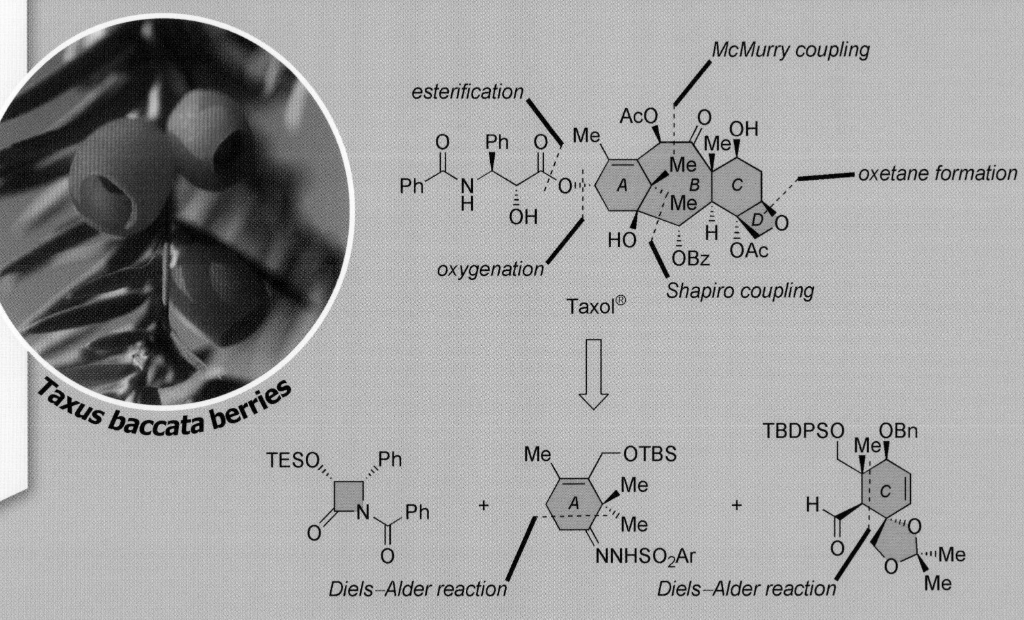

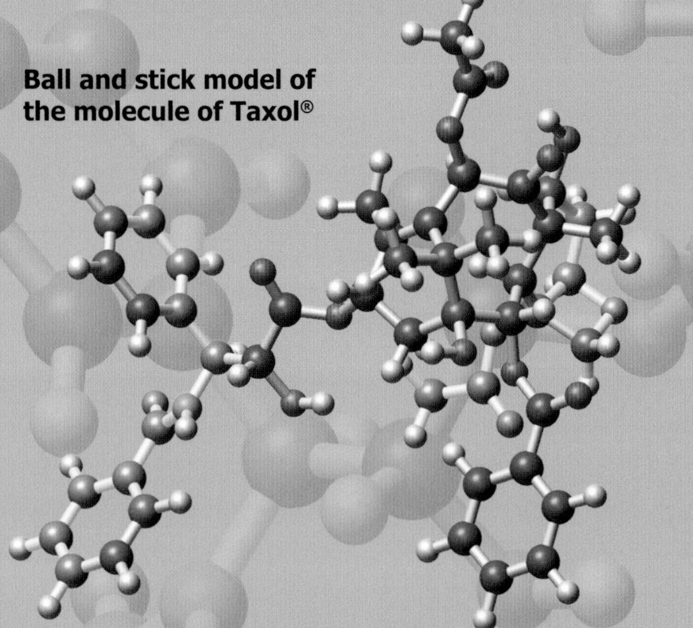

Ball and stick model of the molecule of Taxol®

Space filling model of the molecule of Taxol®

As these groundbreaking investigations were gradually revealed, scientists became increasingly excited about Taxol® because of its emerging portrayal as a long sought-after ally in the fight against cancer. In the 1980s and 1990s, the media was buzzing with stories of a miraculous cure and a drug that could change the face of cancer treatment forever. However, before these dreams could be realized, passage around a number of difficult logistical obstacles still had to be navigated. The most pressing of these issues was that of supply. The Pacific yew is a slow growing tree and has a tendency to grow in dispersed microsites hidden in ecosystems containing mostly older growth and larger tree species. Prior to the discovery of Taxol®, these trees were regarded as scrub and were, for the most part, ignored in primary logging operations, instead being destroyed in the slash and burn disposal fires that followed. Large numbers of mature specimens concentrated in sizeable and harvestable regions were therefore not available. The extraction of Taxol® itself was also a tedious process requiring large quantities of tree bark to yield relatively small amounts of the compound. For example, 38,000 *Taxus brevifolia* trees were sacrificed to obtain just 25 kg of Taxol®, enough to treat 12,000 patients (that is more than three trees per person) in one of the initial clinical studies of this new medication. Such short supplies were hampering the progress of these early clinical trials, and yet the problem was complicated even further by environmental concerns. Indeed, there was a major controversy and a lively debate surrounding the harvest of so many trees from the delicate forest environments. The concern was voiced that the success of Taxol® would drive the Pacific yew to the verge of extinction and cause irreparable damage to other members of these fragile ecosystems, including the endangered Northern Spotted owl.

The frustrated medicinal promise that Taxol® held in these heady days created unprecedented interest among synthetic organic chemists, who were motivated by the recognition that a chemical synthesis could potentially provide a solution to the supply problem. This urgency spurred additional teams of researchers to follow the many groups, from a host of countries, who had already embarked on the quest for a total synthesis of Taxol®. The lure was not only the issue of the limited stocks, but also the intrigue and challenge presented by the highly complex and unique architecture of this molecule. The structure of Taxol® consists of a diterpene core that is distinguished by its ring arrangement, a periphery densely populated with oxygen functionalities, and nine asymmetric centers. Appended to this core, at a relatively hindered site, is an ester side chain bearing two additional asymmetric carbon atoms. Taxol® became the Holy Grail of total synthesis, and the race to reach it took on many twists and turns until it finally ended in success in 1994.

The first total synthesis of Taxol® to be published was that emanating from the Nicolaou laboratories at The Scripps Research Institute in La Jolla, California (USA), reported in *Nature* early in 1994. This achievement, together with another

Box 4 Highlights of the Nicolaou group's total synthesis of Taxol®

PhB(OH)₂ Diels–Alder reaction → boronate cleavage → lactone migration → Shapiro coupling → McMurry coupling → Taxol®

total synthesis which appeared at almost the same time and was accomplished in the laboratories of Robert A. Holton at Florida State University in Tallahassee (USA), ended nearly two decades of recalcitrance exhibited by Taxol®. The Nicolaou group's success relied on the implementation of a highly convergent strategy devised by retrosynthetic analysis (Box 3). Thus, five strategic bonds had been identified and retrosynthetically disconnected, disassembling the molecule of Taxol® into three key building blocks: a β-lactam corresponding to the side chain, a hydrazone encoding the A-ring, and an aldehyde containing the necessary carbon framework and functionality to complete rings C and D. It was anticipated that ring B, which perhaps posed the thorniest problem of the synthesis, might be cast through an intermolecular Shapiro reaction followed by an intramolecular McMurry coupling process. Finally, the venerable Diels–Alder reaction was to be called upon twice, to assemble each of the two requisite cyclohexene rings.

This strategy was successfully implemented, more or less the way it was designed, after a relentless campaign that included many twists and turns and dramatic moments, as summarized in Box 4. Thus, the six-membered C-ring was assembled through the application of an innovative boron-tethered Diels–Alder reaction, developed by Koichi Narasaka (University of Tokyo, Japan), which caused a switch of regioselectivity from the unwanted natural bias of the reactants towards the desired isomer. The A-ring was prepared through a different Diels–Alder reaction, and was then coupled to the C-ring in a subsequent Shapiro reaction. Between each of these key stages adjustment of oxidation levels and functional group manipulations steered the growing molecule towards its final destination. Pleasingly, the challenge of cyclization to furnish the highly strained and crowded 8-membered B-ring was met, as projected, by the McMurry reaction. This remarkable process proved capable of uniting the two aldehyde groups to close the highly strained ring, yielding the tricyclic framework of Taxol® molecule, despite the significant opposing forces. A more advanced intermediate exhibiting all of the structural motifs of the Taxol® was then reached after a number of sensitive and frequently treacherous chemical modifications. Included among these stringent tests were the generation of the highly strained oxetane ring and the regioselective cleavage of the cyclic carbonate to form the desired hydroxy benzoate moiety. To complete the Herculean task of Taxol®'s total synthesis, the hindered ester side chain was finally installed by the reaction of the completed tetracyclic core with an appropriate electrophilic β-lactam ring, according to the

Robert A. Holton

Taxus baccata

The Holton Taxol® team

Box 5 Highlights of the Holton group's total synthesis of Taxol®

(–)-camphor

1964 method of G. Büchi

(+)-patchino

epoxy-alcohol fragmentation

Taxol®

Pierre Potier, Françoise Guéritte-Voegelein and Daniel Guénard

Box 6 Structures of 10-deacetylbaccatin III and Taxotere®

10-deacetylbaccatin III

Taxotere®

Holton–Ojima procedure. The final protecting groups were then removed to reveal the trophy - laboratory synthesized Taxol®.

Within days of the *Nature* report describing the Nicolaou synthesis, Robert Holton and his group at Florida State University published their elegant route to Taxol®. Holton had chosen an altogether different strategy to access the taxane skeleton in which he employed a rather unusual and novel epoxy-alcohol fragmentation that he had previously developed especially for this application (Box 5). His choice of camphor as the starting point for his synthesis of Taxol® provides a telling reminder of how far organic synthesis had advanced over the preceding century, for as you will recall from one of our stories (Chapter 5) camphor itself was once an arduously sought-after synthetic target. In Holton's synthesis of Taxol®, camphor served as a cheap and readily available source of chirality, and the start of the synthesis, not its end. These two essentially simultaneous total syntheses from the Nicolaou and Holton camps were followed over the next few years by completed syntheses from the groups of Samuel J. Danishefsky (1995, Sloan–Kettering Institute and Columbia University, USA), Paul A. Wender (1996, Stanford University, USA), Teruaki Mukaiyama (1997, Tokyo University of Science, Japan), and Isao Kuwajima (1998, Tokyo Institute of Technology, Japan). All of these disparate syntheses can be characterized by one unifying feature; they all exhibit the use of bold strategies and novel tactics, which have contributed considerably to the advancement of total synthesis and exemplified both its awesome power and its enabling nature for biology and medicine.

The conquest of Taxol® by chemical synthesis was not the only victory of science and medicine, nor did it signal the end of the story. In 1992, the US Food and Drug Administration (FDA) approved the use of Taxol® for the treatment of ovarian cancer, and, in 1994, the drug was approved for the treatment of recurrent breast cancer. At its peak in 2001, Taxol®, distributed by Bristol-Myers Squibb (BMS), was the world's top-selling anticancer drug with sales approaching $2 billion. Furthermore, the clinical applications of Taxol® are constantly expanding, opening up a plethora of new opportunities for the treatment of a wider variety of cancers, especially with new formulations, including exciting and omnipotent new combination therapies in which Taxol® is combined with other anticancer drugs. It should be noted that BMS secured the rights to the name Taxol® for use as a trademark, thereby causing the renaming of the natural product as 'paclitaxel', despite the fact that scientists had been using the former name since 1971. To reach a position from which Taxol® could dominate the therapy of breast and ovarian cancer, BMS had to find a synthesis capable of meeting the large demand for such a drug. This task was facilitated by the findings of Françoise Guéritte-Voegelein, Pierre Potier, Andrew E. Greene, and coworkers in France, who discovered that the foliage and other renewable biomass of the more common European yew, *Taxus baccata*, contained significant quantities of a related compound, 10-deacetylbaccatin III (Box 6), that could be converted into Taxol® through a short synthetic sequence. Although three thousand kilograms of needles from *Taxus baccata* were needed to obtain one kilogram of the drug precursor, unlike the isolation of Taxol® from *Taxus brevifolia* bark, the harvesting did not signal the death of the tree. Thus a sustainable supply of Taxol® could be guaranteed. Today, Taxol® and related compounds are produced industrially from 10-deacetylbaccatin III through the attachment of the side chain to the hindered C-13 hydroxyl group in a semi-synthesis which is related to the protocols originally developed by Holton and Iwao Ojima. As we shall see next, the discovery of 10-deacetylbaccatin III led to yet another chapter in the tale of Taxol® and a new advancement in cancer chemotherapy.

Box 7 Selected anticancer drugs

Generic name [trade name] (source)	Structure or description	Mode of action	Marketer (worldwide sales 2005, US $, in millions)	Indication
paclitaxel [Taxol®] (natural product)	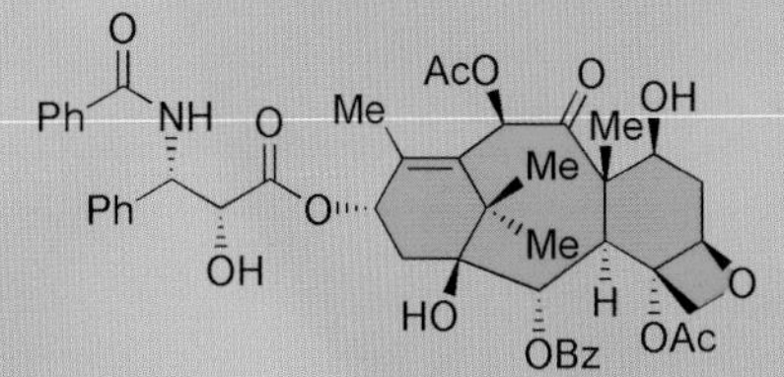	microtubule stabilization	**Bristol-Myers Squibb (747)**	ovarian and breast cancers
docetaxel [Taxotere®] (semi-synthetic)	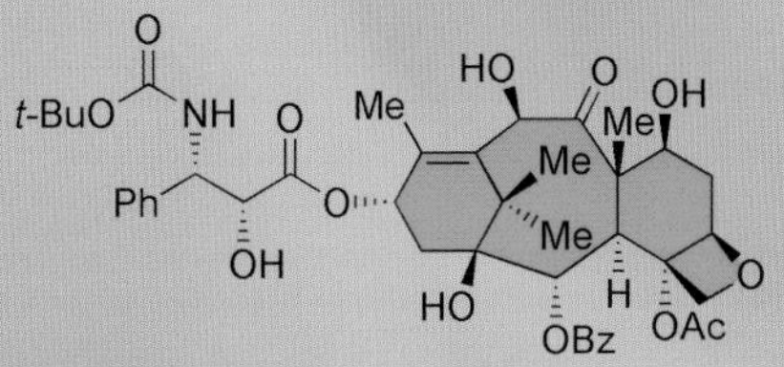	microtubule stabilization	**Sanofi-Aventis (1,609)**	breast and lung cancers
bevacizumab [Avastin®] (biotechnology)	anti-VEGF (vascular endothelial growth factor) antibody	angiogenesis inhibitor	**Genentech and Roche (1,333)**	metastatic colorectal cancer (in combination with 5-fluorouracil)
rituximab [Rituxan®] (biotechnology)	first monoclonal antibody approved for use in treatment of malignant disease in the USA (chimeric murine/human antibody)	targets CD20 antigen on surface of normal and malignant B lymphocytes	**Genentech and Biogen Idec (1,831)**	non-Hodgkin's lymphoma
goserelin [Zoladex®] (synthetic analogue)	decapeptide mimicking natural hormone gonadorelin	causes androgen hormone depletion	**AstraZeneca (1,004)**	palliative treatment of prostate cancer
leuprolide acetate [Leuplin®] (synthetic analogue)	decapeptide mimicking natural hormone gonadorelin	causes androgen hormone depletion	**Takeda and TAP Pharmaceuticals (1,782)**	palliative treatment of prostate cancer
gemcitabine hydrochloride [Gemzar®] (synthetic cytosine nucleotide analogue)	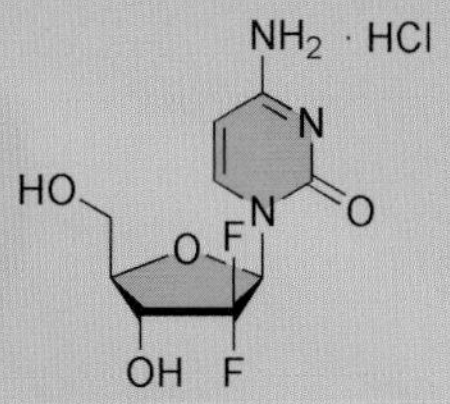	inhibits enzyme producing nucleotides for DNA, also competes with cytosine for incorporation into DNA. Leads to cell death in S-phase and blocks to progression from G1/S-phase	**Eli Lilly (1,335)**	lung cancer
oxaliplatin [Eloxatin®] (synthetic compound)	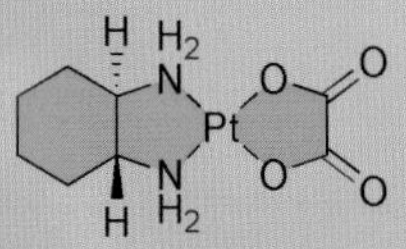	prodrug - hydrolysis gives active compound that produces interstrand DNA crosslinks, cell-cycle non-specific	**Sanofi-Aventis (1,947)**	colorectal cancer (in combination with 5-fluorouracil and leucovorin)

(continues on next page)

Box 7 Selected anticancer drugs (continued)

Generic name [trade name] (source)	Structure or description	Mode of action	Marketer (worldwide sales 2005, US $, in millions)	Indication
tamoxifen citrate [Nolvadex®] (synthetic compound)		potent anti-estrogen, binds competively to estrogen receptors in breast tissue	**AstraZeneca (114)**	breast cancer
irinotecan hydrochloride [Camptosar®] (semi-synthetic analogue)	· HCl	inhibits topoisomerase I	**Pfizer (910)**	metastatic colorectal cancer
bicalutamide [Casodex®] (synthetic compound)	racemate - (*R*)-enantiomer active, (*S*)-enantiomer inactive	non-steroidal anti-androgen, binds to cytosol androgen receptors	**AstraZeneca (1,123)**	palliative treatment of prostate cancer
trastuzumab [Herceptin®] (biotechnology)	recombinant DNA-derived monoclonal antibody	antibody binds to human epidermal growth factor 2 (HER2) protein, whose overexpression is common in primary breast cancers	**Genentech and Roche (1,722)**	breast cancer
epirubicin hydrochloride [Ellence®] (natural product)		forms a complex with DNA by intercalation, thus inhibiting DNA and RNA synthesis	**Pfizer (367)**	solid tumors, particularly in the lung and breast

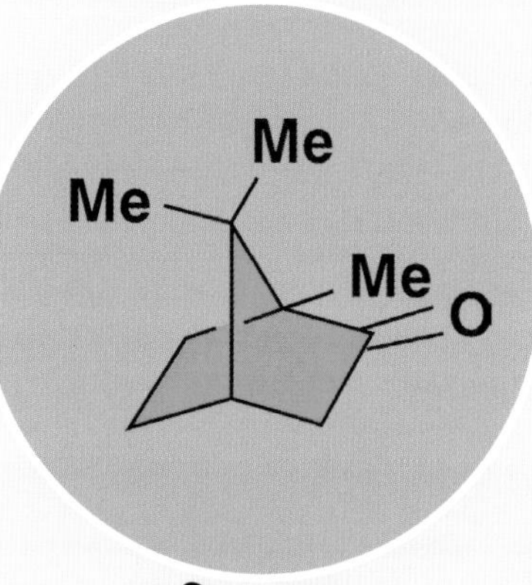

Camphor

The poor solubility of Taxol® in water was recognized as a serious impediment in early clinical studies. The drug originally had to be formulated with ethanol and Cremophor EL (a castor oil derivative) to alleviate this problem; however, the latter component induced major hypersensitivity reactions in some patients. Some claim that this obstacle delayed the clinical trials by as much as five years. Whatever the true timescale, this problem undoubtedly invigorated the search for Taxol® analogues (variations of the drug) with improved pharmacological profiles. One of the earliest and most successful analogues of Taxol® was originally discovered by Guéritte-Voegelein and Potier at Gif-sur-Yvette in France. This analogue was later developed and distributed by Rhône-Poulenc, now Sanofi-Aventis, under the trademark Taxotere®. This anticancer agent possesses a *tert*-butoxycarbonyl group instead of a benzoyl group on the nitrogen atom and a free hydroxyl moiety at C10 rather than an acetoxy group as in Taxol® (Box 6). Taxotere® is today a highly successful anticancer drug in its own right, with sales figures that have well surpassed those of its elder sibling. In early 2005, the FDA approved

Abraxane® for use in the treatment of metastatic breast cancer. This new treatment consists of nanoparticles of albumin-bound paclitaxel, thus overcoming the insolubility problem and offering an improved administration profile with fewer side effects than Taxol® itself.

Chemotherapy using Taxol® and Taxotere® proved to be a decisive advance in the struggle against cancer, but there are other drugs with their own long and rich chemical and clinical histories (Box 7). As the taxoid drugs show continuing success alongside the many other cancer chemotherapies available, we can view their development as an instructive illustration of what mankind can achieve when academe, industry, and government join forces to solve seemingly intractable and pressing problems that seriously affect society. To be sure, there were skirmishes, competitions, and controversies, but the number of lives saved and improved by Taxol® and Taxotere® make these incidents pale in comparison to the overall accomplishment. Chemists should be particularly proud for the crucial contributions that their science has made to the successful outcome of this legendary project.

***Taxus baccata*, 'Standishii', flank the entrance to a house**

Teruaki Mukaiyama

Paul A. Wender

Samuel J. Danishefsky

Isao Kuwajima

Iwao Ojima

K. C. Nicolaou conferring with students

Taxus baccata

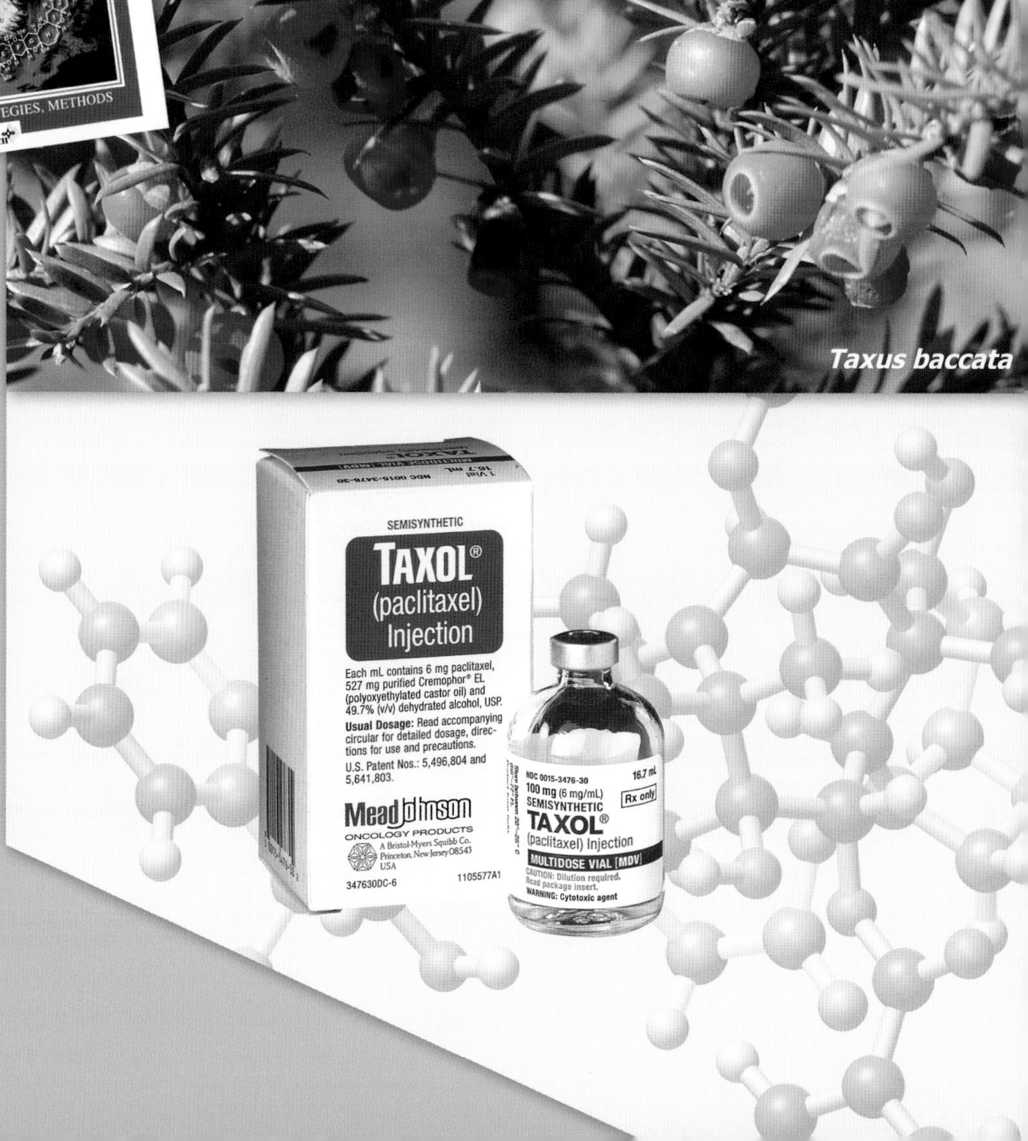

Further Reading

P. B. Schiff, J. Fant, S. B. Horwitz, Promotion of Microtubule Assembly *In Vitro* by Taxol, *Nature* **1979**, *277*, 665–667.

K. C. Nicolaou, Z. Yang, J. J. Liu, H. Ueno, P. G. Nantermet, R. K. Guy, C. F. Claiborne, J. Renaud, E. A. Couladouros, K. Paulvannan, E. J. Sorensen, Total Synthesis of Taxol, *Nature* **1994**, *367*, 630–634.

R. A. Holton, C. Somoza, H.-B. Kim, F. Liang, R. J. Biediger, P. D. Boatman, M. Shindo, C. C. Smith, S. Kim, H. Nadizadeh, Y. Suzuki, C. Tao, P. Vu, S. Tang, P. Zhang, K. K. Murthi, L. N. Gentile, J. H. Li, First Total Synthesis of Taxol. 1. Functionalization of the B Ring, *J. Am. Chem. Soc.* **1994**, *116*, 1597–1598.

R. A. Holton, H.-B. Kim, C. Somoza, F. Liang, R. J. Biedisger, P. D. Boatman, M. Shindo, C. C. Smith, S. Kim, H. Nadizadeh, Y. Suzuki, C. Tao, P. Vu, S. Tang, P. Zhang, K. K. Murthi, L. N. Gentile, J. H. Liu, First Total Synthesis of Taxol. 2. Completion of the C and D Rings, *J. Am. Chem. Soc.* **1994**, *116*, 1599–1600.

K. C. Nicolaou, W.-M. Dai, R. K. Guy, Chemistry and Biology of Taxol, *Angew. Chem. Int. Ed. Engl.* **1994**, *33*, 15–44.

K. C. Nicolaou, R. K. Guy, The Conquest of Taxol, *Angew. Chem. Int. Ed. Engl.* **1995**, *34*, 2079–2090.

K. C. Nicolaou, R. K. Guy, P. Potier, Taxoids: New Weapons Against Cancer, *Sci. Am.* **1996**, *274*, 94–98.

J. Mann, Natural Products in Cancer Chemotherapy: Past, Present and Future, *Nat. Rev. Cancer* **2002**, *2*, 143–148.

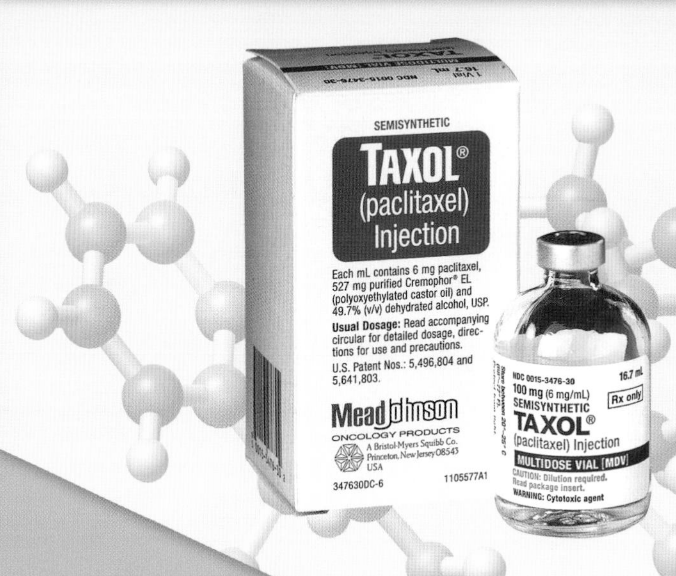

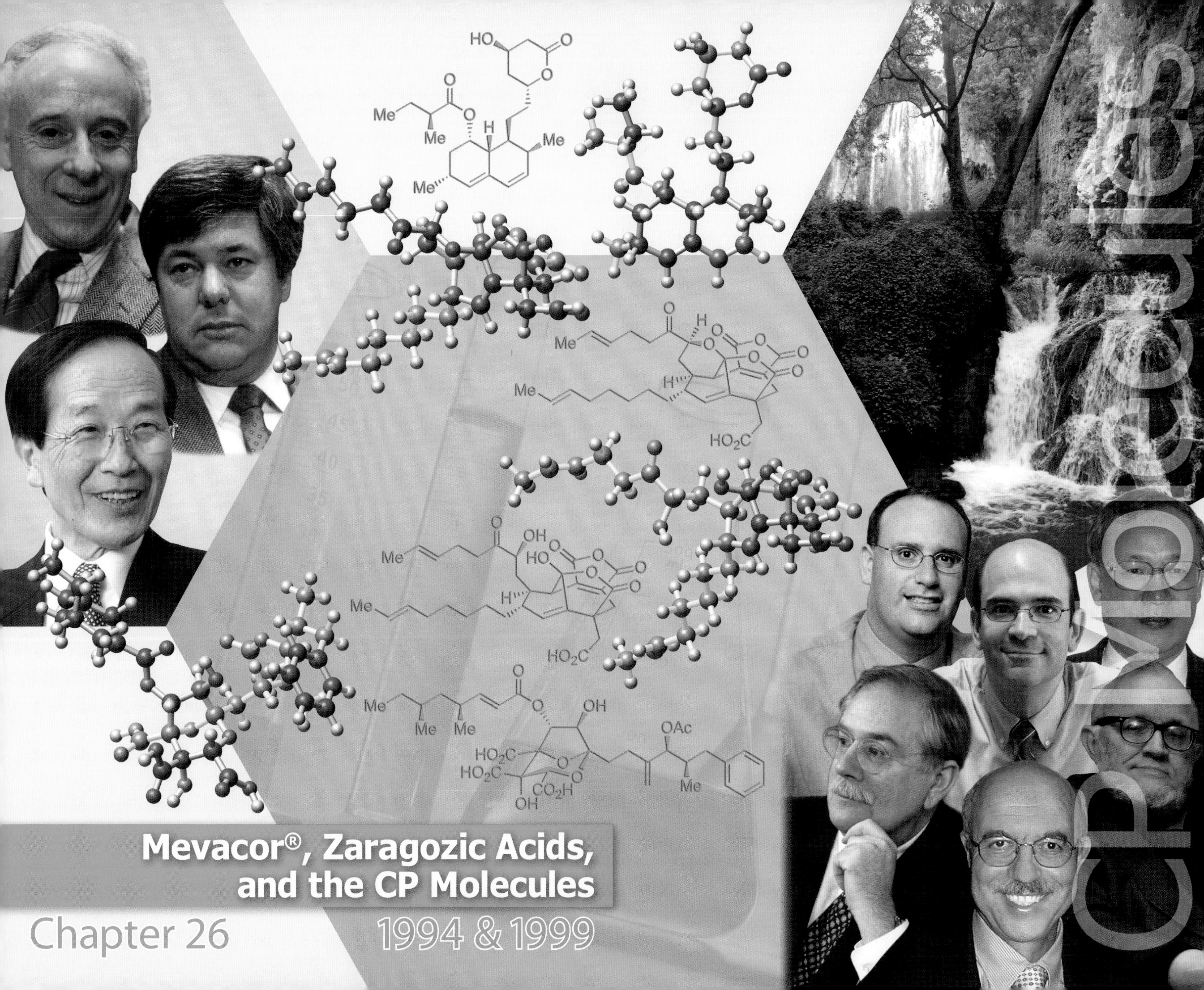
Mevacor®, Zaragozic Acids,
and the CP Molecules
Chapter 26
1994 & 1999
CP Molecules

lovastatin [Mevacor®]

zaragozic acid A (squalestatin S1)

CP-225,917 (phomoidride A)

CP-263,114 (phomoidride B)

Following decades of public health education, the need for society to adopt a healthier diet and reduce blood cholesterol levels has become widely known. Beyond this simple message, however, there is widespread confusion, and the advice given by health authorities has changed significantly over the years. What is clear is that an elevated blood cholesterol level can lead to serious health problems in later life.

The biochemistry of the human body is highly intricate, so the basis for the effects of cholesterol is rather complex. An imbalance in cholesterol metabolism is a primary promoter of atherosclerosis, along with a number of other genetic and environmental factors (Box 1). Atherosclerosis is the underlying cause of both cardiovascular disease (CVD) and cerebral stroke, conditions that account for approximately 50 % of all deaths in industrialized nations. Since elevated cholesterol is the underlying cause of atherosclerosis, its treatment could allow the prevention of numerous deaths. The treatment of elevated cholesterol has received considerable attention from pharmaceutical companies and academic scientists alike. In this chapter, we will introduce three classes of natural products that offer opportunities to better understand and treat high cholesterol. The first class includes Mevacor®, an inhibitor of cholesterol biosynthesis developed as a drug by the pharmaceutical company Merck. We will also discuss the zaragozic acids and the CP molecules. These natural products also inhibit cholesterol biosynthesis, and their complex and intriguing chemistry has proved inspirational to a number of chemists, culminating in the landmark achievements of their total syntheses.

The history of cholesterol chemistry spans some 250 years, since its original discovery in human gallstones. It is a fat that fulfills a number of important roles and is essential to the smooth functioning of the body. In the 1920s, scientists devoted huge efforts to investigating cholesterol, and studies directed towards determining its structure earned the German chemist Adolf O. R. Winhaus the 1928 Nobel Prize in Chemistry. Unfortunately, within three years of this prestigious award, the structure Windhaus had proposed was shown to be erroneous by English scientist John D. Bernal, who used the then new technique of x-ray crystallography to correctly elucidate the gross structure. Within the body, cholesterol forms a structural component of cell membranes and is the precursor to bile acids. It is also related to the steroid hormones, which include the male and female sex hormones (see Chapter 11). Cholesterol exists in only one form, however the popular discussion surrounding its physiological role has led to the terms "good" and "bad" cholesterol. Cholesterol is produced and stored in the liver, and, because it is insoluble in water, it must be transported in the blood by carrier proteins, of which there are several

Adolf O. R. Winhaus

John D. Bernal

types. These lipoproteins carry varying quantities of a range of different lipids, cholesterol being just one component. The different lipids they carry endow the lipoprotein complexes with varying densities, by which they are categorized. High density lipoproteins (HDLs) remove cholesterol from the bloodstream, and return it to the liver. Low density lipoproteins (LDLs) perform the opposite function. The terms "good" and "bad" cholesterol have become associated with these two categories of lipoproteins in reference to their respective functions.

Monocytes

High blood serum levels of LDL lead to a sequence of events culminating in vascular injury and occlusion. The processes involved in such conditions are highly complex and begin in early childhood. By the age of ten most children already have fatty streaks, comprised of cholesterol-engorged macrophages (foam cells), within their aorta. These lesions are not yet in themselves clinically significant, but they are precursors to damaging atherosclerotic plaques. By the age of thirty, fatty streaks have appeared in many more arterial tissues, and their form has begun to change, taking on the shape of a problematic atherosclerotic plaque. Atherosclerosis is not simply an inevitable degenerative consequence of aging, but rather a progressive chronic inflammatory condition with similarities to asthma (see Chapter 15). To understand this connection, we must first learn how the seemingly innocuous fatty streaks form, and how they transform into life-threatening atherosclerotic plaques. Our understanding of this process is based on studying animals that have been fed a diet high in cholesterol and other fats. The first observable change occurs in the artery wall, where lipoproteins and their aggregates accumulate (Box 2). The immune system responds, and monocytes (a type of white blood cell) adhere to the walls of the blood vessel (the endothelium). The monocytes then migrate across the arterial lining, through the collagen and proteoglycan layers, to reach the smooth muscle that constitutes the middle layer of vascular tissue. At this point, the monocyctes proliferate and differentiate into macrophages (large immune cells that collect debris) and begin to consume lipoproteins, eventually becoming foam cells. With time, the foam cells die and the middle of the lesion becomes a necrotic core. Meanwhile, the proximal smooth muscle cells secrete fibrous elements which bind the dead matter. As the lesion grows, the immune system reacts more aggressively and mononuclear cells from the blood enter and multiply, perpetuating the cycle. The lesion can calcify (harden) and ulceration of the luminal surface of the blood vessel may occur, spurring yet more activity from the immune system. Although the legions themselves can grow so large that the blood vessel becomes occluded, the acute clinical events of myocardial infarction (heart attack) and cerebral stroke predominantly occur when vascular ulceration causes plaque rupture, prompting platelets to gather rapidly at the site, where they induce formation of a thrombus (blood

Box 1 Factors associated with development of atherosclerosis

Genetic factors	Environmental factors
• Elevated blood serum levels of LDL and VLDL (low density and very low density lipoproteins, respectively)	• High saturated fat content of diet
• Reduced blood serum levels of HDL (e.g. antioxidant vitamins, flavanoids)	• Low antioxidant levels in diet (high density lipoprotein)
• Elevated blood serum levels of lipoprotein a	• Smoking
• Elevated blood pressure (hypertension)	• Lack of exercise
• Elevated levels of homocysteine	• Infectious agents (viruses and bacterial infection can promote arterial damage, e.g. Chlamydia pneumoniae)
• Family history	• Low socioeconomic status
• Diabetes and obesity	
• Elevated levels of haemostatic factors (fibrinogen, platelet activity, etc.)	
• Depression	
• Gender – males at higher risk	
• Systemic inflammation (e.g. arthritis)	
• Metabolic syndrome	

Box 2 The development of atherosclerosis

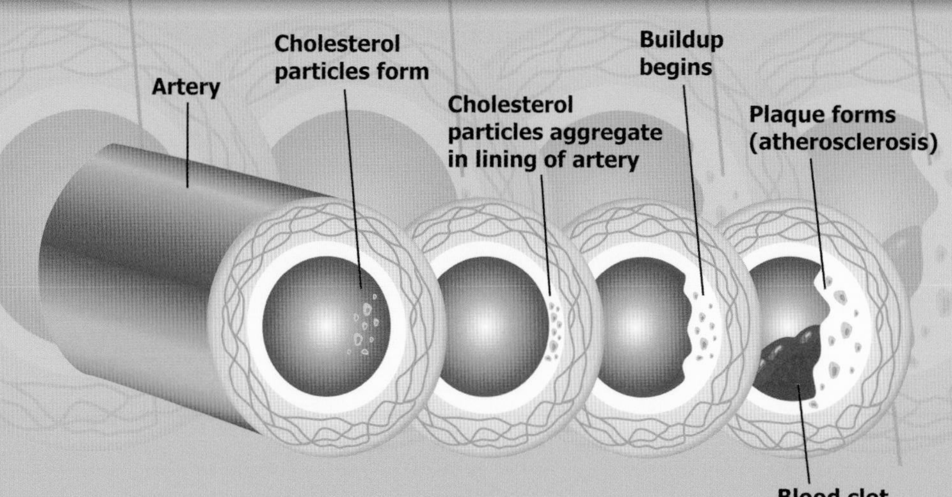

Box 3 Selected drugs used to treat cardiovascular disease

Drug class	Pharmacological description	Leading examples generic name, [trade name] (marketer, worldwide sales in million US $, 2005)
angiotensin converting enzyme (ACE) inhibitors	ACE is a peptidase enzyme that catalyses the conversion of angiotensin I to the vasoconstricting substance angiotensin II (which also stimulates aldosterone secretion). Reduction of angiotensin II levels exerts a negative feedback effect on renin secretion. The combined effects of suppressing the renin-angiotensin-aldosterone system are antihypertensive.	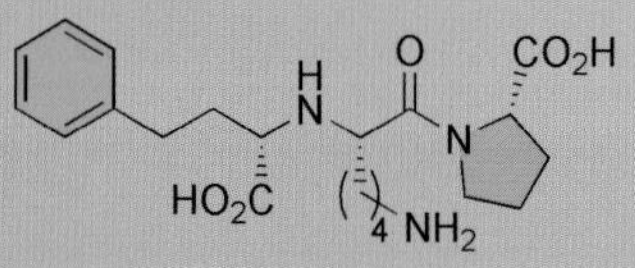 lisinopril [Zestril®] (AstraZeneca, 332) 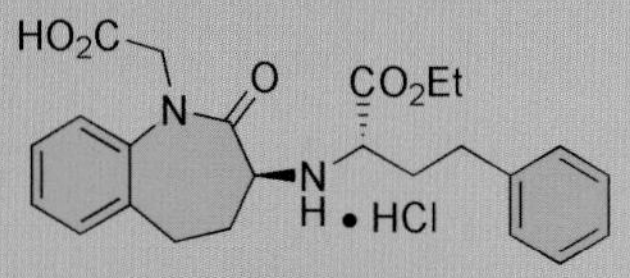 benazepril hydrochloride [Lotensin®] (Novartis, 433)
angiotensin II receptor antagonists	These drugs block the vasoconstrictor and aldosterone-secreting effects of angiotensin II by selectively blocking the binding of angiotensin II to the AT_1 receptor found in many tissues (e.g. vascular smooth muscle and adrenal gland).	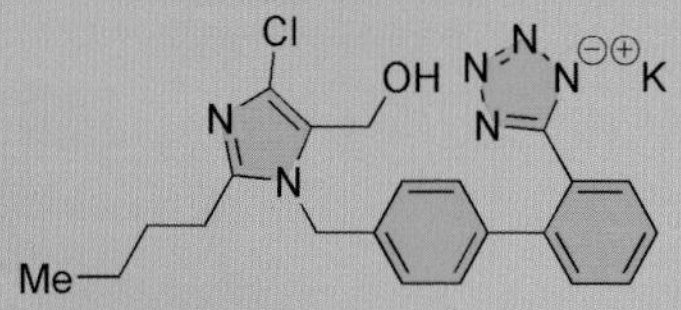 losartan potassium [Cozaar®] (Merck, 3,037) 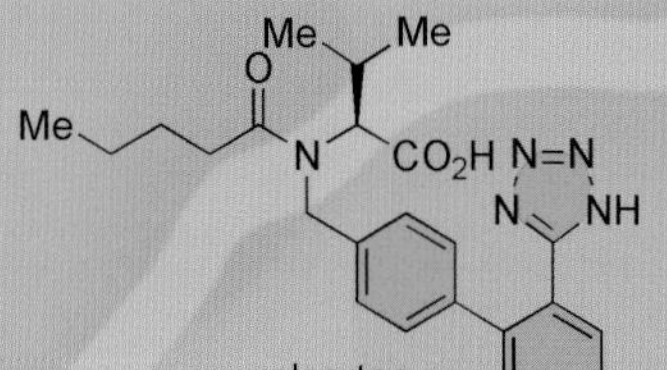 valsartan [Diovan®] (Novartis, 3,676) 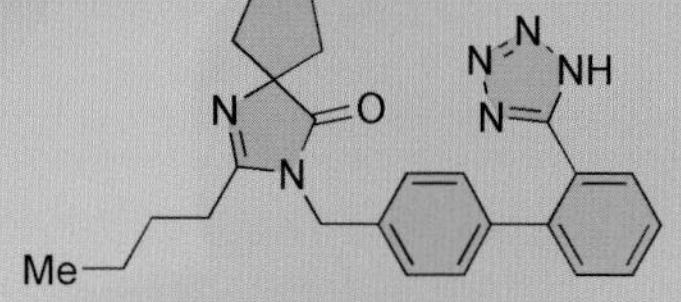 irbesartan [Avapro®] (Bristol-Myers Squibb/ Sanofi-Aventis, 2,050)

clot). The intricate cascade of reactions involved in blood clot formation is discussed briefly in Chapter 15. Parts of the clot can break away, and are swept through the bloodstream. If they are not broken down the clot fragments may eventually reach a small coronary or cerebral blood vessel, creating a blockage. This blockage causes the catastrophic shortage of oxygen in the heart muscle or brain that we experience as a heart attack or stroke. As mentioned above, deaths resulting from these traumas are very common, but the effects of this condition are felt much more widely. The illness and fatigue of atherosclerosis and CVD lead to significant healthcare costs and lost productivity, in addition to the personal suffering of the patients. Means to prevent, or at least minimize, the impact of atherosclerosis have therefore commanded significant attention from clinicians and the pharmaceutical industry.

Currently, treatment strategies for CVD generally combine two or more types of drug in a multi-pronged regime. These include β-blockers, angiotensin converting enzyme (ACE) inhibitors, angiotensin II receptor antagonists, calcium channel blockers, and platelet aggregation inhibitors (Box 3). Although these drugs have saved many lives, they are used only to alleviate symptoms (such as high blood pressure), once damage to the arteries has already set in. The earlier long-term suppression of cholesterol levels (in the form of LDLs) presents the opportunity to intervene before damage occurs, especially when targeted at high risk individuals who are genetically prone to cholesterol-mediated CVD. If patients are treated early enough, serious arterial damage can be prevented, and the onset of high blood pressure delayed. In patients already suffering from CVD, such treatments can slow the rate of arterial damage, and postpone acute events such as heart attacks and strokes.

Cholesterol, though essential for the healthy function of the body, is not a necessary component of our diets because it is biosynthesized endogenously. Some clinicians had previously proposed that eliminating dietary cholesterol, by minimizing the consumption of foods such as eggs, red meats, and shellfish, was sufficient to lower blood cholesterol levels. It is now argued that the body, when deprived of dietary cholesterol, compensates for the loss by producing more itself. Thus, the lower blood levels achieved initially cannot always be sustained by using this simple dietary approach. Much of the detail regarding the mechanism by which the body regulates blood cholesterol was elucidated by Michael S. Brown and Joseph L. Goldstein of the University of Texas, Dallas (USA). They discovered receptors for LDLs on the surfaces of connective tissue cells, and noted that patients with familial hypercholesterolemia (genetic high cholesterol levels) had fewer of these LDL docks than other individuals or, in some cases, none at all. They went on to investigate the fate of LDL packages delivered to these receptors, showing how they were crucial in the regulation of cholesterol biosynthesis. This work earned these pioneering scientists the 1985 Nobel Prize in Physiology or Medicine.

The latest dietary advice to emerge from this research suggests that an ideal diet would minimize the intake of cholesterol and saturated fat, while adding unsaturated lipids along with supplementary foods

Michael S. Brown

Joseph L. Goldstein

Box 3 Selected drugs used to treat cardiovascular disease (continued)

Drug class	Pharmacological description	Leading examples generic name, [trade name], (marketer, worldwide sales in million US $, 2005)
β-blocker	Inhibitors primarily of β_1-adrenergic receptors, especially those in cardiac tissue. These exert antihypertensive properties by reduction of heart rate at rest and during excercise, reduces systolic blood pressure, inhibits tachycardia, and has a central effect of reduced sympathetic outflow to the periphery.	atenolol [Tenormin®] (AstraZeneca, 352) 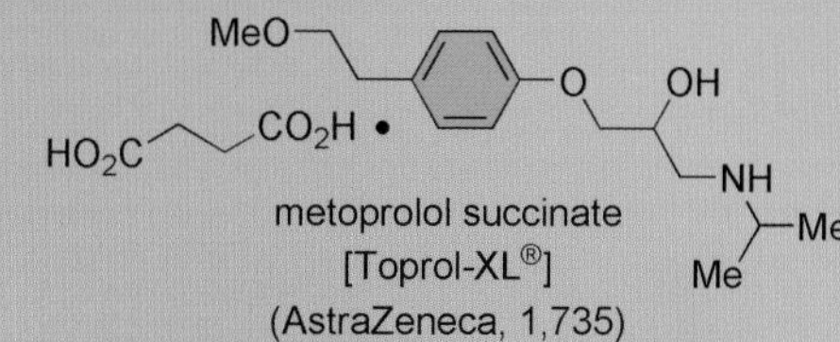 metoprolol succinate [Toprol-XL®] (AstraZeneca, 1,735)
calcium channel blocker	These molecules are dihydropyridine calcium antagonists that inhibit the transmembrane influx of calcium ions through specific ion channels into vascular smooth muscle and cardiac muscle, thereby preventing their contraction. This action is antihypertensive, restoring blood flow to coronary arteries.	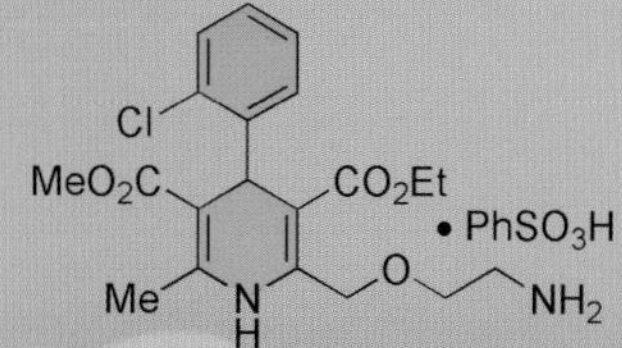 amlodipine benzenesulfonate [Norvasc®] (Pfizer, 4,706) 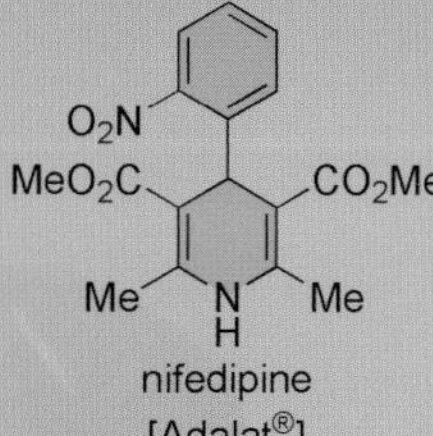 nifedipine [Adalat®] (Bayer, 659)
platelet aggregation inhibitor	Plavix® inhibits ADP-induced platelet aggregation by preventing the binding of adenosine diphosphate (ADP) to its receptor and the subsequent ADP-mediated activation of the glycoprotein GPIIb/IIIa complex. Platelet aggregation is key to thrombus (blood clot) formation. Thrombus-induced blockage of coronary or cerebral blood vessels causes myocardial infarction or cerebral stroke, respectively.	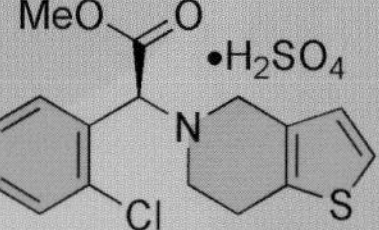 clopidogrel bisulfate [Plavix®] (Bristol-Myers Squibb/ Sanofi-Aventis, 6,250)

that contain natural LDL-lowering agents. Curiously, current advice tends to reflect recommendations made in the United States in the 1960s and 1970s. These were later revoked, or at least diluted, because the message was considered to be too complicated. The USA experienced a significant dip in the incidence of CVD during the 1980s, since the standard American diet had seen an approximate doubling in polyunsaturated fatty acids content during the preceding decade. However, this progress was not sustained as the simple "all fat is bad" message gained popularity.

'Saturated' and 'unsaturated' are broad chemical terms which have been commandeered for common use. Saturated fats and oils are those with no carbon–carbon double bonds (Box 4). This category includes hydrogenated vegetable oils since a molecule of hydrogen has been added across each double bond present in the original natural oil through a process called hydrogenation (see Chapter 18). Hydrogenation makes the oils more stable to a range of conditions, including the elevated temperatures needed for baking and frying, such that these fats polymerize and oxidize less readily than their natural unsaturated counterparts. Naturally occurring saturated fats and oils are found in meats (especially red meat), dairy products, coconut oil, and palm oil; these, along with the refined vegetable oils, are responsible for raising LDL serum levels disproportionately with respect to HDL levels. Conversely, unsaturated fats, such as those found in olive, soybean, and fish oils, contain one or more carbon–carbon double bonds. Through a mechanism that is not yet fully understood, these oils lower LDL (and raise HDL) blood serum levels. The geometry of the double bond in these oils is also crucial. Natural unsaturated fats have the *cis* (Z) configuration, but industrial or biological processing (e.g. in a cow's rumen) can lead to isomerization to the *trans* (E) isomer. This change alters the characteristic chemical and physical properties of these fats. For example, such alterations can convert an oil into a solid fat, as in the manufacture of margarine. Fats with *trans* double bonds also induce unhealthy increases in blood serum LDL levels. Furthermore, *trans*-fatty acids have been shown to raise the blood serum levels of lipoprotein a, another atherogenic lipoprotein. A note of caution must be added to the advice favoring unsaturated fats. The *cis* double bonds of these fatty acids are susceptible to the formation of damaging organic peroxides upon reaction with environmental radical species. The consumption of antioxidants, such as vitamin E, β-carotene, and flavonoids (present in fruits, vegetables, and red wine), is one way to minimize the potential danger of organic peroxides.

Three groups of supplementary foods that may help prevent the progression of atherosclerosis include the B vitamins (folic acid and niacin), foods high in soluble fiber (such as oats), and plant sterols. Folic acid is converted to tetrahydrofolate in the body, which is a vital co-enzyme in the synthesis of the amino acid methionine from homocysteine. This reaction depletes homocysteine in the blood, which is desirable because elevated homocysteine levels have been shown to lead to cumulative

Box 4 Structures of selected dietary fats

Saturated fats

tetradecanoic acid (myristic acid) - butterfat, palm and coconut oils

dodecanoic acid (lauric acid) - palm and coconut oils

hexadecanoic acid (palmitic acid) - animal products

Unsaturated fats (natural)

eicosapentaenoic acid (EPA) - fish oil

docosahexaenoic acid (DHA) - fish oil

octadecatrienoic acid (α-linolenic acid) - sunflower, flax, safflower, and olive oils

cis-9-octadecenoic acid (oleic acid) - olive oil

trans **Unsaturated fat (man made)**

trans-9-octadecenoic acid (elaidic acid) - *trans* isomer of oleic acid, formed by industrial refining process

damage of the cells lining the arteries. High homocysteine levels are also known to interfere in the proper function of clotting factors, and in the oxidation of LDLs, which also promote vascular disease. Excessive homocysteine levels have also been implicated in cognitive disorders such as Alzheimer's disease. Niacin, also known as nicotinic acid or vitamin B_3, affects the production of blood fats in the liver, lowering levels of LDL and triglycerides, while raising HDL levels.

Manufacturers of oat-based foods often advertise the health benefits of such cereals, including their effect on blood cholesterol. Much of the fiber consumed in our diet is insoluble, however, oats contain high levels of soluble fiber. Other sources of soluble fiber include eggplant (aubergine), okra, and barley, but not wheat. Soluble fiber sequesters bile acids as it passes through the gut, preventing their reabsorption. To replace bile acid supplies, the liver uses up stocks of cholesterol, leading to a reduction in blood LDL levels. Plant sterols have the same skeletal framework as cholesterol and other human steroids, differing only in the nature of their substituents. Studies have shown that incorporating plant sterols into the diet can help lower LDL levels by 10–15 %. They presumably act by mimicking cholesterol in our own feedback control mechanisms, thus preventing its release into the bloodstream. The matter is complicated by the lack of long-term data, and the quantity of plant sterols needed for a beneficial effect to be seen is disputed. However, several manufacturers have begun producing foods with added plant sterols, the most well known being the Becel®, Benecol® and Take Control® ranges.

A healthy diet should consist of low levels of saturated fats, but with balance being the key consideration. However, in the words of American writer Mason Cooley, "moderation in all things is best, but it's pretty hard to get excited by it." As this quote indicates, it is easier to suggest a balanced healthy diet than to keep to one, particularly given the hectic lifestyles of many in industrialized nations. In recent years, eating patterns have developed in which meals are prepared quickly and eaten in brief breaks within busy schedules. Meals increasingly include a high proportion of heavily processed foods which are laden with saturated fats, preservatives (including salt), and refined sugars. The inevitable result is more patients with persistently high LDL levels. These patients, along with those who cannot lower their blood cholesterol through dietary and lifestyle changes alone, require alternative or additional means of protecting themselves from arterial disease. In the 1980s, pharmaceutical companies began searching for compounds that could be developed as drugs to fill the emerging market niche for a cholesterol-controlling treatment. The fact that any treatment would be taken for prolonged periods by large segments of society provided added incentive, in the form of the enormous profitability of a successful drug. The cholesterol-lowering drugs that ultimately resulted from this search, such as Zocor® (Merck) and Lipitor® (Pfizer), are amongst the highest selling pre-

Box 5 Selected drugs for the treatment of high cholesterol

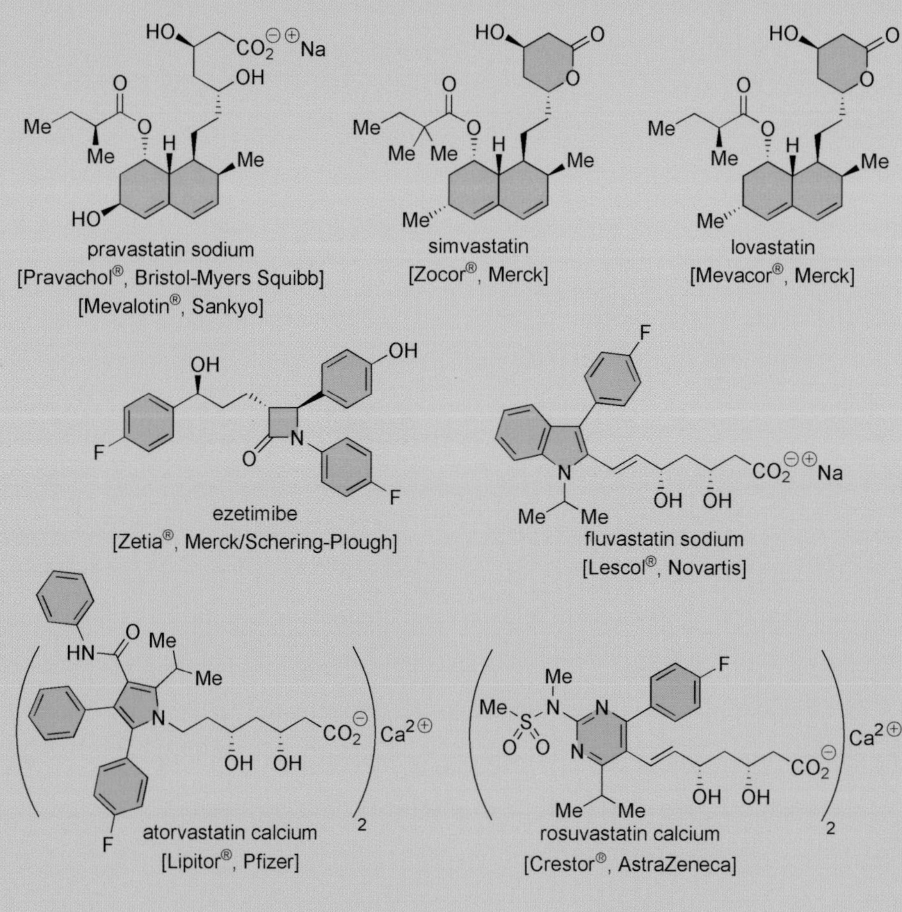

pravastatin sodium [Pravachol®, Bristol-Myers Squibb] [Mevalotin®, Sankyo]

simvastatin [Zocor®, Merck]

lovastatin [Mevacor®, Merck]

ezetimibe [Zetia®, Merck/Schering-Plough]

fluvastatin sodium [Lescol®, Novartis]

atorvastatin calcium [Lipitor®, Pfizer]

rosuvastatin calcium [Crestor®, AstraZeneca]

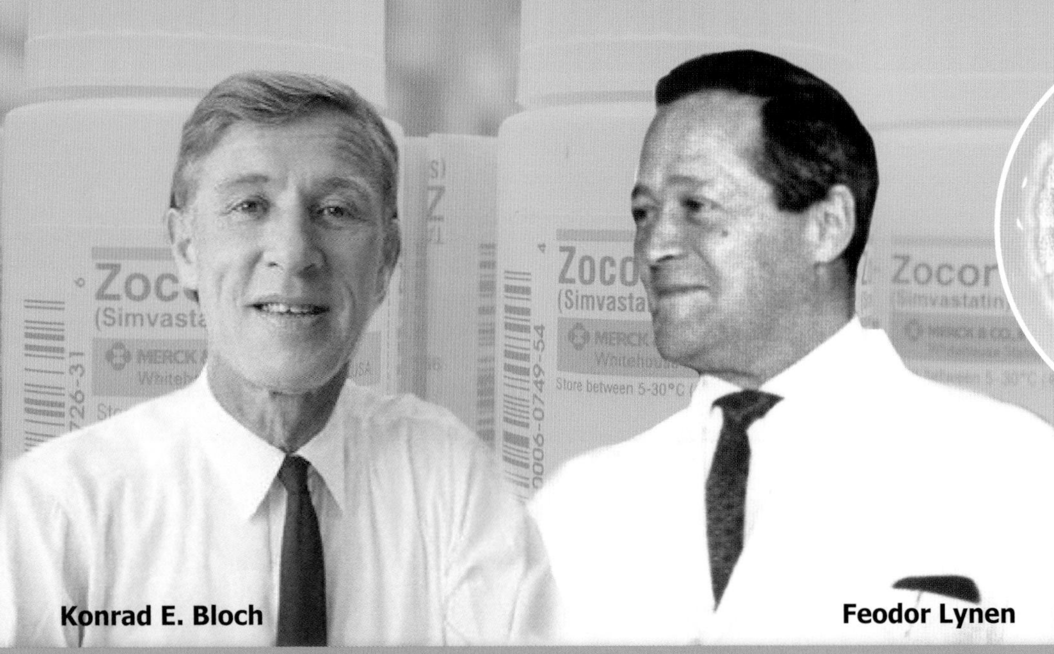

Konrad E. Bloch

Feodor Lynen

Penicillium citrinum

scription drugs ever developed (Box 5).

The objective was to find compounds that would inhibit cholesterol biosynthesis, as at least half of the cholesterol in our bodies is made endogenously. Cholesterol is produced through a complex enzyme-catalyzed process known as the mevalonate pathway (Box 6), which begins with acetyl coenzyme A. This molecule is gradually built upon, proceeding through mevalonic acid, eventually giving the long-chain hydrocarbon squalene. This linear precursor molecule undergoes several more enzymatic reactions, including cyclization to form the steroid skeleton, before furnishing cholesterol. Prior to the formation of squalene, the pathway branches, allowing the generation of other essential biomolecules, a feature with important consequences for drugs that interfere at a point before this branching. The 1964 Nobel Prize in Physiology or Medicine was awarded to two German-born scientists, Konrad E. Bloch (Harvard University, USA) and Feodor Lynen (Max Planck Institute, Münich, Germany), for their part in uncovering many of the details of this critical biosynthetic pathway. Scientists at Merck first examined inhibiting the biosynthesis at a late stage, and began to search for small molecules that could inhibit the conversion of lanosterol into cholesterol, but with little success. Fortunately, work in various academic laboratories had identified an earlier, rate-limiting step in the pathway; inhibition of HMG CoA reductase would prevent cholesterol biosynthesis and lead to a significant drop in blood serum cholesterol levels.

The Japanese pharmaceutical company Sankyo was the first scientific establishment to pinpoint a lead for such an intervention in the form of a natural product. A team led by Akira Endo at their Tokyo laboratories began to search for cholesterol biosynthesis inhibitors in fungal cultures. After a painstaking search they isolated several fungal metabolites from a *Penicillium citrinum* broth that were found to inhibit HMG CoA reductase and effect a reduction of blood serum cholesterol levels in animal models. Despite initial reluctance on the part of his company, Endo and a physician at the Osaka University Hospital, Akira Yamamoto, conducted a small clinical trial in patients with genetic hypercholesterolemia with striking results. This first study led to the development of the statin drugs by Sankyo and other companies. Sankyo named their primary lead mevastatin, although the same compound was also named Compactin® by Beecham Pharmaceuticals (UK), who had isolated it during routine screening for antifungal agents. Merck subsequently isolated a very similar compound from the broth of *Aspergillus terreus*, another fungal species. The Merck compound, named Mevacor® (but also known as lovastatin), was found to be a potent inhibitor of HMG CoA reductase, and was quickly and successfully developed into a drug treatment, gaining approval in the lucrative US market in late 1987. Since then, Mevacor® has shown considerable success, helping to control cholesterol levels in millions of patients. Its success has led to several semi-synthetic analogues (Box 5), including pravastatin (sold as Pravachol® by Bristol-Myers Squibb, and Mevalotin® by Sankyo) and simvastatin (Zocor®, also sold

Box 6 Outline of cholesterol biosynthesis – the mevalonate pathway

acetyl co-enzyme A [Ac CoA] → hydroxymethylglutaryl coenzyme A [HMG CoA] → (HMG CoA reductase) → mevalonic acid → farnesyl pyrophosphate (FPP) → many other lipid biomolecules

farnesyl pyrophosphate (FPP) → squalene → (squalene epoxidase) → squalene epoxide → (cyclase) → lanosterol → (enzymatic) → cholesterol

squalene epoxide → steroid hormones

cholesterol → bile acids

by Merck). Both Mevacor® and Zocor® are in fact prodrugs, meaning that the molecules administered are not the active compounds, but precursors which are transformed to the active drug upon enzymatic degradation (hydrolysis) within the body (Box 7). A number of organic chemists have devised elegant synthetic routes to these natural products, including Charles J. Sih (University of Wisconsin, USA), Masahiro Hirama (then at Suntory Institute of Bioorganic Research, Osaka, Japan), and Paul A. Grieco (Indiana University, USA), as well as Narindar N. Girotra and Norman L. Wendler at Merck (Rahway, New Jersey, USA). These studies facilitated the preparation of analogues and the investigation of their biochemical interactions; however, it has proven more viable to produce Mevacor® and related compounds by fermentation techniques. Lipitor® (sold by Pfizer) and Lescol® (Novartis) are also competitive inhibitors of HMG CoA reductase, although in this case the drugs are fully synthetic molecules, designed and made by chemists in the laboratory (Box 5). Lipitor® was the top-selling drug in 2005, with worldwide sales of $12.2 billion. Zetia®, from Merck/Schering-Plough, is a new type of cholesterol drug which acts by preventing cholesterol uptake from the gut, and can be used in combination with a statin (drugs that inhibit cholesterol biosynthesis) to give a greater overall effect.

Masahiro Hirama

Charles J. Sih

As indicated earlier, HMG CoA reductase catalyzes an early step in the biosynthesis of cholesterol, and therein lies a potential flaw with the blockbuster drugs that we have just introduced (with the exception of Zetia®). While these compounds effectively lower cholesterol levels, they also decrease, or prevent completely, the biosynthesis of numerous other molecules essential to the proper functioning of our bodies (Box 6). This feature, combined with the profitability of cholesterol-lowering drugs, spurred the search to continue in the late 1980s for new cholesterol-lowering agents. Researchers began to seek inhibitors of the enzyme squalene synthase, which is involved in cholesterol biosynthesis after a branch point leading to many other essential compounds. Between 1991 and 1993, three groups from Merck (in the USA

Paul A. Grieco

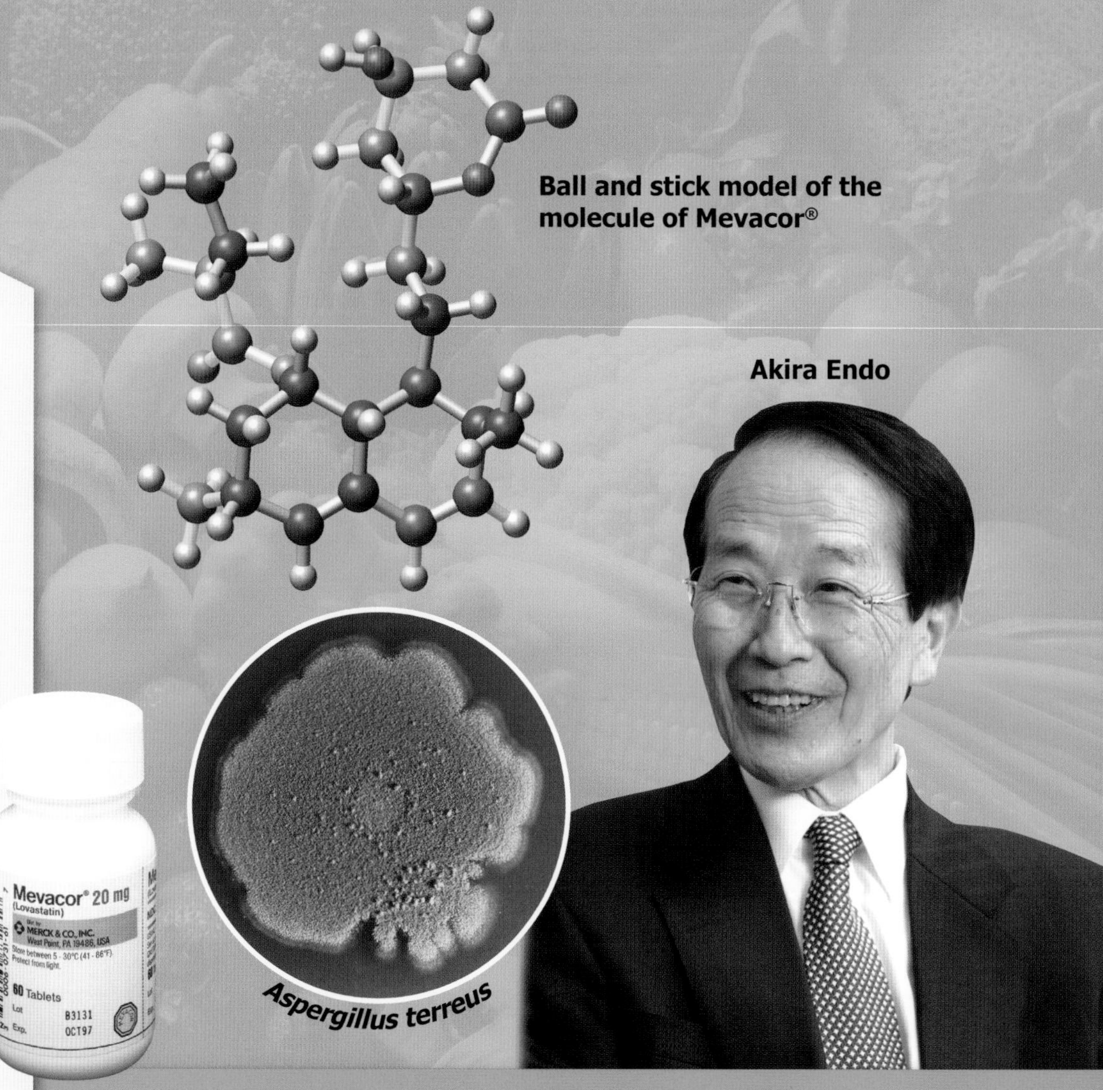

Ball and stick model of the molecule of Mevacor®

Akira Endo

Aspergillus terreus

Box 7 The naturally occurring statins are prodrugs

lovastatin [Mevacor®] *prodrug* or mevastatin [Compactin®] *prodrug* —*hydrolysis in vivo*→ (R = Me or H) *active drug*

Box 8 **Highlights from the Nicolaou group's total synthesis of zaragozic acid A**

and Spain), Glaxo Pharmaceuticals (UK), and Tokyo Noko University (Japan), the latter in collaboration with the Mitsubishi Kasei Corporation (Japan), announced their independent discovery of a new class of natural products. These compounds exhibited potent inhibitory action against squalene synthase. Merck named their compounds the zaragozic acids because the fungal broth from which they were isolated was derived from a sample culture collected from the Jalón River in the Zaragoza region of Spain. Meanwhile, Glaxo named their discoveries the squalestatins, in reference to their activity. Their samples were isolated from fungal broths of the *Phoma* genus, collected in Armação de Pera in Portugal. Despite coming from quite different fungal samples, the two groups of molecules are close structural relatives. Indeed, zaragozic acid A and squalestatin S1 are identical compounds.

Many members of the zaragozic acid family have now been isolated from different fungal species, originating from a range of sites around the world, including Spain, the USA, Kenya, and the Philippines. There is some structural diversity, however all members of the class share the same highly oxygenated bicyclic core, to which is appended two variable lipophilic side chains. The unusual core poses a host of challenges to any total synthesis. Not only are there practical difficulties associated with the handling of such polar compounds, and the reactivity of the highly oxygenated skeleton, but the bicyclic system includes significant stereochemical complexity. This core structure alone incorporates six contiguous stereogenic centers, including three quaternary carbon atoms.

During the early 1990s, these challenges attracted the attention of numerous research groups, culminating in several beautiful total syntheses. The first successes came from the groups of Erick M. Carreira (then at the CalTech, USA, now at the ETH, Zürich, Switzerland) and K. C. Nicolaou (The Scripps Research Institute and the University of California at San Diego, USA), who accomplished total syntheses of zaragozic acids C and A, respectively, in 1994. The same year also saw the completion, with characteristic finesse, of zaragozic acid C by the team led by David A. Evans (Harvard University). More syntheses followed and, while each of them has its own engaging characteristics, a certain commonality can be seen in many aspects. Many rely upon the acid-mediated ketalization of an advanced precursor to form the bicyclic system. For this reason, we shall summarize only the first total synthesis of zaragozic acid A. The pivotal steps of the Nicolaou team's synthesis of zaragozic acid A are highlighted in Box 8. The key stereoselective step in this synthesis employed the Sharpless asymmetric dihydroxylation reaction (Box 12) to provide a chiral diol. The formation of all the remaining stereocenters in the bicyclic core of zaragozic acid A was controlled by these two centers.

Ball and stick model of the molecule of zaragozic acid A

While the zaragozic acids were being synthesized and their structure-activity relationships examined, new fungal metabolites with promising activity were still being sought, often from fungal broths. These broths are popular sources for the isolation of bioactive natural products for a number of reasons. Fungal species are relatively easy to culture and grow, and they can be collected easily from diverse geographical and ecological environments. In addition, fungi produce a range of defense chemicals, which they use to defend their territory and compete for resources. The final characters in this account, the CP molecules, are also fungal metabolites, isolated from a fungus found on juniper twigs in Texas (USA). These compounds are also known by different names. The first names, CP-225,917 and CP-263,114, are the identification codes given to the molecules by their discoverers, led by Takushi Kaneko, at Pfizer (the initials being those of Charles Pfizer, the founder of the company). They are also known as the phomoidrides, a contraction of the fungal genus from which they were isolated (*Phoma*), and nonadride, denoting a nine-membered ring bearing adjacent carboxylic acid groups fused together to afford an anhydride, a key structural feature of these secondary metabolites. From fifteen liters of fermentation broth, the Pfizer chemists initially isolated just 31 mg of CP-225,917, 18 mg of CP-263,114, and 8 mg of zaragozic acid A using HPLC (high pressure liquid chromatography). Such small quantities proved sufficient for structural studies, and initial biological testing.

Squalene synthase catalyzes the coupling of two molecules of farnesyl pyrophosphate, and the zaragozic acids inhibit this enzyme by mimicking the transition state of this transformation. Farnesyl pyrophosphate is also the substrate for another enzyme, ras farnesyl transferase (RFT), which catalyzes the transfer of a farnesyl group to the ras protein. The attachment of a farnesyl group is necessary for the membrane translocation of this protein, which, once inside the cell, acts as a switch to turn on cell growth. Inhibition of RFT has been proposed as a possible strategy for controlling the abnormal growth of cancerous cells promoted by a common mutation of the gene that encodes the ras protein (leading to a deviant form named ras p21). In addition to being inhibitors of squalene synthase, the CP molecules are potent RFT inhibitors, which led to

Ball and stick models of the CP molecules

The Nicolaou CP molecule team

Takushi Kaneko

Box 10 Construction of the maleic anhydride unit by a cascade reaction

(1) MsCl, Et_3N
(2) K_2CO_3

C_5H_9 =
C_8H_{15} =

β-elimination then 5-exo-dig cyclization

air, $(CO_2H)_2$
tautomerization

stepwise addition of $^3O_2$

[a]intermediates have been simplified by removing substituents for clarity

H_2O
$- NH_3$

Box 11 The completion of the CP molecules by the Nicolaou group

aq. LiOH

NaH_2PO_4

$MeSO_3H$
method of Kaneko *et al.*

(+)-CP-263,114

(–)-CP-225,917

(enantiomers of natural products)

Tohru Fukuyama **Matthew D. Shair** **Samuel J. Danishefsky**

great excitement about these natural products and their potential as anticancer agents.

As with the zaragozic acids, the CP molecules have an oxygen-rich polycyclic core bearing a lipophilic side chain. However, the core of the CP molecules is considerably more complex, and includes a highly strained carbon–carbon double bond at a bridgehead position within a bicyclodecane system. In the early part of the twentieth century, K. Julius Bredt performed a systematic study of the camphor and pinane bicyclic ring systems (see Chapter 5), and discovered that double bonds at bridgehead positions are strongly disfavored, a situation that is now described by Bredt's rule. It has subsequently been found that as the ring sizes of the bicyclic system increase the additional flexibility does allow for bridgehead double bonds (see, for e.g., calicheamicin γ_1^I and Taxol®, Chapters 23 and 25, respectively), although frameworks analogous to the CP molecules are exceptionally rare in nature.

Four groups have risen to the challenge posed by the total synthesis of these molecules. The group led by K. C. Nicolaou completed the first total synthesis in 1999, and in the following year their accomplishment was joined by similar feats from the groups of Tohru Fukuyama (University of Tokyo, Japan), Matthew D. Shair (Harvard University), and Samuel J. Danishefsky (Sloan-Kettering Institute and Columbia University, USA). Each one of these syntheses is a daring adventure that took skill, determination, and ingenuity from the protagonists to bring it to a successful conclusion. We will focus here on the Nicolaou team's synthesis, as it is most familiar to us, but the reader is strongly encouraged to explore the original literature for full details of all these outstanding achievements.

The final strategy followed by the Nicolaou group was based on a retrosynthetic analysis that made use of information from the isolation report regarding chemical means by which to convert one target (CP-225,917) into the other (CP-263,114), before disconnecting the lactone and anhydride rings, and the side chain, leaving the bicyclic core as the central challenge. The core

K. C. Nicolaou

Box 12 The Sharpless asymmetric dihydroxylation (SAD) reaction

In 1874, Dutchman Jacobus H. van't Hoff and Frenchman Joseph-Achille Le Bel independently discovered that carbon atoms having four groups, each attached by a single bond, arrange these bonds so they point to the corners of a tetrahedron. This arrangement gives rise to the phenomenon of chirality in organic compounds. If the four groups attached to a carbon atom are all different, two different arrays can exist, which are termed enantiomers. These enantiomers are mirror images of one another and cannot be superimposed. Many chemical and physical properties of enantiomers are identical. However, they differ in their interactions with chiral reagents or environments. Biological molecules are found in only one enantiomeric form, thus biological environments are chiral, and two enantiomeric molecules will interact differently with biological systems. In the late 1950s, pregnant women were administered a new drug, thalidomide, to prevent morning sickness. Both enantiomers of the drug were present in these preparations, and while one enantiomer prevented nausea, the other caused severe birth defects. In an effort to avert similar tragedies, regulatory authorities now generally require drugs to be supplied as single enantiomers. In rare instances where racemates are administered, the effects of both enantiomers must be fully investigated. Thus, the preparation of compounds as single enantiomers is of great importance.

The two reactions discussed here, the Sharpless asymmetric dihydroxylation and the Sharpless–Katsuki asymmetric epoxidation (see Box 13), have made an enormous contribution to our ability to synthesize organic molecules in one enantiomeric form. K. Barry Sharpless, the leading scientist involved in the discovery of both these processes, shared the 2001 Nobel Prize in Chemistry for "...his work on chirally catalysed oxidation reactions" with Ryoji Noyori and William S. Knowles, who were rewarded "...for their work on chirally catalysed hydrogenation reactions" (see Chapter 28).

Milestones in the development of the asymmetric dihydroxylation:

1908 – O. Makowka reports that olefins can be *cis*-dihydroxylated using stoichiometric (one equivalent of reagent for every equivalent of substrate) amounts of osmium tetroxide (OsO_4).

olefin — NMO, cat. OsO_4, acetone/H_2O (*Upjohn dihydroxylation*) → [intermediate proposed by Criegee] → *cis*-diol

1912 – K. A. Hofmann reports the use of metal chlorates ($MClO_3$) as co-oxidants, thereby allowing the amount of osmium employed to be reduced. This represented a major advance, as osmium is a rare and expensive metal and osmium tetroxide is volatile and toxic.

1936 – Rudolf Criegee notes that the addition of amines, such as pyridine, to the reaction mixture accelerates its rate. Criegee also proposes a mechanism for the dihydroxylation reaction.

Rudolf Criegee

1976 – K. Barry Sharpless introduces *tert*-butyl hydroperoxide (*t*-BuOOH) as co-oxidant. This innovation was rapidly superseded by a procedure reported by a group of chemists at the Upjohn Pharmaceutical Company (USA), using *N*-methylmorpholine-*N*-oxide (NMO) as co-oxidant. This protocol is still very popular today, even though it delivers both enantiomers.

1980 – K. Barry Sharpless, inspired by Criegee's observation, publishes a protocol for the stoichiometric osmium tetroxide dihydroxylation using chiral amine additives. Two pseudo-enantiomeric ligands, DHQ and DHQD, give promising results. These natural product-derived ligands are not exact mirror images of one another, but give opposite selectivity in the dihydroxylation reaction.

The subsequent decade saw many refinements of the asymmetric dihydroxylation protocol, with the introduction of potassium ferricyanide as co-oxidant, and the use of catalytic quantities of non-volatile potassium osmate ($K_2OsO_4(H_2O)_2$) in place of OsO_4. All the components can now be purchased as a convenient pre-mixed solid reagent, containing either of the ligands (AD-mix α and AD-mix β). Addition of these mixtures to a solution of an alkene gives reliable and predictable enantioselective dihydroxylation. The predictability of the reaction has led to its widespread use in organic synthesis.

$(DHQD)_2PHAL$, $K_2OsO_4(H_2O)_2$, $K_3Fe(CN)_6$, K_2CO_3, $MeSO_2NH_2$, *t*-BuOH/H_2O — *Sharpless asymmetric dihydroxylation* — 99 : 1

$(DHQ)_2PHAL$ ligand

$(DHQD)_2PHAL$ ligand

was then disconnected to an acyclic triene, and finally to relatively simple building blocks. It should be noted that this was by no means the first route explored. Indeed, it is almost impossible to convey to those who have not experienced it first hand the ingenuity, skill, and, perhaps most of all, perseverance required to complete such a project in the face of countless setbacks. The forward synthesis was completed first in racemic form, however, this feat was followed shortly thereafter by a synthesis of a single enantiomer of the CP molecules, by the sequence highlighted in Boxes 9–11. In the event, the Nicolaou group prepared the unnatural enantiomer of the CP molecules, thus determining their absolute configuration, which was unknown at the outset of the campaign. The synthesis began with the assembly of a linear triene precursor to the core skeleton (Box 9). Under Lewis acid catalysis, this triene underwent the planned intramolecular Diels–Alder reaction to form the bicyclic core of the targets, including the bridgehead alkene, which was carried intact throughout the remainder of the sequence. Notably, a single stereocenter on the Diels–Alder substrate controlled the facial selectivity in this key reaction, setting the other stereocenters in their proper configuration. The side chain was then introduced in preparation for the formation of the anhydride ring. While in simpler substrates the installation of such a group would not be problematic, in the setting of the CP molecules this task proved challenging. The team employed an intricate cascade sequence to overcome this difficulty, as shown in Box 10. Beginning from a dihydroxy nitrile, the cascade proceeded through an epoxide and an aminofuran, which underwent addition of oxygen to generate the anhydride motif. Following this cascade, a number of protecting group and oxidation state manipulations led to the completion of the total synthesis of the CP molecules (Box 11).

The CP molecules provided an inspirational challenge to chemical synthesis. In addition to the total synthesis achievements themselves, these endeavors have provided a greater understanding of the reactivity of such systems and the motivation for the development of numerous synthetic methods. Such programs also provide rigorous education and training for numerous students of organic chemistry. While these molecules have been successfully conquered by synthesis, the full spectrum of their biological properties remains unexplored. The biological secrets that may come to light through future studies promise to open new avenues for further biomedical breakthroughs.

K. Barry Sharpless

Tsutomu Katsuki

Box 13 The Sharpless–Katsuki asymmetric epoxidation (SAE) reaction

Since the beginning of the twentieth century, chemists have known that alkenes can be converted to epoxides by reaction with a peracid. The reaction is faster between electron rich carbon–carbon double bonds and electron deficient peracids. Such a reaction is employed frequently by organic chemists to prepare epoxides, often using *meta*-chloroperbenzoic acid (*m*-CPBA) as the oxidant.

In the early 1970s, Sharpless found that certain peroxy vanadates and molybdates (oxides of the metals vanadium and molybdenum) were efficient catalysts for producing epoxides from allylic or homoallylic alcohols, in conjunction with *tert*-butyl hydroperoxide as co-oxidant. In 1980, Sharpless and his then student, Tsutomu Katsuki, extended this technique with the development of "...the first practical method for asymmetric epoxidation." This reaction involves a chiral titanium catalyst to deliver an oxygen atom to the alkene of an allylic alcohol. Cheap and readily available tartrate esters are used as the chiral ligands on the titanium, leading to high levels of selectivity in the epoxide products. Again, the reliability and predictability of the Sharpless–Katsuki epoxidation has made it one of the most widely used and important enantioselective reactions in synthesis.

$Ti(Oi\text{-}Pr)_4$, t-BuOOH; L-(+)-diethyltartrate — $Ti(Oi\text{-}Pr)_4$, t-BuOOH; D-(−)-diethyltartrate

Further Reading

A. W. Alberts, Discovery, Biochemistry and Biology of Lovastatin, *Am. J. Cardiol.* **1988**, *62*, 10J–15J.

A. Nadin, K. C. Nicolaou, Chemistry and Biology of the Zaragozic Acids (Squalestatins), *Angew. Chem. Int. Ed. Engl.* **1996**, *35*, 1622–1656.

A. J. Lusis, Atherosclerosis, *Nature* **2000**, *407*, 233–241.

K. C. Nicolaou, P. S. Baran, The CP Molecule Labyrinth: A Paradigm of How Endeavors in Total Synthesis Lead to Discoveries and Inventions in Organic Synthesis, *Angew. Chem. Int. Ed.* **2002**, *41*, 2678–2720.

W. C. Willett, M. J. Stampfer, Rebuilding the Food Pyramid, *Sci. Am.* **2003**, *288*, 64–71.

K. C. Nicolaou, S. A. Snyder, *Classics in Total Synthesis II*, Wiley-VCH, Weinheim, **2003**, pp. 381–421.

D. Krummel, in *Krause's Food Nutrition and Diet Therapy*, L. K. Mahan, S. Escott-Stump, W. B. Saunders, Eds., Philadelphia, **2003**, pp. 558–595.

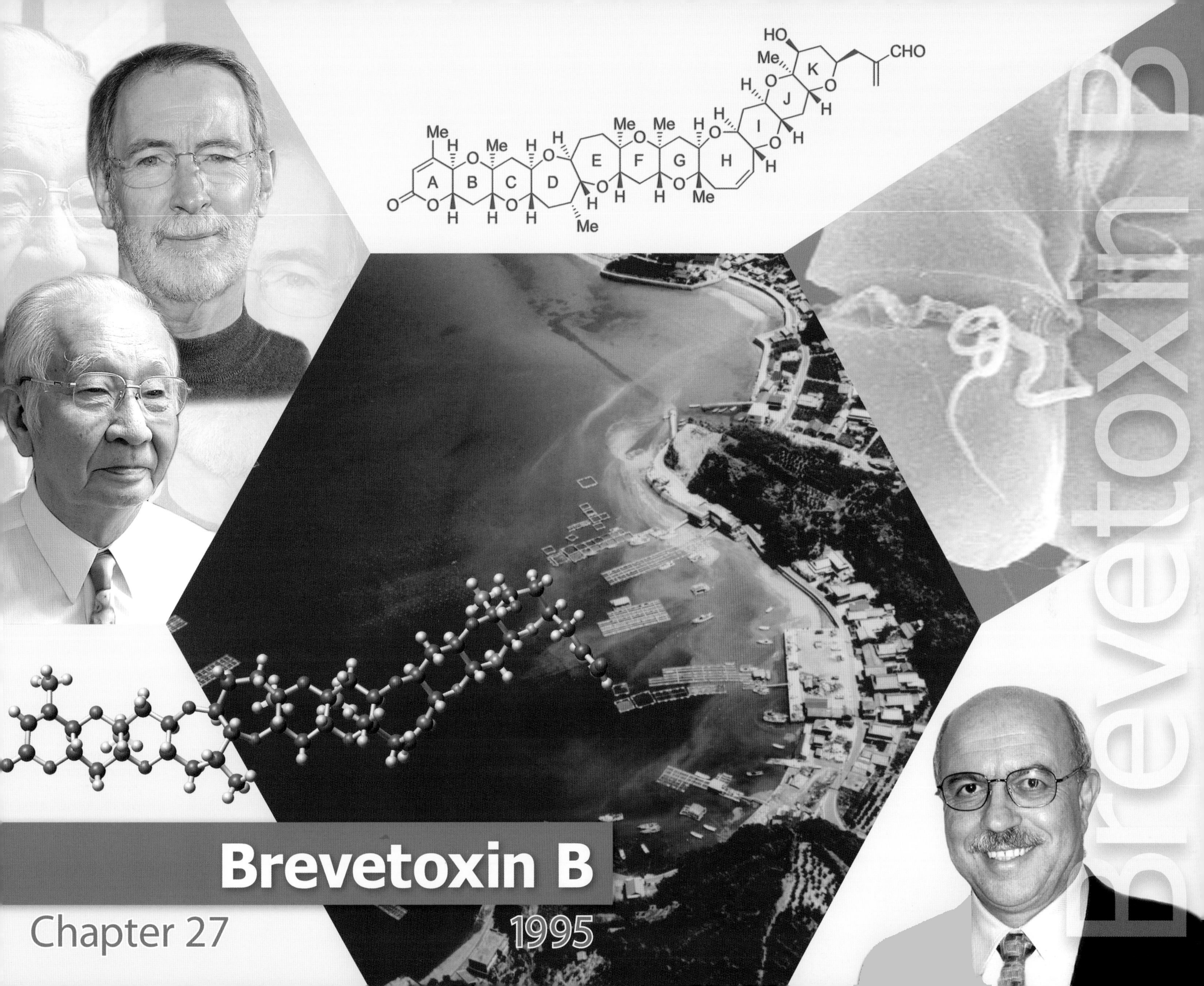
Brevetoxin B
Chapter 27
1995
Brevetoxin B
HO
CHO
Me
H
O
K
A
B
C
D
E
F
G
H
I
J

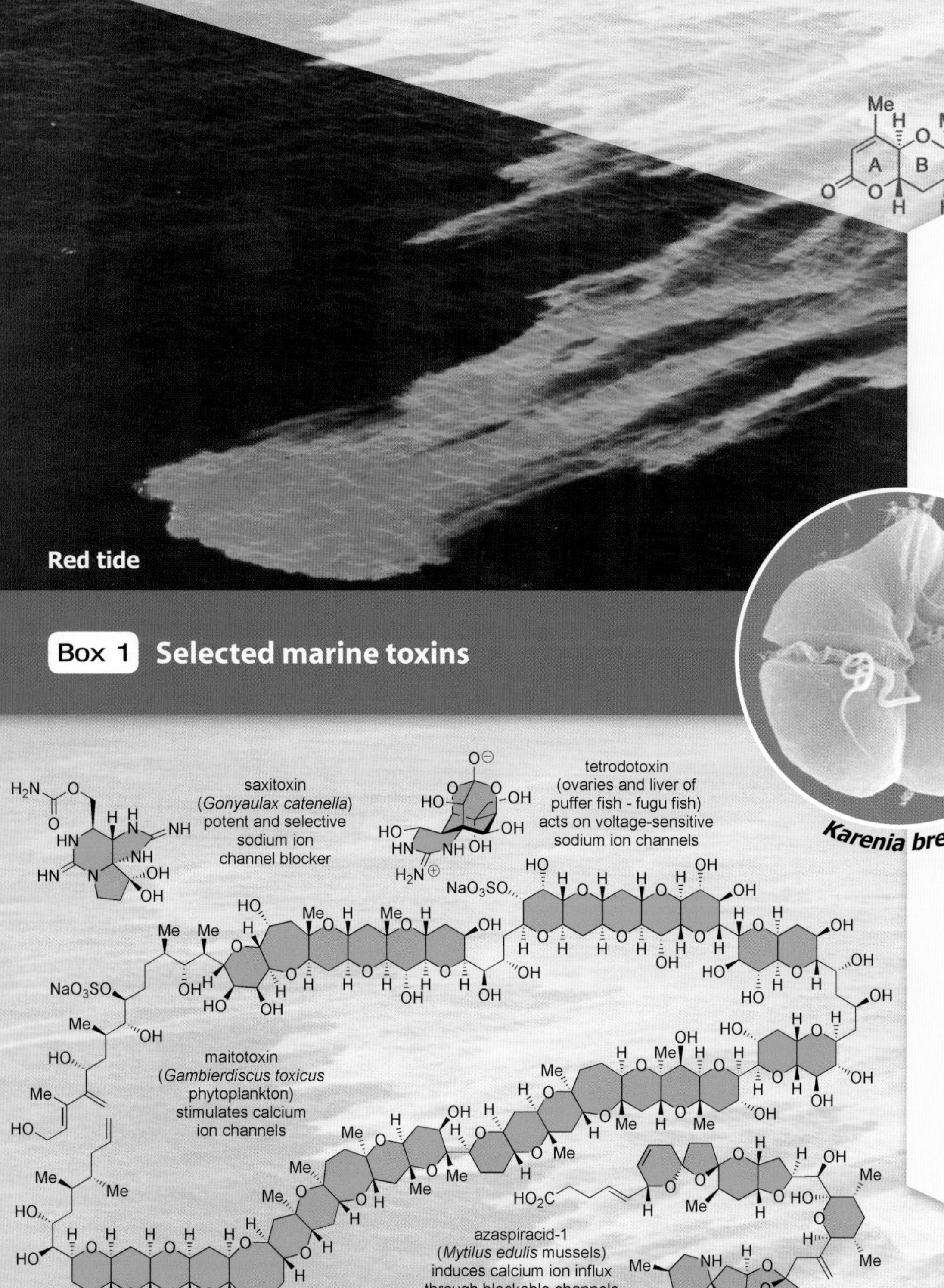

brevetoxin B

Karenia brevis

"...behold, I will smite with the rod that is in mine hand upon the waters which are in the river, and they shall be turned to blood. And the fish that is in the river shall die, and the river shall stink; and the Egyptians shall loathe to drink of the water of the river."

Exodus 7:17-18

Above is a short excerpt from what some believe is one of the first descriptions of an occurrence known around the world today as a red tide phenomenon. The quote comes from the *Bible*, where the story of the opening plague executed at the command of God against the Egyptians who had enslaved Moses and the Israelites is recounted. 'Red tide' is the modern name commonly given to the vast blooms of single-celled phytoplankton occurring sporadically in coastal waters, which derive a red hue from the dense growth of algae containing the carotenoid pigment peridinin. The Red Sea is thought to have acquired its name from its susceptibility to such epidemics. It should be noted that not all toxic algal blooms are red; various other colors are also seen, and outbreaks even occur with little or no associated color. Oceans tinged with a bright crimson stain present spectacular opportunities for photography, but, as the biblical quote ominously implies, the red tide also carries a sinister threat. A small number of the algae causing these outbreaks produce deadly defense toxins that can bring catastrophe to marine life within the bloom. These toxins persist and can reach dangerous concentrations throughout the food chain, so humans are also at risk if they consume contaminated seafood. There are no geographical restrictions to these algal blooms; they occur frequently throughout the world. The members of one class of powerful neurotoxic molecules isolated from the waters associated with such events were named the brevetoxins after the organism responsible for producing the deadly compounds, a dinoflagellate called *Karenia brevis* (formerly *Gymnodinium breve*). The first structure to be determined from this class, that of brevetoxin B, was reported in 1981. The story of the conquest of brevetoxin B by total synthesis constitutes the main theme of this chapter, but before we hear of the incredible twelve-year odyssey that it took to construct brevetoxin B in the laboratory, let us delve further into the extraordinary history and effects of red tides.

K. C. NICOLAOU, E. J. SORENSEN

CLASSICS IN TOTAL SYNTHESIS

TARGETS, STRATEGIES, METHODS

VCH

In 1793, the English seafaring captain George Vancouver became the first European to navigate and explore the coastal waters around British Columbia (present-day Canada). During this expedition, he landed at a place now called Poison Cove. Vancouver noted in his diary that the native Indians would not eat the shellfish from the colored waters of the bay because they were considered too dangerous. It now seems reasonable to attribute this historical episode, which gave Poison Cove its name, to a red tide event. As we move forward in time, we can be more certain of the incidences of this phenomenon. For example, in 1972, a massive bloom stretching along the northeast coast of the United States from Maine all the way to Massachusetts was definitively identified as a red tide. The incident followed a hurricane that had originated further south in the warm waters of the Gulf of Mexico and which had presumably been responsible for relocating the toxic algae to new waters as the storm moved northwards. 1972 was also a disastrous year for the Japanese fishery industry due to a severe red tide outbreak in the Seto Inland Sea killing caged yellowtail tuna worth an estimated \$500 million. The tiny unicellular algae responsible for the blooms are impressive killers, for they can destroy even the giants of the sea. Between 1987 and 1988, fourteen humpback whales died off Cape Cod, and 740 bottlenose dolphins were washed up on the Atlantic coast of North America from Florida in the south to New Jersey in the north. Both of these tragedies were later traced back down the food chain to algal bloom toxins. In the case of the whales the poisoning was traced to mackerel contaminated by feeding on toxic algae called *Alexandrium tamarense*, and with the dolphins to fish containing lethal brevetoxins from *Karenia brevis*. In 1991, a red tide incident was blamed for hundreds of sick and dying pelicans found on the beaches of Monterey, California (USA). Humans have also become ill as a result of such blooms. In 1987, a mass poisoning was reported in Canada with patients complaining of vomiting, diarrhea, disorientation, and abdominal cramps. In this instance, the culprits were identified as mussels from Prince Edward Island that were apparently contaminated by marine toxins from the red tide bloom of a *Pseudonitzschia* species. For this very reason, folklore in Nova Scotia warns people not to eat shellfish in the months whose name contains a letter "r".

Puffer fish (fugu)

The prevalence of red tide blooms has increased in recent times as population, pollution, and maritime transport have all swelled to previously unimagined levels. During the period between 1976 and 1986, when the population of Hong Kong grew six-fold, this startling expansion was matched in nearby

Box 2 K. C. Nicolaou's thoughts on seeing the structure of brevetoxin B

"It was love at first sight! I was intrigued, challenged and excited by brevetoxin's exquisite and fascinating [structural] regularity...but worries crept into my mind...It [the synthesis] would clearly be a treacherous and risky project..."

K. C. Nicolaou

Box 3 Selected cytotoxic marine natural products

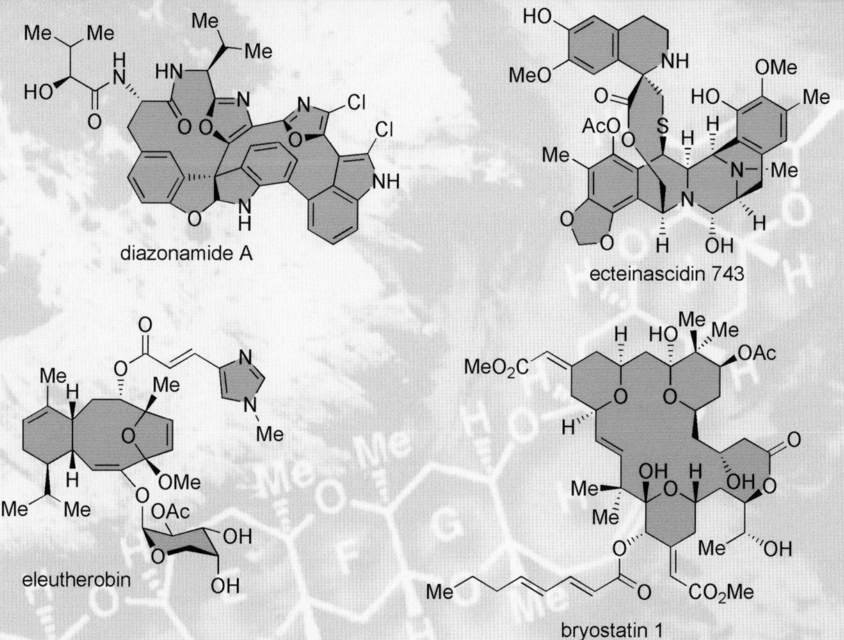

Box 4 **Nicolaou's strategic bond disconnections and retrosynthetic analysis of brevetoxin B**

hydroxy dithioketal cyclization
methylenation
Wittig reaction
oxygenation
brevetoxin B
intramolecular Horner–Wadsworth–Emmons olefination
$NiCl_2/CrCl_2$ coupling
intramolecular conjugate addition
intramolecular conjugate addition
hydroxy epoxide cyclization
hydroxy epoxide cyclization

Eleutherobia grayi coral

Tolo Harbor by an eight-fold rise in the occurrence of red tides. Pollutants from the land in the form of agricultural run-offs and sewage provide algae with the requisite nutrients to grow, sometimes out of control, triggering a bloom. In Florida (USA), high levels of chelated iron compounds in the Peace River effluent have been demonstrated to cause outbreaks in the immediate vicinity. Gibberellic acid, a plant hormone often washed into our waterways, has also been implicated as a red tide promoter. Long distance shipping by oil tankers, cargo vessels, and naval fleets has also facilitated the uncontrolled distribution of algal colonies. A number of these species have the ability to form hardened cysts that travel the oceans in a dormant state only to be reawakened when they encounter conditions that are conducive to their growth and reproduction (temperature range 10–33 °C, salinity levels 27–37 %, and appropriate duration and intensity of light). It is unfortunate that certain human activities are somewhat responsible for this growing hazard. The damage to marine ecology and human health, and the considerable worldwide economic losses due to red tides, all suggest that we ought to take this issue more seriously and look for prophylactic solutions.

The search for understanding of these catastrophic phenomena began sometime ago and it led to the identification of several biotoxins from poisoned waters. Amongst the most notorious marine toxins identified thus far are tetrodotoxin (associated with the fugu or puffer fish, a delicacy in the Far East), saxitoxin, palytoxin, maitotoxin, ciguatoxin, azaspiracid-1, and brevetoxins A and B (Boxes 1 and 6). One should not, however, associate the oceans only with harmful chemical compounds, for certain marine natural products can be tamed and exploited to provide powerful leads for treating disease, particularly cancer. Indeed, static or slow-moving aquatic creatures, such as sponges, algae, and anemones, have survived evolutionary pressures by producing powerful toxins to fend off potential predators. While several of these molecules have already been found and are providing clues towards new chemotherapeutic approaches to cancer treatment, it is certain that many more remain to be discovered. Included among the stunning known molecular architectures from the marine environment are those of the antitumor agents ecteinascidin 743, eleutherobin, diazonamide A, and bryostatin 1 (Box 3). Exploring, in environmentally benign ways, the largely untapped molecular diversity stored in the oceans is bound to lead to new beginnings in science and medicine, and therefore will ultimately be of great benefit to mankind.

Brevetoxin B, the potent and highly complex neurotoxin associated with red tides, was the first structure of its kind to be unraveled and as such it holds a special place within the annals of natural products chemistry. The exact structure was elucidated by Koji Nakanishi (Columbia University, USA), in collaboration with Yong-Yeng Lin (University of Texas, USA) and Jon C. Clardy (Cornell University, USA), who carried out the x-ray crystallographic analysis of brevetoxin B. The mode of

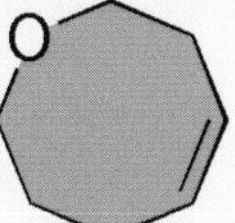

oxocene ring

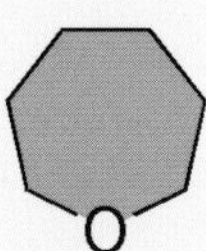

oxapane ring

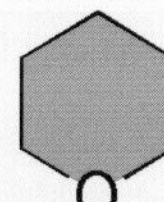

tetrahydropyran ring

K. C. Nicolaou

action of the brevetoxins involves activation of voltage-dependent sodium channels embedded within neurons. When these channels are activated by brevetoxin A or B they open, allowing a continuous influx of sodium ions, rather than the normal regulated flow, thus disabling the neuron. Symptoms in humans include a tingling sensation in the mouth and digits, disrupted coordination, hot/cold reversal of temperature sensitivity, dilated pupils, brachycardia, and diarrhea.

The amazing molecular architecture of brevetoxin B is breathtaking to organic chemists (see Box 2). Its fifty carbon, fourteen oxygen, and seventy hydrogen atoms are woven together in a stunning array of eleven contiguous rings locked into a rigid ladder-like structure. Despite its complex nature, this magnificent structure shows remarkable regularity. Thus, all eleven rings are *trans*-fused and each one contains a single oxygen atom.

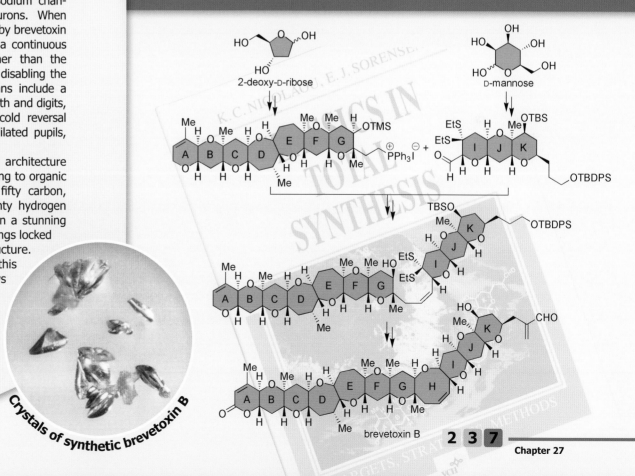

Crystals of synthetic brevetoxin B

The final brevetoxin B team

Box 5 Completing the odyssey: highlights of the Nicolaou brevetoxin B total synthesis

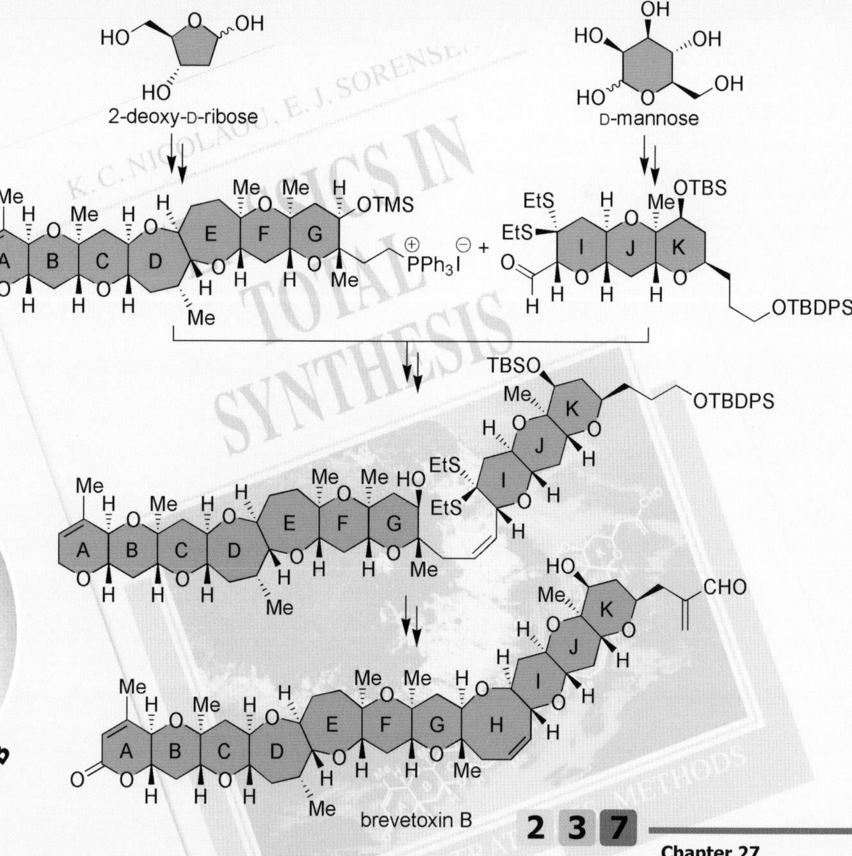

Box 6 **Molecular structures of brevetoxin A and ciguatoxin CTX1B**

brevetoxin A

ciguatoxin (CTX1B)

All the ring oxygen atoms are separated from their closest neighboring ring oxygens by two carbons and each is flanked by two *syn*-disposed hydrogen (or methyl) substituents, except for the first (ring A) which carries a carbonyl group to its 'left', and the last (ring K) which is flanked by two *anti*-orientated hydrogen atoms.

The first total synthesis of brevetoxin B was accomplished by the group of K. C. Nicolaou in 1995, following an arduous campaign. This twelve-year "synthetic odyssey" required tremendous stamina and skill, not to mention ingenuity and imagination on the behalf of those working at the front line. After several abortive attempts based on a myriad of diverse strategies, brevetoxin B finally fell, with the details of the conquest reported in the *Journal of the American Chemical Society*. This celebrated total synthesis was accompanied by the discovery and invention of much new chemistry. Included among the methods endowed by this program were new means to construct certain types of oxygen-containing rings, such as tetrahydropyrans, oxepanes, and oxocenes, methods that today remain as widely used tools in organic synthesis.

The final and successful strategy employed in the Nicolaou total synthesis of brevetoxin B is shown retrosynthetically in Box 4. While experts will not be surprised, many readers may be interested to know that this strategy toward brevetoxin B was not the one originally designed (it very rarely is in complex molecule construction), but rather it was one that had evolved over the duration of the campaign. It involved eighty-six steps in its longest linear sequence (from 2-deoxy-D-ribose) and traversed a terrain full of surprises, puzzles, and fascinating sites of exploration. The convergent synthesis began with 2-deoxy-D-ribose and D-mannose, two readily available carbohydrate building blocks which were separately elaborated towards two advanced intermediates (Box 5). These key fragments were then united using a Wittig reaction, and the product advanced to brevetoxin B through a carefully orchestrated synthetic sequence. Brevetoxin A (Box 6), which presents its own unique synthetic hurdles, was also synthesized by the Nicolaou group, in 1998. Brevetoxin A is the most potent neurotoxin isolated from *Karenia brevis*, and its structure had been elucidated as the result of a collaboration, undertaken during the early 1980s, between Yuzuru Shimizu (University of Rhode Island, USA) and Jon C. Clardy.

The ciguatoxins are a class of polyether marine toxins produced by another dinoflagellate, *Gambierdiscus toxicus*, which lives on macro-algae. The ciguatoxins are also potent neurotoxins and, through accumulation up the aquatic food chain, are thought to be responsible for the poisoning of over 20,000 people annually in subtropical and tropical regions of the world. The parent member of this natural product class,

Ball and stick model of the molecule of azaspiracid-1

CTX1B (Box 6), was first isolated by Paul J. Scheuer and his group (University of Hawaii, USA) in 1967. Through the heroic efforts of Takeshi Yasumoto and co-workers (Tohoku University, Japan), using a mere 0.35 mg of CTX1B extracted from 4000 kg of moray eels (*Gymnothorax javanicus*), the structure of this molecule was finally elucidated in 1989. In a similar feat of skill and endurance to that accomplished earlier by the Nicolaou brevetoxin teams, ciguatoxin (CTX1B, Box 6) was synthesized by Masahiro Hirama (Tohoku University, Japan) and his group in 2006. The Hirama group has also completed the total syntheses of other members of this challenging class.

Another accomplishment in this area is the total synthesis, by the Nicolaou group, of azaspiracid-1 (Box 1), a neurotoxin which contaminates mussels causing human poisoning in Europe. Originally isolated by the Yasumoto team from poisoned mussels (*Mytilus edulis*) collected from Killary Harbor (Ireland), this molecule inspired a campaign towards its synthesis that became a high-drama detective story once the first synthesis proved that the originally proposed structure was incorrect. The scientists then re-designed their strategy so as to decipher the true structure of azaspiracid-1, a formidable task that was finally accomplished by opportunistic intelligence gathering and total synthesis. As with the story of strychnine (Chapter 12), this tale exemplifies another value of chemical synthesis, that of providing the ultimate proof of a natural product structure.

Completion of all these syntheses may be hailed as a testament to the awesome power of modern chemical synthesis. Yet, if we compare the overall efficiency of the processes involved in these syntheses to the facile biosyntheses of these marine toxins, we must admit our inferiority and conclude that we still have a long way to go before we can claim equal status with nature. Incidentally, while both saxitoxin and tetrodotoxin (Box 1) have also been synthesized in the laboratory, the total synthesis of maitotoxin (Box 1), the largest secondary metabolite yet isolated, remains a challenge – perhaps to be met one day by one or more of the young students reading this book!

Azaspiracid-1

The Nicolaou azaspiracid-1 team

Box 7 Highlights of the Nicolaou group's endeavor in the brevetoxin area

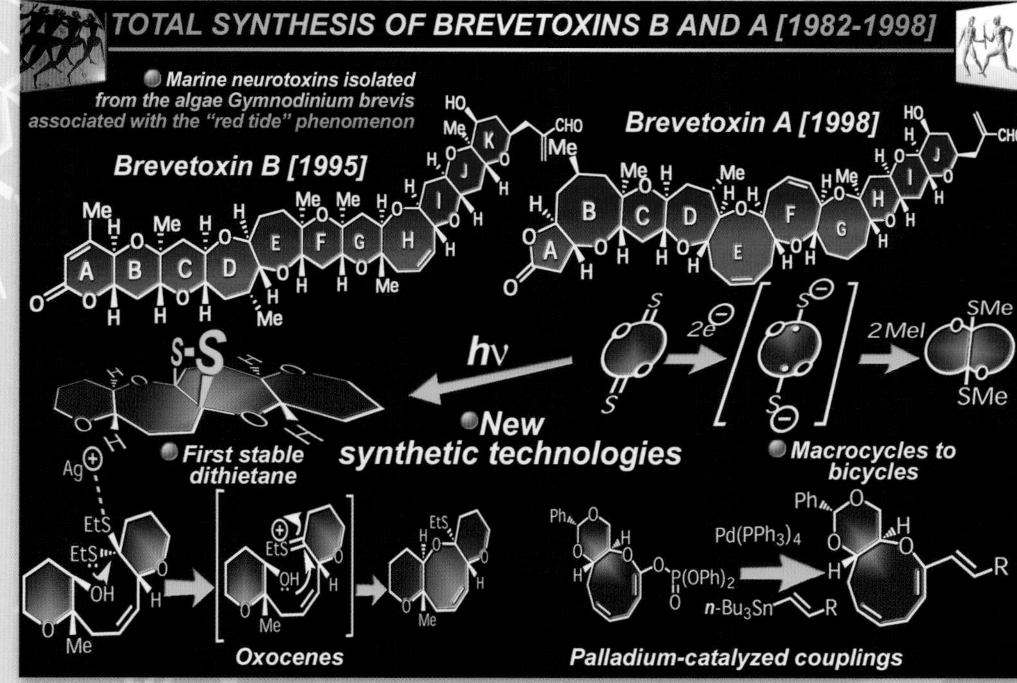

Ball and stick model of the molecule of brevetoxin B

Further Reading

K. C. Nicolaou, F. P. J. T. Rutjes, E. A. Theodorakis, J. Tiebes, M. Sato, E. Untersteller, Total Synthesis of Brevetoxin B. 3. Final Strategy and Completion, *J. Am. Chem. Soc.* **1995**, *117*, 10252–10263.

K. C. Nicolaou, The Total Synthesis of Brevetoxin B: A Twelve-Year Odyssey in Organic Synthesis, *Angew. Chem. Int. Ed. Engl.* **1996**, *35*, 589–607.

K. C. Nicolaou, E. J. Sorensen, *Classics in Total Synthesis*, Wiley-VCH, Weinheim, **1996**, pp. 731–786.

K. C. Nicolaou, Z. Yang, G.-Q. Shi, J. L. Gunzner, K. A. Agrios, P. Gärtner, Total Synthesis of Brevetoxin A, *Nature* **1998**, *392*, 264–269.

M. Inoue, K. Miyazaki, Y. Ishihara, A. Tatami, Y. Ohnuma, Y. Kawada, K. Komano, S. Yamashita, N. Lee, M. Hirama, Total Synthesis of Ciguatoxin and 51-HydroxyCTX3C, *J. Am. Chem. Soc.* **2006**, *128*, 9352–9354.

K. C. Nicolaou, T. V. Koftis, S. Vyskocil, G. Petrovic, T. Ling, Y. M. A. Yamada, W. Tang, M. O. Frederick, Structural Revision and Total Synthesis of Azaspiracid-1. Part 2: Definition of the ABCD Domain and Total Synthesis, *Angew. Chem. Int. Ed.* **2004**, *43*, 4318–4324.

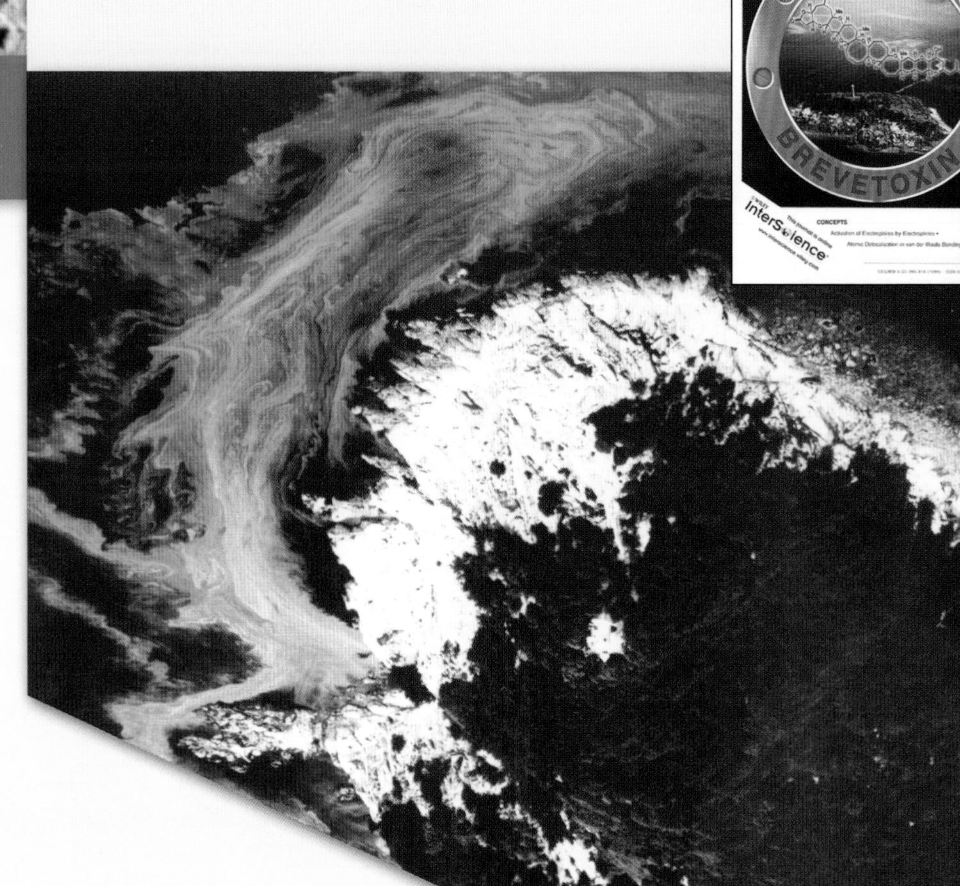

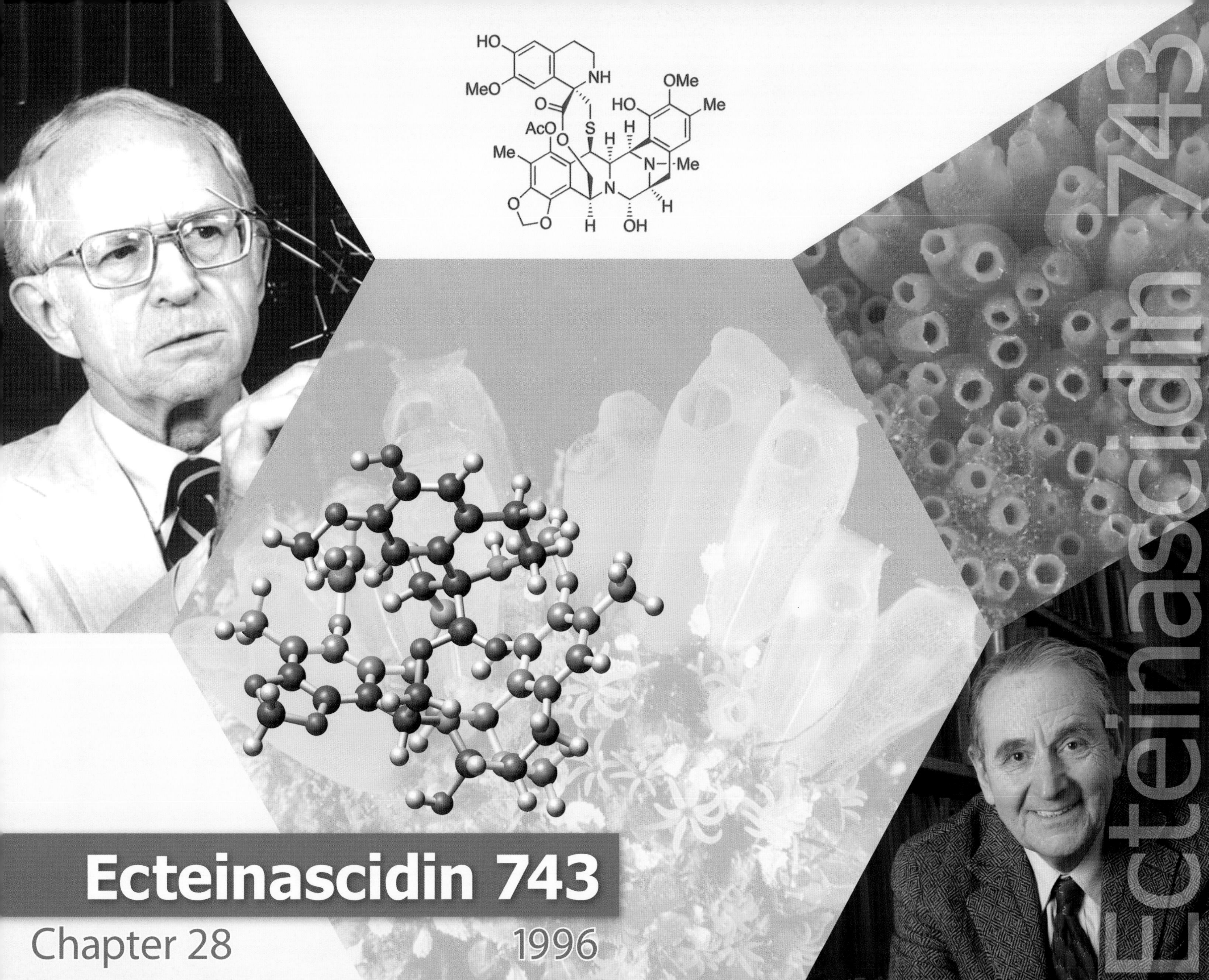

Ecteinascidin 743

Chapter 28

1996

Box 1 The mechanism of action of ecteinascidin 743

ecteinascidin 743

activation

reaction

DNA damage

CELL DEATH

DNA-ecteinascidin adduct
[DNA distortion]

ecteinascidin 743

The oceans and seas, especially the shallow warm waters of the tropics, are home to many soft-bodied invertebrates. These strange and resplendent creatures include soft corals, sponges, starfish, and sea anemones. The story of ecteinascidin 743 begins in 1969 when a preparation containing a mixture of compounds was extracted from one such marine creature, a sea squirt, collected during a scientific diving expedition in the Caribbean Sea.

Sea squirts (or ascidians) have sack-like bodies into which they draw water so that nutritional microorganisms can be absorbed; the water is then squirted out, hence their rather comical name. Sea squirts contain an array of organs within their body wall, including a circulatory system driven by a heart which, curiously, can change the direction of its pumping. The mixture extracted following maceration of the entire body of this particular ascidian, *Ecteinascidia turbinata*, was propelled into the scientific limelight because it showed an extraordinary level of lethal activity against P388 murine leukemia cells. This cancer cell line has been specially engineered and can be cultivated in the laboratory to test for anticancer properties in new compounds. Usually, the next steps in such a process would have been separation of the mixture, analysis and identification of its individual components, and their respective activities, followed by a full-scale investigation pursuing the constituents that had been identified as the most biologically promising. In this instance, however, such work was stymied for more than fifteen years due to the instability of the components. Irrespective of the care employed, almost any attempt to purify the extracted mixture led to the decomposition of its constituent compounds. During this frustrating period this medley of compounds continued to tantalize scientists, owing to the fact that it was the most potent marine natural product tested by the National Cancer Institute (USA) in the eight years to 1980. The development of a number of new chromatographic techniques finally allowed the composition of the mixture to be deconvoluted in 1986 by Kenneth L. Rinehart and his colleagues at the University of Illinois (USA). With pure materials now available, the full structural elucidation was also completed shortly thereafter (1990), unveiling the intricate molecular architectures of several ecteinascidins. The most abundant member of this new class of natural products was ecteinascidin 743, and experts were anxious to understand the origins of its impressive biological activity, having become convinced that a novel mode of action was at work.

It was not long before investigators found that ecteinascidin 743 (Et743) binds tightly in the minor groove of the DNA double helix, showing a preference for a specific sequence of bases (guanine followed by cytosine). Upon docking in the minor groove, the ecteinascidin 743 molecule reacts chemically with the DNA, forming a covalent bond with a nearby DNA functional group and converting the Et743 molecule into a highly reactive chemical species (Box 1). It has been suggested that the DNA–Et743 adduct absorbs oxygen from its environment, generating a reactive hydroxyl radical. This radical then slices the backbone of the DNA strands in the locality. This mechanism is thought to be the manner by which the ecteinascidins (and another family of structurally related molecules, the saframycins, see Box 5) destroy the genetic material that is essential to living cells. However, our cells, cancerous or normal, are not so weak as to submit to this violent attack without resistance; enzymes exist within cells whose sole purpose is repairing lesions in DNA. They shuttle up and down DNA duplexes in search of stretches requiring maintenance, like methodical train track engineers fixing railroads. These enzymes evolved because DNA is constantly being damaged under the ordinary circumstances of daily wear and tear. For example, exposure to ultraviolet light can promote dimerization reactions within DNA to form cyclobutane rings that distort the helix. One of these enzymes, called nucleotide excision repair (NER) enzyme, removes lesions in damaged helices by excising them out and then, in conjunction with other cellular machinery, incorporates new nucleotides and stitches the DNA back up to restore a perfect helix. A recent investigation offered an interesting new insight into the mode of DNA cleavage by Et743, which accounts for the correlation between the cytotoxicity of Et743 and the presence of intact NER enzyme in a cell. In this hypothesis, the kink made in DNA by its reaction with Et743 is recognized by the NER enzyme, which then proceeds to snip out the damaged region. However, Et743 has the last word, since it disables the enzyme afterwards, such that the gaping hole in the helix is not mended following this surgery. With damage this severe, the cell dies rapidly. Since it has been established beyond doubt that Et743 is capable of cleaving DNA, it may be that both the mechanistic hypotheses mentioned above are valid and that there is more than one cellular target for Et743, with multiple modes for DNA interference operating simultaneously. At a molecular level, fine details such as these are always difficult to tease apart. What is beyond dispute is that ecteinascidin 743 exerts its effects through a new mode of

Leukemia cells

Double-stranded DNA structure

Kenneth L. Rinehart

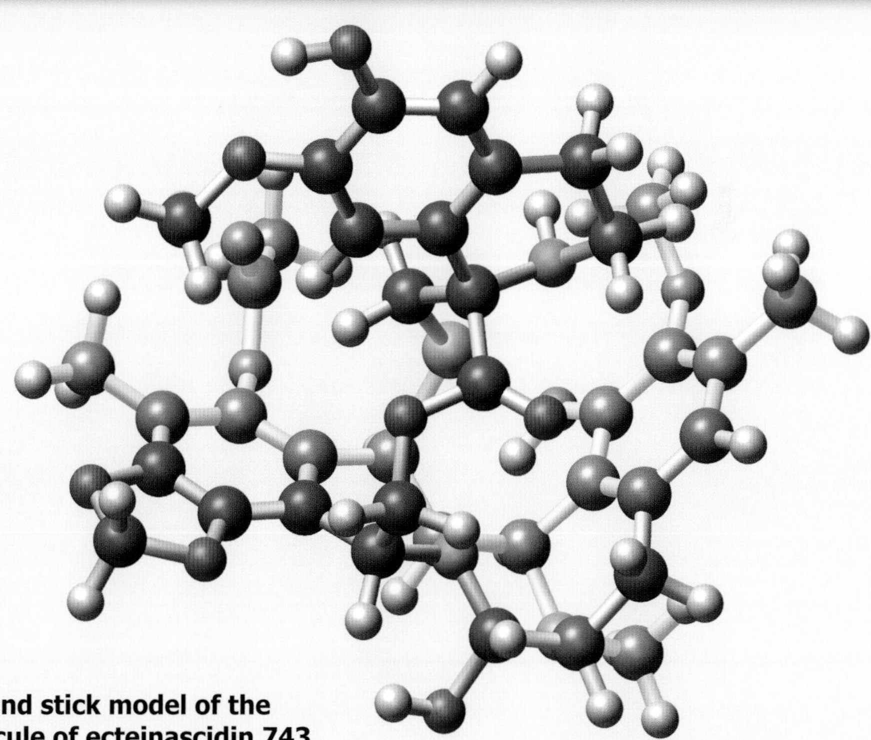

Ball and stick model of the molecule of ecteinascidin 743

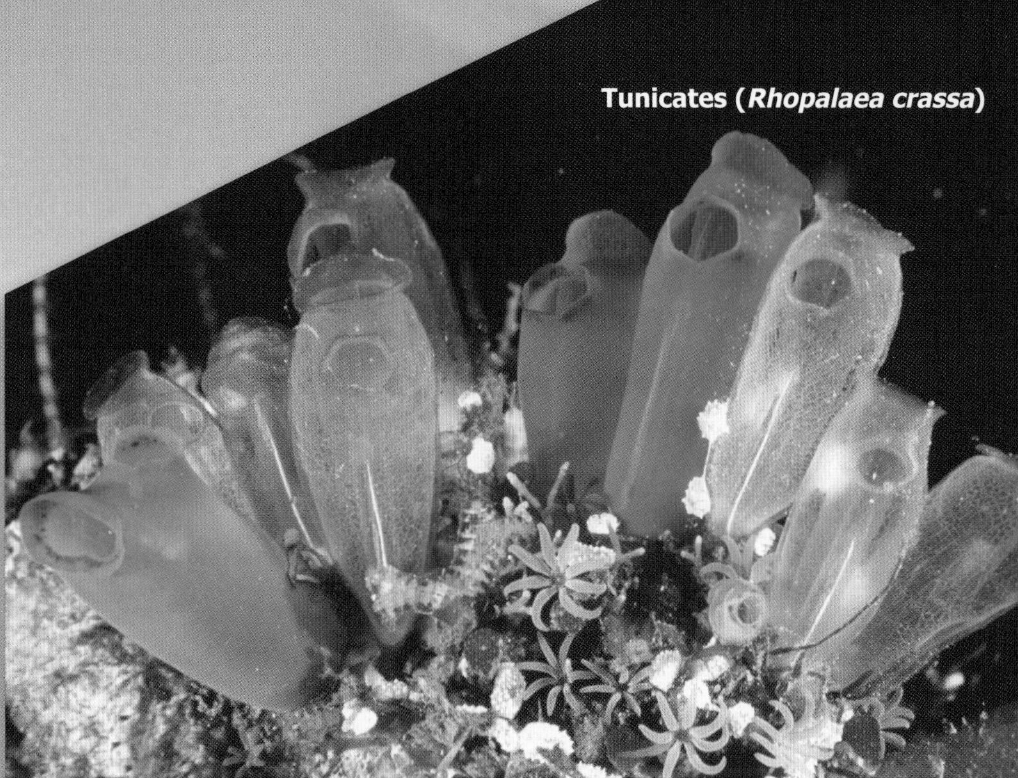

Box 2 Corey's retrosynthetic analysis of ecteinascidin 743

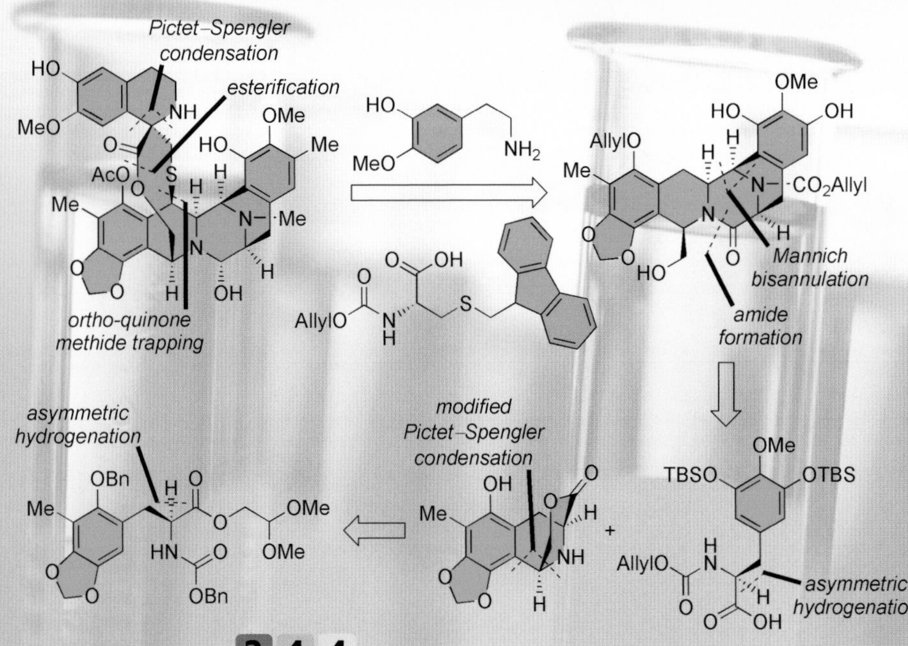

action. Thus, it is hoped that Et743 may become a powerful new weapon in our fight against cancer, provided of course that it makes it through the final stages of clinical trials. It should be emphasized that approval is not automatic, once a drug candidate has entered clinical trials. This period is one of the most arduous in the whole process of drug discovery and development, and may not ultimately end with the launch of a new medicine. All clinical drug candidates are required to pass certain safety and efficacy criteria in humans, and also to demonstrate advantages over existing therapies before they receive the official stamp of ratification from the appropriate regulatory authorities.

Before the ecteinascidin 743 clinical trials could commence, a pressing question had to be answered, that of supply. A single dose of Et743 would require the harvesting and processing of such vast numbers of the rare and inaccessible sea squirts that this approach is not remotely practical. A chemical synthesis, carried out by E. J. Corey and his group at Harvard University in 1996, was to provide the solution, and allow the exploration of this potential drug in the clinic. The first clues towards solving this synthetic challenge were suggested in the hypothesis regarding the biosynthetic origins of the ecteinascidins proposed by Rinehart and his colleagues, the same group of scientists that had succeeded in isolating and structurally characterizing the individual ecteinascidins. They proposed a sequence of transformations beginning with the dimerization of the amino acid tyrosine, that eventually led to the ecteinascidin skeleton. In nature, of course, this construction is a trivial operation when compared to a laboratory synthesis because of the enzymatic machinery that helps to organize and accelerate the reactions within the producing organism.

Ecteinascidin 743 has three tetrahydroisoquinoline units tightly bound together by a series of complex ring systems, both bridged and fused, including one unusual ten-membered ring bearing an intrinsic sulfide moiety. Corey's enantioselective total synthesis incorporated a number of elegant biomimetic features and impressive 'one-pot' sequences. In addition, and in the tradition of the Corey school, the employed strategy (Box 2) cleverly overcame the hurdles thrown up by the molecule along the way. Highlights of this synthesis include two asymmetric hydrogenations (see Box 6) employing state-of-the-art technologies to furnish the desired stereochemistry, intra- and intermolecular Pictet–Spengler reactions, and an intramolecular Mannich bisannulation process (Box 3). This elegant sequence allowed the researchers to construct an advanced hexacyclic intermediate quickly and efficiently, setting the scene for a dramatic cascade reaction to install the ten-membered sulfide-containing ring (Box 4). This daring reaction involved the generation of an *ortho*-quinone methide, the unmasking of a pendant thiol group, and the intramolecular union of these two highly reactive groups to form the large ring. A further cascade sequence to forge the final ring system, and two deprotection operations completed the total synthesis. The Corey synthesis proved suitable for large-scale production, and was later improved further in a second-generation route from the Corey group. The Spanish pharmaceutical firm PharmaMar has since reported a semi-synthesis of Et743 from cyanosafricin B (Box 5), a simpler natural product available in large quantities from bacterial culture. These triumphs of synthetic organic chemistry have overcome the supply problem, facilitating the clinical trials that are now underway around the world to evaluate ecteinascidin 743 (also known in this regard by the names trabectidin and Yondelis®).

In the year 2002, Tohru Fukuyama and his

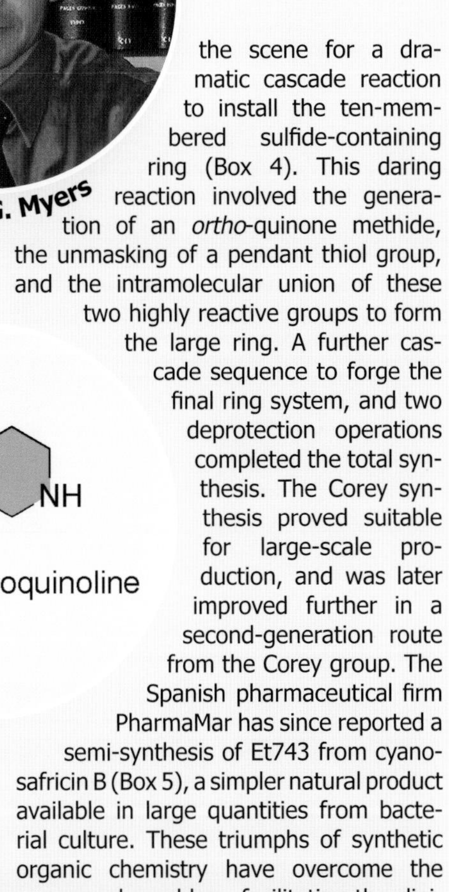

Box 3 Highlights from the early stages of Corey's total synthesis of ecteinascidin 743

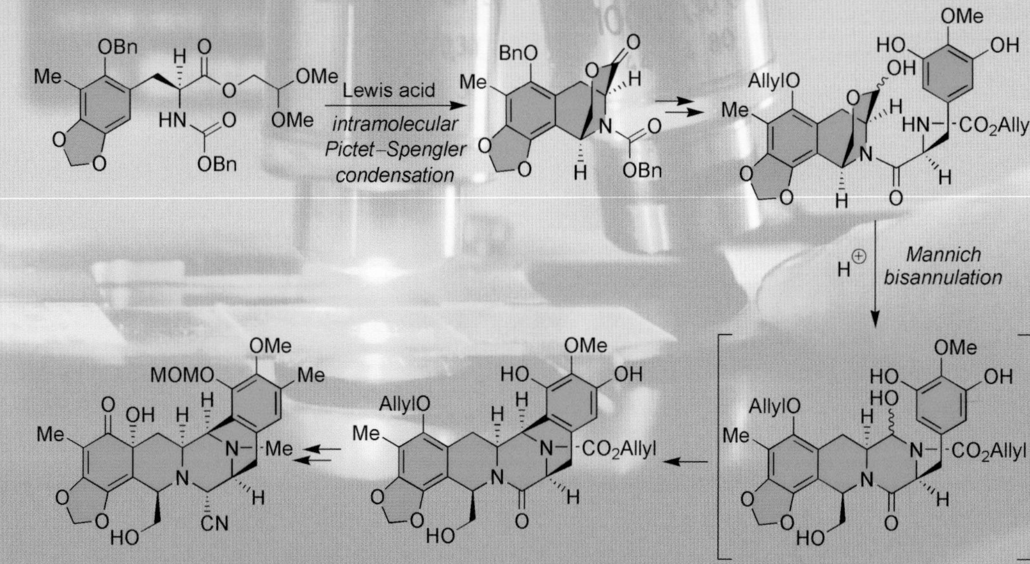

Box 4 The late stages of Corey's total synthesis of ecteinascidin 743

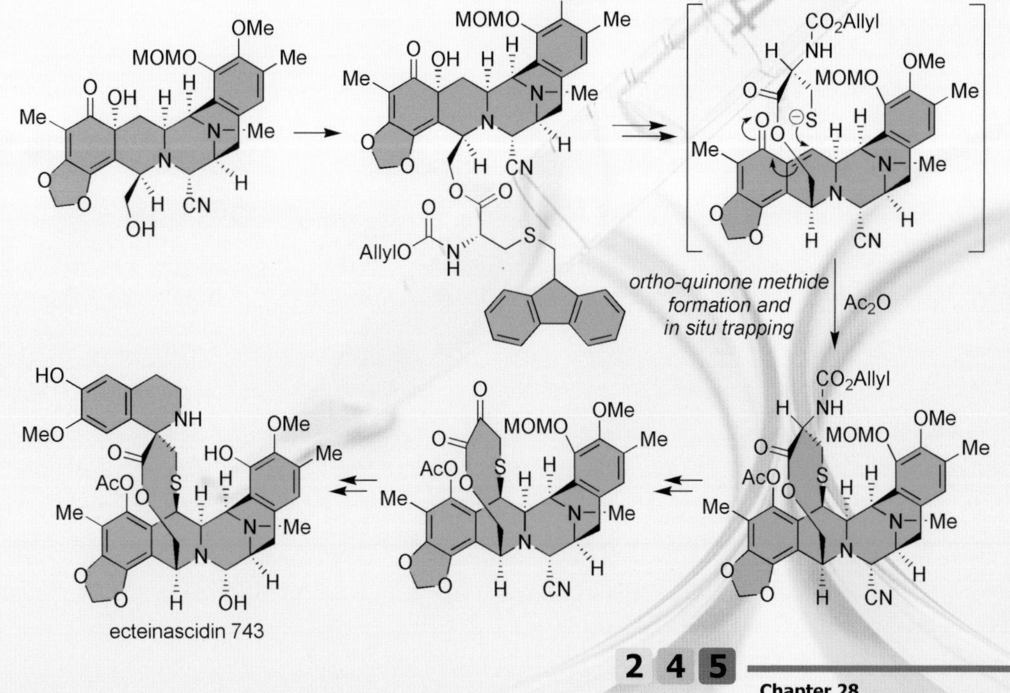

Box 5 Selected cytotoxic marine natural products

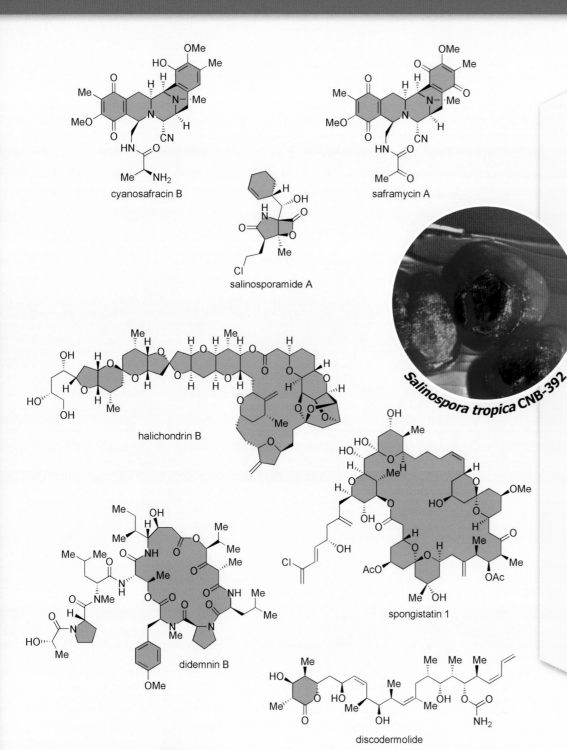

cyanosafracin B

saframycin A

salinosporamide A

halichondrin B

spongistatin 1

didemnin B

discodermolide

Salinospora tropica CNB-392

group at the University of Tokyo (Japan) published their elegant total synthesis of ecteinascidin 743, which was joined in 2006 by a total synthesis from the laboratory of Jieping Zhu (CNRS, Cedex, France) and a formal total synthesis from Samuel J. Danishefsky and co-workers (Columbia University and the Sloan-Kettering Institute for Cancer Research, USA).

The saframycin family of natural products, which includes saframycin A (Box 5) and cyanosafricin B, has also been the subject of many synthetic studies, including the recent synthesis of large numbers of analogues using solid phase techniques (see Chapter 29) by the group of Andrew G. Myers (Harvard University).

In 2004, Corey and co-workers reported the efficient total synthesis of another potential anticancer marine natural product, salinosporamide A (Box 5). This highly potent protease inhibitor had been isolated only a year earlier by the group of William H. Fenical at the Scripps Institution of Oceanography (La Jolla, USA). Fenical has been a leading investigator of marine natural products, isolating an impressive number of fascinating chemical structures with a range of important biological activities. D. John Faulkner, also at the Scripps Institution of Oceanography, was another pioneer in the field of marine natural products chemistry, with many important discoveries to his credit.

William H. Fenical

A great many cytotoxic secondary metabolites have been isolated from marine organisms and ascribed the loose collective title of cytotoxic marine natural products. While they share a common

Hyrtios erecta

Ecteinascidia turbinata

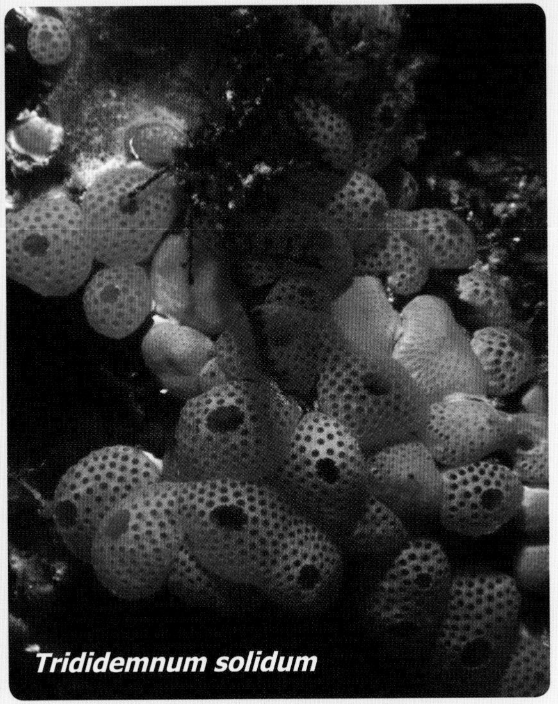

Trididemnum solidum

environmental origin and broadly defined biological activity, the compounds within this group have little else in common. Marine natural products display a wealth of structural diversity and often include intricate arrangements of heteroatoms not seen in terrestrial compounds. These features, combined with the constant need to develop new anticancer drugs, both inspire and entice synthetic organic chemists. Some of the examples shown in Box 5 (halichondrin B, didemnin B, spongistatin 1, and discodermolide) have been among the most sought after synthetic targets of recent years. Yoshito Kishi and his group (Harvard University) completed the total synthesis of halichondrin B in 1990 and the Japanese pharmaceutical company Eisai is currently investigating the potential of truncated halichondrin B analogues as drugs. The didemnins were isolated and synthesized by the Rinehart group, and have been the subject of many biological and synthetic investigations, notably by the group of Madeleine M. Joullié (University of Pennsylvania, USA), with other total syntheses reported by Ulrich Schmidt (Stuttgart University, Germany) and Takayuki Shioiri (Nagoya City University, Japan). The spongistatins are a family of potent anticancer polyketides isolated from marine sponges. The scarcity of these compounds from natural sources has prompted a number of synthetic programs, culminating in seven total syntheses to date. These emanated from the groups of Yoshito Kishi, David A. Evans (Harvard University), Amos B. Smith, III (University of Pennsylvania), Ian Paterson (University of Cambridge, UK), Michael T. Crimmins (University of North Carolina at Chapel Hill, USA), Clayton H. Heathcock (University of California, Berkeley, USA), and Steven V. Ley (University of Cambridge). Finally,

D. John Faulkner

Ian Paterson

Amos B. Smith, III

Clayton H. Heathcock

David A. Evans

Yoshito Kishi

Madeleine M. Joullié

Michael T. Crimmins

Steven V. Ley

the potent anticancer activity of discodermolide has also inspired numerous total syntheses, including the recent preparation of large quantities by the Novartis pharmaceutical firm, whose hybrid synthesis borrows from the work of both Smith and Paterson.

In closing this chapter, describing some of the many medicinal leads and potential drugs discovered among marine natural products, we must reflect on the modern tendency to neglect our oceans and seas. In past times, the ocean was revered; the ancient Greek philosopher–scientists were awed by its powerful currents and tides, and the medieval European explorers of its vast and mysterious domains were heralded as heroes, brave pioneers of this frightening and majestic environment. In the Vedic writings from India (3,000 B.C.), the *sagara* (sea) is regarded as a vault of riches, including *Amrita*, the tonic of immortality. These spiritual writers may have been closer to the truth than we care to acknowledge, for if the oceans contain a myriad of new anticancer medicines, antibiotics, and other much needed drugs, then they may indeed have a profound influence on the mortality of Man.

Discodermia dissoluta

Box 6 Catalytic asymmetric hydrogenation

Hydrogenation is a chemical reaction wherein a molecule of hydrogen is added to a substrate, commonly across an alkene double bond to give the corresponding alkane, via a process often mediated by a metal catalyst. This process is an extremely useful transformation in chemical synthesis, for it provides an entry to many intermediates and new compounds required for further investigations. If the alkene is unsymmetrically substituted, then during the hydrogenation one, or even two, new chiral centers can be generated. If the stereochemistry of these centers is appropriately controlled by employing a suitable chiral catalyst, then the reaction, now elevated to the status of catalytic asymmetric hydrogenation, becomes even more powerful; its usefulness being due to the fact that most naturally occurring substances and active pharmaceutical components are enantiomerically pure, that is to say of one handedness.

Catalytic hydrogenation

metal

H--H

cis-addition of hydrogen

enantiomeric products possible

Catalytic asymmetric hydrogenation

H_2 (3 atm)

(100 % yield, 96 % *ee*)

The following milestone events trace the evolution of this important chemical process:

■ **1912** – Paul Sabatier (University of Toulouse, France) is awarded the Nobel Prize in Chemistry "for his method of hydrogenating organic compounds in the presence of finely disintegrated metals."

■ **1956** – S. Akabori (Osaka University, Japan) reports the use of palladium on silk in asymmetric hydrogenation reactions. Interestingly, the silk had to be wild, not the cultivated type, and the enantiomeric excesses (ee) were low with poor reproducibility.

■ **1965** – The 1973 Nobel Laureate in Chemistry, Geoffrey Wilkinson (Imperial College, London, UK), discovers the rhodium catalyst system [$RhCl(PPh_3)_3$] which bears his name, and which could be employed in homogeneous hydrogenation reactions, unlike the heterogeneous metal catalysts employed previously.

■ **1968** – William S. Knowles, from the Monsanto Company (USA), publishes a reliable enantiopure catalyst system that favors one enantiomeric product over the other in hydrogenation reactions. Knowles used a variant of Wilkinson's catalyst, in which he had replaced a non-chiral triphenylphosphine ligand with a chiral analogue. He later refined his chiral catalyst to achieve practical levels of enantiomeric enrichment. This process found applications in the industrial production of both L-DOPA, a drug used to treat Parkinson's disease, and NutraSweet®, the non-calorific sugar substitute.

■ **1971** – Henri B. Kagan (Orsay University, France) develops a rhodium catalyst with a DIOP ligand (a diphosphine derived from tartaric acid) that induced impressive asymmetric induction in hydrogenation reactions of dehydro amino acid precursors. This C2 symmetric ligand provided impetus for further developments in chiral catalyst design, with major implications for asymmetric synthesis.

■ **1980** – Ryoji Noyori and his associates at Nagoya University (Japan) publish the first asymmetric reactions induced by BINAP, a chiral phosphine ligand. This remarkable ligand and its variants have since assumed a prominent role in asymmetric catalysis with important applications in the production of enantiomerically pure compounds, through a wide range of reactions.

■ **2001** – The groundbreaking works of Knowles and Noyori are recognized by the Nobel Foundation with the award of the Nobel Prize in Chemistry "for their work on chirally catalyzed hydrogenation reactions," which they shared with a third leader in the field of asymmetric catalysis, K. Barry Sharpless (see Chapter 26).

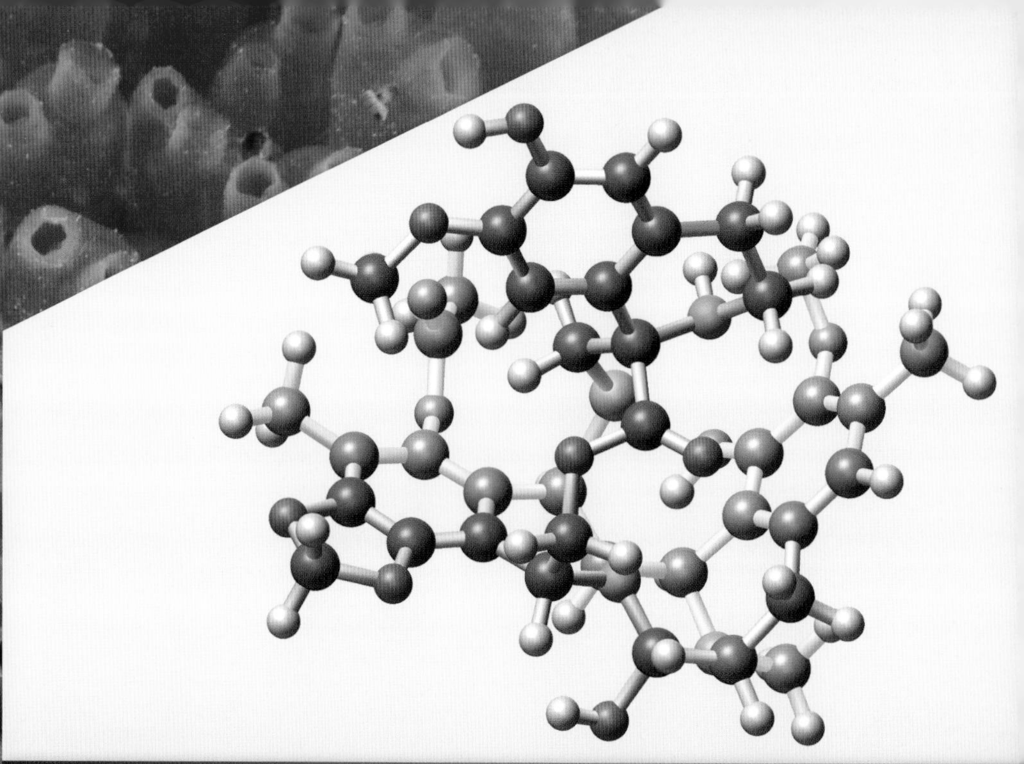

Further Reading

K. L. Rinehart, T. G. Holt, N. L. Fregeau, P. A. Keifer, G. R. Wilson, T. J. Perun, Jr., R. Sakai, A. G. Thompson, J. G. Stroh, L. S. Shield, D. S. Seigler, L. H. Li, D. G. Martin, C. J. P. Grimmelikhuijzen, G. Gäde, Bioactive Compounds from Aquatic and Terrestrial Sources, *J. Nat. Prod.* **1990**, *53*, 771–792.

E. J. Corey, D. Y. Gin, R. S. Kania, Enantioselective Total Synthesis of Ecteinascidin 743, *J. Am. Chem. Soc.* **1996**, *118*, 9202–9203.

E. J. Martinez, E. J. Corey, A New, More Efficient, and Effective Process for the Synthesis of a Key Pentacyclic Intermediate for Production of Ecteinascidin and Phthalascidin Antitumor Agents, *Org. Lett.* **2000**, *2*, 993–996.

A. Endo, A. Yanagisawa, M. Abe, S. Tohma, T. Kan, T. Fukuyama, Total Synthesis of Ecteinascidin 743, *J. Am. Chem. Soc.* **2002**, *124*, 6552–6554.

J. D. Scott, R. M. Williams, Chemistry and Biology of the Tetrahydroisoquinoline Antitumor Antibiotics, *Chem. Rev.* **2002**, *102*, 1669–1730.

Drugs from the Sea, N. Fusetani, Ed., S. Karger AG, Basel, **2000**.

K. C. Nicolaou, S. A. Snyder, *Classics in Total Synthesis II*, Wiley-VCH, Weinheim, **2003**, pp. 109–136.

HO
Me
Me
Me
Me
O
HO
O
O
H
O
Me
S
N
Me
Epothilones
Chapter 29
1996, 1997
Epothilones

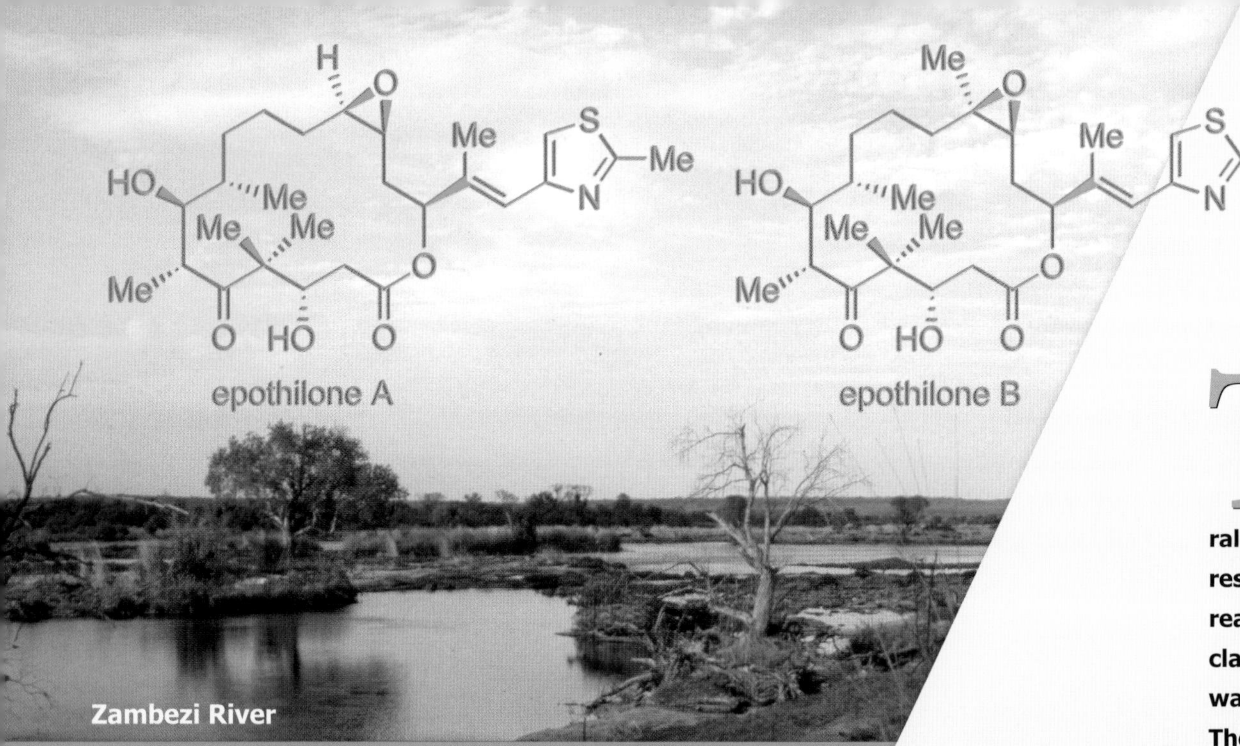

epothilone A

epothilone B

Zambezi River

Hans Reichenbach and Gerhard Höfle

The search for anticancer drugs is a major driving force for the isolation and screening of new natural products. Just as the frenzied research on Taxol® (Chapter 25) was reaching its peak in the 1990s, a new class of potential anticancer drugs was, quite literally, being unearthed. The story of these cytotoxic agents, the epothilones, began slowly, but developed rapidly after biological testing indicated the potential of these compounds as medicines to fight drug-resistant cancers.

Epothilones A and B were first isolated in the early 1990s from the culture extracts of the myxobacterium *Sorangium cellulosum* (strain So ce90) which had been found in soil samples taken from the banks of the Zambezi River in southern Africa. The discoverers, Gerhard Höfle, Hans Reichenbach, and their collaborators at the Gesellschaft für Biotechnologische Forschung (Braunschweig, Germany), initially sought to probe the antifungal activity of these compounds with a view to employing them as agricultural fungicides. However, field trials proved that these compounds were too toxic against plants for this application, and their full medicinal potential remained undiscovered for a while. Meanwhile, scientists at the pharmaceutical company Merck (USA) independently isolated both epothilones A and B and, by 1995, had discovered that these polyketide secondary metabolites were highly effective antitumor agents. Indeed, in many of the cell lines used in typical first round tests for cytotoxicity, the epothilones outperformed Taxol®. Perhaps most importantly, further studies revealed that the new substances were

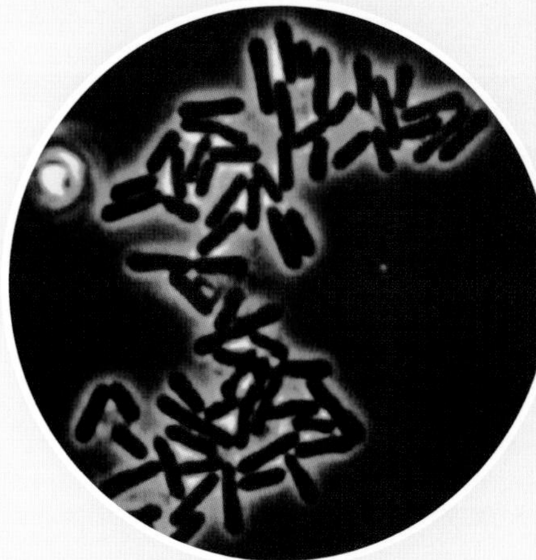

The epothilone-producing myxobacterium *Sorangium cellulosum*: growing cells (left) and spore capsules (right)

remarkably potent against Taxol®-resistant tumor cells. Intriguingly, they were also found to operate by a mechanism identical to that of Taxol®, even binding to the same site on the same protein target. Specifically, the epothilones were found to induce tubulin polymerization and to stabilize microtubules, and thereby irreversibly damaging replicating cells, driving them to their death (see diagram below and pictures on the right). As a result of this highly significant observation, a great deal of excitement and interest spread amongst the scientific community as hopes were raised about the possibility of developing new and superior agents for use in cancer chemotherapy. In a curious twist, a series of miscalculations by the major pharmaceutical players meant that the epothilones were left without patent protection for their potential anticancer application. Fortunately, this feature did not curb their march to the clinic for long, and soon a new wave of investigators opted to pursue them as promising leads.

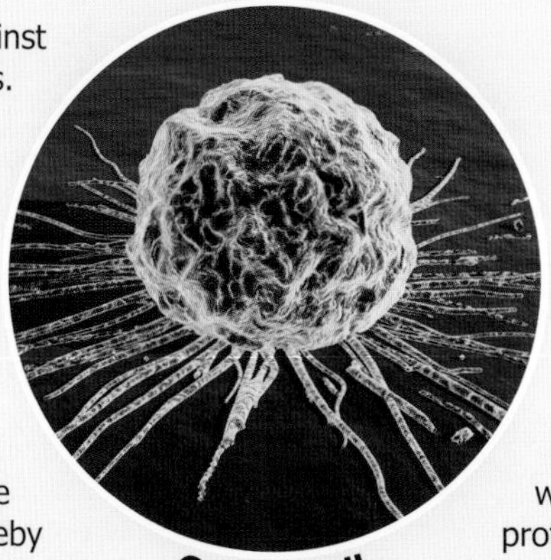

Cancer cell

Despite the fact that the original patent describing the epothilones was filed in early 1990, it was not until July 1996 that the German scientists published the full details of their molecular structures.

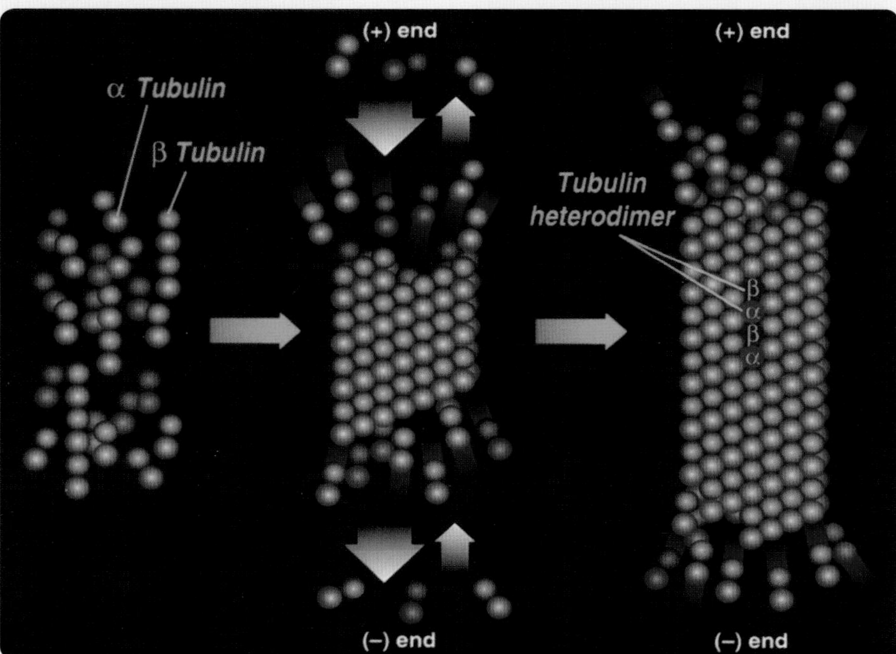

Polymerization of tubulin to microtubules

Effect of epothilone B on *Potorous tridactylis* (PtK_2) cells

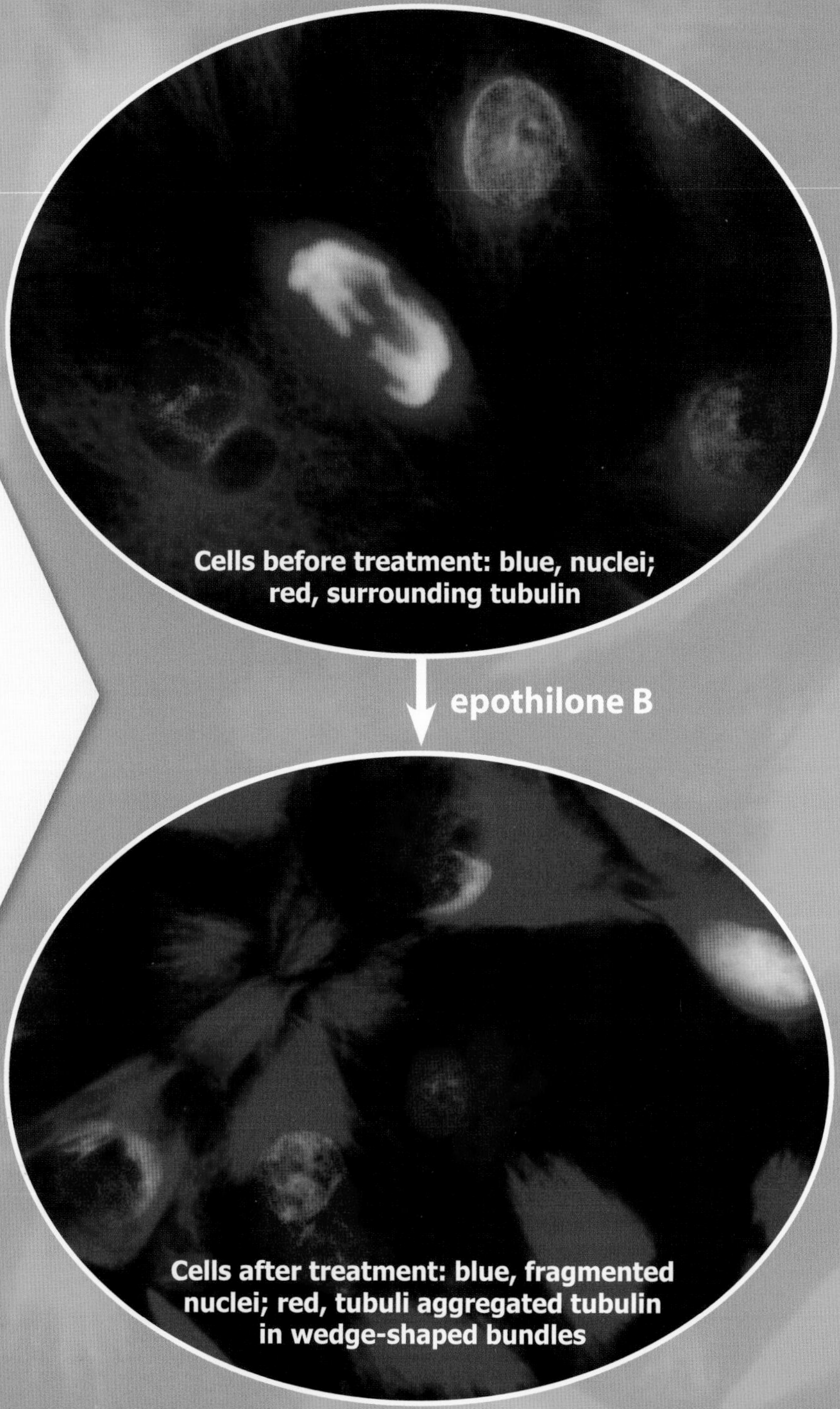

Cells before treatment: blue, nuclei; red, surrounding tubulin

epothilone B

Cells after treatment: blue, fragmented nuclei; red, tubuli aggregated tubulin in wedge-shaped bundles

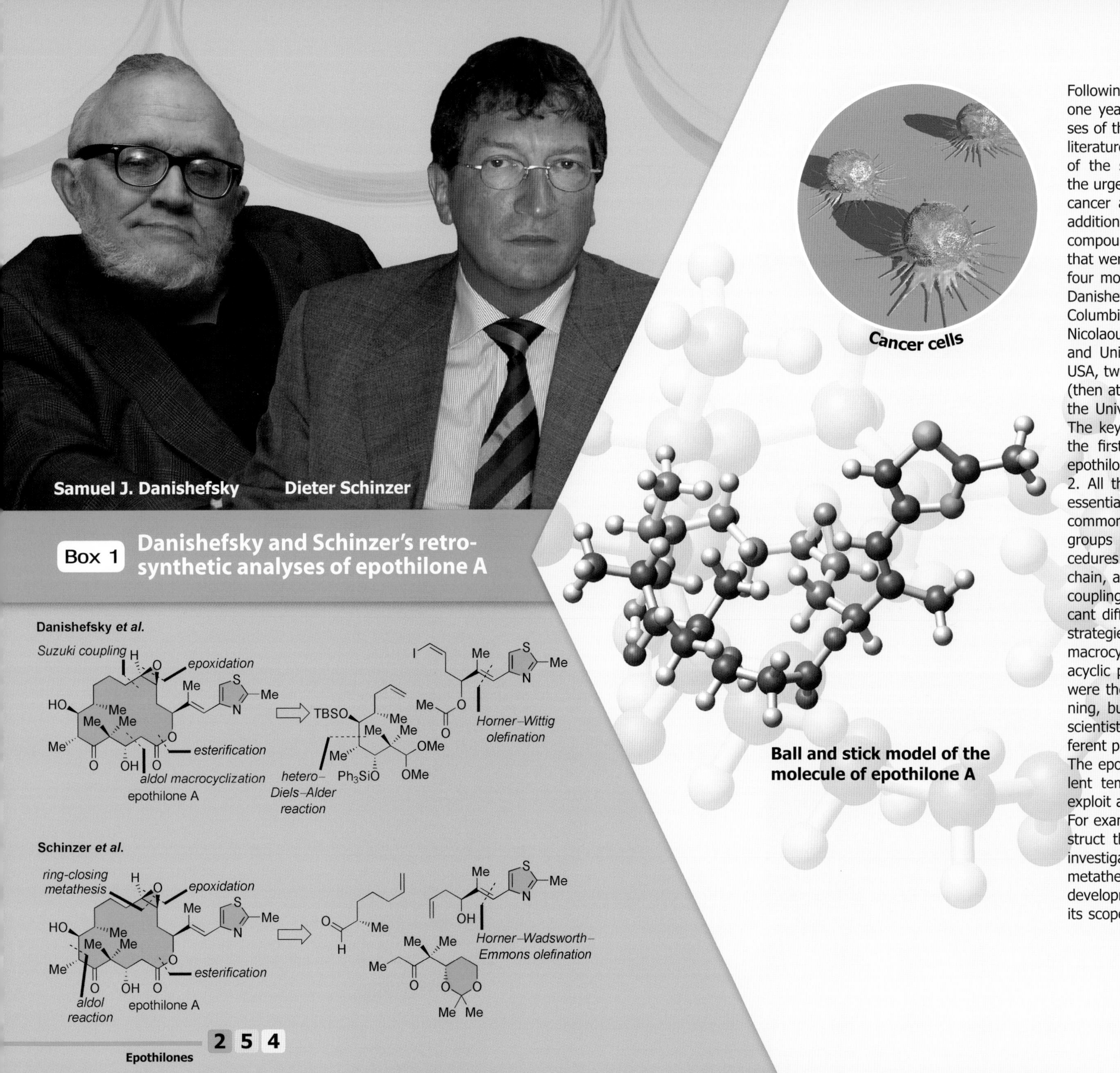

Box 1 Danishefsky and Schinzer's retrosynthetic analyses of epothilone A

Ball and stick model of the molecule of epothilone A

Following this revelation, it took less than one year for the first laboratory syntheses of these compounds to appear in the literature, providing testament to the skill of the synthetic chemists involved and the urgency of the need to find new anticancer agents. Since then, over twenty additional synthetic routes toward these compounds have joined the original four that were published over a period of only four months by the groups of Samuel J. Danishefsky (Sloan-Kettering Institute and Columbia University, New York, USA), K. C. Nicolaou (The Scripps Research Institute and University of California, San Diego, USA, two strategies), and Dieter Schinzer (then at Braunschweig University, now at the University of Magdeburg, Germany). The key retrosynthetic disconnections of the first four successful approaches to epothilone A are shown in Boxes 1 and 2. All these syntheses were carried out essentially simultaneously, and some common features are evident. All three groups chose related olefination procedures to attach the heterocyclic side chain, although the exact conditions and coupling partners varied. However, significant differences are found in the overall strategies employed, both for forming the macrocyclic ring and for constructing the acyclic precursor. All of these sequences were the result of careful scientific planning, but the individual creativity of the scientists involved is reflected in the different paths followed to the same target. The epothilones also served as an excellent template for chemists to test and exploit a broad range of new techniques. For example, the problem of how to construct the epothilone macrocycle fueled investigations into the scope of olefin metathesis (see Box 6), accelerating its development and significantly expanding its scope and utility. Today, this powerful reaction has become an indispensable tool to the organic chemist, often providing novel solutions to synthetic puzzles of considerable complexity.

The first total synthesis of epothilone A to use an olefin metathesis approach came from the Nicolaou group and was published early in 1997. Although the molecular architectures of the epothilones are considerably less complex than that of Taxol®, the construction of a sixteen-membered macrolide ring, the installation of seven stereocenters around its periphery, and the attachment of a thiazole side chain still presented serious obstacles. The chemists knew that a better understanding of the biological interactions and structure-activity relationships of the epothilones could potentially lead to the discovery of clinically useful therapeutic agents. Driven by this goal, the Nicolaou group sought to develop a flexible and convergent synthetic strategy that would deliver not only the naturally occurring substances, but also a diverse collection of designed analogues. Thus, contrary to the objective of most other synthetic plans where the controlled installation of stereochemical elements is a major concern, the goal in this instance was to gain access to a range of isomers and related compounds in order to explore which structural motifs were crucial for biological activity. The result of this plan was the olefin metathesis-based approach shown in Box 2. The acyclic precursor was to be formed from three building blocks of roughly equal size via aldol and esterification reactions.

K. C. Nicolaou

This plan was successfully executed to furnish epothilone A. The key olefin metathesis reaction was accomplished using a ruthenium catalyst developed by Robert H. Grubbs at the California Institute of Technology (USA). This remarkable macrocyclization reaction gave rise to a mixture of double bond isomers, slightly favoring the *cis*-olefin, thereby supplying the precursor to the desired final product. Following completion of this synthesis, which was undertaken using conventional solution-phase reactions, the Nicolaou team adapted it for use with solid phase synthesis technology. This facilitated the synthesis of several hundred epothilone analogues for biological screening.

The solid phase total synthesis of epothilone A (Box 4) and its application to the construction of a collection of designed epothilone analogues (described as an epothilone library) was based on principles pioneered by R. Bruce Merrifield at The Rockefeller University (New York, USA) for the synthesis of peptides, for which he received the 1984 Nobel Prize in Chemistry. In this approach, the growing molecule is chemically attached to a polymer support, or resin, which comes in the form of small beads, usually made of polystyrene. During the synthesis, reactions are driven to completion (with consequentially higher yields) by the addition

The Nicolaou epothilone team

Box 2 Nicolaou's retrosynthetic analyses of epothilone A

Nicolaou *et al.* – ring-closing metathesis approach

ring-closing metathesis; epoxidation; esterification; aldol reaction; epothilone A; asymmetric allylation; Wittig reaction; asymmetric allylation

Nicolaou *et al.* – macrolactonization approach

Wittig reaction; epoxidation; macrolactonization; aldol reaction; epothilone A; Wittig reaction; asymmetric allylation

Box 3 The split-and-pool strategy for combinatorial synthesis

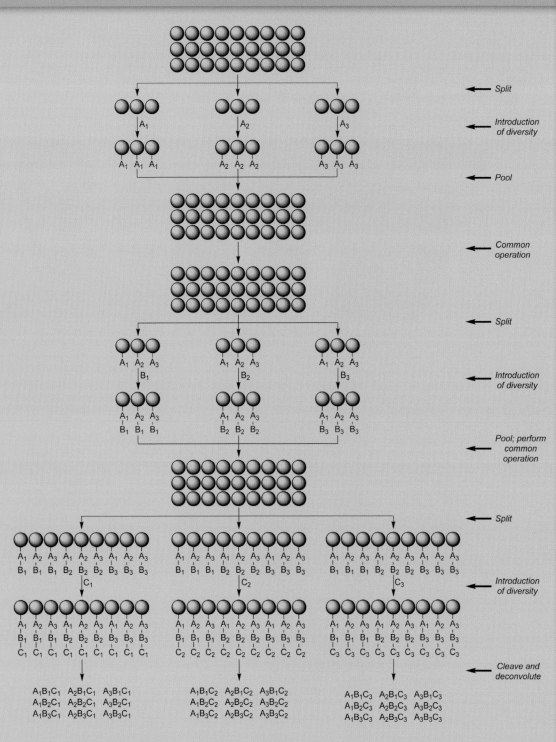

Split

Introduction of diversity: A_1, A_2, A_3

$A_1\ A_1\ A_1$ — $A_2\ A_2\ A_2$ — $A_3\ A_3\ A_3$

Pool

Common operation

Split

$A_1\ A_2\ A_3$ — $A_1\ A_2\ A_3$ — $A_1\ A_2\ A_3$

Introduction of diversity: B_1, B_2, B_3

$A_1B_1\ A_2B_1\ A_3B_1$ — $A_1B_2\ A_2B_2\ A_3B_2$ — $A_1B_3\ A_2B_3\ A_3B_3$

Pool; perform common operation

Split

Introduction of diversity: C_1, C_2, C_3

Cleave and deconvolute

$A_1B_1C_1$ $A_2B_1C_1$ $A_3B_1C_1$
$A_1B_2C_1$ $A_2B_2C_1$ $A_3B_2C_1$
$A_1B_3C_1$ $A_2B_3C_1$ $A_3B_3C_1$

$A_1B_1C_2$ $A_2B_1C_2$ $A_3B_1C_2$
$A_1B_2C_2$ $A_2B_2C_2$ $A_3B_2C_2$
$A_1B_3C_2$ $A_2B_3C_2$ $A_3B_3C_2$

$A_1B_1C_3$ $A_2B_1C_3$ $A_3B_1C_3$
$A_1B_2C_3$ $A_2B_2C_3$ $A_3B_2C_3$
$A_1B_3C_3$ $A_2B_3C_3$ $A_3B_3C_3$

R. Bruce Merrifield

of reagents in large excess. The leftover reagent can easily be removed after each step by washing the product-containing beads with appropriate solvents. This method has the advantage of avoiding costly and time-consuming purification procedures at the end of each operation. Furthermore, the characteristics of solid phase chemistry mean that it is particularly well suited for use in combinatorial chemistry. The purpose of combinatorial chemistry is the assembly of collections (libraries) of many compounds using combinations of building blocks that come in sets with specified variations. The objective is met by forging such fragments into compounds using all possible or logical combinations, thereby building up the maximum number of analogous compounds. The development of solid phase techniques also allowed for the implementation of the so-called split-and-pool combinatorial strategy, which allows the construction of large compound libraries very rapidly (Box 3). To achieve this goal, the resin is divided into lots and a different variant of the first building block is introduced to each lot. The resin is then pooled so that common operations can be undertaken on the entire batch (saving time and money). These operations range from simple washings to reactions such as deprotections and functional group manipulations. The resin is subsequently split into lots again and the synthesis is continued by the addition of the next building block, with a different variant employed for each newly divided lot of resin. This iterative process is repeated until the library construction is complete, at which time the products are cleaved from the resin and purified to provide the desired library of compounds. Even with relatively small numbers of each unit, this approach allows for the construction of very large libraries and requires less time and effort than the conventional synthesis of individual compounds. The structural information regarding the growing molecules that could potentially be lost at the pooling stages may be saved by using encoding techniques, such as bar coding or radio-frequency tagging of the resin cartridges. The epothilone library was synthesized using MicroKan® resin micro-reactors, developed by the IRORI company, bearing barcodes that could be read by a scanner in order to track the synthetic history of each compound.

Microkan® reactors

The elegance of the solid phase combinatorial synthesis of the epothilones (Box 4) lies in the use of the olefin metathesis not only to bring about macrocyclization, but also as a tactic to cleave the product from the resin in a single process. This so-called cyclorelease strategy has the added advantage of leaving no trace of the resin attachment site in the final products. The final two synthetic steps were carried out in solution and furnished a diverse library of epothilone analogues as discrete and pure compounds ready for biological evaluation.

Today, solid phase chemistry can be applied to the synthesis of both natural and designed molecules, either in a one-at-a-time fashion or as relatively small compound libraries (focused or designed libraries) for screening in various biological assays. Notable developments in the field include the use of solid-supported reagents, and the use of scavenger resins equipped with chemical groups to remove excess reagents and impurities from reaction mixtures, thus facilitating the advancement of the growing synthetic intermediate to the next stage. Steven V. Ley (University of Cambridge,

Steven V. Ley

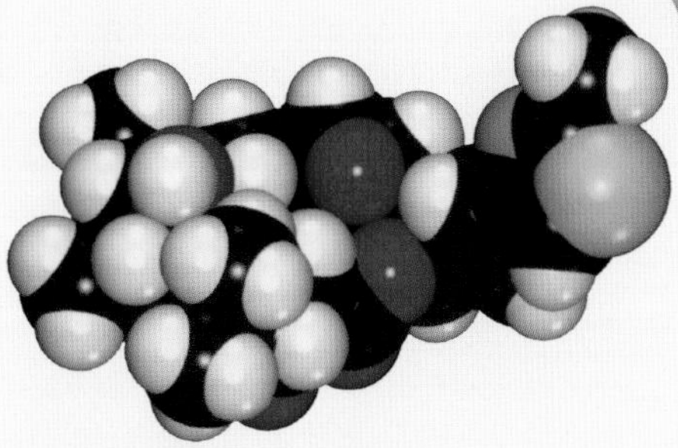

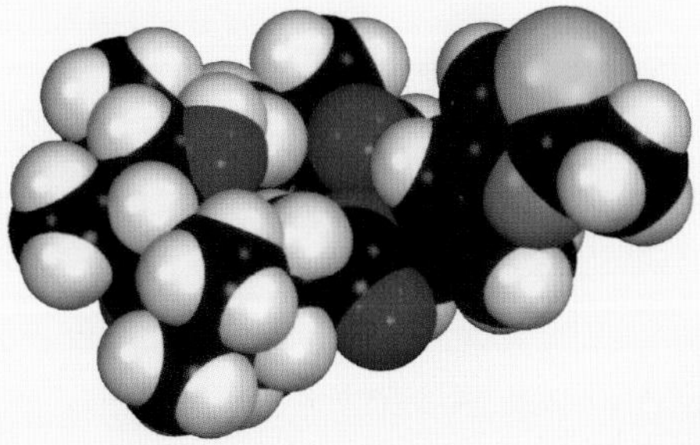

Space filling models of epothilones A (top) and B (bottom)

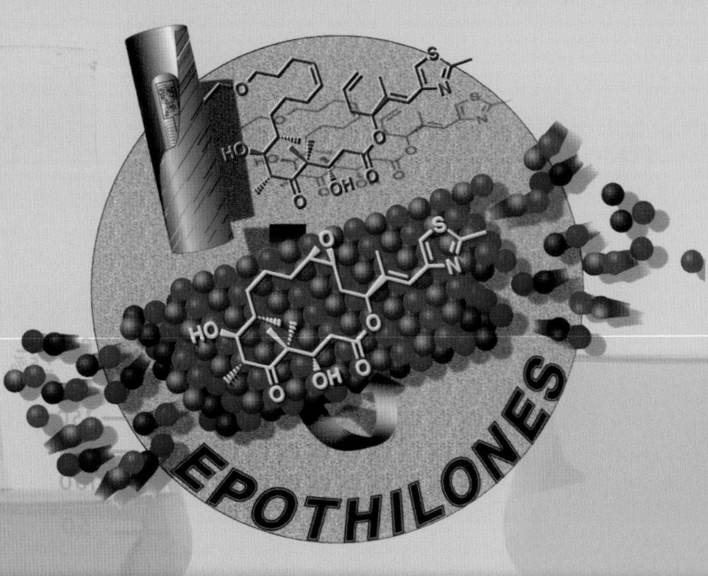

Box 4 Nicolaou's solid phase synthesis of epothilones A and C

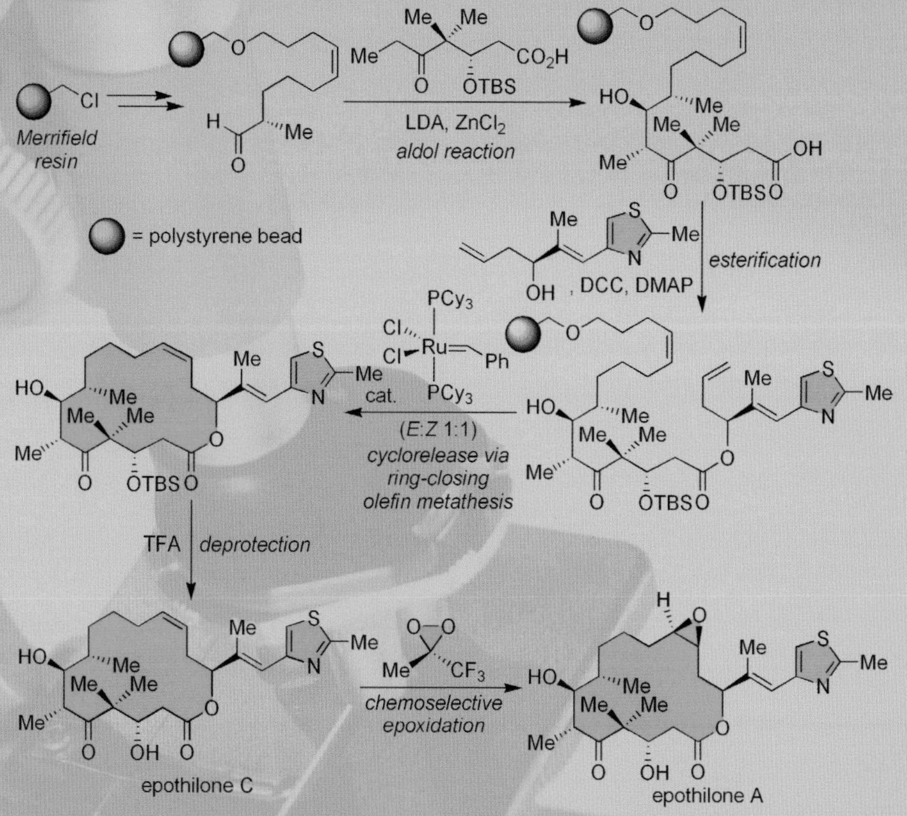

Merrifield resin; = polystyrene bead; LDA, $ZnCl_2$ aldol reaction; esterification, DCC, DMAP; cat. $Cl_2Ru(=CHPh)(PCy_3)_2$ (E:Z 1:1) cyclorelease via ring-closing olefin metathesis; TFA deprotection; chemoselective epoxidation; epothilone C; epothilone A

Box 5 SAR results for the epothilones and selected designed analogues

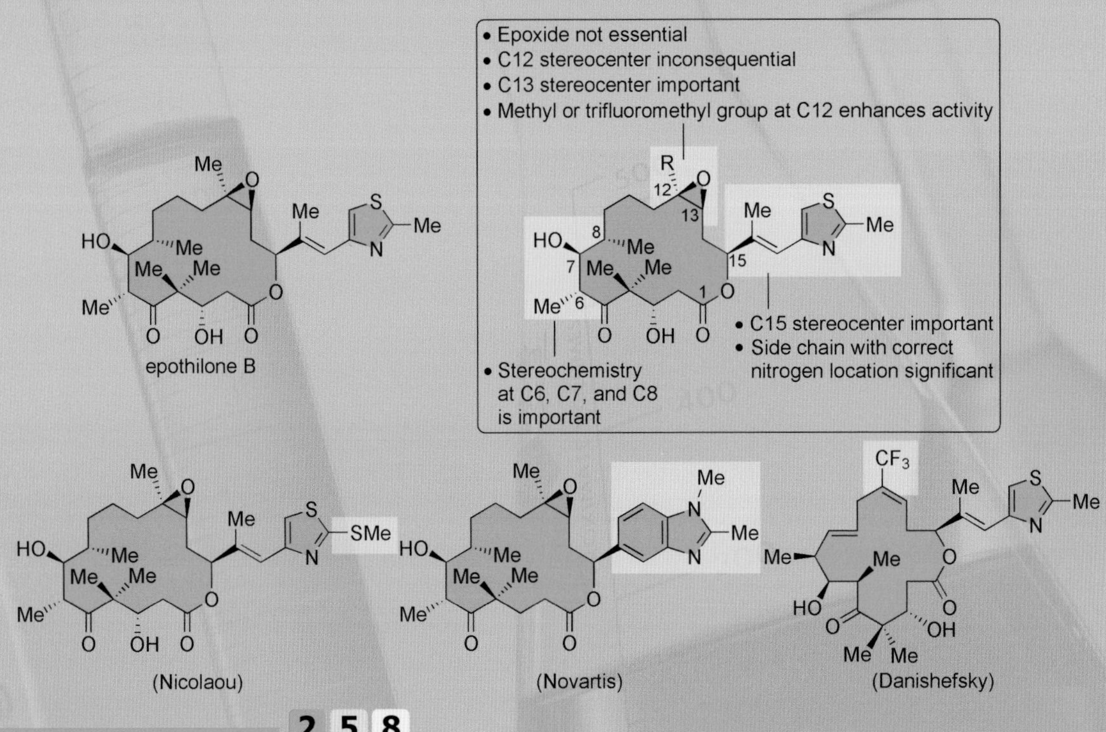

UK) has been a pioneer in developing such solid phase reagents and strategies, now finding their way into common laboratory practices.

Epothilone B, the most potent of the naturally occurring epothilones, has been synthesized in the laboratory by several groups, including those of Danishefsky and Nicolaou. The worldwide efforts in the chemical synthesis of the epothilones enabled extensive chemical biology investigations and allowed their structure-activity relationships (SARs) to be established (Box 5). Today, several epothilones are undergoing late-stage clinical trials, spearheaded by companies such as Novartis, Bristol-Myers Squibb, Schering AG, and Hoffmann-La Roche, with the hope that at least some will soon be approved as medicines to help patients afflicted by the dreaded disease of cancer.

Nissim Calderon

Yves Chauvin

Richard R. Schrock

Thomas J. Katz

Robert H. Grubbs

Box 6 The olefin metathesis reaction

Metathesis reactions, such as those used in the synthesis of the epothilones, revolutionized organic synthesis in the 1990s, but the roots of these processes can be traced back nearly fifty years. There are three main variations of the metathesis reaction (see adjacent): alkene metathesis, alkyne metathesis, and enyne metathesis. Alkene metathesis involves the formal cleavage of the alkene reactants, and the interchanging and recombining of the alkylidene fragments to give two new alkenes. The intermolecular cross-metathesis process is used in the petrochemical industry for the production of linear alkenes. If the two alkenes are within the same molecule, a ring-closing metathesis (RCM) reaction results, usually with the expulsion of a small alkene by-product. Metathesis reactions can also be used to prepare polymers from either cyclic alkenes or acyclic dienes. This process is widely used in both academic and industrial laboratories.

Alkyne metathesis, using metal alkylidyne catalysts, works in an analogous manner. Enyne metathesis involves the formal cleavage of the alkene bond, and the addition of the fragments to the ends of the alkyne, forming a diene product. Again, this reaction can be employed in both cross- and ring-closing variants, however, unlike the alkene and alkyne metatheses, there is no by-product expelled in this process.

The olefin metathesis reaction was recognized by Nissim Calderon of the Goodyear Tire and Rubber Company, who coined the term from the Greek for 'change places'. The generally accepted mechanism for alkene metathesis is based upon the suggestions made by the French chemist Yves Chauvin, with key experimental evidence later provided by the groups of Thomas J. Katz (Columbia University, USA) and Robert H. Grubbs (then at the University of Michigan, now at California Institute of Technology, USA). In 1976, Katz and his group described the first well-defined metathesis catalyst, a tungsten Fischer carbene. This milestone was followed in 1990 by the introduction of a molybdenum-based catalyst by Richard R. Schrock (Massachusetts Institute of Technology, USA), and in 1995 and 1999 by Grubbs' first- and second-generation ruthenium-based catalysts. Today, chemists have a large number of catalysts at their disposal, most frequently based on ruthenium or molybdenum, developed by a number of investigators. Each catalyst has its own characteristics, and the choice of which to utilize is dictated by the particular reaction being considered. Grubbs, Schrock, and Chauvin were awarded the 2005 Nobel Prize in Chemistry for their work on the development of the metathesis reaction as a tool for organic synthesis.

As mentioned above, metathesis processes are used in the petrochemical industry for processing hydrocarbons. The cross-metathesis process for the production of polymers was one of the first examples of olefin metathesis observed and is still widely used today. The use of RCM in the pharmaceutical industry is beginning to emerge. For example, Boehringer Ingelheim Pharma in Germany used RCM in the preparation of almost 400 kg of a drug candidate for the treatment of hepatitis C, BILN 2061 ZW.

Alkene metathesis

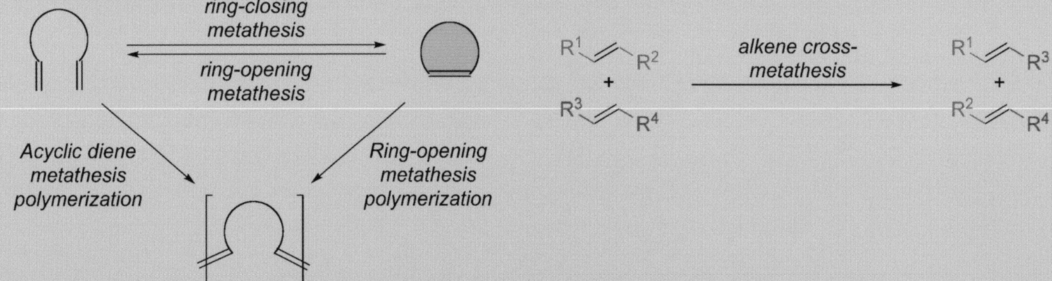

Enyne metathesis

Alkyne metathesis

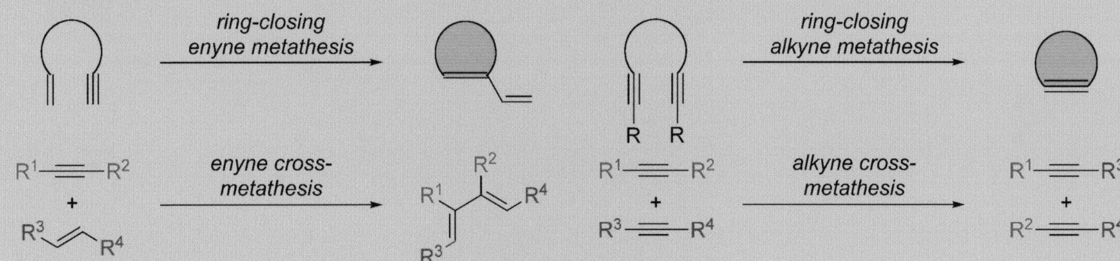

Selected milestones in the development of metathesis catalysts

Katz (1976)

Schrock (1990)

Grubbs (1995)

Grubbs (1999)

alkyne metathesis catalyst Schrock (1982)

Industrial application of the olefin metathesis reaction

cat. ring-closing metathesis

BILN 2061 ZW

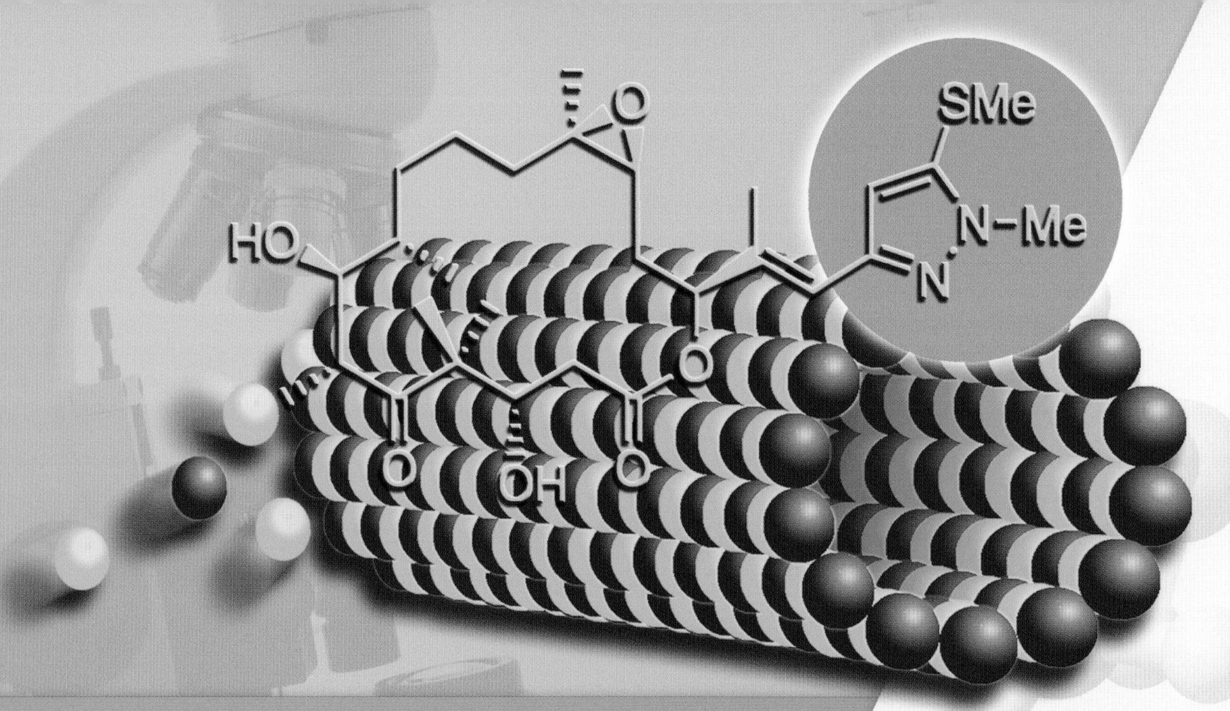

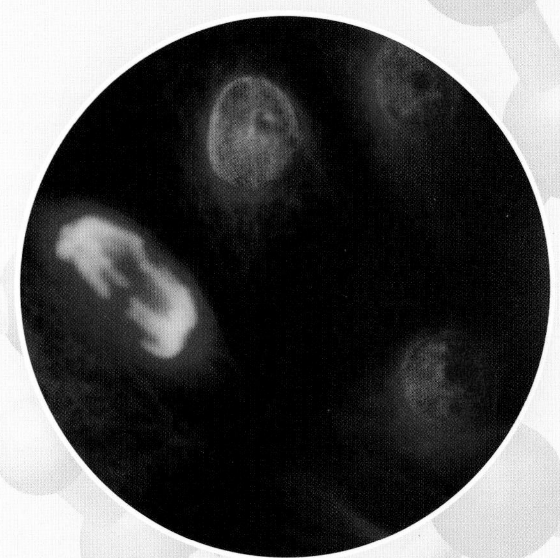

Further Reading

D. M. Bollag, P. A. McQueney, J. Zhu, O. Hensens, L. Koupal, J. Liesch, M. Goetz, E. Lazarides, C. M. Woods, Epothilones, a New Class of Microtubule-Stabilizing Agents with a Taxol-Like Mechanism of Action, *Cancer Res.* **1995**, *55*, 2325–2333.

G. Höfle, N. Bedorf, H. Steinmetz, D. Schomburg, K. Gerth, H. Reichenbach, Epothilone A and B – Novel 16-Membered Macrolides with Cytotoxic Activity: Isolation, Crystal Structure, and Conformation in Solution, *Angew. Chem. Int. Ed. Engl.* **1996**, *35*, 1567–1569.

K. C. Nicolaou, F. Roschangar, D. Vourloumis, Chemical Biology of Epothilones, *Angew. Chem. Int. Ed.* **1998**, *37*, 2014–2045.

K. C. Nicolaou, A. Ritzén, K. Namoto, Recent Developments in the Chemistry, Biology and Medicine of the Epothilones, *Chem. Commun.* **2001**, 1523–1535.

Handbook of Combinatorial Chemistry, Vols. 1 & 2, K. C. Nicolaou, R. Hanko, W. Hartwig, Eds., Wiley-VCH, Weinheim, **2001**.

K. C. Nicolaou, S. A. Snyder, *Classics in Total Synthesis II*, Wiley-VCH, Weinheim, **2003**, pp. 161–210.

A. Rivkin, T.-C. Chou, S. J. Danishefsky, On the Remarkable Antitumor Properties of Fludelone: How We Got There, *Angew. Chem. Int. Ed.* **2005**, *44*, 2838–2850.

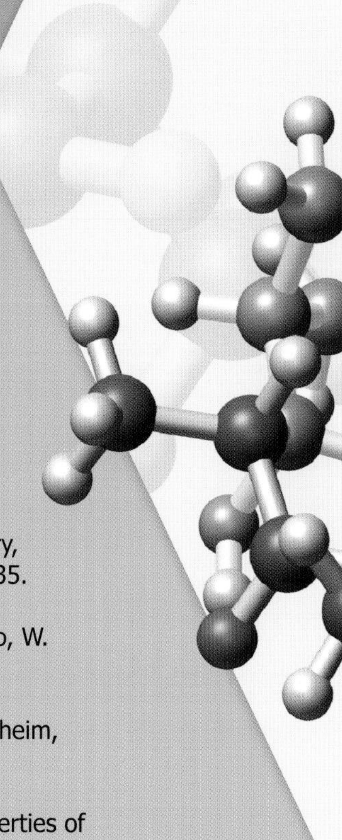

Resiniferatoxin
Resiniferatoxin
Chapter 30
1997
Me
O
O
Me
H
H
O
Me
OMe
OH
O
O
OH

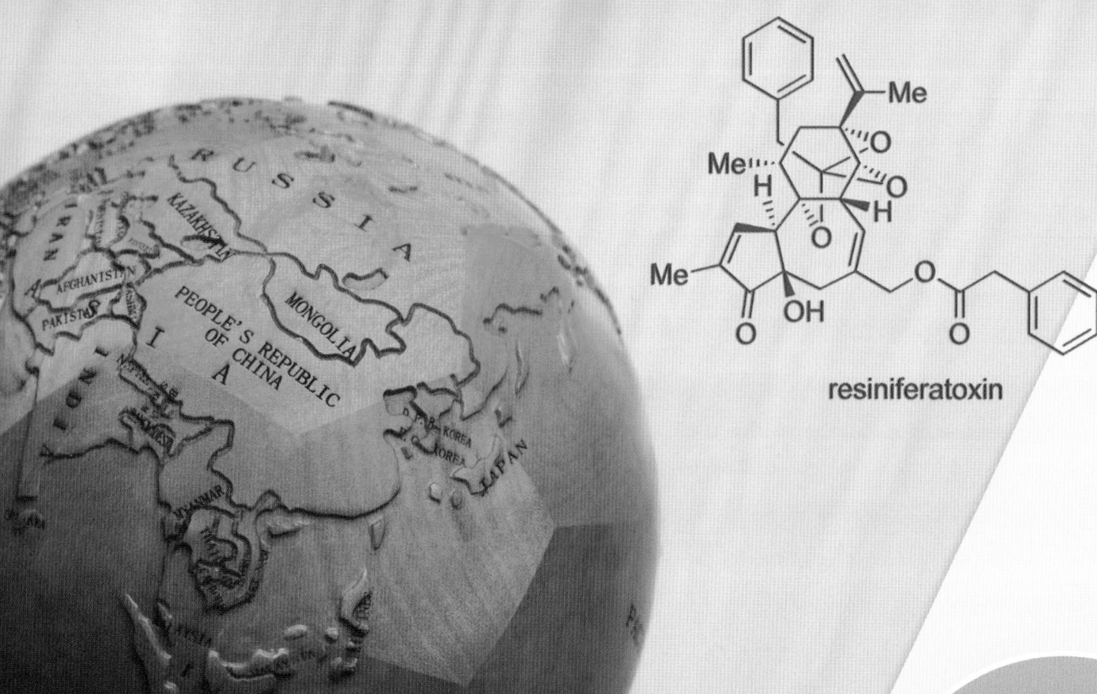

Box 1 Molecules with characteristic flavors

linalool (tea)

benzaldehyde (almond)

furaneol (strawberries and pineapple)

α-sinensal (mandarin, orange)

(+)-nootkatone (grapefruit)

1-*p*-menthene-8-thiol (grapefruit)

2-*trans*-6-*cis*-nonadienal (cucumber)

4-phenyl-2-butanone (raspberries)

Most taste experiences arise from a complex blend of chemicals interacting subtly to create the overall symphony of flavor

Globalization is thought to have had a profound impact on almost every aspect of our lives in recent years. However, the process of globalization is by no means an exclusively modern phenomenon. In fact, it has been shaping civilization since the beginning of recorded history. In this chapter, we will explore one early manifestation of globalization, namely the spice trade. The rich history of the spice trade began in ancient times. It fuelled the earliest cross-border trade, laid the foundations for international banking and credit systems, and gave rise to the first global corporations. This story will also lead us to the discovery of resiniferatoxin, a relative of capsaicin (the active component of chili peppers), which holds great medicinal promise. We will also describe how the active molecules contained within spices have extended our understanding of the biochemical basis of flavor, and given us lead compounds for the development of drugs.

Flavor has three equally important sensory components: taste (gustation), tactile sensation (termed mouthfeel), and smell (olfaction). Richard Axel and Linda B. Buck discovered the complex processes involved in the olfactory system in the early 1990s, and were awarded the 2004 Nobel Prize in Physiology or Medicine in recognition of this work. Much research has recently been undertaken to determine how the interaction of the taste buds with flavor molecules is translated into neuronal activity and interpreted in the brain as taste. Selected examples of the molecules responsible for some common flavors are shown in Box 1. Taste is biochemically and physiologically the least understood of all our senses, but recent advances have shown that certain parallels can be drawn between the mechanisms of taste perception and the processes involved in vision. Flavor molecules were traditionally divided into four general categories, sweet, bitter, sour, and salty (Box 2). Each is received by taste cells within a taste bud via one of two types of protein. Sweet and bitter flavors dock within a G-protein at a receptor site on the surface of the cell. A signal is then transmitted across the membrane into the cell cytoplasm where a cascade of signaling chemicals (similar to those involved in vision) relay the message to neurons, which then transmit it to the brain. Sour and salty tastes are usually caused by molecules bearing formal charges (at least under physiological conditions), such as acetic acid (as in vinegar), and these compounds interact with ion channels in the taste cell membrane. Ion channels are large proteins that form a pore in the cell membrane to facilitate the flow of ions. Once inside the cell, these charged flavor molecules trigger their own neuronal response. In recent times, a fifth taste, termed umami, has been recognized. The sodium salt of the amino acid glutamic acid (MSG, Box 2) and the disodium salts of the 5′-inosine monophosphate and 5′-guanosine monophosphate are responsible for this taste that characterizes certain vegetable (e.g., tomatoes, mushrooms), dairy (e.g., milk, cheese), and meat products (e.g., ham, veal). Recent studies suggest that neurons can respond to more than one dissimilar flavor stimulus, with the brain giving an overall verdict on the nature of each taste. Once again this parallels vision, where color perception depends on a pattern of activity across a group of photoreceptor cells in the eye.

The ancient Egyptians were perhaps the first civilization to leave a record of their use of spices (Box 3), which they imported from Asia having defined the first spice trade routes as their explorations proceeded eastwards. When King Khnum Khufwy oversaw the construction of the Great Pyramid of Cheops over five thousand years ago, he is said to have fed his laborers with doses of Asian spices to fuel their strength. The Egyptians also used spices such as cassia, cinnamon, cumin, and anise in the embalming process. The spices served two roles; they helped to preserve the bodies and they gave them a fine fragrance (see also Chapter 6). Throughout history, spices have been used for their fragrance and medicinal properties, and also for food preservation in a logical evolution of the ancient Egyptians' mummification techniques. In the plants from which they originate, spice chemicals often act as defense chemicals to protect the plant from animals, insects, and microbes, making them ideal for food preservation. Furthermore, spices could be used to disguise food that was beginning to rot. From these utilitarian roots, spices have developed into essential ingredients for cooking. A correlation can be seen between the prevailing temperature of a country, and thus the rate at which food becomes tainted by bacterial or fungal activity, and the amount of spice employed in traditional cuisine. In warmer nations such as Mexico, India, and Thailand, standard dishes are famously spicy, whilst colder states, like England, tout a much blander menu. Furthermore, traditional vegetable recipes often contain fewer spices than meat dishes from the same region. This may be due to the enhanced susceptibility of meats to bacterial and fungal invasion.

Taste
•Sweet •Bitter
•Sour •Salty
•Umami

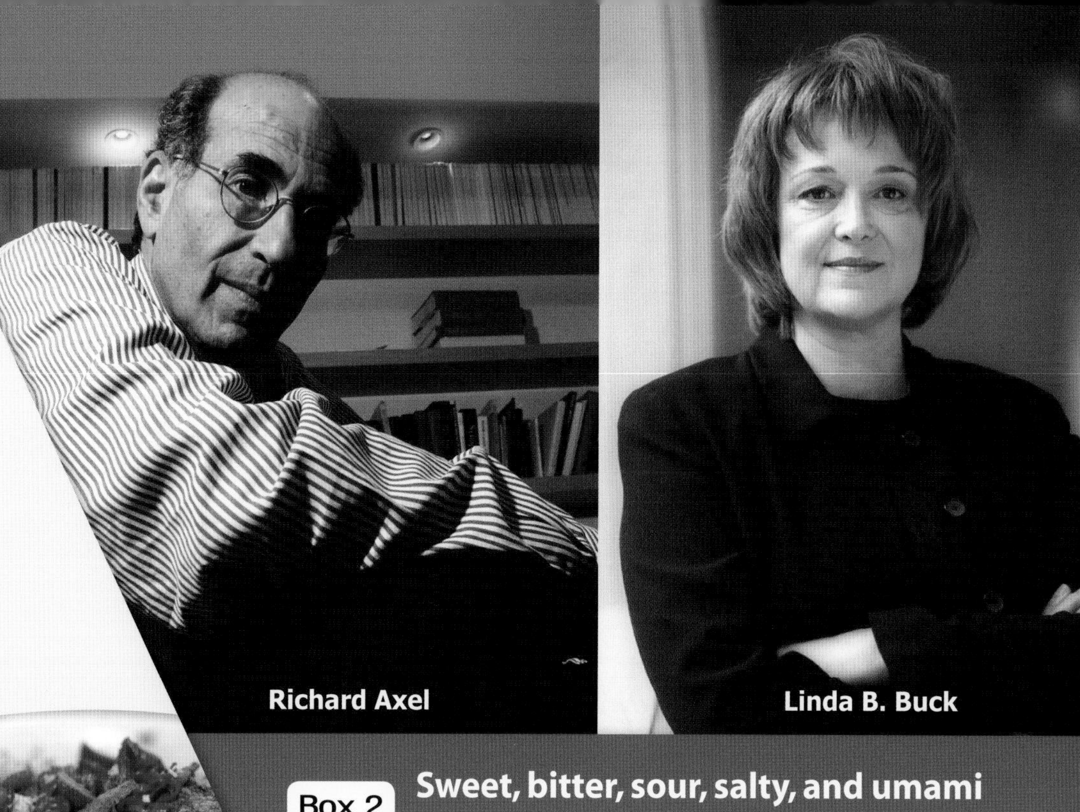
Richard Axel — Linda B. Buck

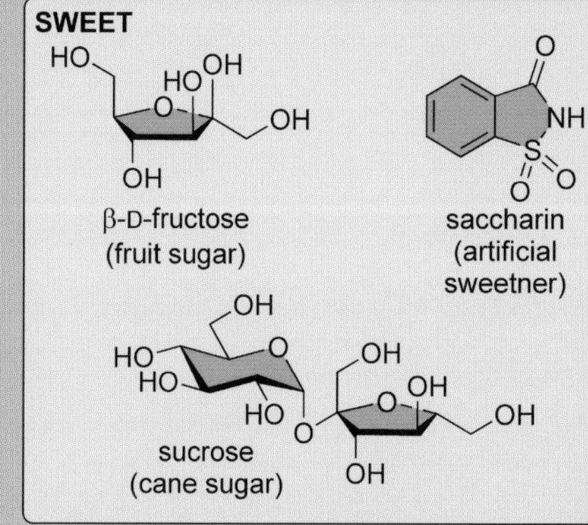

Box 2 **Sweet, bitter, sour, salty, and umami tasting molecules**

SWEET

β-D-fructose (fruit sugar)

saccharin (artificial sweetner)

sucrose (cane sugar)

BITTER

quercetin (red wine)

caffeine (coffee, tea)

SOUR

acetic acid (vinegar)

citric acid (citrus fruit)

SALTY

$Na^{\oplus}\,^{\ominus}Cl$

sodium chloride (salt)

UMAMI

glutamic acid (as monosodium glutamate, MSG) (cheese, tomatoes, soy sauce)

Pliny the Elder

Herodotus

Box 3 Spice molecules

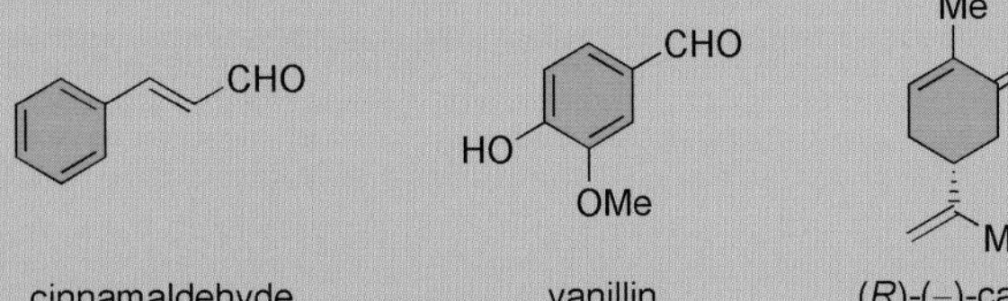

cinnamaldehyde (cinnamon and cassia)

vanillin (vanilla)

(*R*)-(–)-carvone (spearmint)

(*S*)-(+)-carvone (caraway, dill)

eugenol (oil of cloves, also bay leaves, cassia, and allspice)

isoeugenol (nutmeg, mace)

anethole (oil of aniseed, also fennel and tarragon)

The ancient Syrians also made frequent use of spices, including cloves, nutmeg, and mace. These grew only on small and remote islands in the Indonesian archipelago. The Arabs had discovered that the monsoon winds in early summer could be harnessed to take their ships quickly eastwards from the Red Sea coast towards the Indian subcontinent. The mid-season direction change could then be used to facilitate their voyage home. This early knowledge gave the Middle East (ranging from Alexandria in Egypt in the south, to Baghdad in modern-day Iraq, and Constantinople in modern-day Turkey in the north) control over the supply of spices to Europe over an astonishing period spanning four millennia. These cities gained incredible wealth from their control over spice trade, and the merchants jealously guarded their sources in order to maintain their monopoly. In the fifth century B.C., the Greek historian Herodotus was fooled by Syrian merchants, who told him that cinnamon grew only on the top of a single mountain in Arabia, guarded by huge birds of prey that had to be distracted with donkey meat to allow the harvesting of the spice. Later, Pliny the Elder, another famous historian and writer of antiquity, noted the price inflation caused by the flow of gold from the Roman Empire to the Arab spice merchants. The imbalance in the supply of spices had two consequences, both of which impacted the region for at least five hundred years and might be considered to have echoes in modern geopolitics. The first was a series of wars between the European spice consumers and the Arabian traders. The second was the impetus the Arabian monopoly gave European leaders to seek new routes to the spice producing regions. This drove many explorations and inventions, and eventually led to the discovery of the Americas by the Europeans. Another consequence of the Arabian spice trade was the spread of Islam. The prophet Muhammad was a spice trader, and his business activities facilitated the spread of his ideology eastwards to Indonesia and Malaysia.

The Romans were particularly extravagant in their use of spices; the rich slept on pillows of saffron and legionnaires went into battle wearing spice-rich perfumes. This provided the incentive for the Romans' attempt to conquer Arabia in 24 B.C. The failure of this campaign served to further the Arab stranglehold on the supply of spices, with no effective challengers for several centuries to come. Later, during the siege of Rome in 408 A.D., the attacking Visigoths demanded a bounty of

Cinnamon

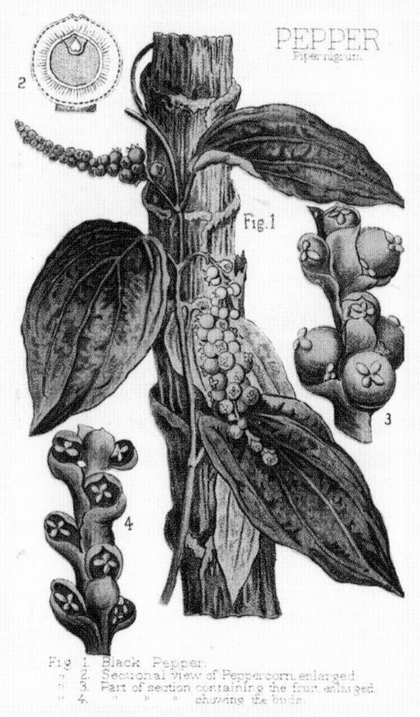

Piper nigrum

pepper, along with the more conventional riches of gold and silver.

In medieval Europe, the Italian port of Venice, originally made rich by the sale of salt from its lagoons, became increasingly powerful and dominated its own empire around the Mediterranean Sea. The power and affluence of Venice was derived from its collaboration with Arab merchants, through which it acted as the gatekeeper to Europe. The market for spices in Europe was ever increasing, with enormous demands for pepper, cloves, nutmeg, and other fine spices. The Europeans' demands were fuelled by the exposure of the people to spices brought back by the Crusaders on their return from the Holy Lands.

When the Ottoman Turks finally seized Constantinople (now known as Istanbul) in 1453, they secured the last overland route to the spice-rich eastern continents. This caused a vast increase in spice prices in Europe and led to the generous financing of expeditions in search of alternative routes to the Orient. The competition was fierce and resulted in an era of far-reaching sea exploration and expansionism by various European powers. The Portuguese explorer Vasco da Gama succeeded in accomplishing the first sea voyage from Europe to India by rounding the Cape of Good Hope, eventually finding Calicut (Southwest India) and its bountiful supply of peppercorns. The *Piper nigrum* vines are native to the neighboring region of Kerala, where their growth is sustained by the annual monsoon rains. All attempts to transplant the vines back to Europe met with failure, since no land could match the nourishing combination of rain and sun at Kerala. European sailors that safely reached India became accustomed to carrying peppercorns in their purses rather than gold coins, for the former commanded a greater and more universal value. Ferdinand Magellan, a Portuguese-born explorer funded by Spain, led the first successful expedition to circumnavigate the world. He was killed in the Philippines during the return leg of his voyage; however, one of his captains, Juan Sebastian del Cano, assumed command and completed the momentous voyage, cementing Magellan's fame. The Portuguese narrowly beat their Spanish rivals to claim the port of Malacca in 1511, a triumph that afforded them the intelligence needed to find the Banda Islands, the world's only source of nutmeg and mace. Both of these spices are derived from the same tree, *Myristica fragrans*. In Lisbon, Portugal, the price of spices plummeted and merchants flocked to the new source of these prized commodities. Today, nutmeg continues to play a role in globalization as a key ingredient in Coca-Cola®.

On one particular return journey from the Banda Islands, a Portuguese ship laden with nutmeg and mace ran aground beside a remote island. The sailors were brought before the Sultan of Ternate, and courted him with gifts. Thus, the Portuguese gained control over the source of cloves,

Cloves

Myristica fragrans

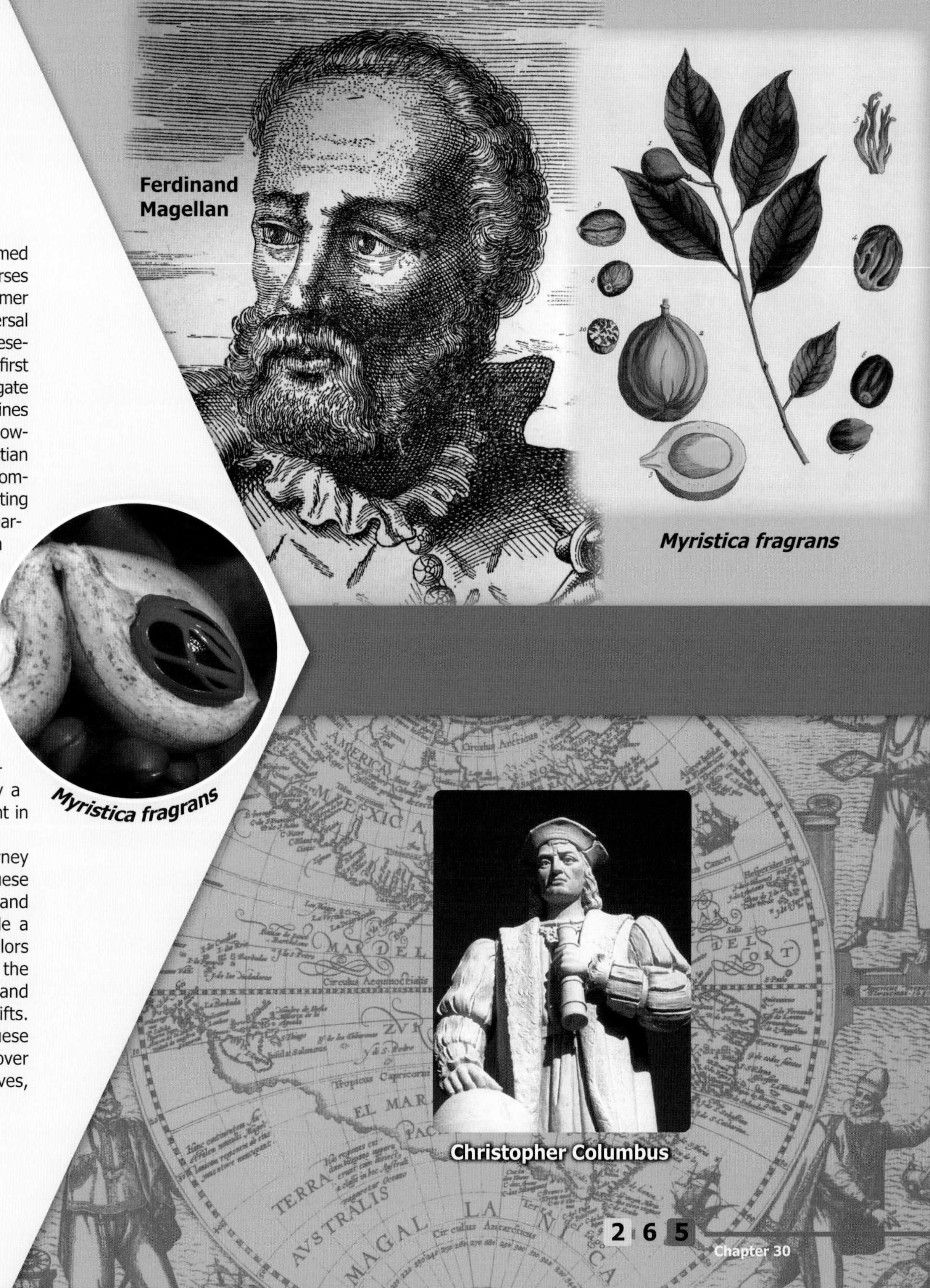

Ferdinand Magellan

Myristica fragrans

Christopher Columbus

from the tree *Eugenia aromatica*. Their dominance, however, was short-lived, as feuds with Spain began to distract the Portuguese. This gave the Dutch an opportunity to seize control over the spice trade. In 1602, a group of Dutch merchants formed the Dutch East India Company (Vereenigde Oostindische Compagnie, VOC), a trade association dedicated to the conquest of the spice trade. Such was their dedication that, at one point, they traded their American colony of New Amsterdam (now known as Manhattan!) for the deserted spice island of Run. By 1670, their ruthless tactics had made Dutch East India Company a truly global corporation, and the richest in the world, employing more than fifty thousand people. It owned some two hundred ocean-going vessels and its interests were protected by thirty thousand mercenary fighters who brutally enslaved and manipulated native populations in the spice growing tropics, using torture and violence to maintain their control. To insure their ships and cargoes over the course of the risky voyages, the company would allow merchants to bid for a share of produce before the expedition ever set sail. This process evolved into the modern practice of trading stocks and shares in a company, which is now used universally to fund corporations.

Meanwhile, the Italian explorer Christopher Columbus had discovered the Americas. Columbus, who was funded by the Spanish King and Queen, departed from Spain with the intention of discovering a western passage to the Indies, and thereby wrestle control of the spice trade from the Portuguese. Instead, of course, he landed on a Caribbean island on October 12, 1492. At first he believed he had landed in China, which had been described by Marco Polo a century earlier. When he sailed on further to new islands, he believed he had reached India. In an effort to convince his benefactors of the worth of his excursions, Columbus brought back many products to the Spanish court, including a red chili pepper that was sacred to the local people of the islands. Chili peppers, which had until this point been native solely to regions of Central and South America, were easily cultivated,

Ball and stick model of the molecule of capsaicin

unlike the extremely temperamental *Piper nigrum* vine, and as a result these plants spread quickly around the world. Today, many people still refer to chilies as red peppers, a confusion propagated by Columbus, who substituted dried and powdered chilies for black pepper in an effort to provide Spain with a plentiful source of pepper. To add to the confusion, black peppercorns in fact begin life as red berries, which blacken on ripening. Within fifty years, red chili peppers were a sought after commodity in places as far apart as Western Europe, the Indian subcontinent, and the bustling ports of China. Columbus was also responsible for the introduction of vanilla and allspice into Europe.

The molecule that causes the ferocious heat of chili peppers is called capsaicin, and it has remarkable structural similarities to zingerone and piperine, the active components of ginger and pepper (Box 4). Capsaicin is a defense chemical evolved by the chili plants to ward off grazing animals. Birds, which are much better vectors for spreading seeds, have no receptors for capsaicin and are, therefore, immune to the vicious effects experienced by mammals. This feature has been harnessed commercially to produce squirrel-proof birdseed. Curiously, capsaicin is also found in much lower doses in the herbs oregano, cinnamon, and cilantro (coriander). Many people enjoy the strong taste of chilies, and it is said that the cult-like following of this spice by some is an indication of its mildly addictive nature. Capsaicin causes a persistent burning sensation in the mouth even at concentrations as low as ten parts per million. This is because the trigeminal cells located in the mouth, nose, and throat are activated and irritated, leading to the relay of a pain message. Capsaicin induces the opening of a trans-membrane ion channel, causing an influx of calcium ions. The depolarization caused by the ion flux is directly transmitted by the primary afferent sensory neurons (nociceptors) to the brain. These nerve cells are responsible for sensing noxious chemical, thermal, and mechanical stimuli. In turn, this event triggers the release of endorphins (natural painkillers) to subdue the perceived pain. It is the natural euphoric high produced by the endorphins that is said to cause addiction to hot spicy chili-containing foods. At the site of capsaicin stimulus, inflammation-mediating chemicals are released and bodily functions such as heart rate, metabolism, gastrointestinal tract activity, salivation, and perspiration are increased. The ability of capsaicin to induce pain and tears has been employed for many years as a deterrent against would-be assailants. The Mayan Indians burned chilies to create an acrid and stinging smoke screen when they fought in battles. Gourds were also thrown that contained powdered chilies suspended in water to blind attackers. Today we use this very same

weapon in the urban environment, where citizens and law enforcers carry pepper sprays containing capsaicin to protect themselves against or subdue criminals.

Originally, clinicians were drawn to chilies and capsaicin because repeated application of this irritant leads to desensitization. This feature was regarded as having significant potential for the treatment of conditions such as arthritis and diabetic neuropathy, which cause chronic pain. At high concentrations capsaicin kills nociceptor cells, whilst sustained lower concentrations cause reversible changes in the nociceptors such that they no longer function in the transmission of pain signals. In addition, certain alterations to brain chemistry contribute to the process of desensitization. Firing of the nociceptors causes the release of a number of neuropeptides, including tachykinin substance P (t-SP), within the central nervous system (CNS). When this firing becomes too frequent, the levels of t-SP can become depleted, hindering the perception of pain. However, patients attempting to use capsaicin treatments had to tolerate such extreme initial discomfort that its use to treat moderate pain has more or less discontinued. Fortunately, other compounds from the vanilloid class, to which capsaicin belongs, presented more favorable activity profiles and became candidates for use in pain management.

David Julius

The vanilloid molecules usually, but not always, include a homovanillyl moiety (Box 4). All of them trigger pain receptors that had historically been named the vanilloid receptors (VR), leading to the slightly confusing nomenclature. The existence of the vanilloid receptor and its relation to capsaicin were predicted as early as the 1960s by the Hungarian pharmacologist Nicholas Jancsó. His work succeeded in promoting capsaicin from a curiosity into a neuropharmalogical tool, though it was not until 1990 that direct evidence for the receptors was uncovered by Peter M. Blumberg and Arpad Szallasi (National Cancer Institute, Bethesda, USA). By the time the terminology of 'vanilloid receptor' had gained acceptance, it had also emerged that these biomolecules could recognize and bind a wider variety of compounds, including those without vanillyl groups. The VRs have since been found in a number of additional tissues (including the brain and bladder), and exist with distinct well-defined subtypes. They respond to multiple stimuli, such as mechanical damage and heat. This is why the brain interprets the effect of eating chili peppers as a sensation of heat. Interestingly, an analogous process is responsible for the cold sensation caused by compounds such as menthol. The search for alternative vanilloid stimulants, which

Peter M. Blumberg

could be employed in pain management without the severe discomfort associated with capsaicin, led to the discovery of resiniferatoxin. This work was inspired by medicinal folklore surrounding euphorbium, the dried latex from the cactus *Euphorbia resinifera*, found in mountainous regions of Morocco. Resiniferatoxin is a highly potent capsaicin analogue, and the active component of this traditional medicine.

King Juba II of Mauretania (50 B.C. to 23 A.D.), who was married to Antony and Cleopatra's daughter, Selene, wrote extensive academic texts including one of the first pharmacological monographs, *On Latex*. He described a material named euphorbium, most probably named after his personal physician, but perhaps relating to the shape of the cactus, as the Greek translation can mean 'well fed'. The text has become fragmented with time, so the applications of euphorbium recommended by Juba remain vague. Later, the writers Dioscorides (Greek) and Pliny the Elder (Roman) both promoted the use of euphorbium as a sternutative (an agent provoking sneezing) and vesicant. Pliny also states that poisons, including those introduced by snake, spider, or scorpion bites, are simply dealt with by an incision into the skull and insertion of euphorbium, regardless of the location of the bite. By renaissance times it was used in popular vesicant plasters such as the infamous *Moche di Milano* (*Flies of Milan*), a plaster that did indeed contain flies, as well as euphorbium, turpentine, and storax. Euphorbium was also recommended as a purgative. Many physicians, including Étienne François Geoffroy, disapproved this application because of the drastic outcome. Geoffroy instead proposed the use of euphorbium as an excellent treatment for bone cavities and nerve pains. In Transylvania, around the same period, it was a fashionable medication for toothache. However, its irritant characteristics were so unpleasant that few pharmacists would pulverize the material themselves, preferring instead to leave it to more junior staff. Euphorbium powder was also a favorite amongst practical jokers. One tale tells of a ballroom floor dusted with euphorbium powder, such that all the revelers sneezed without rest for days after attending the grand celebrations of a member of the French aristocracy!

Coin of King Juba II

Despite the reputation of euphorbium as an agent for the management of pain, it had almost disappeared from medical literature by the beginning of the nineteenth century, and only gained renewed interest in recent decades. By the time Peter M. Blumberg (then at Harvard University) became interested in the pharmacological potential of resiniferatoxin in the early 1980s, euphorbium collection and sales had become so rare that he was forced to enlist the help of United States Senator Edward Kennedy, whose intervention led to per-

Box 5 The related molecular structures of capsaicin, phorbol, and resiniferatoxin

capsaicin (chili pepper)

phorbol

resiniferatoxin

Box 6 Wender's retrosynthetic analysis of resiniferatoxin

resiniferatoxin

Stanford University

mission for staff at the American Consulate in Morocco to collect samples. The major active principle, resiniferatoxin, had been isolated from euphorbium and identified in 1975 by W. Adolf and coworkers at the Institute of Biochemistry in Heidelberg (Germany). Since then, the pharmacology of resiniferatoxin has been extensively investigated by several researchers, including Peter Blumberg, Arpad Szallasi, and David Julius (University of California, San Francisco, USA). These studies, which significantly expanded the knowledge of vanilloid receptors, spurred great interest in the chemistry and biology of resiniferatoxin (now registered as RTX®). One of the findings of this research was that resiniferatoxin desensitizes, rather than overstimulates, the neurons, making it an ideal neuropathic agent bearing none of the painful side effects of capsaicin. Ultimately, this vigorous research activity led to RTX® entering clinical trials in the late 1990s as an agent for the management of pain and the control of various overactive bladder disorders. Recent laboratory results show that resiniferatoxin can also kill tumor cells, and thus may be of use in the fight against cancer, although more work is needed to determine its full potential in this regard.

Synthetic organic chemists have also been involved in the development of this molecule. Along with its structural relationship to capsaicin, resiniferatoxin shares a degree of structural homology with another group of molecules, including phorbol (Box 5). The phorbol family includes the most potent tumor promoters known. Fortunately, resiniferatoxin shares only certain architectural features with these harmful compounds, and has been shown to lack the carcinogenic activity of phorbol. Both these natural products contain a similar arrangement of three fused rings with a number of pendant oxygen-based functional groups. However, in the case of resiniferatoxin, three such oxygen substituents are tied together in a delicate orthoester group that significantly complicates the assembly of this already challenging motif. The only total synthesis of resiniferatoxin completed to date has come from the group of Paul A. Wender at Stanford University (California, USA), who also completed the asymmetric total synthesis of phorbol.

Throughout his respected career Wender has pursued some of the toughest challenges within synthetic organic chemistry and has always been keen to

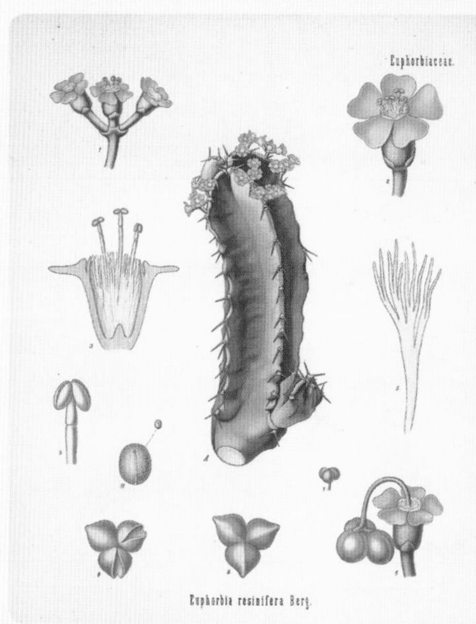

Euphorbia resinifera

Box 7 Highlights of the early stages of Wender's total synthesis of resiniferatoxin

(1) *m*-CPBA
(2) Ac_2O
oxidative rearrangement

DBU
heat

1,3-dipolar cycloaddition

(1) $Cp_2Zr(n\text{-}Bu)_2$
(2) AcOH
enyne ring closure

combine the discovery of unique solutions for these problems with investigations into related aspects of biology and medicine. In their route to resiniferatoxin, the Wender group employed a typically bold and innovative strategy; highlights of their retrosynthetic analysis are shown in Box 6. Their synthetic plan included a proposal to harness both a 1,3-dipolar cycloaddition and a zirconium-mediated enyne ring closure, interspersed in the systematic construction of the remaining architectural features. The result is a well orchestrated showcase of many state of the art modern tools, developed to meet the ever growing challenges posed by newly discovered and increasingly complex natural products.

As illustrated in Box 7, the 1,3-dipolar cycloaddition was used to accomplish the stereoselective formation of the first complex tricyclic intermediate. Later, the zirconium-mediated enyne cyclization proceeded as planned to fuse the final carbocyclic ring onto the existing framework. It should be remembered that with molecules as large and densely functionalized as resiniferatoxin, there are no trivial operations. As the molecule grows, greater demands of selectivity and tolerance are placed on each successive step. In addition, the complexity of such targets can lead to unpredictable behavior, caused by unforeseen interactions between different groups. Such complications can derail even the most ingenious synthetic plan. The team led by Wender successfully overcame all the hurdles encountered in their brave campaign, and completed their total synthesis of this inspiring molecule in 1997 (Box 8).

As this story shows, spices have played a key role in the modern world, and the drive to acquire and understand them has fueled numerous advances in science and technology. In addition to the varied and exciting flavors they give our cooking, we have embraced spices for their medicinal properties for millennia. We are now able to understand these properties at a molecular level and, in turn, the spice molecules themselves have served as the inspiration for developments in the synthesis of complex molecules. Above all, the history of spices demonstrates that an understanding of chemistry is of paramount importance to many aspects of our daily lives, from cooking, flavors, and perfumes, to agriculture and medicine.

Paul A. Wender

Euphorbia resinifera

Box 8 Completion of Wender's total synthesis of resiniferatoxin

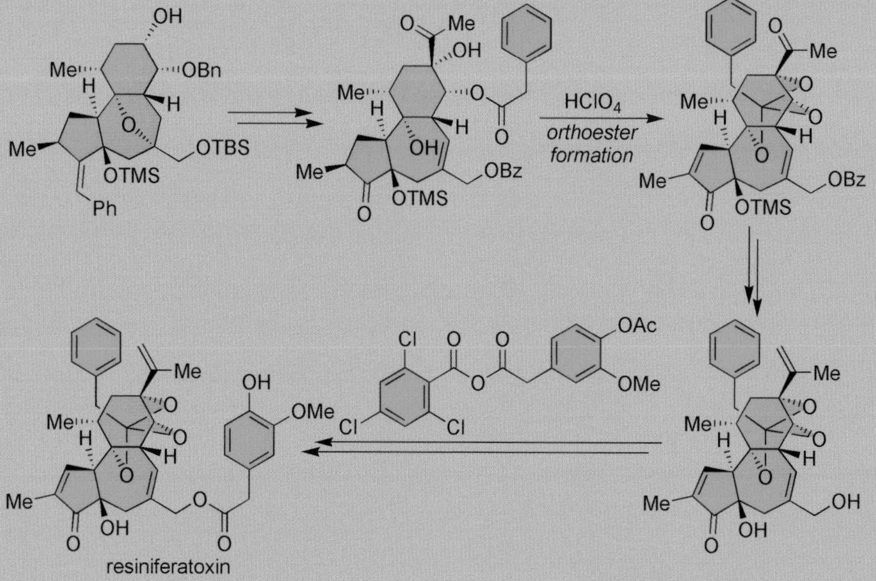

Richard Axel and Linda B. Buck

Further Reading

D. V. Smith, R. F. Margolskee, Making Sense of Taste, *Sci. Am.* **2001**, *284*, 32–39.

C. Corn, *The Scents of Eden: A History of the Spice Trade*, Kodansha International, Tokyo, **1999**.

G. Appendino, A. Szallasi, Euphorbium: Modern Research on its Active Principle, Resiniferatoxin, Revives an Ancient Medicine, *Life Sciences*, **1997**, *60*, 681–696.

R. Axel, Scents and Sensibility: a Molecular Logic of Olfactory Perception, *Angew. Chem. Int. Ed.* **2005**, *44*, 6110–6127.

L. B. Buck, Unravelling the Sense of Smell, *Angew. Chem. Int. Ed.* **2005**, *44*, 6128–6140.

P. A. Wender, K. D. Rice, M. E. Schnute, The First Formal Asymmetric Synthesis of Phorbol, *J. Am. Chem. Soc.* **1997**, *119*, 7897–7898.

P. A. Wender, C. D. Jesudason, H. Nakahira, N. Tamura, A. L. Tebbe, Y. Ueno, The First Synthesis of a Daphnane Diterpene: The Enantiocontrolled Total Synthesis of (+)-Resiniferatoxin, *J. Am. Chem. Soc.* **1997**, *119*, 12976–12977.

K. C. Nicolaou, S. A. Snyder, *Classics in Total Synthesis II*, Wiley-VCH, Weinheim, **2003**, pp. 137–159.

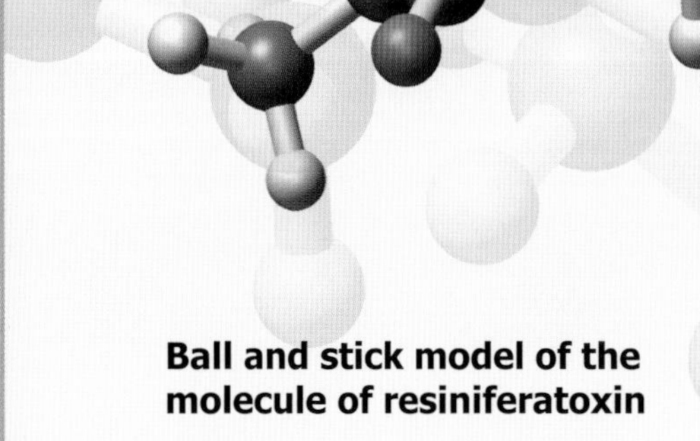

Ball and stick model of the molecule of resiniferatoxin

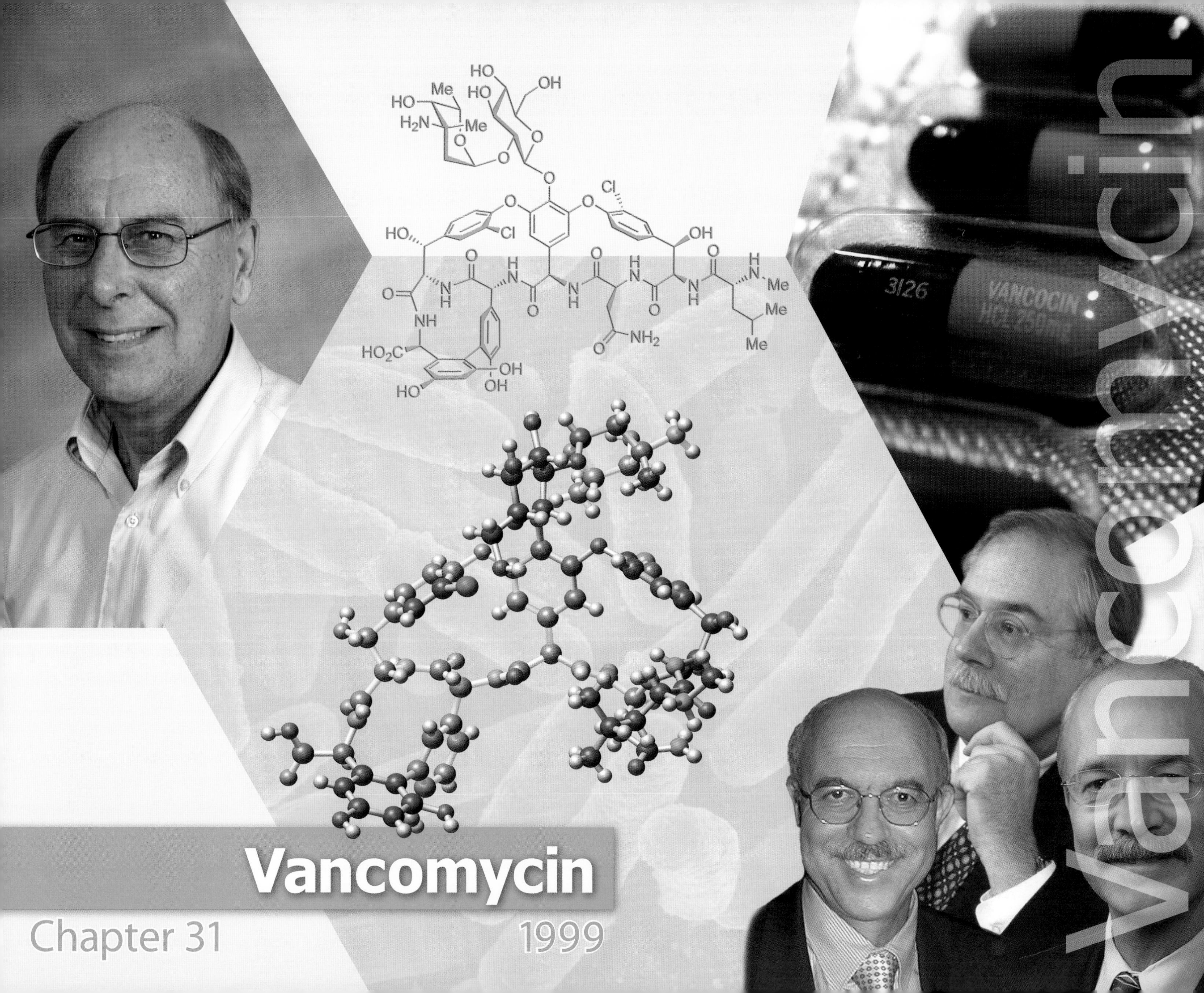
Vancomycin
Chapter 31
1999
3126
VANCOCIN
HCL 250mg
Vancomycin

Sir Alexander Fleming

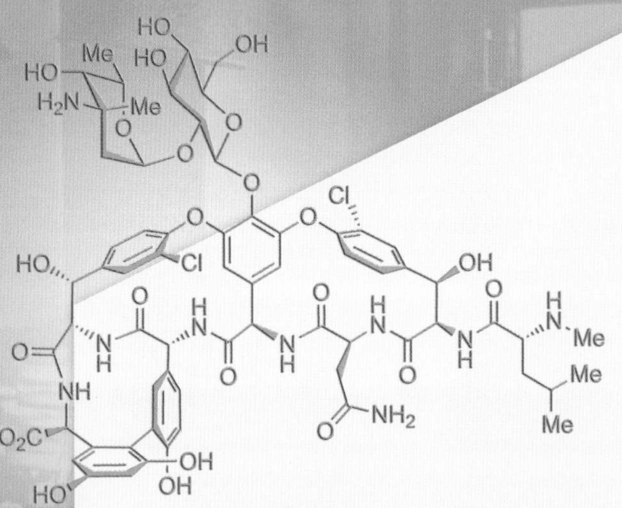

Box 1 Selected naturally occurring antibiotics used to treat infections

streptomycin
[various brands]
(*Streptomyces griseus*)

teicoplanin A_2-1
[Teicoplanin A_2®, Targocid®; marketed as a mixture of at least five compounds varying at the lipophilic side chain]
(*Actinoplanes teichomyceticus*)

erythromycin A
[Erythrocin®, Erythroped®, Erythromid®]
(*Saccharopolyspora erythraea*)

rifampin (semisynthetic)
[Rifadin®, Rimactane®]
(*Amycolatopsis mediterranei*)

The first patient to be treated with penicillin was a forty-three year old policeman in the United Kingdom in 1941. A small sore next to his mouth had developed into a systemic bacterial infection, with abscesses appearing around his eyes and in his lungs. This desperately ill man was injected with penicillin, which led to a pronounced improvement in his condition. Unfortunately, at the time there were not sufficient supplies of this experimental drug available to continue treatment and, despite attempts to recover some penicillin from his urine, this patient would later die from the infection. Prior to the introduction of antibiotics, this was a common fate for those who developed bacterial infections.

Penicillin mold

Cuts, chest disorders, childbirth, or even minor scratches could lead to infections, severe illness, and even death. Penicillin essentially eliminated this scourge, and has saved millions of lives. However, Sir Alexander Fleming, who was instrumental in the discovery of penicillin (Chapter 13), began to notice a potentially hazardous feature of antibiotic therapy. In his laboratory, Fleming found that low concentrations of the antibiotic did not kill all the bacteria. Instead, strong penicillin-resistant bacteria survived and multiplied. The problem of clinical penicillin resistance, foreshadowed by these early experiments, has grown steadily ever since. So far, doctors and scientists have kept just ahead of the problem by modifying the structure of penicillin and discovering new antibiotics from nature (Box 1). Vancomycin is one such post-penicillin discovery and has become the so-called antibiotic of last resort in the fight against certain bacterial infections.

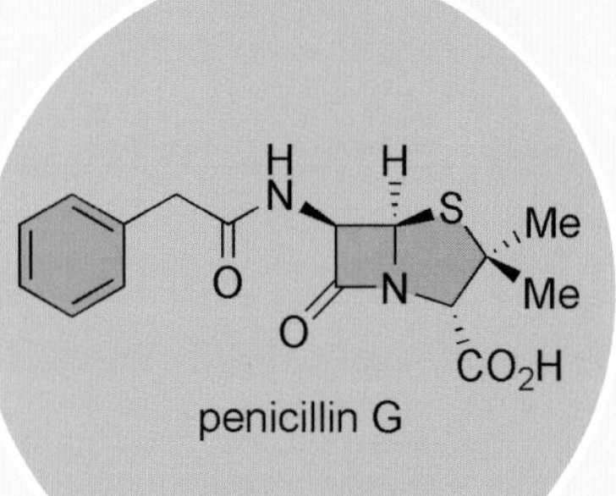

Following the introduction of antibiotics, the treatment of bacterial infections became routine, and society began to forget the dangers formerly posed by such pathogens. However, antibiotic resistance threatens a return to the days of untreatable deadly infections, despite our arsenal of some one hundred fifty antibiotic drugs. Resistance is an inevitable consequence of antibiotic use and, therefore, we must improvise to slow its progression and strive to discover new antibiotics in order to defeat bacterial diseases. Bacteria can develop resistance to drugs in many different ways. Mutations in the proteins to which the antibiotic binds can reduce the effectiveness of the antibiotic by preventing effective binding. Bacteria can also produce enzymes that modify or degrade the antibiotic structure. Bacteria evade penicillin and other β-lactam antibiotics by producing β-lactamase enzymes which hydrolyze the lactam ring before it can disrupt cell wall synthesis. Alternatively, bacteria can develop molecular pumps which actively remove the antibiotic from the bacterial cells, thereby minimizing its impact.

The development of antibiotic resistance to bacteria is driven by exposure to antibiotics, which provides an evolutionary pressure in favor of resistant individuals. This process occurs especially quickly in bacteria because these microbes grow and reproduce rapidly. Bacteria can increase the rate at which random mutations take place in response to stress (such as exposure to antibiotics), increasing the chances of a beneficial mutation. They can also pass genetic information between individuals through a number of mechanisms (see Box 2). Genes can be passed directly from one bacterium to another (A); viruses that infect bacteria can transfer genes from one bacterial host to the next (B); and bacteria can also absorb sections of DNA from their environment and incorporate them into their own chromosomes (C). Thus, antibiotic resistance can spread quickly throughout a bacterial population. The way in which we use antibiotics can further promote bacterial resistance. Since their initial development, antibiotics have been vastly overused, both in human populations and in agriculture. Antibiotics are demanded for minor infections and, sometimes, even for viral diseases, against which they are completely ineffective. Indeed, researchers estimate that as many as one third to one half of all antibiotic prescriptions are unnecessary. Additionally, patients often do not complete their prescribed course of antibiotics, a practice that leaves some stronger bacteria alive. The more that bacteria are exposed to antibiotics, particularly at low concentrations that do not kill the entire population, the more opportunities they have to develop resistance. Resistant bacteria can then thrive and multiply as they have a strong advantage over antibiotic-susceptible competitors. The trend for overusing antibacterial agents is spreading, with many household items and cleaners incorporating such substances, further exacerbating the resistance problem. To counter antibiotic resistance, we must continuously develop new and different antibiotics, so as to destroy resistant strains. However, this process is slow and expensive, and must be complemented by a change in our attitude towards antibiotics

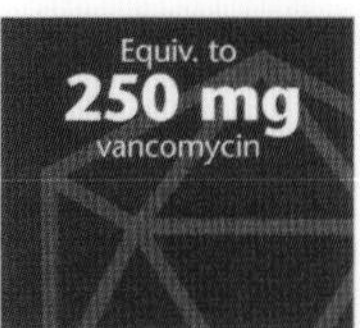

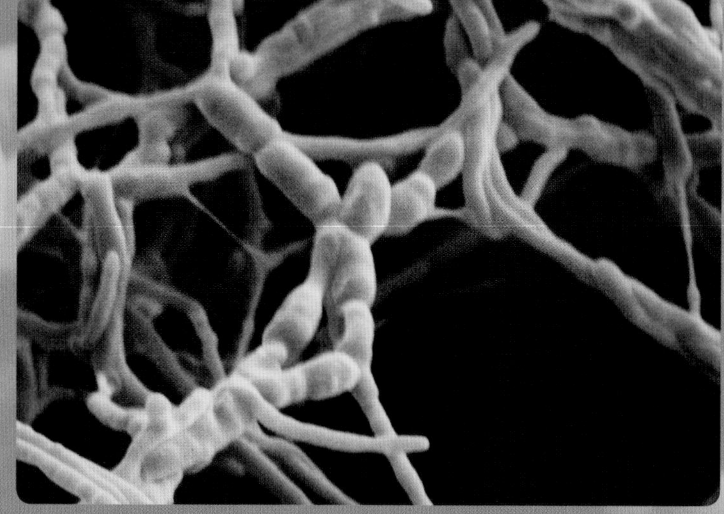

Amycolatopsis orientalis

Box 2 Gene exchange between bacteria leading to drug resistance

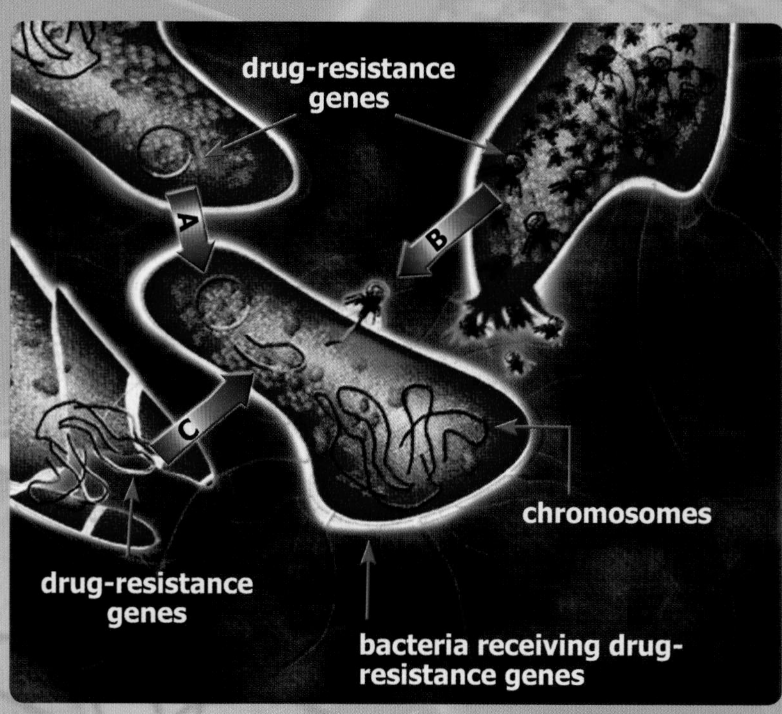

Selman A. Waksman

Enterococcus faecium

Box 3 Disruption of bacterial cell wall synthesis by vancomycin

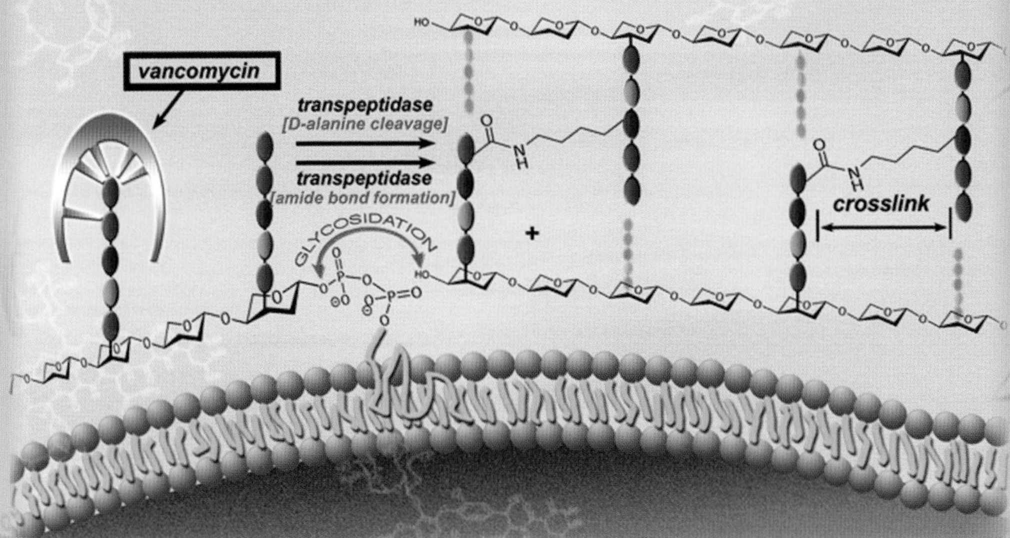

and bacteria in general. Using antibiotics sparingly and only when genuinely needed will certainly help extend the lifetime of existing drugs and provide more time for the development of new ones.

Vancomycin was first isolated in 1956, by scientists at the American pharmaceutical company Eli Lilly, from a soil microbe of the Actinomycete family (first classified as *Nocardia orientalis*, now reclassified as *Amycolatopsis orientalis*). Its original designation, O5865, was soon replaced by a new name, vancomycin, derived from its ability to vanquish bacteria. The soil sample had come from the jungles of Borneo where the ecosystem exhibits a rich biodiversity that invites the harvest of new bioactive natural substances and potential medicines. Soil has been a major focus in the search for new antibiotics ever since the discovery of streptomycin (produced by *Streptomyces griseus*) by the Ukrainian scientist Selman A. Waksman (Rutgers University, USA), in 1943. Soil contains approximately one billion microbes per cubic centimeter, and Waksman was fascinated by the territorial battles within these dense populations of microorganisms. Indeed, he rigorously redefined the term antibiotic to describe the chemicals produced by microbes to wage war upon one another. Inspired by the observation that the tubercule bacillus does not survive in the soil, Waksman isolated streptomycin, the first antibiotic effective against tuberculosis, from a soil bacteria.

Streptomyces griseus

The Food and Drug Administration (FDA) approved vancomycin for clinical use in the USA in 1958. At first, the impure nature of the drug led to some unwanted toxicity and certain negative side effects. Fortunately, these early complications were quickly resolved and the drug has since become widely known for its ability to treat methicillin-resistant *Staphylococcus aureus* (MRSA) infections. MRSA is a bacterial strain that causes systemic infections, particularly in weakened individuals, that are otherwise difficult to treat. The MRSA problem is especially acute in hospital wards, where these communicable bacterial infections find many hosts with impaired immune systems in close proximity. The problem of hospital-acquired infections has increased markedly in recent years and led to a renewed awareness of hygiene in hospitals. Today, vancomycin and the related antibiotic teicoplanin A_2-1 (Box 1) are the drugs of last resort to treat

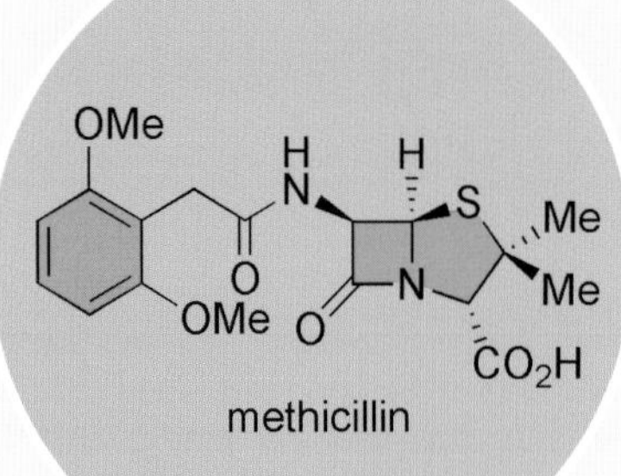

methicillin

patients infected with drug-resistant bacterial strains.

Like penicillin, vancomycin exerts its antibiotic effects by disrupting bacterial cell wall growth, thereby compromising the integrity of this key structural element and causing lysis of the cell (Box 3). Since phospholipid membranes, rather than peptidoglycan-based cell walls associated with bacteria, encapsulate mammalian cells, this attack is selective for bacteria. The bacterial cell wall is composed of cross-linked peptide and sugar (glycan) networks (hence its name, peptidoglycan) that are continuously being renewed. The main polymer consists of an alternating polysaccharide formed by the addition of heterodisaccharide monomers to a growing chain. A short peptide sequence is attached to every second sugar ring. The exact sequence of the peptide varies from species to species, but generally ends with a diamino acid (e.g. lysine in *Staphylococcus* species) and two D-alanine residues. During construction of the peptidoglycan membrane, the peptide chains are cross-linked to provide the rigidity needed to resist the internal

Staphylococcus aureus

Box 4 Vancomycin binding to peptidoglycan models of vancomycin-susceptible (left) and vancomycin-resistant (right) bacteria

X-Ray-derived model of the complex of vancomycin with diacetyl-L-Lys-D-Ala-D-Ala

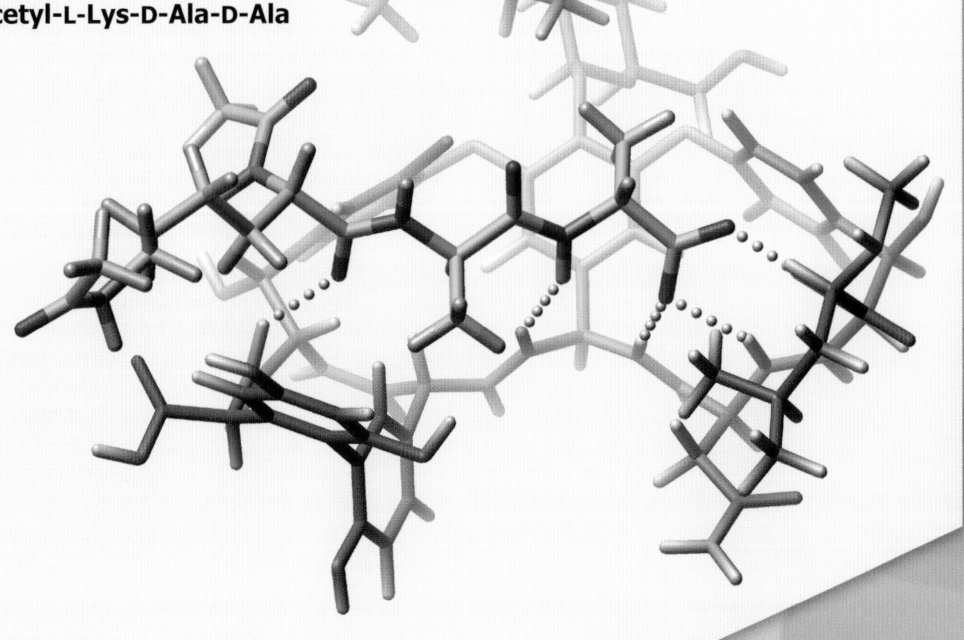

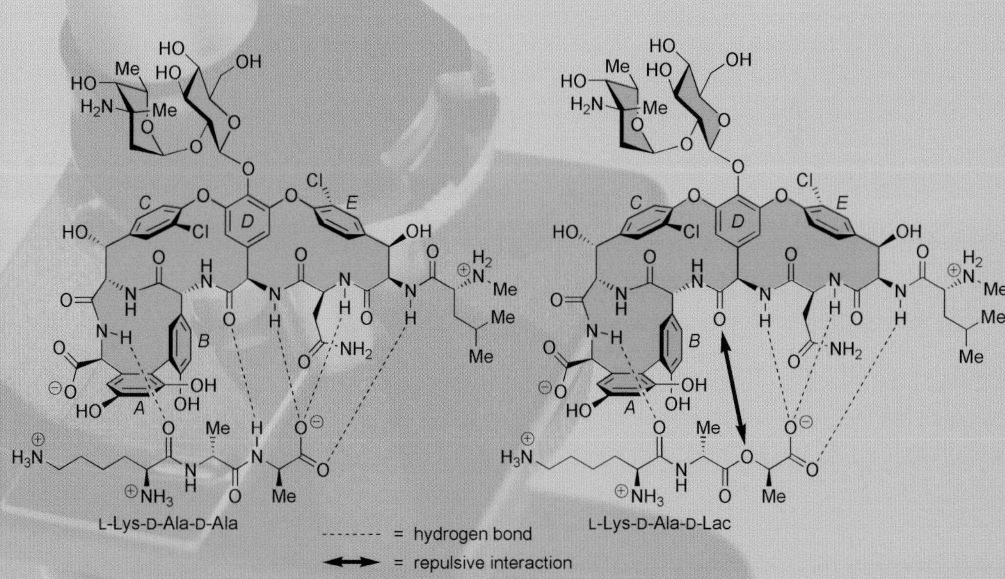

David A. Evans

Box 5 Evans' retrosynthetic analysis of the vancomycin aglycon

nitro-activated S_NAr bisaryl ether macrocyclization

peptide coupling

vancomycin aglycon

nitro-activated S_NAr bisaryl ether formation

peptide coupling

peptide coupling

peptide coupling

oxidative coupling

peptide coupling

peptide coupling

osmotic pressure of the cells, thus preventing them from bursting. This process involves cleavage of the terminal D-alanine residue and coupling of the resulting carboxylic acid to the lysine moiety of another chain via a small intervening peptide. Vancomycin binds to the peptide chain of the monomeric disaccharides (Box 3), preventing their incorporation into the polymer matrix. This leads to a collapse in the strength of the cell wall, and eventual cell lysis (rupture). However, despite the undisputed power of vancomycin, a new threat has recently emerged in the form of vancomycin-resistant strains of pathogenic bacteria.

Resistance to vancomycin among *Staphylococcus aureus* infections was first noted in three distinct geographical locations in 1998. Scientists had originally focused their attention on another bacterium species, *Enterococcus faecalis*, because vancomycin-resistant strains had been known since the late 1980s. They worked to understand the mechanism by which vancomycin-resistant *Enterococcus* (VRE) evades vancomycin with the hope that this would allow the design of strategies to overcome the resistance. At a molecular level, vancomycin binds the pendant pentapeptides in the peptidoglycan through five different hydrogen bonds (Box 4). The mode of binding was uncovered by Dudley H. Williams and his team at the University of Cambridge (UK) using sophisticated nuclear magnetic resonance (NMR) spectroscopic techniques. In resistant bacteria, it was shown that the nature of the terminating amino acids is changed just enough to loosen this grip. The final D-alanine in the peptidoglycan precursors of these bacteria is replaced by the hydroxy acid D-lactate. The net result is the exchange of an N–H unit with an oxygen atom (Box 4). This simple alteration leads to the loss of one hydrogen bond with a repulsive interaction taking its place, thus loosening the grip of vancomycin on the peptide chain and giving the bacteria leeway to squeeze free, resulting in a 1,000-fold reduction in the activity of the drug. The genetics and enzymology of vancomycin resistance were unraveled in a series of studies beginning in the early 1990s by the groups of Patrice Courvalin (Institut Pasteur, France) and Christopher T. Walsh (Harvard University, USA). Less is known about the origin of vancomycin resistance in *Staphylococcus aureus* strains, although it does not involve the same gene cluster as in the *Enterococci* species. Instead, the resistant *Staphylococci* are thought to produce thicker peptidoglycan cell walls with substantially less cross-linking.

The potent antibiotic activity and novel molecular architecture of vancomycin (which had finally been elucidated by Dudley Williams and his group in 1981) lured synthetic chemists to contemplate its total synthesis. Such an endeavor, it was thought, might lead to a better understanding of the chemistry of vancomycin and facilitate the design and synthesis of superior antibiotics, including some exhibiting the much sought after ability to kill vancomycin-resistant bacteria. These chemists recognized a number of challenging obstacles that had to be overcome if a total synthesis of vancomycin

Enterococcus faecalis

was to be accomplished. These included the preparation of the novel amino acids and sugars that make up the structure, the formation of the glycoside and amide bonds associated with the sugar and amino acid units, and the construction of the three strained macrocyclic rings. The latter consideration also required the control of atropisomerism of the macrocycles. Atropisomerism is an interesting stereochemical phenomenon that arises when rotation about a single bond is restricted due to the presence of adjacent bulky groups, leading to two different isolable isomers. It is most common in biaryl systems with at least three *ortho*-substituents other than hydrogen. The groups cannot pass by each other at ambient temperature, so two different spatial arrangements can exist. Vancomycin has three such axes, which called for special consideration in any design of its chemical synthesis.

Following several years of arduous campaigning, K. C. Nicolaou and coworkers at The Scripps Research Institute (USA) finally accomplished the total synthesis of vancomycin in 1999. The aglycon of vancomycin (i.e. the core structure without the sugar groups) had been synthesized a year earlier by both the groups of Nicolaou and David A. Evans (Harvard University, USA). The group of Dale L. Boger (The Scripps Research Institute) subsequently matched this feat in 1999. The synthesis of such a complex molecular structure demanded the invention of new strategies and reactions. Indeed, both the Evans and Nicolaou syntheses highlighted in this chapter are rich in new chemistry (Boxes 5–8).

The Evans total synthesis of the vancomycin aglycon followed the retrosynthetic strategy shown in Box 5. This plan was devised following detailed model studies for the construction of the various ring systems. After assembling the required building blocks, the three macrocycles were forged sequentially (Box 6). The first was formed by an oxidative biaryl coupling to give the AB macrocycle. This was followed by two nucleophilic aromatic substitution ring closures, facilitated by electron-withdrawing nitro groups on the aromatic rings, to form the biaryl ether rings. Subsequent functional group manipulations and deprotection reactions then furnished the vancomycin aglycon.

The Nicolaou group's assembly of the vancomycin aglycon was predicated on an entirely different retrosynthetic plan (Box 7), which relied upon a new technology specifically developed for the construction of the macrocyclic bisaryl ethers. The AB biaryl axis was formed first using a diastereoselective Suzuki coupling. The CD biaryl ether macrocycle was then closed using the specially developed triazene-driven ether formation and the AB macrocycle was closed via a macrolactamization. This was followed by the coupling of the left hand fragment, and a second triazene-

Dale L. Boger

Box 6 Highlights from the Evans group's synthesis of the vancomycin aglycon

VOF_3, $BF_3\bullet OEt_2$, $AgBF_4$, CF_3CO_2H; then $NaBH(OAc)_3$

oxidative coupling

Ddm =

vancomycin aglycon

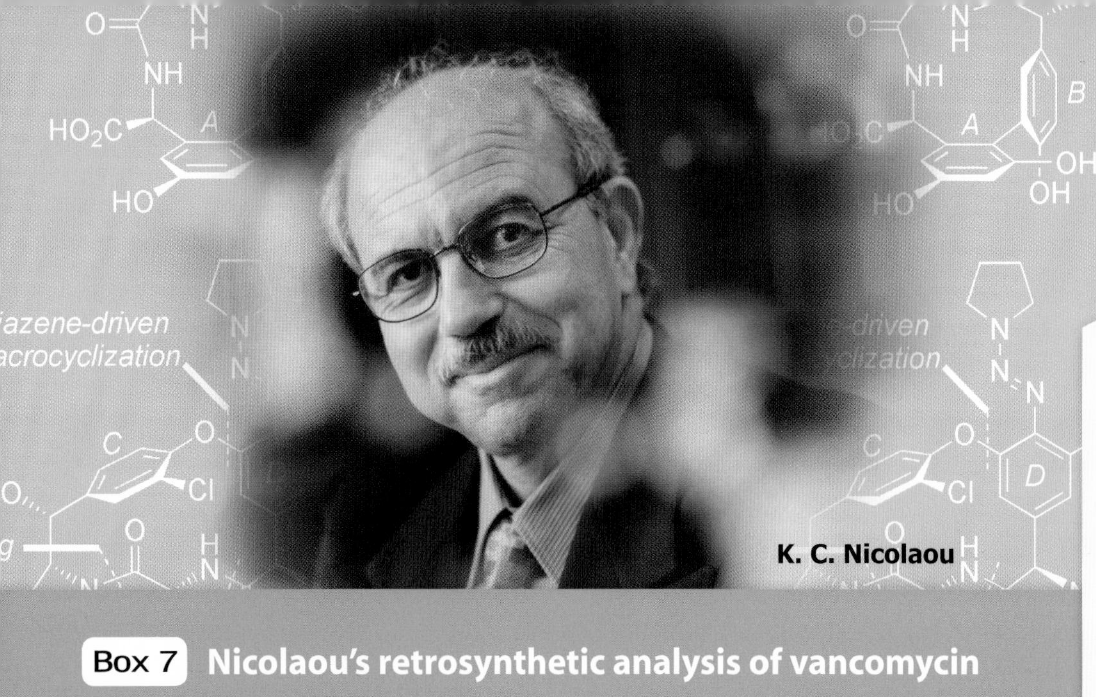

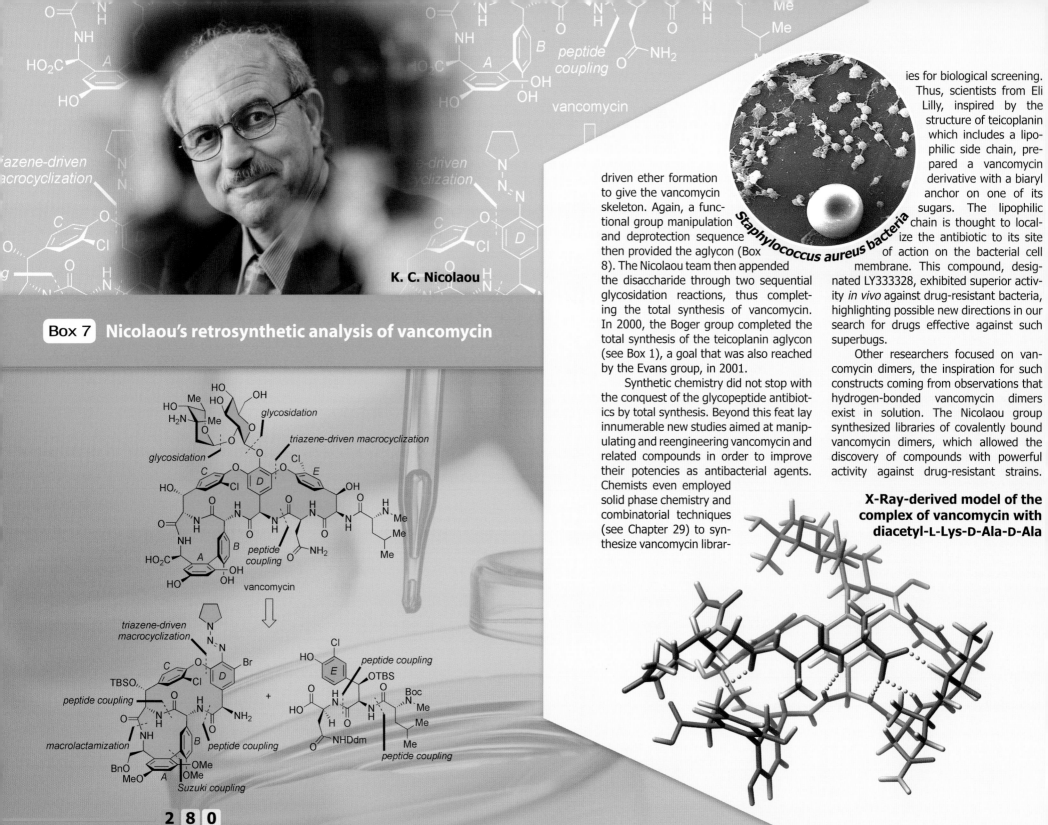

K. C. Nicolaou

Box 7 Nicolaou's retrosynthetic analysis of vancomycin

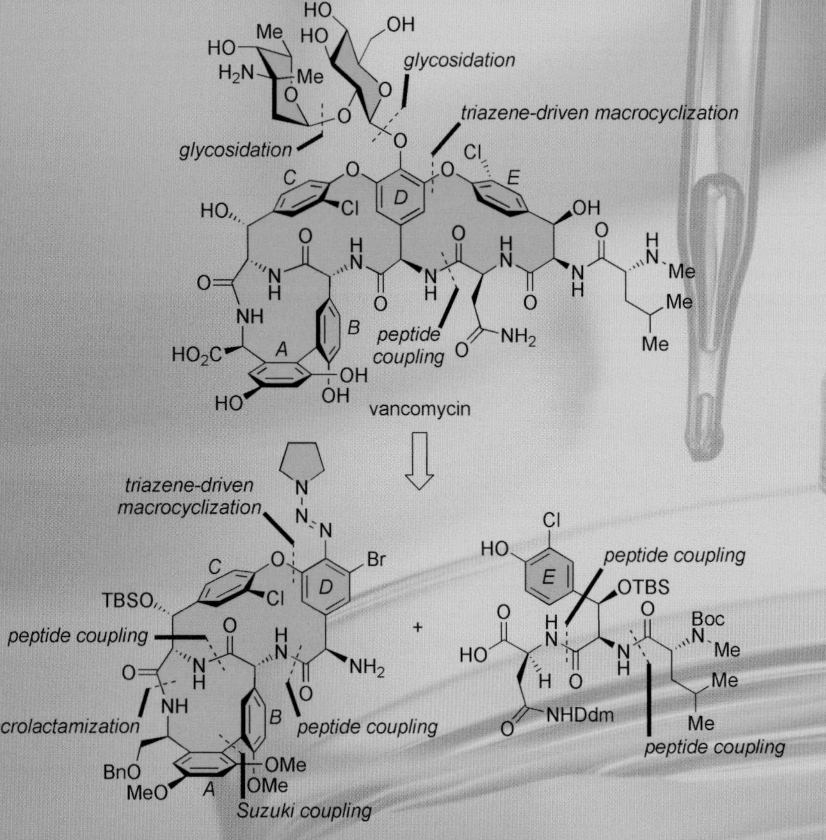

driven ether formation to give the vancomycin skeleton. Again, a functional group manipulation and deprotection sequence then provided the aglycon (Box 8). The Nicolaou team then appended the disaccharide through two sequential glycosidation reactions, thus completing the total synthesis of vancomycin. In 2000, the Boger group completed the total synthesis of the teicoplanin aglycon (see Box 1), a goal that was also reached by the Evans group, in 2001.

Synthetic chemistry did not stop with the conquest of the glycopeptide antibiotics by total synthesis. Beyond this feat lay innumerable new studies aimed at manipulating and reengineering vancomycin and related compounds in order to improve their potencies as antibacterial agents. Chemists even employed solid phase chemistry and combinatorial techniques (see Chapter 29) to synthesize vancomycin libraries for biological screening. Thus, scientists from Eli Lilly, inspired by the structure of teicoplanin which includes a lipophilic side chain, prepared a vancomycin derivative with a biaryl anchor on one of its sugars. The lipophilic chain is thought to localize the antibiotic to its site of action on the bacterial cell membrane. This compound, designated LY333328, exhibited superior activity *in vivo* against drug-resistant bacteria, highlighting possible new directions in our search for drugs effective against such superbugs.

Staphylococcus aureus bacteria

Other researchers focused on vancomycin dimers, the inspiration for such constructs coming from observations that hydrogen-bonded vancomycin dimers exist in solution. The Nicolaou group synthesized libraries of covalently bound vancomycin dimers, which allowed the discovery of compounds with powerful activity against drug-resistant strains.

X-Ray-derived model of the complex of vancomycin with diacetyl-L-Lys-D-Ala-D-Ala

Dimerization is thought to improve the activity of vancomycin in two ways. The first involves cooperative binding wherein, after the first molecule binds a peptidoglycan chain, the second vancomycin unit is held in the correct location to bind further peptides. The second mode of activity enhancement involves a so-called allosteric effect, which results from the hydrogen bonds that hold the dimer together. This hydrogen bonding polarizes the peptide bonds of the vancomycin framework in such a way as to increase their affinity for the peptidoglycan target. Interestingly, this effect also works in reverse such that, once a ligand is bound to vancomycin, it shows a greater propensity for dimer formation. The Nicolaou group made use of this feature in a target-accelerated combinatorial approach to the discovery of potent dimers. Thus, they used a model of the natural ligand (i.e. Ac_2-L-Lys-D-Ala-D-Ala) to aid the assembly of modified vancomycin units into dimers, which were then joined by a chemical reaction to form a permanent dimer. In this way, the dimers with the strongest affinity for the target are formed selectively. Recently, Boger has developed a reengineered vancomycin aglycon analogue, in which one of the amide carbonyl groups is replaced by a methylene (CH_2) group. This leads to a hundred-fold increase in affinity for the peptide chain found in VRE by eliminating the destabilizing interaction between the D-Ala-D-Lac ester and the vancomycin backbone (see Box 4). The modified structure is active against VRE strains, while retaining activity against vancomycin-susceptible strains.

Vancomycin is a stunning and intricate molecule and the chemistry developed to complete its total synthesis demonstrates once again the vibrancy of synthetic organic chemistry. Additionally, the chemical modification of the glycopeptide structures to increase their activity exemplifies the influence of organic chemistry on biology and medicine. However, we must remain vigilant and continue to discover and develop new antibacterial agents, for bacteria will always evolve to attain drug resistance; therefore, whether a decisive final victory over them can ever be found remains in doubt.

Ball and stick model of the molecule of vancomycin

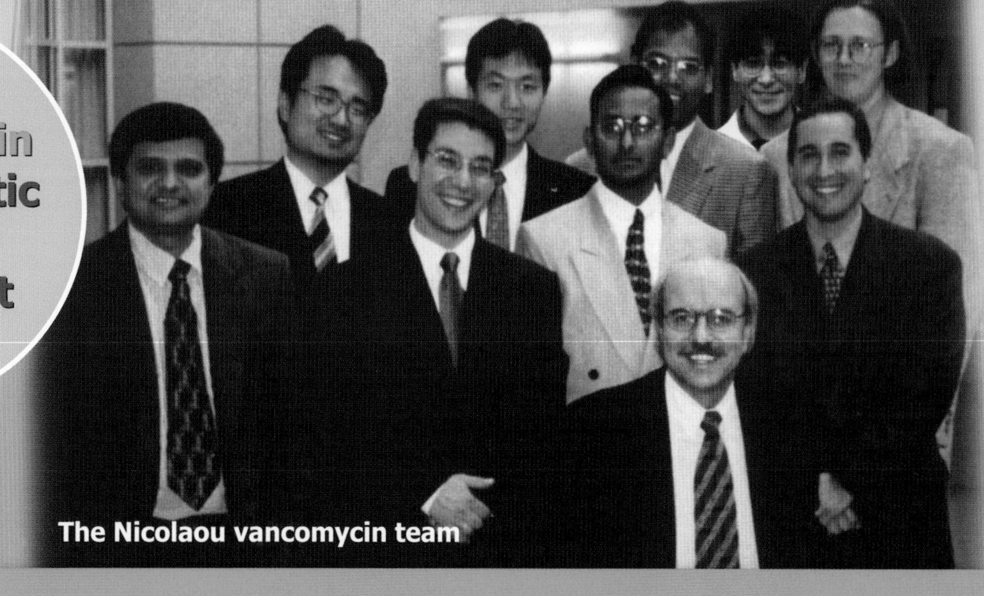
The Nicolaou vancomycin team

Box 8 **Highlights from the Nicolaou group's synthesis of vancomycin**

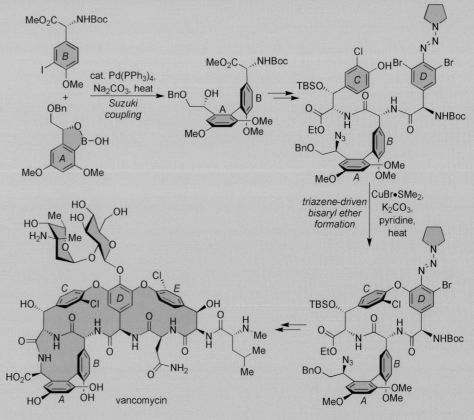

cat. $Pd(PPh_3)_4$, Na_2CO_3, heat
Suzuki coupling

triazene-driven bisaryl ether formation
$CuBr{\bullet}SMe_2$, K_2CO_3, pyridine, heat

vancomycin

Figure 1. The structure of swinholide A

swinholide A

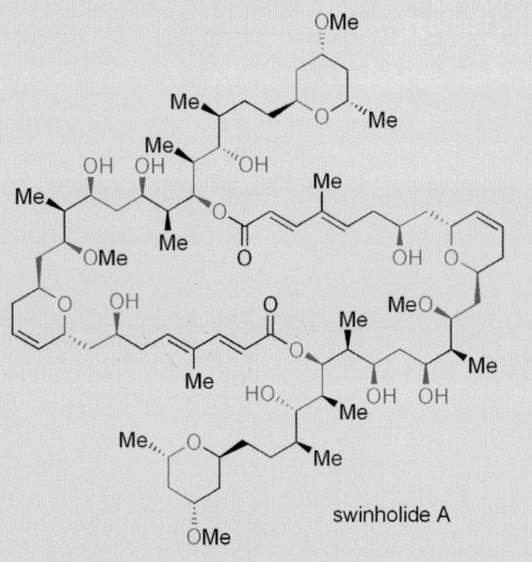

Illustrated above is the impressive molecular structure of swinholide A. This fascinating marine natural product was isolated from a Red Sea sponge and showed potential for the treatment of cancer. However, in this context, we are more concerned with its structure. The molecular architecture of swinholide A contains many repeats of an oxygen-carbon-carbon-carbon-oxygen unit, which are highlighted in red. This particular structural motif is very common in natural products, and is particularly evident in polyketides such as swinholide A. When chemists target this type of structure, they will often turn to the aldol reaction, which involves the combination of an en***ol*** or en***o-l***ate of a carbonyl compound, with an ***ald***ehyde or ketone. As shown in Figure 2, the aldol reaction gives a β-hydroxy carbonyl product (aldol moiety), which is a versatile building block for the preparation of several structural motifs and compound types. Indeed, examples of the aldol reaction can be found throughout this book, and many natural products are biosynthesized through a series of aldol reactions, with subsequent steps producing highly diverse molecular frameworks. For example, the aromatic natural product alternariol (Figure 2) is formed from a polyketide precursor through aldol and dehydration reactions.

Figure 2. Recognizing aldol products

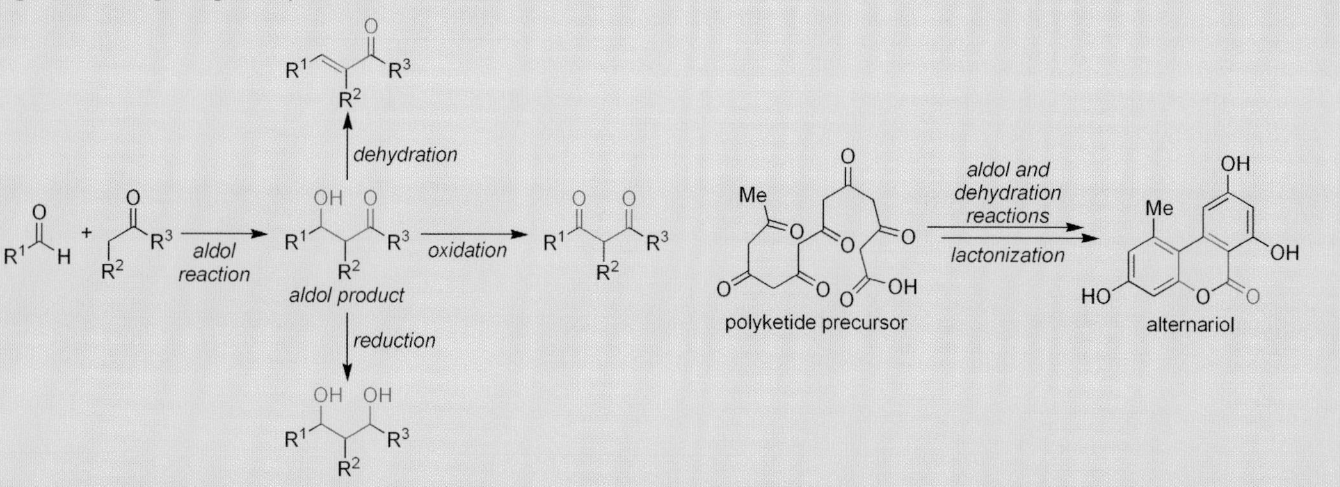

aldol reaction; aldol product; dehydration; oxidation; reduction

polyketide precursor — aldol and dehydration reactions, lactonization → alternariol

Figure 3. Stereochemical course of aldol reactions

Aldol reactions can be catalyzed by either acid or base, and a considerable body of research has transformed them into sharp tools for chemical synthesis. In addition to forming a new carbon–carbon bond, the aldol reaction can generate up to two new chiral centers. Modern techniques allow chemists to control the geometry of the products with very high selectivity.

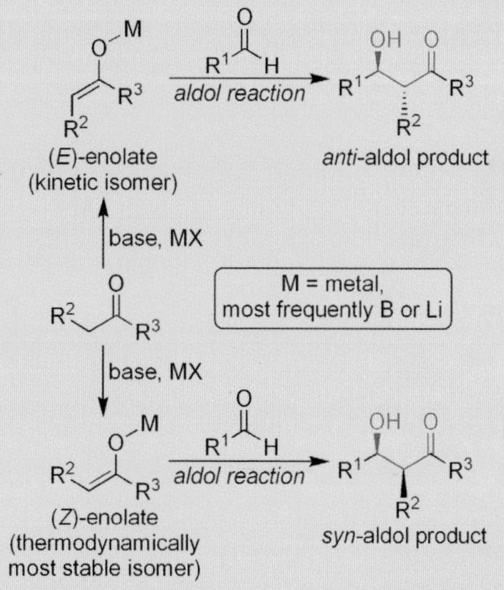

(*E*)-enolate (kinetic isomer) — aldol reaction → *anti*-aldol product

(*Z*)-enolate (thermodynamically most stable isomer) — aldol reaction → *syn*-aldol product

base, MX

M = metal, most frequently B or Li

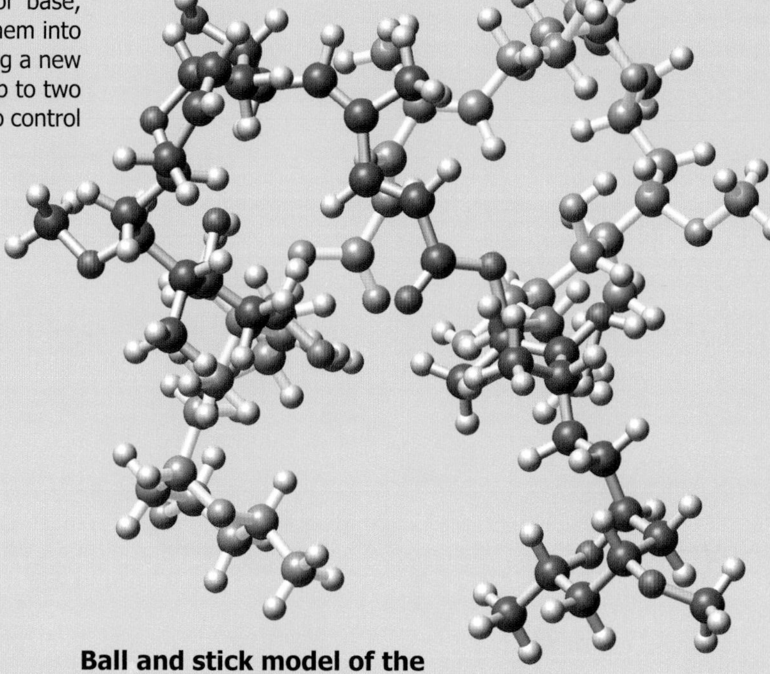

Ball and stick model of the molecule of swinholide A

Figure 4. Silyl enol ethers and lithium enolates in aldol reactions

$PhCHO$ + silyl enol ether ($OSiMe_3$) — $TiCl_4$, *Mukaiyama aldol reaction* → aldol product; cyclohexanone — LDA → Li enolate (OLi) — PhCHO → aldol product

Teruaki Mukaiyama (University of Tokyo, Japan) developed the Lewis acid-mediated aldol reaction of silyl enol ethers. The Mukaiyama aldol reaction, as this process has come to be known, is a very useful tool for organic synthesis, and finds new applications every day. Another pioneer in the field was Clayton H. Heathcock (University of California Berkeley, USA), who studied the behavior of lithium enolates in the aldol reaction. Heathcock's research led to greater understanding of the selectivity of these processes and his findings have found useful applications in the construction of complex molecules.

Figure 5. Use of one of David A. Evans' oxazolidinone auxiliaries in an aldol reaction

Satoru Masamune (Massachusetts Institute of Technology, USA) introduced the use of boron enolates as coupling partners in the aldol reaction. The predictability of the boron aldol reaction makes it one of the most versatile and reliable processes in organic chemistry. In the late 1970s and early 1980s, David A. Evans (Harvard University), a key protagonist in the story of vancomycin, began to make his mark on organic synthesis. Perhaps the most valuable of his many contributions is an asymmetric version of the boron aldol process, which is now known as the Evans aldol reaction. He developed a series of readily available chiral auxiliaries, based on the oxazolidinone framework, which give excellent selectivities when used in aldol reactions, as well as a host of other transformations. He and many other synthetic chemists have employed these auxiliaries in the total synthesis of countless natural products and designed molecules. The example below is taken from Evans' total synthesis of another polyketide, roxaticin.

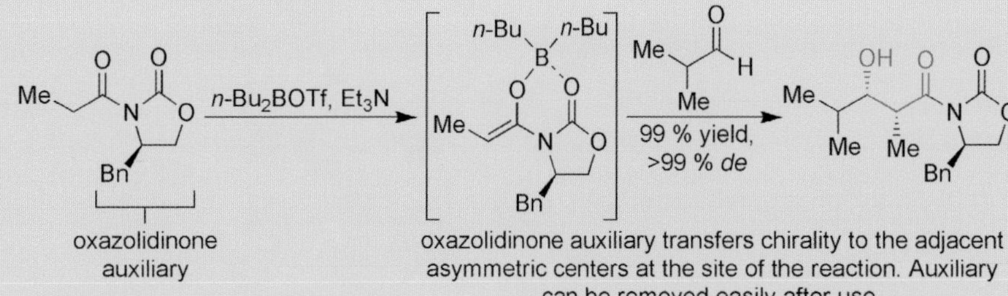

The striking molecule of swinholide A with which we began this box brings us to another pioneer of aldol chemistry. In 1994, Ian Paterson and his group (University of Cambridge, UK) reported their total synthesis of swinholide A, in which they made significant use of boron aldol reactions. Paterson's studies in boron-mediated aldol reactions have led to a number of important developments which he and others have exploited cleverly in many challenging total syntheses. Swinholide A was later conquered again, in 1996, this time by the Nicolaou group, who also used several aldol reactions during their synthesis.

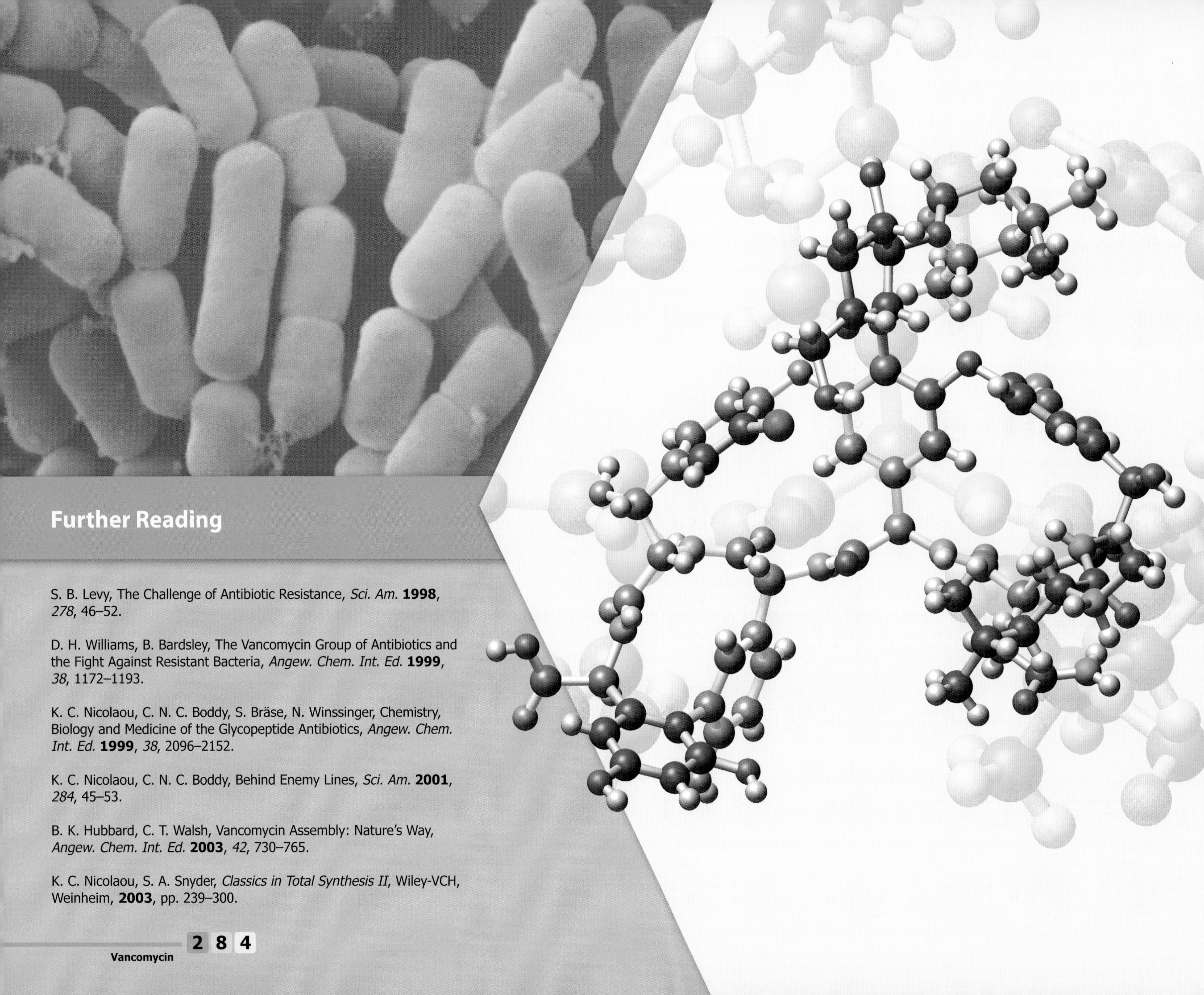

Further Reading

S. B. Levy, The Challenge of Antibiotic Resistance, *Sci. Am.* **1998**, *278*, 46–52.

D. H. Williams, B. Bardsley, The Vancomycin Group of Antibiotics and the Fight Against Resistant Bacteria, *Angew. Chem. Int. Ed.* **1999**, *38*, 1172–1193.

K. C. Nicolaou, C. N. C. Boddy, S. Bräse, N. Winssinger, Chemistry, Biology and Medicine of the Glycopeptide Antibiotics, *Angew. Chem. Int. Ed.* **1999**, *38*, 2096–2152.

K. C. Nicolaou, C. N. C. Boddy, Behind Enemy Lines, *Sci. Am.* **2001**, *284*, 45–53.

B. K. Hubbard, C. T. Walsh, Vancomycin Assembly: Nature's Way, *Angew. Chem. Int. Ed.* **2003**, *42*, 730–765.

K. C. Nicolaou, S. A. Snyder, *Classics in Total Synthesis II*, Wiley-VCH, Weinheim, **2003**, pp. 239–300.

Thiostrepton
Thiostrepton
Chapter 32
2004

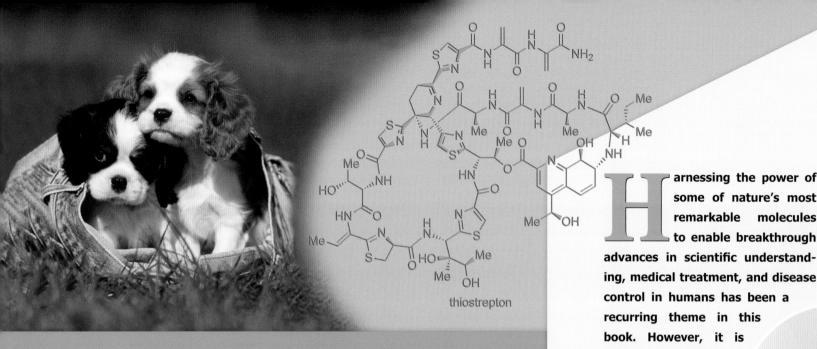

Box 1 **The constituents of Panolog® and Animax®, veterinary medicines for dermatologic disorders in cats and dogs**

thiostrepton (antibacterial)

neomycin B - major component of neomycin antibiotic complex (antibacterial)

nystatin A_1 (antifungal)

triamcinolone acetonide (corticosteroid, anti-inflammatory)

Harnessing the power of some of nature's most remarkable molecules to enable breakthrough advances in scientific understanding, medical treatment, and disease control in humans has been a recurring theme in this book. However, it is not only the health of the human population which has benefited from these processes, as impressive strides have also been made in veterinary science and medicine. Many of the same life saving or enhancing treatments developed for humans, such as insulin injections, hip replacements, and even pacemakers, are now available to animals, and therapeutic drugs are no exception in this regard. Many of the pharmaceuticals used by people are employed to treat the corresponding ailments in animals, although, of course, the vast array of physiologies encountered in the latter means that there are also many more species-specific medicines.

A problem commonly encountered in veterinary medicine is one of precise diagnosis (or, often more accurately, the relative cost of establishing a correct diagnosis). Therefore, a number of prescribed treatments for animals are rather all-encompassing in that they target several different possible causes of the ailment at once. Examples include Panolog® and Animax®, used as topical treatments to manage dermatological disorders in cats and dogs. Each of these medications contains no fewer than four active ingredients (see Box 1). Nystatin A_1 is an antifungal agent active against *Candida albicans*, whilst triamcinolone acetonide is a semi-synthetic corticosteroid which reduces inflammation and pruritis. The last two components combine to provide a comprehensive therapy against bacterial infection; they are neomycin sulfate and thiostrepton, the latter molecule being the focus of this chapter.

Thiostrepton is the flagship member of a remarkable class of natural products known as the thiopeptide antibiotics. As this name might suggest, this family of compounds, which numbers in the dozens, is comprised of a variety of sulfur-rich peptide-based structures, the majority of which exhibit activity against Gram-posi-

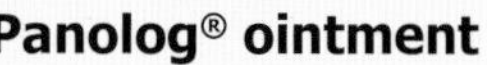

Panolog® ointment

tive bacteria. Besides thiostrepton, a number of other representative members of the thiopeptide antibiotics are illustrated in Box 2. A common structural feature among them is the presence of a polysubstituted nitrogen heterocycle (e.g. a dehydropiperidine or pyridine ring system), onto which is appended a variety of modified amino acid residues, themselves contained within a larger macrocyclic framework. These modified residues may take the form of thiazole or oxazole rings, dehydroamino acid units, or one of several other structural motifs resulting from enzymatic post-translational modification of the parent amino acid residue.

Despite being one of the first members of this class of compounds to be discovered, more than half a century ago, thiostrepton remains the jewel in the crown of the thiopeptide antibiotics. It is the most structurally complex and most extensively studied, and the only one yet to have reached commercial application as an antibacterial agent. The story of thiostrepton begins in 1954 in the desert of New Mexico, USA, with the collection of a soil sample by a team of scientists from the Squibb Institute for Medical Research, New Jersey, USA (now part of the Bristol-Myers Squibb company). Inhabiting this innocuous pile of dirt was a strain of bacteria (*Streptomyces azureus*) which the scientists cultivated in the laboratory. Subsequent extraction of these cultures resulted in the isolation of a novel substance, named thiostrepton, that displayed potent antibiotic activity against a variety of Gram-positive bacteria, in particular against a strain resistant to penicillin, the leading antibiotic of the era. However, hamstrung by the absence of the high resolution spectroscopic techniques that chemists today take for granted, the Squibb Institute team was unable to speculate as to the structure of this mysterious and promising new antibiotic, other than to identify its polypeptidic nature and perform a rudimentary analysis of its physicochemical properties. The latter included the interesting observations that the antibacterial activity of thiostrepton was "... not appreciably decreased in twenty-four hours by incubation at room temperature in the presence of artificial intestinal juice, artificial gastric juice, ten per cent solution of human feces, or in water."

An arduous, and at times rather confusing, international quest to unravel the mystery regarding the structure of thiostrepton unfolded over subsequent years. Painstaking classical degradative experiments, carried out largely by the groups of Miklos Bodanszky (Squibb Institute) and George W. Kenner (University of Liverpool, UK), revealed the nature of several of the amino acid components, but left unanswered the question of their connectivity. Concurrently, several varying estimates of the molecular weight and empirical formula of thiostrepton were made by other groups. Matters were complicated further by the isolation, in 1955, of a purportedly novel thiopeptide antibiotic, ini-

Box 2 Selected examples of thiopeptide antibiotics

berninamycin C

promothiocin A

nocathiacin I

Box 3 The molecular structure of thiostrepton

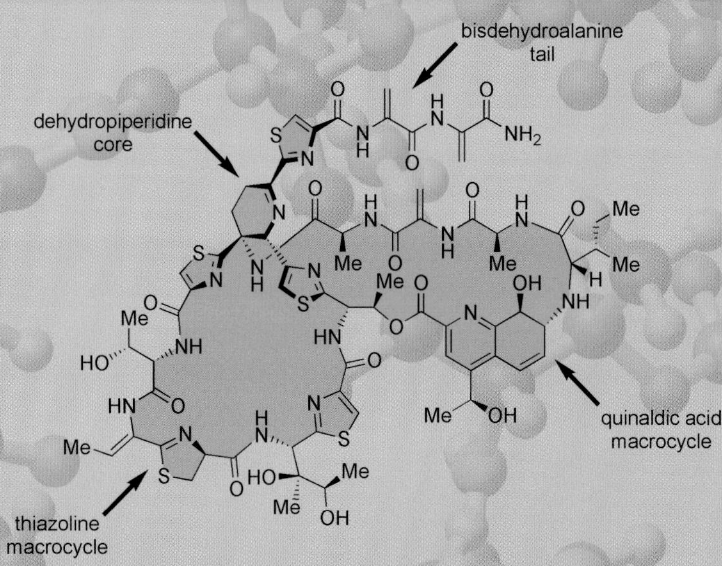

Ball and stick model of the molecule of thiostrepton

tially given the trade name bryamycin but later renamed thiactin, by Bernard Heinemann and co-workers (Bristol Laboratories, USA) from cultured extracts of soil bacteria (*Streptomyces hawaiiensis*) collected in Hawaii. This substance was in fact shown to be identical to thiostrepton by the Bodanszky group in 1963, but the somewhat unfortunate use of three different names (thiostrepton, bryamycin or thiactin) to describe the same compound persists to this day. As events transpired, the major breakthrough in dispelling the clouds of uncertainty surrounding the structure of thiostrepton would not arrive until 1970, when Dorothy Crowfoot Hodgkin (of penicillin and vitamin B_{12} fame, see Chapters 13 and 16, respectively) and co-workers at the University of Oxford published their x-ray crystallographic structure of thiostrepton. This study established the stereochemistry and atom connectivity of most, but, significantly, not all of the molecule. Although x-ray crystallography is generally considered to be the final arbiter of structure determination, exceptions can occur, as in this case where disorder in the region containing the side chain prevented the unambiguous assignment of that domain of the molecule. In 1976, ^{13}C-NMR (Carbon-13 Nuclear Magnetic Resonance) studies by Kazuo Tori and co-workers at the Shionogi Research Laboratory in Osaka, Japan, identified the side chain as being comprised of two dehydroalanine units. Thus, more than twenty years after its isolation, the complete structure of thiostrepton was finally revealed (Box 3). Confirmation of this assignment has since been provided by further NMR studies, x-ray analysis on an alternative crystal polymorph, and as we shall describe later in this chapter, total synthesis.

Bernard Weisblum

As shown in Box 3, the extraordinary polycyclic architecture of thiostrepton can be divided into four domains, namely the twenty-six-membered thiazoline macrocycle, the twenty-seven-membered quinaldic acid macrocycle, the dehydropiperidine core, and the bisdehydroalanine tail. The first three domains form a roughly globular structure, with one macrocycle folded over the top of the other. This compact conformation is stabilized by numerous hydrogen bonds, including several contacts to a water molecule which, although not chemically bound to thiostrepton, is held within a hydrophilic cleft in the interior of its structure.

The year 1970 proved to be seminal in the history of thiostrepton, for not only was the first x-ray crystal structure reported, but also the first indications as to the molecular basis of its antibiotic activity were revealed. A study by

Dorothy Crowfoot Hodgkin

Bernard Weisblum and Veronika Demohn at the University of Wisconsin Medical School (USA) demonstrated that thiostrepton inhibits bacterial protein biosynthesis by blocking the function of the 50S ribosome subunit, triggering a sequence of events which ultimately lead to bacterial cell death. A later study by Yanyan Xing and David E. Draper (Johns Hopkins University, Maryland, USA) in 1996 more accurately pinpointed the binding site of thiostrepton on the bacterial ribosome as being at a region known as the L11 binding domain.

Roger D. Kornberg

The majority of known antibiotics function by targeting and inhibiting the bacterial cell machinery required for one of the four following processes: (1) cell wall biosynthesis, (2) folic acid biosynthesis (a precursor to DNA/RNA), (3) inhibition of DNA/RNA replication or repair, and (4) inhibition of protein biosynthesis on the ribosome. The former two pathways have no parallels in eukaryotic cells, whereas the latter two clearly occur in both prokaryotic (i.e. bacterial) and eukaryotic (e.g. animal, human) cells. Fortuitously, however, there are sufficient structural differences between the enzymes required for protein and DNA biosynthesis in prokaryotic and eukaryotic cells such that selective inhibition is possible.

The expression of the genetic information encoded within DNA occurs (in the broadest possible terms) in two stages: transcription, in which messenger RNA (mRNA) is synthesized from a DNA template, and translation, in which the mRNA serves as the template for protein synthesis. In prokaryotic organisms, these two events are closely coupled, while in eukaryotic cells these events are separated in both space and time; transcription occurs in the nucleus while translation occurs in the cytosol. In eukaryotic cells, the enzyme RNA polymerase II transcribes DNA to synthesize precursors of mRNA (in addition to a number of non-coding RNA molecules). In 2001, Roger D. Kornberg and his group at Stanford University (USA) reported an x-ray crystallographic snapshot of an actively transcribing complex of RNA polymerase II in complex with DNA (Box 4). The significance of this, and other work from the Kornberg laboratories, in enhancing our understanding of the structural basis of transcription was recognized with the awarding of the 2006 Nobel Prize in Chemistry to Kornberg "for his studies of the molecular basis of eukaryotic transcription."

The ribosome is the site of protein biosynthesis in the cell, functioning to translate the genetic information encoded within mRNA to assemble amino acids into polypeptides (Box 5), and is the largest and most complex of all the enzymes. All ribosomes display a high degree of structural and functional homology. Nevertheless, several key differences between prokaryotic and eukaryotic ribosomes enable selectivity in antibiotic action. It is these structural differences that result in thiostrepton inhibiting bacterial protein biosynthesis over one hundred

Box 4 X-Ray crystallographic structure of the yeast RNA polymerase II transcribing complex

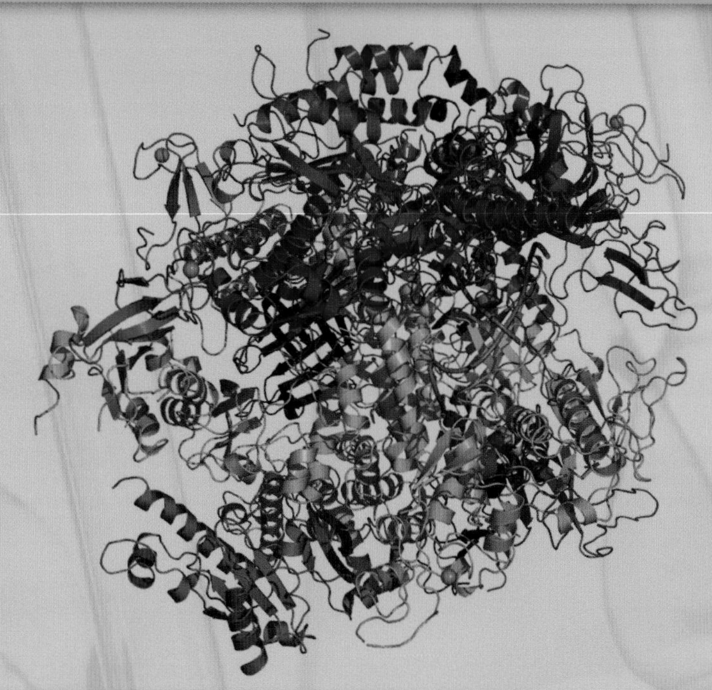

Box 5 Protein synthesis on the bacterial ribosome

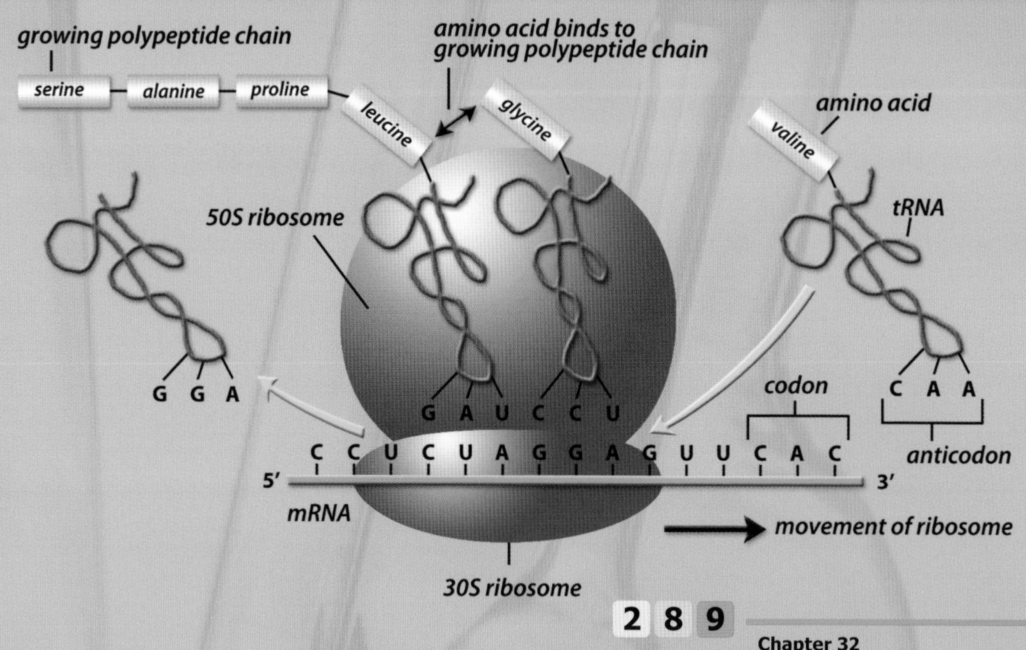

Box 6 Selected drugs that target the bacterial ribosome

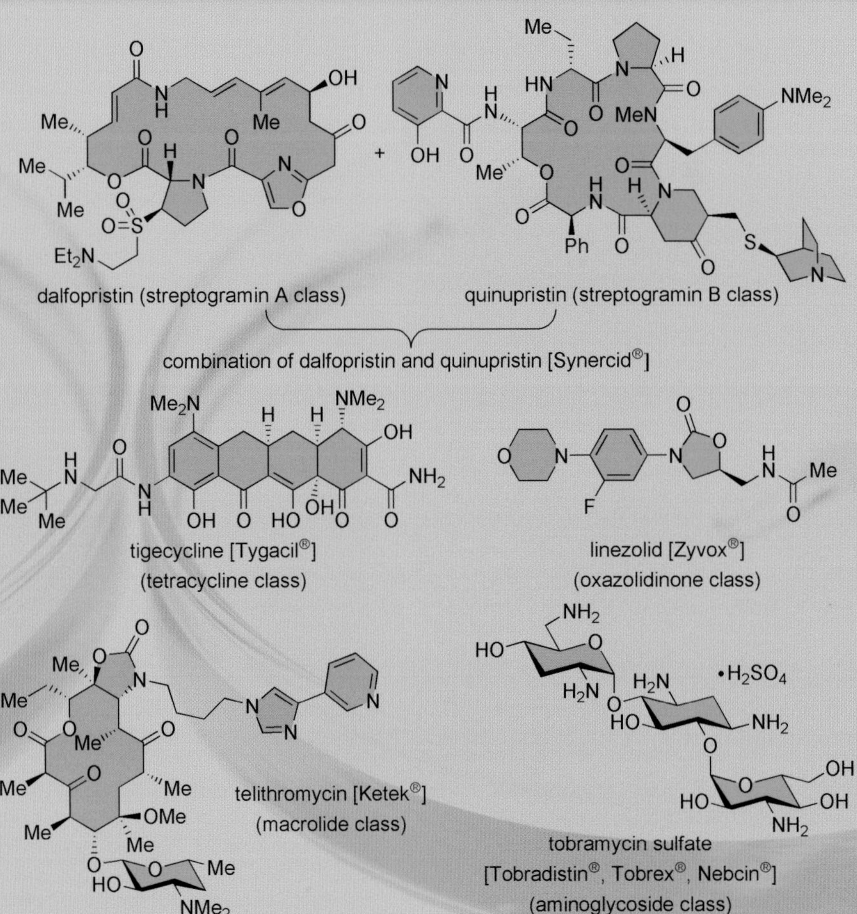

dalfopristin (streptogramin A class) + quinupristin (streptogramin B class)

combination of dalfopristin and quinupristin [Synercid®]

tigecycline [Tygacil®] (tetracycline class)

linezolid [Zyvox®] (oxazolidinone class)

telithromycin [Ketek®] (macrolide class)

tobramycin sulfate [Tobradistin®, Tobrex®, Nebcin®] (aminoglycoside class)

Methicillin-resistant *Staphylococcus aureus*

times more potently than eukaryotic protein biosynthesis. Bacterial ribosomes, which are about 30 % smaller than in eukaryotes, are composed of two subunits of unequal size, known as the 30S and 50S subunits. These subunits are roughly two-thirds ribosomal RNA (rRNA) and one-third protein by weight, and associate through noncovalent interactions. The larger 50S subunit is itself a conglomerate of approximately thirty individual proteins and two strands of rRNA, the latter known as the 23S and 5S strands. Binding of thiostrepton to a specific site of the 50S subunit composed of protein L11 and 23S rRNA (the L11 binding domain) prevents conformational changes of the ribosome required to drive movement of the ribosome along the mRNA. Protein biosynthesis is thus stalled, ultimately leading to the death of the bacterial cell.

Although thiostrepton has been a mainstay topical antibiotic in veterinary medicine for many years, its application in humans has been limited by its low solubility in aqueous systems and its poor bioavailability (the proportion of an administered dose of medication that reaches the systemic circulation) due to low absorption from the gastrointestinal tract. These factors result in the antibiotic inducing resistance to itself in the proliferating bacteria before its concentration becomes therapeutically significant. There are, however, a number of clinically important classes of antibiotics relevant to human health which act by inhibiting various steps of protein biosynthesis on the bacterial ribosome (Box 6). A 7:3 mixture of the macrolactone dalfopristin and the depsipeptide quinupristin (semi-synthetic derivatives, with improved water solubility, of naturally occurring streptogramins A and B, respectively), marketed as Synercid®, was granted accelerated approval by the Food and Drug Administration, USA, in 1999 for the treatment of serious or life threatening infections associated with vancomycin-resistant *Enterococcus faecium* (VRE) bacteremia. The two components act synergistically to inhibit bacterial protein biosynthesis, the combination proving more effective than either component individually. Macrolides, such as erythromycin A (Chapter 17) and its analogues, form arguably the major class of ribosome-targeted antibiotics. Telithromycin (Ketek®, Sanofi Aventis) is one of the most recent semi-synthetic erythromycin derivatives approved for use in humans, and is representative of the latest generation of macrolide antibiotics that have been optimized for potency, stability to gastric acids in the stomach, and activity against macrolide-resistant pathogens.

The aminoglycosides, which can trace their esteemed lineage back to the discovery of streptomycin in 1944, are potent drugs against Gram-negative bacteria, tobramycin sulfate being one of the most prominent members in contemporary clinical use. The tetracy-

Enterococcus faecium

Box 7 Retrosynthetic analysis of thiostrepton

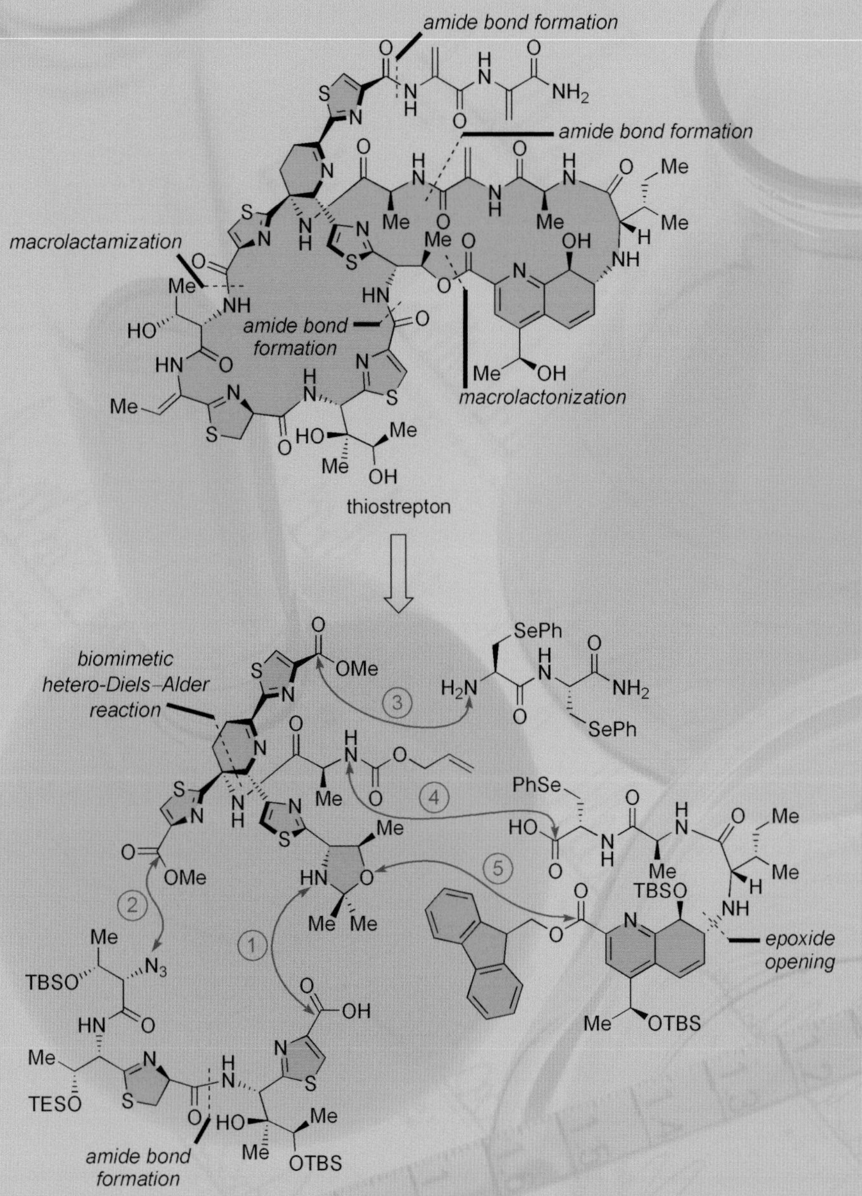

cline class of antibiotics has suffered a diminishing role as a front line therapy in recent decades due to emerging bacterial resistance, but this trend may start to be reversed following the fast track approval by the FDA of tigecycline (Tygacil®, Wyeth) in 2005. Substitution of the aromatic ring with a glycine amide functionality has been found to restore activity against tetracycline-resistant *Escherichia coli* and *Staphylococcus aureus*, as well as against methicillin-resistant *Staphylococcus aureus* (MRSA). Linezolid (Zyvox®, Pfizer) is the first, and to date only, member of the oxazolidinone class of antibiotics to be approved for use in humans, and is also currently the only totally synthetic (as opposed to being derived from or inspired by a natural product) antibiotic in clinical use. Its use is usually reserved for serious Gram-positive bacterial infections where older antibiotics have failed due to bacterial resistance. Linezolid is also novel in that it inhibits the initiation of bacterial protein biosynthesis, a stage of protein synthesis previously unexplored as a target, by binding to the 50S subunit and preventing its association with the 30S subunit and mRNA. In addition to the small selection described here, a number of other new antibiotics, which target a variety of aspects of the bacterial cell machinery, are currently in late-stage clinical development. Thus, the outlook for combating the spread of bacterial resistance, and the concurrent perceived lack of new antibiotics, is not as bleak as the periodic sensationalism from the media and other sources might have one believe.

Christopher J. Moody

The structural complexity and biological properties of the thiopeptide antibiotics have made them attractive targets for many synthetic practitioners keen to probe and expand the boundaries of their field. Christopher J. Moody and his group (then at the University of Exeter, now at the University of Nottingham, UK) reported the landmark first total synthesis of a member of the thiopeptide family in 2000, namely that of promothiocin A (Box 2). It was also about this time that Nicolaou and co-workers at The Scripps Research Institute (La Jolla, California, USA) began their march towards the total synthesis of thiostrepton, an endeavor which forms the basis for the rest of this discussion.

The retrosynthetic analysis for what would ultimately be the successful strategy towards thiostrepton is illustrated in Box 7, in which the molecule was dissected into smaller fragments, corresponding to each of the domains of the target. Also illustrated is the order of coupling of these fragments, namely generation of the thiazoline macrocycle, followed by attachment of a masked form of the dehydroalanine tail, and finally the appendage of the quinaldic acid macrocycle. It should be noted that this order of bond construction was

Box 8 Proposed biogenesis of the dehydropiperidine core of thiostrepton

Box 9 Biomimetic approach to the dehydropiperidine core of thiostrepton

not entirely arbitrary, but, rather, arose from a series of systematic reevaluations of the synthetic route, which, in turn, resulted from encountering numerous pitfalls and roadblocks on early paths to the coveted prize.

Of all the prominent structural features encompassed within thiostrepton, it is undoubtedly the dehydropiperidine core, which lies at the heart of the molecule, that is the most striking from a synthetic perspective. One of the most pressing problems facing the Nicolaou group was, therefore, to find an efficient (and, ideally, elegant) method to forge this sector of the target. Fortunately, a combination of clues provided by nature and insightful studies by a number of researchers appeared to offer a solution. In 1978, Barrie W. Bycroft and Maxim S. Gowland at the University of Nottingham suggested that the dehydropiperidine core of thiostrepton may be derived biogenetically "from the interaction of two dehydroalanine units in a single peptide chain," although no mention of the nature of this interaction was put forward. This proposal was further refined into the scheme depicted in Box 8 by Heinz G. Floss and co-workers (University of Washington, USA) during the course of isotopic labeling experiments on several members of the thiopeptide family aimed at probing their biosynthesis. The cyclization event in this scheme represents a formal Diels–Alder cycloaddition, albeit with slightly esoteric reaction partners. With this hypothesis as both their inspiration and guide, Nicolaou and co-workers investigated whether this type of process could be replicated in the laboratory. Indeed, they found that under the appropriate conditions a thiazolidine precursor could be coaxed into following the reaction cascade illustrated in Box 9, via the fleeting intermediacy of the aza-diene intermediate shown, to furnish the desired dehydropiperidine core structure.

The final stages in the total synthesis of thiostrepton are shown in Box 10. Following several agonizing near misses, it was found necessary to install the reactive dehydroalanine moieties once all the ring systems had been constructed. Formation of the dehydroalanines was achieved through the selective oxidation of the three selenium atoms, followed by the spontaneous elimination of the resulting selenoxide species. The final step involved removal of the silyl protecting groups under suitable conditions, an event which was accompanied, gratifyingly, by the selective dehydration of just one of the hydroxyl groups, to install the final piece in the puzzle, namely the trisubstituted (*Z*)-olefin. Synthetic thiostrepton was in hand on June 29, 2004,

Barrie W. Bycroft

Heinz G. Floss

Box 10 Completion of the total synthesis of thiostrepton by the Nicolaou team

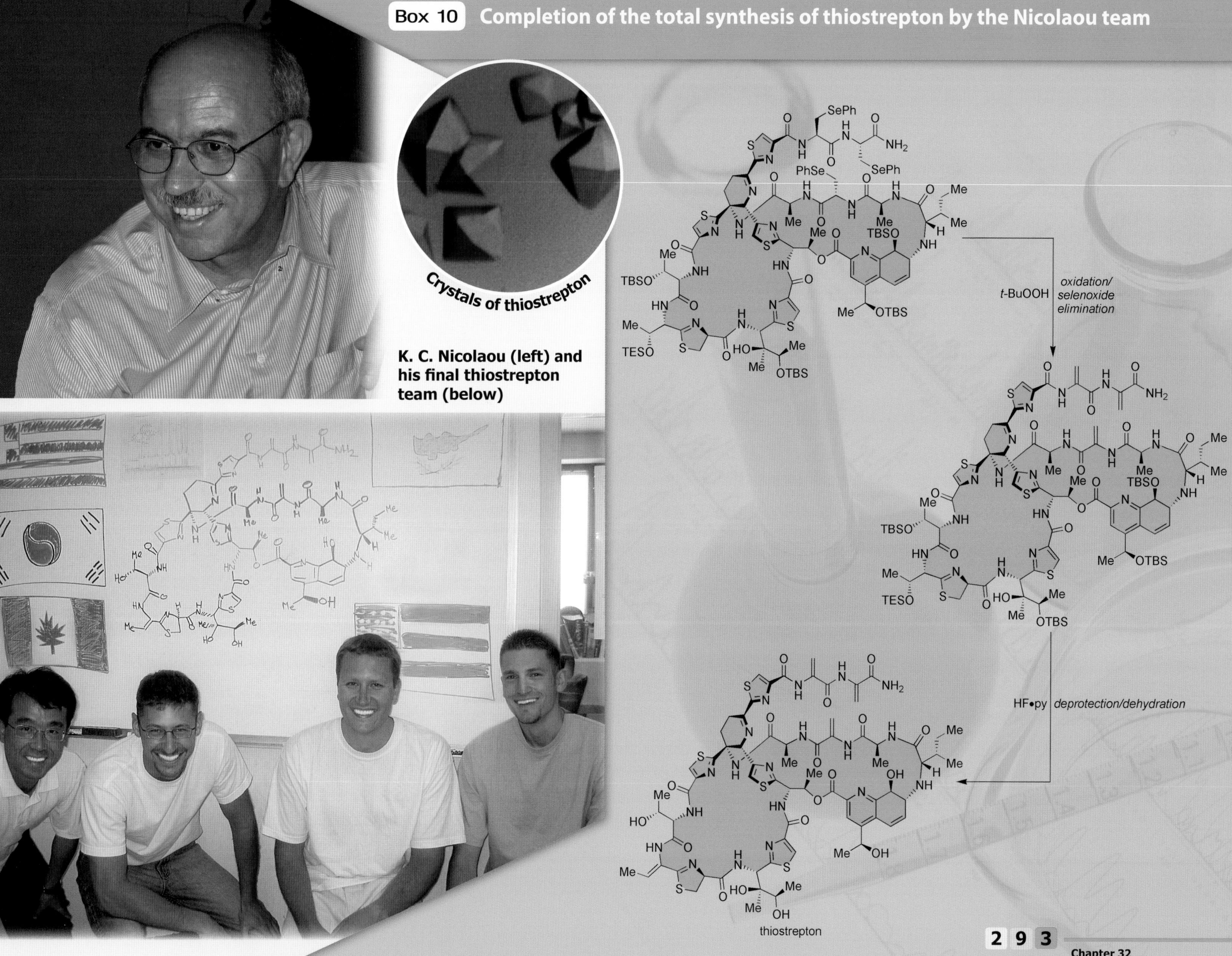

K. C. Nicolaou (left) and his final thiostrepton team (below)

delightfully ending a five-year synthetic odyssey.

The total synthesis of thiostrepton is representative of the state of the art of chemical synthesis as this creative discipline moves into the new millennium. It also serves to illustrate just how far the field has advanced since its formative years in the nineteenth century. Much of this advancement has come within the last few decades; thirty years ago a synthetic target such as thiostrepton would have been considered unattainable, yet today it lies conquered at the hands of synthetic chemists, a feat only made possible by gains in our understanding of fundamental concepts such as mechanism, reactivity, spectroscopy, and biosynthetic pathways, to name but a few. It is with this knowledge that chemistry, and science as a whole, can be driven forward, to address the challenges and opportunities facing humanity in the twenty-first century and beyond.

Further Reading

M. C. Bagley, J. W. Dale, E. A. Merritt, X. Xiong, Thiopeptide Antibiotics, *Chem. Rev.* **2005**, *105*, 685–714.

C. J. Walsh, *Antibiotics: Actions, Origins, Resistance*, ASM Press, Washington, **2003**.

M. C. Bagley, K. E. Bashford, C. L. Hesketh, C. J. Moody, Total Synthesis of Promothiocin A, *J. Am. Chem. Soc.* **2000**, *122*, 3301–3313.

K. C. Nicolaou, B. S. Safina, M. Zak, S. H. Lee, M. Nevalainen, M. Bella, A. A. Estrada, C. Funke, F. J. Zécri, S. Bulat, Total Synthesis of Thiostrepton. Retrosynthetic Analysis and Construction of Key Building Blocks, *J. Am. Chem. Soc.* **2005**, *127*, 11159–11175.

K. C. Nicolaou, M. Zak, B. S. Safina, A. A. Estrada, S. H. Lee, M. Nevalainen, Total Synthesis of Thiostrepton. Assembly of Key Building Blocks and Completion of the Synthesis, *J. Am. Chem. Soc.* **2005**, *127*, 11176–11183.

Small Molecule Drugs

Chapter 33

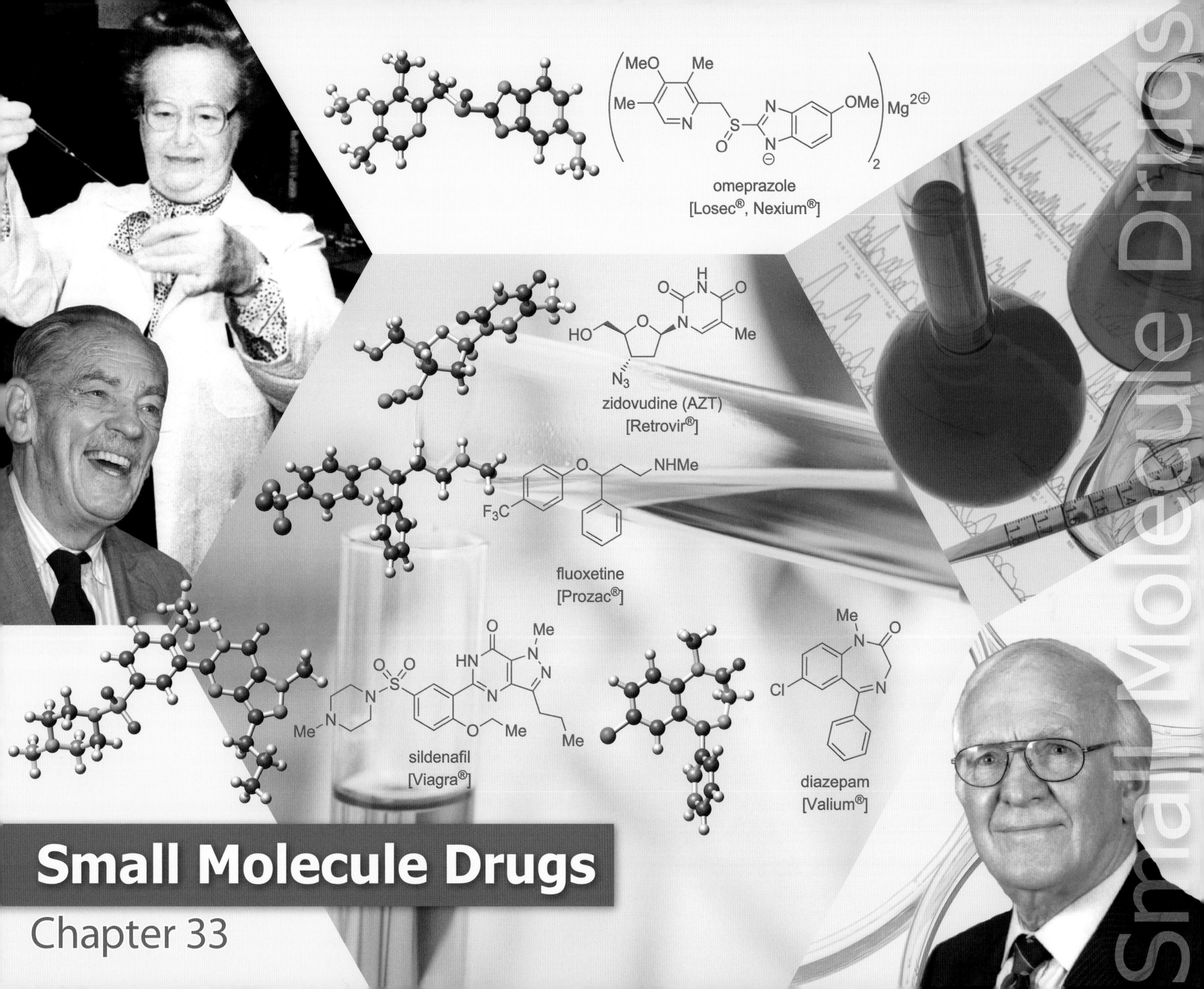

fluoxetine
[Prozac®]

omeprazole
[Losec®, Nexium®]

diazepam
[Valium®]

cimetidine
[Tagamet®]

sildenafil
[Viagra®]

zidovudine (AZT)
[Retrovir®]

propranolol
[Inderal®]

Box 1 The drug discovery and development process

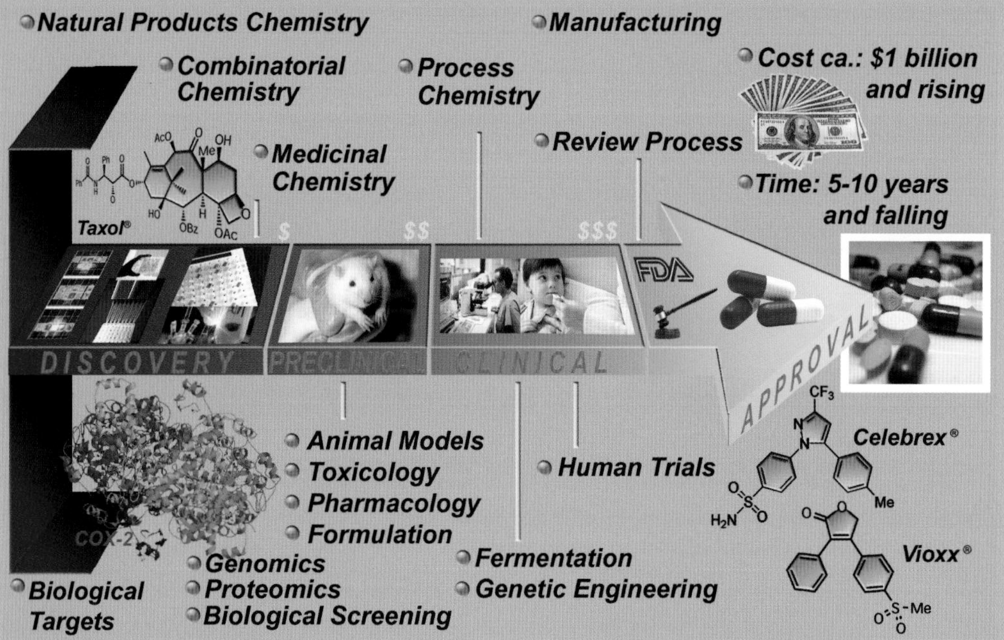

In this chapter, we will focus not on a story of a natural product, but rather on some important classes of designed small molecule drugs. As in so many other fields, chemical synthesis plays a vital role in drug discovery and development. We will begin with a brief overview of the modern drug discovery and development process, highlighting the different roles played by chemists, and then relate the stories of various medications to treat mental illnesses, viral infections, gastrointestinal disorders, heart conditions, and sexual dysfunction.

Modern Drug Discovery and Development

While the practice of drug discovery and development in the modern pharmaceutical industry may vary from company to company, it follows the general pattern described below, consisting of five distinct phases. The overall process is challenging, time-consuming, and very expensive (Box 1).

Determination of the disease pathogenesis and development of an assay. Before chemists become involved, the drug discovery process begins with biologists and physicians, who must unravel the complex pathways that lead to a specific disease. Ideally, this research will identify a particular target such as an enzyme or a receptor, modulation of which is expected to cure or manage the disease. Biochemists must then develop an assay to determine the effects of small molecules on the target. The complexity of these assays can vary dramatically, from simple biochemical tests, to cultures of viruses, bacteria, or tumor cells. Simpler assays are easier to perform and can be used to screen hundreds of thousands of compounds quickly; more complex systems may give a more reliable indication of the efficacy of a compound under physiological conditions. Validation of the biological target must be carried out as early as possible in the development process. This involves confirming that interference with the selected biological target will indeed result in benefits to the patient.

Finding a lead compound. With an assay available, chemists can begin to search for compounds as potential leads for further refinement. Such preliminary screening often looks for a single property in the compounds tested, selecting, for example, only those compounds that exhibit significant activity in the biological assay. As we have seen in the preceding chapters, identifying leads from nature can be highly rewarding, and some believe natural products offer the best opportunities for drug discovery, although isolating and determining their structures can be a laborious and expensive process. Alternatively, chemists can design and synthesize any number of compounds, often with the help of robotics and combinatorial techniques (see Chapter 29). This latter strategy has the advantage that the structures of the molecules are already well-defined, although the compounds prepared in this manner are often of limited structural complexity. Over the last few decades, pharmaceutical companies, as well as specialist suppliers, have amassed vast libraries of compounds which are screened to provide leads for developing new drugs.

Optimizing a lead compound. Once a suitable lead compound has been identified, chemists can begin refining its chemical structure to improve both its activity in the assay and its physicochemical properties in a process called rational drug design. The relationship between the structure of the molecule and its biological activity is explored by designing, synthesizing, and testing a series of analogues. The resulting structure-activity relationships (SARs) are then used to design the next generation of compounds in an iterative process. Computer modeling can be useful at this stage, particularly if structural information about the biological target is available (e.g. from x-ray crystallography). X-Ray snapshots of ligands bound to their targets are invaluable to the drug designer as they reveal the interactions between the compound and its receptor at the molecular level. Those regions of the molecular structure found to be essential to the binding and biological activity of the compound are identified and optimized, while other regions can be fine-tuned to give the most desirable physical and chemical properties. These properties include factors such as bioavailability, solubility and stability, towards chemical and biological degradation and other types of metabolism *in vivo*. These features are as important to the success of the drug as the biological activity. For example, solubility is essential for proper absorption and distribution of the drug within the body and, if a drug is to be administered orally, it should ideally be stable to the acidic environment of the stomach. If ideal physical properties cannot be installed in the drug candidate, considerable effort is required at the formulation stage to address this problem.

Developing a drug. Once a drug candidate has been identified, it progresses to the next stage of development. From this point on, considerably larger quantities of material are required. Initially, batches of up to a few kilograms must be prepared for use in further tests, such as toxicology

•Chemistry
•Biology
•Medicine

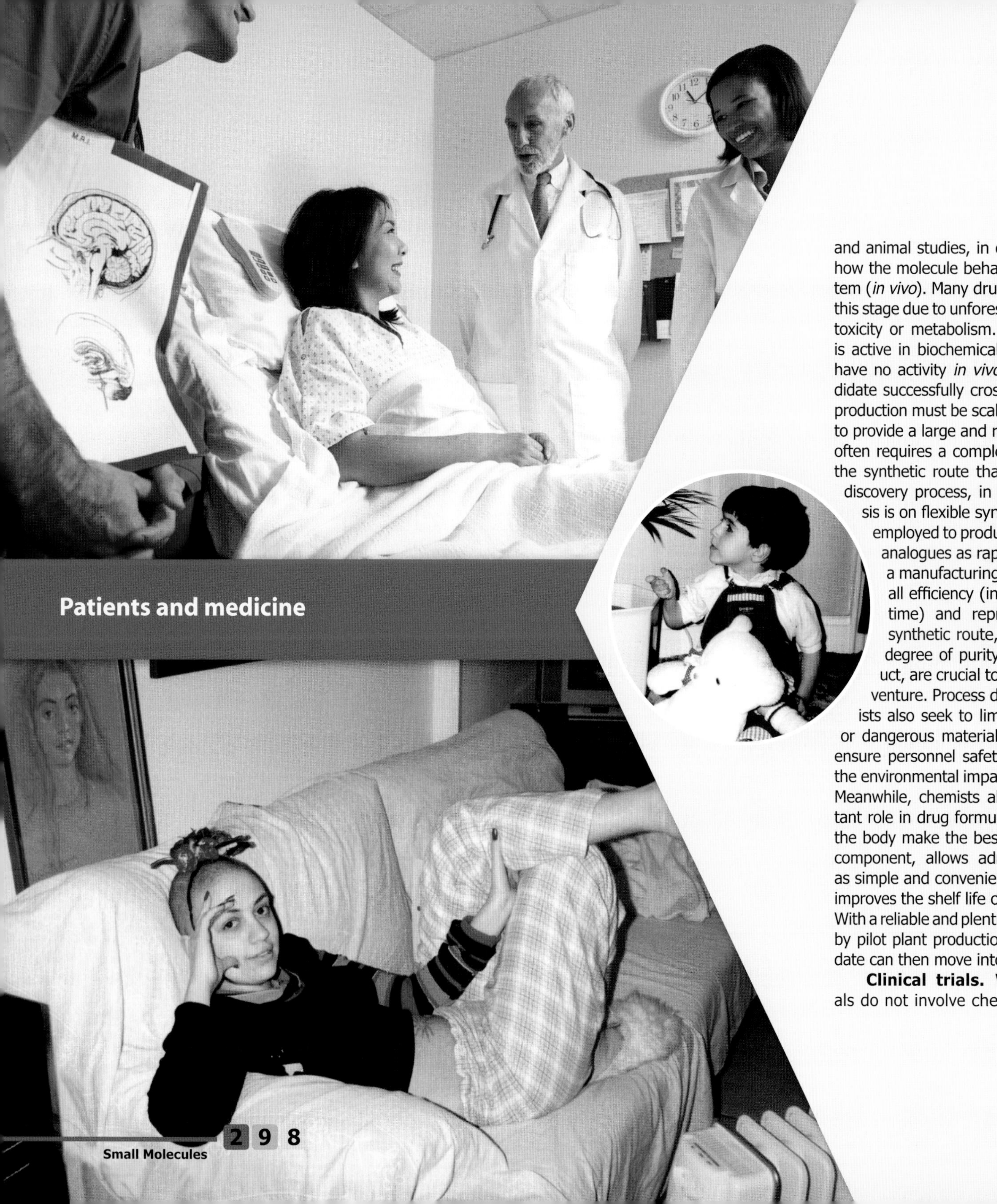

and animal studies, in order to determine how the molecule behaves in a living system (*in vivo*). Many drug candidates fail at this stage due to unforeseen problems with toxicity or metabolism. A compound that is active in biochemical assays may even have no activity *in vivo*. If the drug candidate successfully crosses these hurdles, production must be scaled-up even further to provide a large and reliable supply. This often requires a complete reevaluation of the synthetic route that was used in the discovery process, in which the emphasis is on flexible syntheses that can be employed to produce many structural analogues as rapidly as possible. In a manufacturing process, the overall efficiency (in terms of cost and time) and reproducibility of the synthetic route, together with the degree of purity of the final product, are crucial to the success of the venture. Process development chemists also seek to limit the use of toxic or dangerous materials and solvents, to ensure personnel safety and to minimize the environmental impact of the synthesis. Meanwhile, chemists also play an important role in drug formulation, which helps the body make the best use of the active component, allows administration to be as simple and convenient as possible, and improves the shelf life of the final product. With a reliable and plentiful supply provided by pilot plant production, the drug candidate can then move into clinical trials.

Clinical trials. While clinical trials do not involve chemists directly, we will provide a brief outline here for the sake of completeness. These trials must produce objective evidence of the safety and efficacy of the new drug, as well as its advantages over existing treatments, before regulatory authorities can approve it for general clinical use. There are three phases of clinical trials. In Phase I trials, the drug is given to a small group of healthy subjects to determine safe and tolerable dosage levels and investigate drug metabolism. Phase II clinical trials typically involve several hundred patients, and are used to establish effective doses. If the trials so far have been successful, the candidate drug then enters Phase III clinical trials, usually involving several thousand patients at various sites. The new treatment is compared to existing treatments and sometimes to a placebo, usually in double-blind studies in which even the doctor does not know which treatment is being used. Following analysis of all the data collected during trials, the regulatory authorities, such as the Federal Drug Administration (FDA) in the United States, decide whether to approve the new drug. Many treatments fail during clinical trials, a rather unfortunate and costly occurrence, considering the years of effort that have already been invested to discover and advance the drug candidate to this stage by the pharmaceutical company. Those that get through, however, can make all the difference, often between life and death, for the patients who receive them. Post-marketing studies (Phase IV clinical trials) may also be a condition of product approval, but are typically undertaken by companies even in the absence of a regulatory mandate. Phase IV studies provide information on the effects of the drug over a much larger patient population and longer timescale than is possible during Phase II and III trials. Adverse side effects detected during Phase IV studies can result in the restriction or withdrawal of a drug.

Drugs for the Mind

Mental illnesses are particularly difficult to diagnose and treat. Unlike other types of disease, such as infections and cancer, significant controversy surrounds the diagnosis and treatment of many mental health conditions. The exact nature of depression and the point at which anxiety stops being a normal response to stress and becomes a clinical problem are examples of the ambiguities surrounding mental health. To further complicate matters, medical opinions can vary among practitioners as to the mental health status of an individual patient, and, on a larger scale, general concensus of diagnosis among practitioners can vary greatly between different countries.

Much of this difficulty stems from our poor understanding of the functioning of the brain, as compared with our relatively advanced knowledge of other organs. The brain is particularly problematic to study, as it is the most fragile and sensitive organ and, unlike the liver for example, its function cannot easily survive physical and chemical investigations such as sampling and analysis. Mental illnesses involving problems such as anxiety, personality disorders, or cognitive function, are also very poorly suited for investigation using animal models. An additional problem for both diagnosis and treatment is the presence of the blood-brain barrier (a physical barrier caused by the low permeability of capillary walls in the blood supply of the brain), which prevents many substances from crossing between the brain and the blood stream. As a result, few mental health problems have a physical diagnosis based on blood tests or scans; even Alzheimer's disease, which currently affects over four million Americans, can be diagnosed definitively only by autopsy.

Having established some of the challenges to treating mental health, we will now trace the development of selected classes of drugs that have been used to alleviate such conditions. From a historical perspective, our ability to treat these illnesses has been ever increasing, as has the specificity of the drugs, giving hope for the future of therapy in this field.

Barbiturates. Our discussion of drugs for the mind begins with the barbiturate class. Barbiturates were the first class of drugs for the mind developed by the fledgling pharmaceutical industry. Prior to their introduction, the only treatments available for psychiatric disorders were opiates and alcohol. The pioneer of barbiturates was German chemist Adolf von Baeyer, who we first encountered

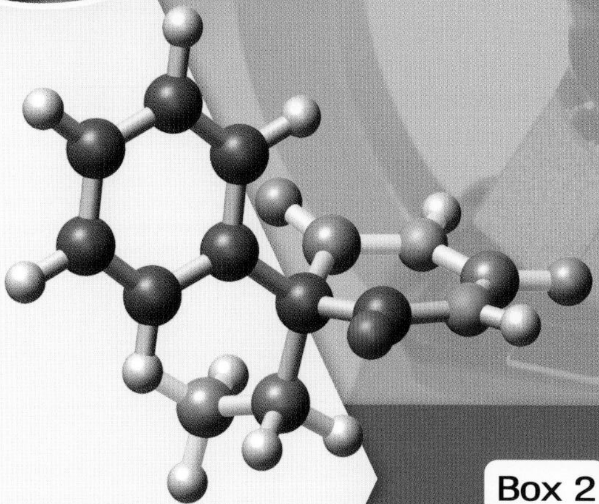

Ball and stick model of the molecule of phenobarbital (Luminal®)

Adolf von Baeyer

Box 2 Selected barbiturates

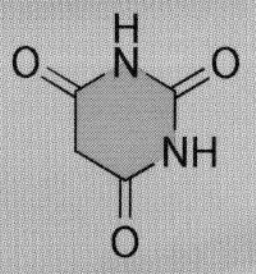

barbituric acid

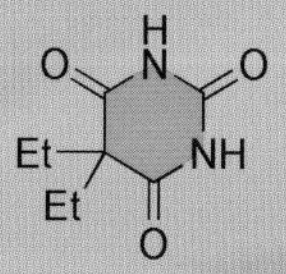

diethylbarbital [Veronal®] Bayer

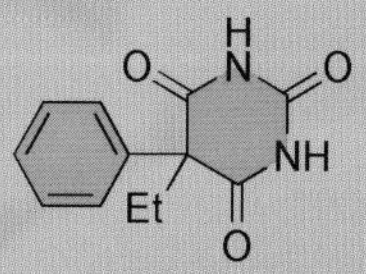

phenobarbital [Luminal®] Bayer

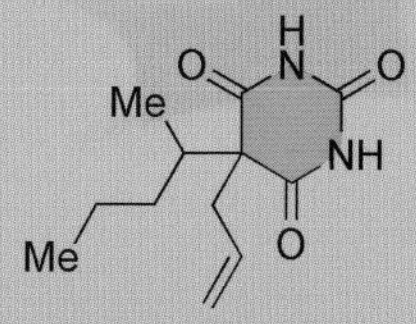

secobarbital [Seconal®] Eli Lilly

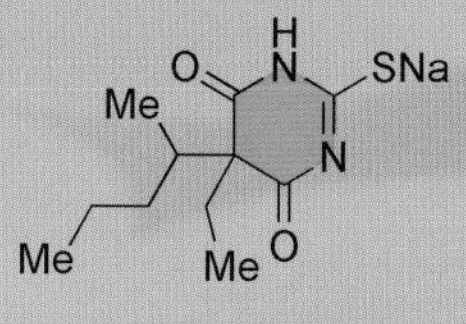

thiopental sodium [Sodium Pentothal®] Abbott

Drugs for the Mind
- **Barbiturates**
- **Phenobarbital**

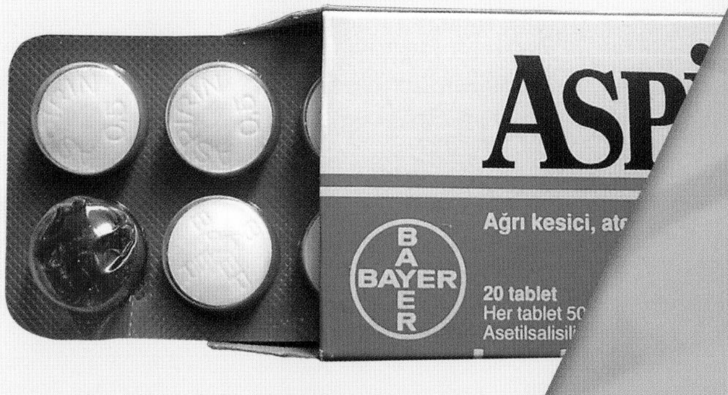

Marilyn Monroe

Jimi Hendrix

in the story of haemin (Chapter 8). von Baeyer prepared barbituric acid (Box 2), the first barbiturate, on December 4, 1863. An enthusiastic researcher with a famous zest for life, he is thought to have named his discovery in honor of St. Barbara, a Christian saint on whose feast day he was working. An alternative explanation is that he named his compound after an acquaintance called Barbara. Baeyer had no therapeutic use in mind for this new molecule, but nevertheless he, and others, prepared a number of derivatives such that when Emil Fischer (see Chapter 3, Glucose) and his colleague Joseph von Mering recognized the medicinal potential of barbiturates in 1903, a number of analogues were already available.

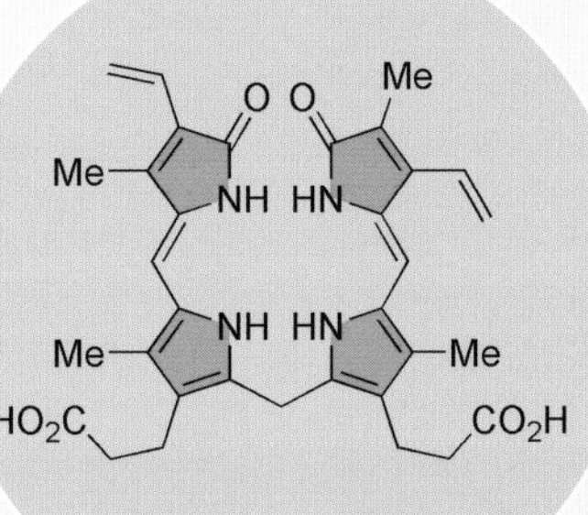

Bilirubin

Fischer and von Mering were studying the effects of diethylbarbital (Box 2) and discovered that it induced sleep in dogs. This led to diethylbarbital, marketed under the trade name Veronal®, being widely used as a sedative and sleeping pill, as well as being the first treatment for anxiety. Many other barbiturates (Box 2) followed in quick succession, the most famous being phenobarbital. Marketed as Luminal®, phenobarbital was introduced by Bayer (Adolf von Baeyer's fledgling pharmaceutical company) in 1912. In addition to its sedative activity, phenobarbital was found to be an effective anticonvulsant and was used to treat epilepsy, a disease for which it is still used today, albeit rarely. Phenobarbital offered considerable improvement over potassium bromide, which had been used for epilepsy therapy since its discovery in 1857 by Sir Charles Locock, as the latter can cause dermatitis and psychosis. Phenobarbital was also used to treat neonatal jaundice. The characteristic yellow color of the jaundiced infant is caused by high levels of bilirubin in the blood. Bilirubin arises as a breakdown product of normal hemoglobin catabolism when red blood cells die. The characteristic colors of urine and feces are due to the excretion of further breakdown metabolites of bilirubin. Although neonatal jaundice is usually harmless and self-correcting, rare complications can lead to neurological damage. This application of phenobarbital has now been superseded by phototherapy, which was discovered accidentally in the 1950s and is much safer than drug therapy. Today, phenobarbital is still used regularly in veterinary medicine to treat convulsions.

In all, some 2,500 different barbiturates have been used over the last century. These drugs have variable potencies and durations of action, with effects ranging from mild sedation to complete anesthesia. Approximately twenty barbiturates are still licensed in the United States as sedatives, anesthetics, and anxiolytics (drugs for anxiety). The barbiturates are central nervous system (CNS) depressants that have effects similar to alcohol. They enhance the activity of the neurotransmitter gamma-aminobutyric acid (GABA) at the $GABA_A$ subgroup of receptors by altering the structure of the receptor so

Box 3 Selected benzodiazepines

chlordiazepoxide
[Librium®]
Hoffmann-La Roche

diazepam
[Valium®]
Hoffmann-La Roche

alprazolam
[Xanax®]
Pfizer

flunitrazepam
[Rohypnol®]
Hoffmann-La Roche

Ball and stick model of the molecule of diazepam (Valium®)

that GABA can bind more easily. This type of interaction is known as allosteric activation. Activation of the $GABA_A$ receptors opens an ion channel in the neuron membrane, allowing chloride ions to enter the cell, interfering with its function. Alcohol operates in part by the same mechanism, hence the cross-tolerance and dependence that is observed in addicts to these drugs.

Despite the benefits brought by the barbiturate drugs, their use is waning due to some serious problems. Barbiturates quickly induce both physical and psychological dependence, and long-term users develop a high tolerance for the drug, eroding its safety margin because greater doses are then needed to produce the same effect. Barbiturates have been implicated in the deaths of a number of high profile celebrities, including actress Marilyn Monroe and guitarist Jimi Hendrix. Accidental barbiturate overdoses are common and frequently occur when the patient also consumes alcohol. The dangers of barbiturate overdose were highlighted by a case of industrial negligence in 1940. The company Winthrop was manufacturing phenobarbital and the antibiotic sulfathiazole in the same plant. Contamination of a batch of sulfathiazole with phenobarbital led to many deaths amongst patients who had unknowingly taken high doses of the barbiturate along with their antibiotic. This disaster led to the introduction of strict new rules governing the manufacture of drug substances.

Benzodiazepines. The benzodiazepines (Box 3) quickly gained popularity in place of barbiturates. During clinical trials, psychiatrists noted that the benzodiazepines could relieve anxiety without significantly affecting cognitive function, which was a significant advantage over earlier treatments. Their action is also more specific than that of barbiturates, leading to fewer side effects and allowing lower doses to be prescribed. As a result, fewer accidental or deliberate overdoses were recorded with these agents. Within fifteen years of the launch of the benzodiazepines, the most popular of these drugs, Valium® (diazepam), was the most widely prescribed pharmaceutical in the world and, by 1987, an astounding 2.8 billion tablets of Valium® were being manufactured each year. Astonishingly, the first member of the benzodiazepine class, Librium®, was nearly discarded without ever being tested!

Leo H. Sternbach, a chemist at the New Jersey (USA) facility of Hoffmann-La Roche, first discovered what eventually became Librium® in 1957. Sternbach had fled his native Poland for Switzerland in the 1930s, joining Hoffmann-La Roche. However, fearing a Nazi invasion, the company opted to evacuate all their vulnerable employees to the United States in 1941. As a postdoctoral scholar in Krakow (Poland), Sternbach had worked on a class of dye molecules called benzheptoxdiazines. In the USA, he sought to exploit his experience with these compounds in his search for new drugs to replace the barbiturates. He found that all the compounds he made contained six-membered rings, rather than the seven-membered ring structures he had supposed, and were inactive in the assays the company was using. When he treated one such compound with methylamine ($MeNH_2$), it underwent a ring expansion to give a crystalline material (Box 4),

Benzodiazepines
- Librium®
- Valium®

Leo H. Sternbach

Box 4 Benzheptoxdiazines — structure and conversion to benzodiazepines

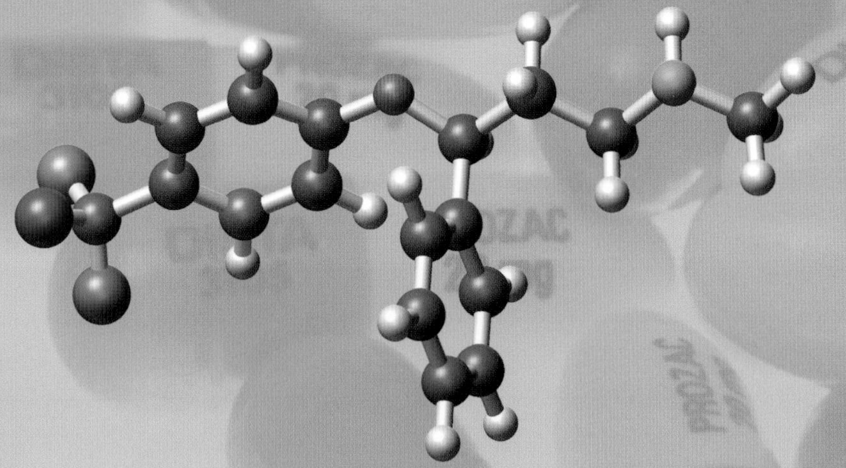

Ball and stick model of the molecule of fluoxetine (Prozac®)

Box 5 Selective serotonin reuptake inhibitors (SSRIs)

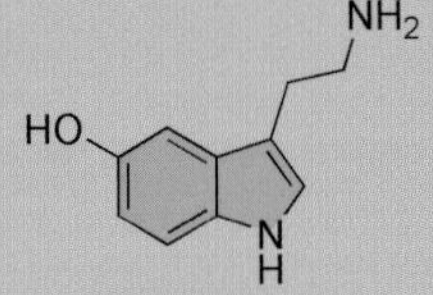

serotonin
(5-hydroxytryptamine, 5-HT)

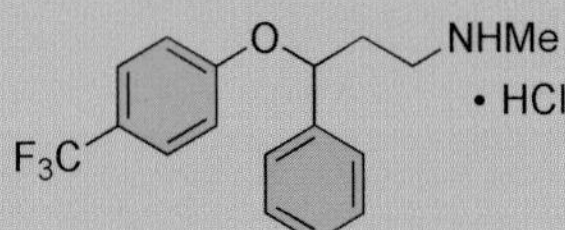

fluoxetine hydrochloride
[Prozac®]
Eli Lilly

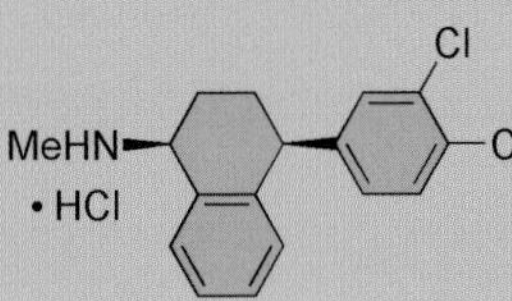

sertraline hydrochloride
[Zoloft®]
Pfizer

paroxetine hydrochloride
[Paxil®]
GlaxoSmithKline

which he labeled Ro 5-0690 and set aside. Eighteen months later, in 1957, the laboratory was being cleared and an assistant asked if a sample of Ro 5-0690 should be sent for biological evaluation or discarded. Fortunately, Sternbach chose the former, and soon received positive results from Lowell O. Randall, the head of pharmacology at Hoffmann-La Roche. During Phase III trials, the full potential of the drug became clear, as patients being treated for anxiety showed marked improvements in their mental health without loss of cognitive function or motor coordination. Hoffmann-La Roche launched Librium® in 1960, followed three years later by Valium®, a more active congener. In 1981, Xanax®, a still better drug, was launched. All three drugs proved highly lucrative for Hoffmann-La Roche and have benefited millions of patients for many years.

Given their similar physiological effects, it is perhaps unsurprising that benzodiazepines have a similar mechanism of action to the barbiturates, although the processes involved are not identical. The benzodiazepines appear to affect only those $GABA_A$ receptors located in the subcortical nuclei, and not those in the brain stem or spinal cord, in accord with the profound tranquilizing effect accompanied by only minor impairment of motor and cognitive function. The benzodiazepines and GABA are thought to augment each other's binding to the $GABA_A$ receptors, although the precise details remain a mystery.

Unfortunately, the benzodiazepines have also become drugs of abuse and, although they were not initially believed to be addictive, they have since been shown to induce a strong physical dependence in patients who take them for prolonged periods. It has been claimed that a large number of alcohol abusers and illicit drug users may also be or have been reliant on benzodiazepines. A lesser known example of an abused benzodiazepine is the so-called 'date-rape' drug, Rohypnol® (Box 3). Despite being illegal in many countries, this highly potent CNS depressant is available on the streets of most major cities. Although more commonly used as a recreational drug, Rohypnol® is also used in the sinister practice in which an unwitting victim is drugged and then robbed or raped. The effects of the drug can prevent the victim from clearly recalling the incident, leading to difficulties in detecting and prosecuting the crime.

The high incidence of benzodiazepine abuse has its roots in misguided prescribing practices during the early life of these drugs. They were often prescribed by family doctors ill-equipped to diagnose pathological anxiety accurately. The widespread use of Valium® led to the drug becoming known as "mother's little helper". As this name implies, addiction to benzodiazepine sedatives was much more common in women than men. Fortunately, more selective, non-addictive anxiolytic drugs have since been developed (see below), and the use of benzodiazepines has shown a marked decline. However, as they are effective and safe for short-term treatment, their continuing use is probably secured, at least for the foreseeable future.

Selective serotonin reuptake inhibitors. Selective serotonin reuptake inhibitors (SSRIs, Box 5) were launched as antidepressants in the late 1980s with the introduction of Prozac® (fluoxetine hydrochloride), manufactured by the pharmaceutical company Eli Lilly. Within two years of its launch, Prozac® had become the most highly prescribed antidepressant ever and, by 1999, two years before its patent expired, it accounted for around 25 % of Lilly's $10 billion annual revenues.

Depressive disorders are common and on the increase, probably due in part to both improved diagnosis and an increase in the numbers of patients seeking help; indeed, 10–20 % of the populations of most Western countries are estimated to take regular medication to combat negative feelings. Between 1991 and 2001, antidepressant prescriptions in the UK increased by fifteen million to twenty-four million per annum, and so much Prozac® is now taken that it has been detected in rivers and groundwater.

The predecessors of the SSRIs were the monoamine oxidase inhibitors (MAOIs) and the tricyclic antidepressants, both of which were introduced in the 1950s. Interestingly, iproniazid (Marsilid®, Box 6), the first MAOI, was originally used to treat tuberculosis before its antidepressant effects were noted. However, these drugs are not selective enough, and patients must endure serious side effects, including hypertension, blurred vision, and headaches. The relative lack of such side effects associated with SSRIs has contributed to their popularity.

While depression is difficult to study, it is believed to be linked to neurotransmitter imbalances in the brain of the patient. Serotonin is stored at nerve junctions (synapses) and released when needed. It is then destroyed by enzymes or taken up by the receiving cell. Those suffering from depressive disorders have unusually low serotonin levels. SSRIs act to inhibit the action of the pump that removes serotonin, leading to a rise in the concentration of the neurotransmitter in the synapse. This mechanism explains the short-term effects of these drugs; however, in the longer term SSRIs appear to induce permanent changes in brain structure, making the patients less prone to depressive events, an effect that is more difficult to explain.

Antidepressants
- **Prozac®**
- **Zoloft®**
- **Paxil®**

Prozac® was discovered in the USA by a team of scientists at Eli Lilly in the 1960s in an early example of rational drug design. Chemists Bryan B. Molloy and Klaus K. Schmiegel synthesized and tested a number of compounds based on the template of the antihistamine drug diphenhydramine hydrochloride (marketed as Benadryl® in the USA) (Box 6). One of these compounds was found to inhibit serotonin uptake specifically, and, after several years of development, Eli Lilly announced their findings in 1974. Branded Prozac®, the drug was launched in 1987. The competitors Zoloft® (Pfizer) and Paxil® (SmithKlineBeecham, now GlaxoSmithKline) were launched in 1991 and 1992, respectively, and have also become hugely successful (Box 5).

The SSRIs have not been free of controversy. Ironically, the lack of side effects associated with Prozac® has been a contributing factor. The apparent safety of Prozac® led to rather liberal prescrib-

Bryan B. Molloy **Klaus K. Schmiegel**

Box 6 Structures of iproniazid and diphenhydramine hydrochloride

iproniazid
[Marsilid®]
Hoffmann-La Roche

diphenhydramine hydrochloride
[Benadryl®]
Pfizer/Johnson & Johnson

Ball and stick model of the molecule of iproniazid (Marsilid®)

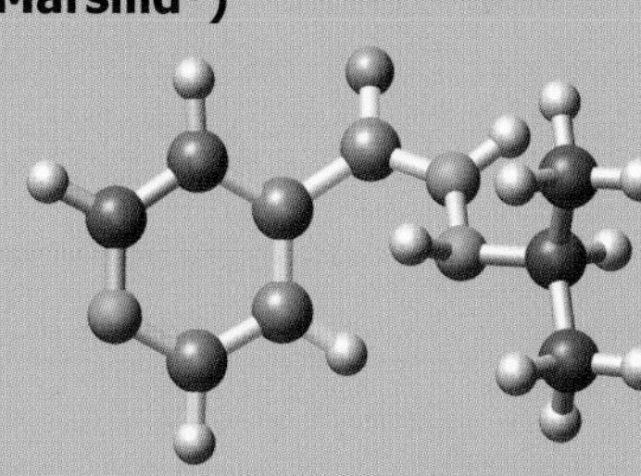

Ball and stick model of the molecule of diphenhydramine (Benadryl®)

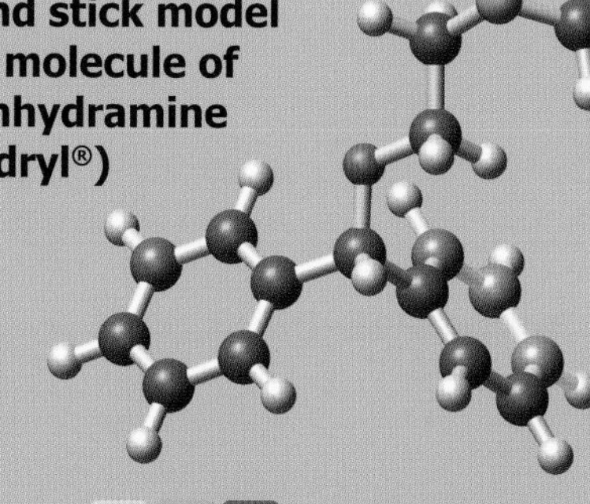

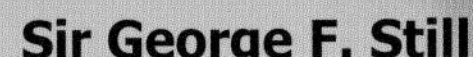

Sir George F. Still

Charles Bradley

Box 7 Ritalin® and selected amphetamines

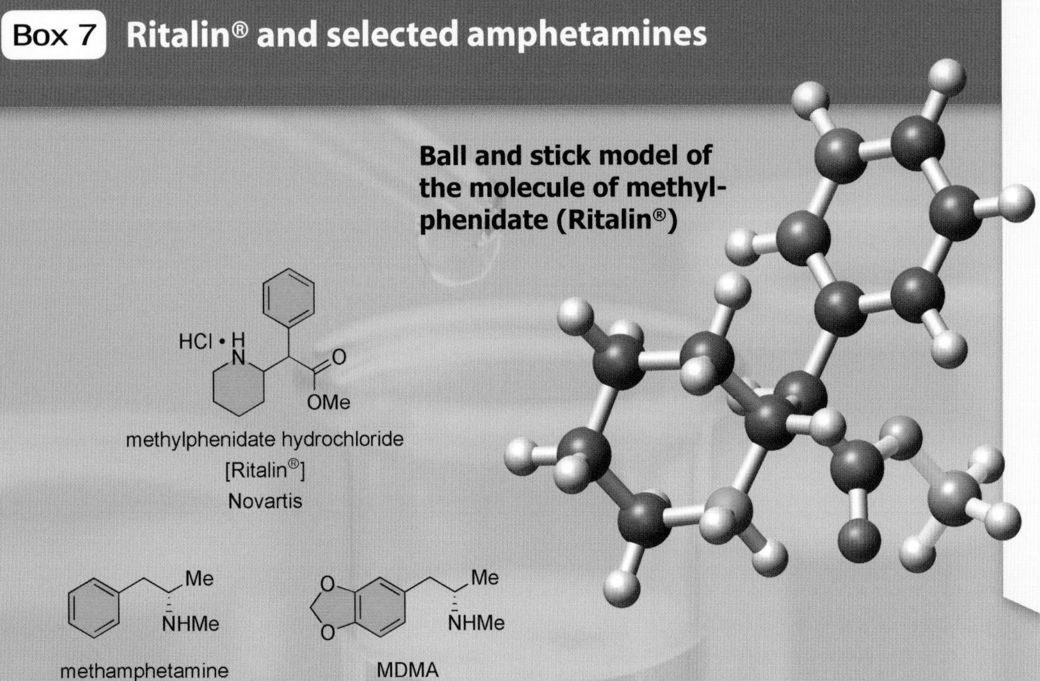

ing practices, and prompted some to voice concerns over its widespread use. One issue was the belief that it was being prescribed to those who were not clinically ill, but rather experiencing a tough period in their life. Other disturbing reports concern changes in character in those who have taken SSRIs, including violent behavior and an increased risk of suicide. This led the FDA to demand that stronger warnings be placed in the drug packaging. Supporters of the drugs argue that untreated depression carries a greater risk of suicide, and Eli Lilly strongly denies any link between Prozac® treatment and suicide or violent behavior. Recently, there have also been reports that some SSRIs can lead to dependence if used for long periods. Despite these controversies, SSRIs remain the first line of treatment for clinical depression.

Amphetamines. Amphetamines (Box 7) are the final class of drugs for the mind discussed in this section. They are stimulants based on the structure of α-methylphenethylamine (amphetamine), the parent member. Many derivatives have been prepared and, unfortunately, several members of the class are drugs of abuse, used recreationally to produce euphoric sensations. They are also used clinically to treat behavioral problems and narcolepsy. Arguably one of the most controversial drugs for the mind is the amphetamine Ritalin®, discussed below.

ADHD •Ritalin®

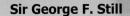

***Ritalin®* (methylphenidate hydrochloride, Box 7).** The difference between true behavioral problems that affect children's social development and learning and their natural propensity to become excited or distracted can be very difficult to define, leading to problems in diagnosing such conditions. In 1902, British pediatrician Sir George F. Still described twenty children in his care as "...passionate, defiant, spiteful..." and lacking in "...inhibitory volition..." and suggested that this behavior was due to neurological damage, rather than bad parenting. This theory gained credence following the viral encephalitis epidemic of 1917–1928. Clinicians noted similar behavioral patterns in children who had been affected by the virus and were assumed to have suffered mild brain damage. An indication of the potential benefits of stimulants came in 1937 when Charles Bradley, working at a Rhode Island (USA) hospital for children with behavioral problems, described how amphetamines he had prescribed to treat severe headaches seemed to cure the youngsters of their behavioral difficulties. By the 1950s, children with the symptoms described above were labeled 'minimally brain dysfunctional', reflecting the prevailing theory on behavioral problems. In 1955, the FDA approved Ritalin®, made by Ciba-Geigy (now part of Novartis), for the treatment of various psychological disorders, including depression, chronic fatigue, and narcolepsy. Following the expiration of their patent in 1967, the company sought to introduce Ritalin® for the treatment of children with minimal brain dysfunction. Ciba-Geigy also sponsored a great deal of research into behavioral disorders in children. By 1975, one million children in the United States had been diagnosed with a behavioral disorder; around half of these were prescribed medication, with some 265,000 receiving Ritalin®. Since then, the number of children diagnosed with what is now known as attention deficit hyperactivity disorder (ADHD) has increased to startling levels, as has the frequency of Ritalin® prescription. Approximately 3.5 million children in the United States have been diagnosed with ADHD, corresponding to about five percent of all children under the age of eighteen; the diagnosis is three times more common in boys. Notably, the rate of ADHD diagnosis in Europe is roughly one tenth of that in the United States, underscoring the differences in prevailing clinical opinion, and, perhaps, culture. ADHD is also becoming an increasingly common diagnosis in adults, with Ritalin® being one of the fastest growing drugs prescribed.

It may seem curious that treating hyperactive individuals with a stimulant should give a calming effect, especially when the same drug induces euphoria and excitement in healthy subjects. The frontal lobes of the brain regulate behavior and are rich in receptors for the neurotransmitters dopamine and norepinephrine. ADHD sufferers have reduced electrical activity and blood flow in these areas, and also low dopamine levels. Ritalin® is known to affect the release and uptake of dopamine and norepinephrine. Induction of an increase in dopamine levels in the frontal lobes may explain its curative effects in ADHD patients, however, the details of the mechanism of action are, as yet, unknown.

Critics of Ritalin® therapy contend that the drug is grossly overprescribed, and has become an easy option for pacifying obstreperous or distracted children who are more in need of better parenting or behavioral therapy. Misdiagnosed ADHD may also mask other serious problems, including neglect or abuse. It is claimed that Ritalin® makes children zombie-like and stifles their creativity and intellect. Many parents of afflicted children disagree and credit Ritalin® for making their home life more bearable.

Ritalin® pills

Numerous other drugs for the mind have been discovered and are in use today; some of these are shown in Box 8.

Drugs to Tackle Viruses

Viruses are amongst the smallest and simplest infective agents. Far from being a handicap, their structural simplicity makes them particularly difficult to fight. Viruses comprise a core of genetic material (DNA or RNA) and a few enzymes enclosed in a protein shell. Some are also surrounded by a lipoprotein envelope, but they have essentially no metabolism. Viruses cause a broad range of diseases in animals, plants and even bacteria. Examples of viruses include the rhinovirus and adenovirus, which cause the common cold and sore throat, respectively, and the herpes family, which cause a variety of conditions from glandular fever (caused by the Epstein-Barr virus) to recurring sores on the skin. Hepatitis, measles, mumps, rubella, polio, and influenza are among the many other examples of viral

Box 8 Other selected drugs for the mind

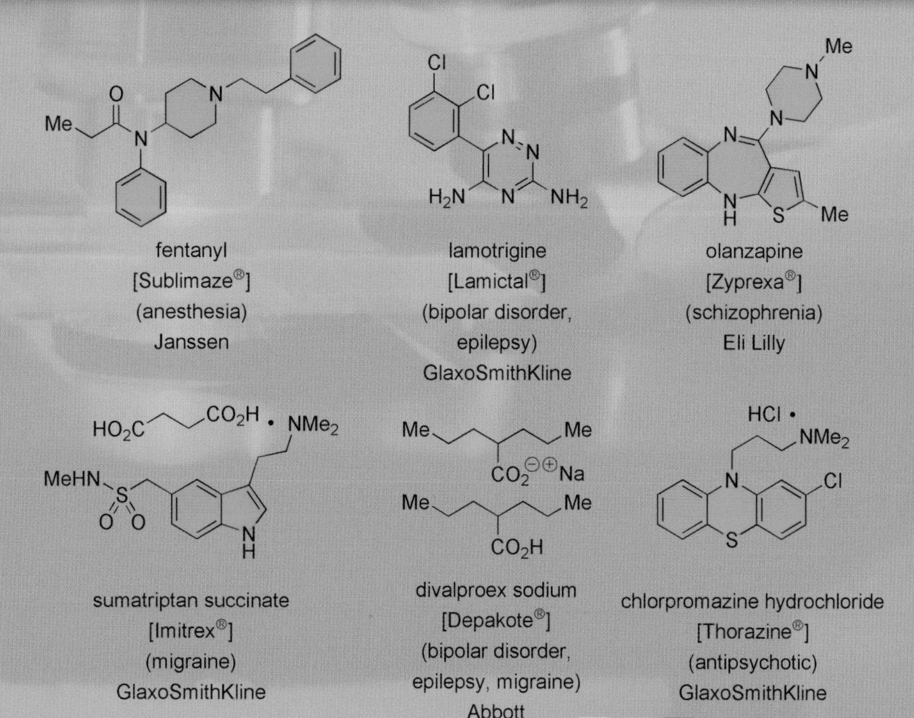

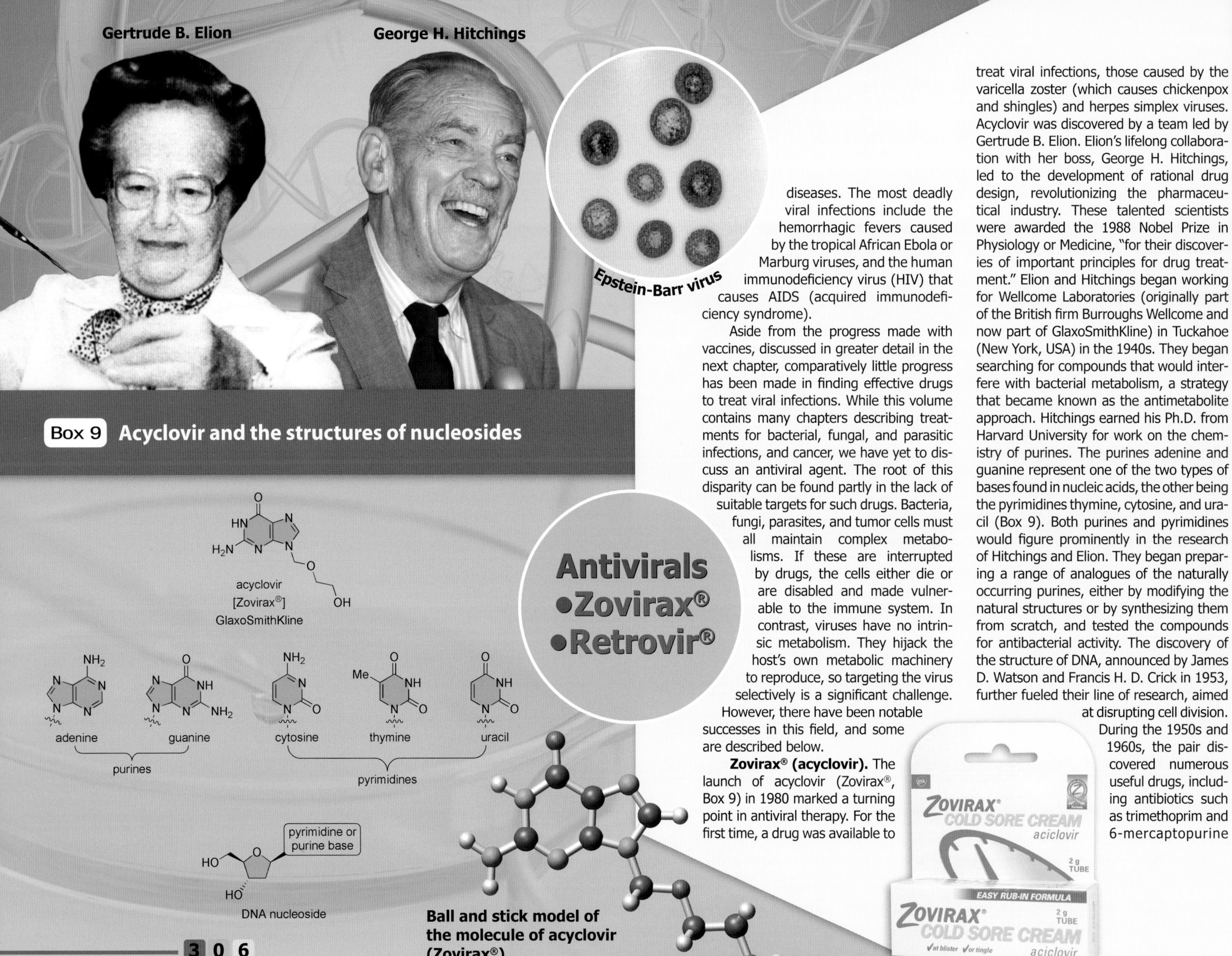

Ball and stick model of the molecule of acyclovir (Zovirax®)

diseases. The most deadly viral infections include the hemorrhagic fevers caused by the tropical African Ebola or Marburg viruses, and the human immunodeficiency virus (HIV) that causes AIDS (acquired immunodeficiency syndrome).

Aside from the progress made with vaccines, discussed in greater detail in the next chapter, comparatively little progress has been made in finding effective drugs to treat viral infections. While this volume contains many chapters describing treatments for bacterial, fungal, and parasitic infections, and cancer, we have yet to discuss an antiviral agent. The root of this disparity can be found partly in the lack of suitable targets for such drugs. Bacteria, fungi, parasites, and tumor cells must all maintain complex metabolisms. If these are interrupted by drugs, the cells either die or are disabled and made vulnerable to the immune system. In contrast, viruses have no intrinsic metabolism. They hijack the host's own metabolic machinery to reproduce, so targeting the virus selectively is a significant challenge. However, there have been notable successes in this field, and some are described below.

Zovirax® (acyclovir). The launch of acyclovir (Zovirax®, Box 9) in 1980 marked a turning point in antiviral therapy. For the first time, a drug was available to treat viral infections, those caused by the varicella zoster (which causes chickenpox and shingles) and herpes simplex viruses. Acyclovir was discovered by a team led by Gertrude B. Elion. Elion's lifelong collaboration with her boss, George H. Hitchings, led to the development of rational drug design, revolutionizing the pharmaceutical industry. These talented scientists were awarded the 1988 Nobel Prize in Physiology or Medicine, "for their discoveries of important principles for drug treatment." Elion and Hitchings began working for Wellcome Laboratories (originally part of the British firm Burroughs Wellcome and now part of GlaxoSmithKline) in Tuckahoe (New York, USA) in the 1940s. They began searching for compounds that would interfere with bacterial metabolism, a strategy that became known as the antimetabolite approach. Hitchings earned his Ph.D. from Harvard University for work on the chemistry of purines. The purines adenine and guanine represent one of the two types of bases found in nucleic acids, the other being the pyrimidines thymine, cytosine, and uracil (Box 9). Both purines and pyrimidines would figure prominently in the research of Hitchings and Elion. They began preparing a range of analogues of the naturally occurring purines, either by modifying the natural structures or by synthesizing them from scratch, and tested the compounds for antibacterial activity. The discovery of the structure of DNA, announced by James D. Watson and Francis H. D. Crick in 1953, further fueled their line of research, aimed at disrupting cell division.

During the 1950s and 1960s, the pair discovered numerous useful drugs, including antibiotics such as trimethoprim and 6-mercaptopurine

(6-MP, which was used to treat childhood leukemia, see Chapter 22, Cyclosporin, FK506, & Rapamycin). These purines inhibit cellular reproduction by mimicking the natural purine DNA bases. The synthetic purines either become incorporated into DNA during replication and thus corrupt it, or inhibit an enzyme involved in DNA synthesis or manipulation, leading to a disruption of cell division. This approach is particularly suited for attacking rapidly dividing cells such as bacteria or cancer cells. Following their successes against bacteria and cancer, the next step was to apply the antimetabolite approach to antiviral therapy. In the 1970s, Hitchings had been promoted away from active research, and Elion, now head of experimental therapy at Wellcome Laboratories (relocated to North Carolina, USA), presided over a flourishing antiviral program. In an extension of the purine and pyrimidine analogue program, the team now began to prepare more sophisticated compounds that mimicked more of the structure of the nucleoside. Nucleosides are composed of a purine or pyrimidine base and a sugar moiety. Acyclovir, an analogue of the guanosine nucleoside, was first synthesized by Howard J. Schaeffer and his team, under Elion's guidance. Acyclovir is actually a prodrug that is converted into the active compound (the corresponding triphosphate) in the body. A viral thymidine kinase enzyme initiates this process, converting acyclovir into the monophosphate derivative; host cell kinases then complete the transformation. The viral kinase is many times more efficient than those in the host, giving the drug its required selectivity. Acyclovir triphosphate inhibits the enzyme DNA polymerase, blocking DNA synthesis and preventing viral replication. The drug candidate quickly progressed through trials and the regulatory process, and was launched in 1980 as Zovirax®. Since then, it has been used to treat Epstein-Barr virus infections, shingles, chickenpox, genital herpes, and cold sores, as well as rarer herpes infections. Today, weaker formulations are also available without prescription, particularly for the treatment of cold sores.

Luc Montagnier

Retrovir® (AZT, zidovudine). In the early 1980s, a viral illness began to claim lives in clusters around the world. Fear spread quickly as many of the victims were apparently fit and healthy young men. The illness was completely new and displayed many varied symptoms. Eventually it became clear that the deaths were not due to the virus directly, but to a variety of opportunistic infections, including PCP (*Pneumocystis carinii* pneumonia), Kaposi's sarcoma (rare forms of pneumonia and skin cancer, respectively), and toxoplasmosis (an infection caused by the protozoan *Toxoplasma gondii*). People with healthy immune systems are normally able to fight off such conditions, but this virus invades and destroys the patient's CD_4 (or T_4) cells, compromising their immune systems. In 1984, Luc Montagnier of the Pasteur Institute (Paris, France) identified a retrovirus (a virus with an RNA genome

Box 10 AZT, its active triphosphate, and thymidine triphosphate

zidovudine (AZT) [Retrovir®] GlaxoSmithKline → phosphorylation in vivo → AZT triphosphate

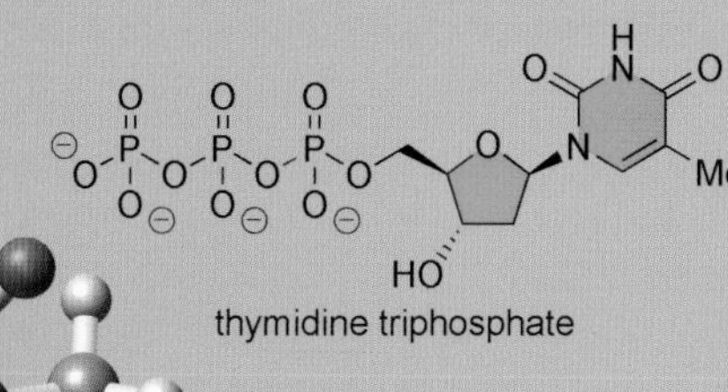
thymidine triphosphate

Ball and stick model of the molecule of AZT (Retrovir®)

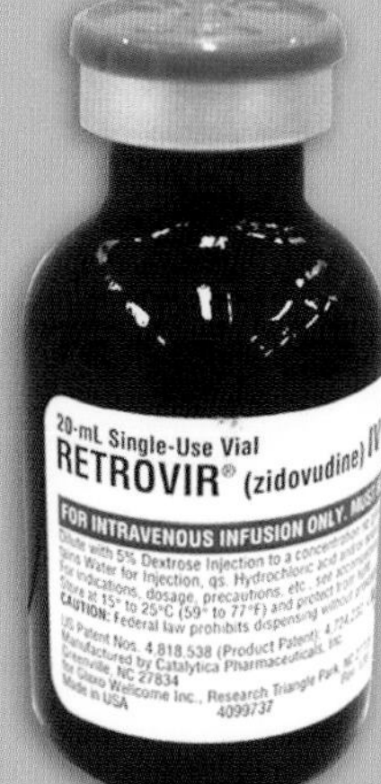

Box 11 Selected nucleoside reverse transcriptase inhibitors (NRTIs)

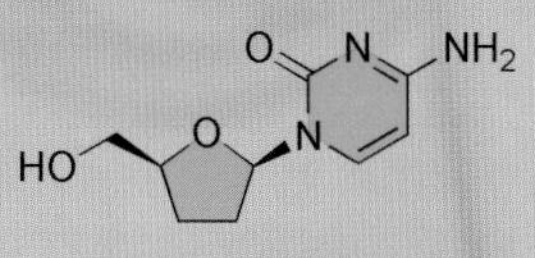

zalcitabine
[Hivid®]
Hoffmann-La Roche

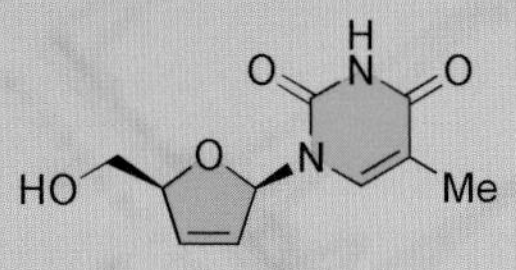

stavudine
[Zerit®]
Bristol-Myers Squibb

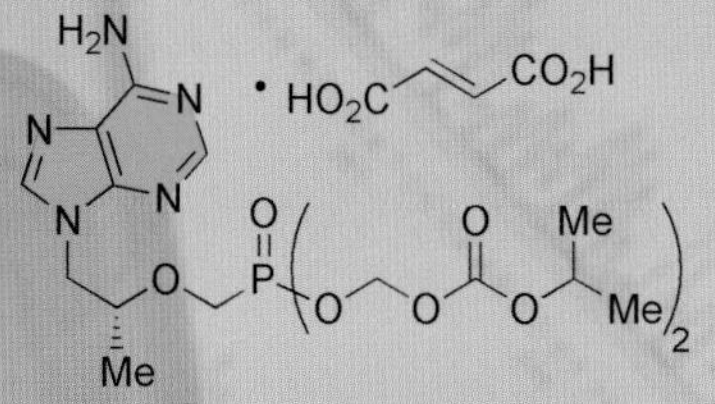

tenofovir disoproxil fumarate
[Viread®]
Gilead Sciences

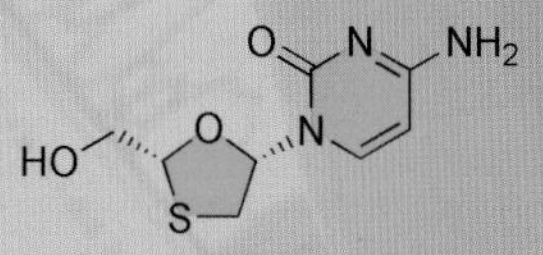

lamivudine
[Epivir®]
GlaxoSmithKline

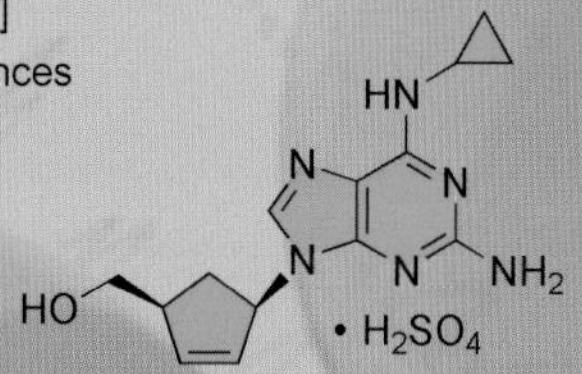

abacavir sulfate
[Ziagen®]
GlaxoSmithKline

Box 12 Selected non-nucleoside reverse transcriptase inhibitors (NNRTIs)

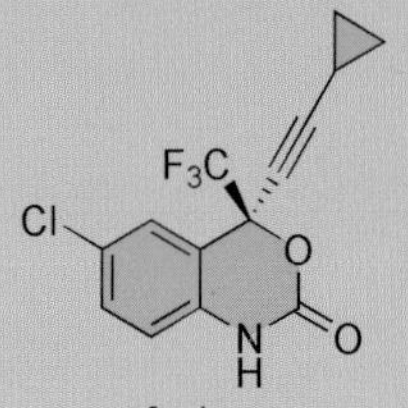

efavirenz
[Sustiva®]
Bristol-Myers Squibb

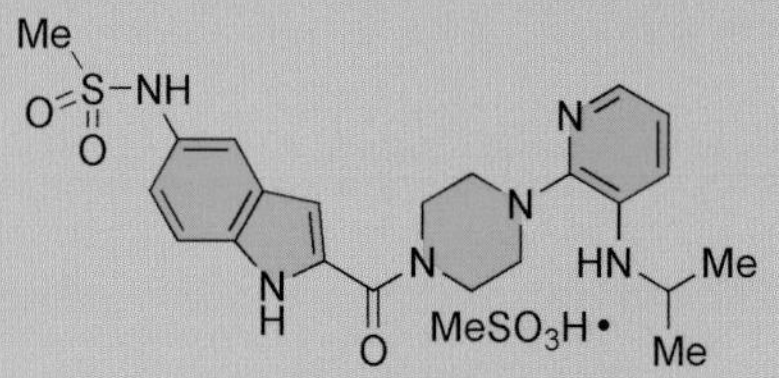

delavirdine mesylate
[Rescriptor®]
Pfizer

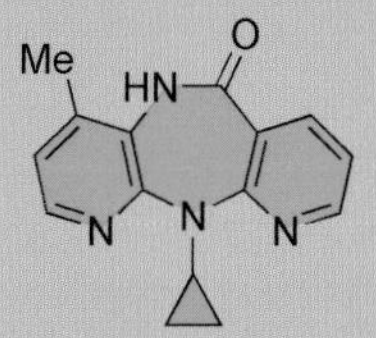

nevirapine
[Viramune®]
Boehringer Ingelheim

rather than a DNA genome) in patients suffering from this new disease. His discovery was followed quickly by similar reports from Abraham Karpas (University of Cambridge, UK), Jay A. Levy (University of California, San Francisco, USA) and Robert C. Gallo (National Cancer Institute, USA). Initially, each group named the virus differently, but a short time later a consensus was reached and it was universally named the human immunodeficiency virus (HIV). Today, researchers know a great deal about HIV infection and the onset of AIDS; however, the worldwide spread of the virus and problems of drug resistance continue to challenge scientists and patients fighting the disease. In 2006, the UNAIDS organization estimated that 38.6 million people are infected with HIV worldwide, and that in the preceding year there were 4.1 million new HIV infections and 2.8 million deaths from AIDS. So far, the epidemic has been concentrated in sub-Saharan Africa. Some drugs have been developed to fight HIV infection, but they have yet to give us the upper hand in eradicating this disease.

The first drug to be launched for the treatment of HIV was azidothymidine (AZT, Retrovir®, or zidovudine). This compound (Box 10), an analogue of the thymidine nucleoside, was first synthesized and tested by the Elion–Hitchings team in the 1960s as an anticancer drug. Although it was unsuccessful in this regard, it was not forgotten, and, in the 1980s, it was found to suppress viral replication. It was approved by the FDA for treating HIV in 1987. AZT works by inhibiting the viral enzyme reverse transcriptase. The HIV genome is encoded in RNA, and must be transcribed into DNA in order to be incorporated into the genome of the host cell. This process is called reverse transcription, as it is literally the reverse of the normal cellular transcription process. As with acyclovir, AZT is converted into its triphosphate and incorporated into the growing DNA chain. This event terminates the growing DNA polymer as AZT has an azide group in place of the hydroxyl group needed for attachment of the next DNA unit. In the early days of HIV therapy, AZT offered a crucial lifeline, but its use is now limited to a few specialist applications because it is relatively toxic and often ineffective due to viral resistance that has since developed. Indeed, the high mutation rate of the HIV virus makes drug resistance an especially challenging problem. A host of other reverse transcriptase inhibitors, both nucleoside- (NRTIs, Box 11) and non-nucleoside-based (NNRTIs, Box 12), have since been developed to replace AZT.

HIV protease inhibitors. In the 1970s, cancer researchers began studying retroviruses because they were known to cause various cancers. They found that retrovirus proteins are not made individually, but in amalgamated polyproteins. To release the individual proteins the viruses possess protein-cleaving enzymes, which are not activated until after the virus particles bud from the host cell. If these enzymes could be inhibited, the virus would be rendered incapable of invading new hosts. The emerging gene tech-

Protease Inhibitors
- Invirase®
- Norvir®
- Crixivan®

nologies of the 1980s allowed scientists to sequence the complete HIV genome, which they used to identify viral proteins. In 1986, several independent groups isolated HIV protease and confirmed that the enzyme was an aspartic protease, meaning that the amino acid residue aspartate occupies a key position in the active site. The enzyme also bore distinct similarities to the digestive enzymes renin and pepsin. In 1988, Irving Sigal, a talented biochemist working for Merck, showed that mutating the HIV protease led to the production of immature viruses incapable of infecting new cells, thus confirming the potential of this enzyme as a drug target. Within a year, two groups led by Manuel Navia (Merck) and Alexander Wlodawer (Frederick Cancer Research Facility, Maryland, USA) independently determined the three dimensional structure of HIV protease by x-ray crystallography, revealing a dimeric enzyme shaped like a walnut, with the active site between the two halves.

Medicinal chemists now had more than enough information to begin searching for HIV protease inhibitors. Indeed, they had a head start as several potent renin inhibitors were already known. They also knew that the role of the enzyme was to cleave the peptide chain at the amide bond between phenylalanine and proline residues, so compounds that mimic the structure of this junction (e.g. peptides or their analogues) could potentially act as inhibitors. However, finding peptide-like molecules that do not undergo degradation by our own proteases was a significant challenge. Aided by the sharp tools of chemical synthesis, structural biology, and computer modeling, medicinal chemists gave this problem their urgent attention. The enormous efforts directed towards this research led to an early triumph for structure-driven drug design with the approval, in December 1995, of the Hoffmann-La Roche drug saquinavir (Invirase®, Box 13), the first HIV protease inhibitor to be approved for clinical use. In the following spring, two further drugs, ritonavir (Norvir®, Abbott Laboratories) and indinavir (Crixivan®, Merck), were launched. Today, many more inhibitors are available (for examples, see Box 13), and these drugs play a vital role in treating those living with HIV. However, they are not without side effects, including lipodystrophy (the redistribution of body fat), kidney stones, and depression.

Combination therapy. Treating HIV was initially found to be extremely difficult because the virus mutates rapidly, enabling it to develop resistance to drugs remarkably quickly. By the mid-1990s, drug resistance was dominating the scientific literature on HIV, but in 1996, at an international AIDS conference in Vancouver (Canada), a number of speakers described a new strategy

Ball and stick model of the molecule of indinavir (Crixivan®)

Box 13 Selected HIV protease inhibitors

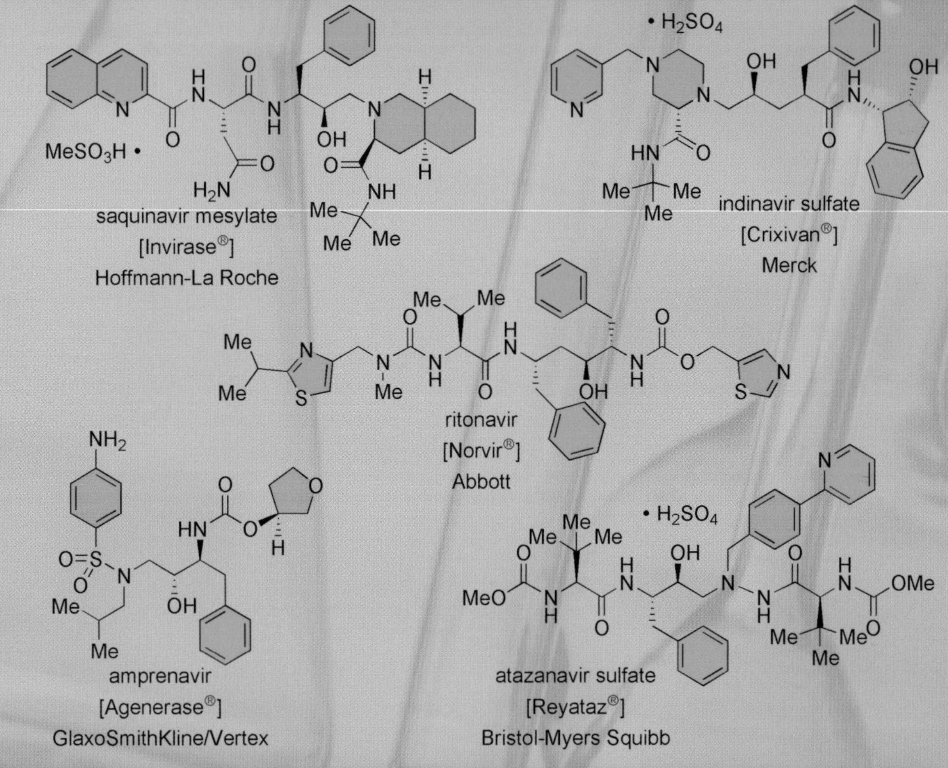

Below: X-Ray crystallographic structure of HIV-1 protease in complex with saquinavir. The small molecule inhibitor is shown in space filling format with hydrogen atoms omitted for clarity (green: carbon, red: oxygen, blue: nitrogen).

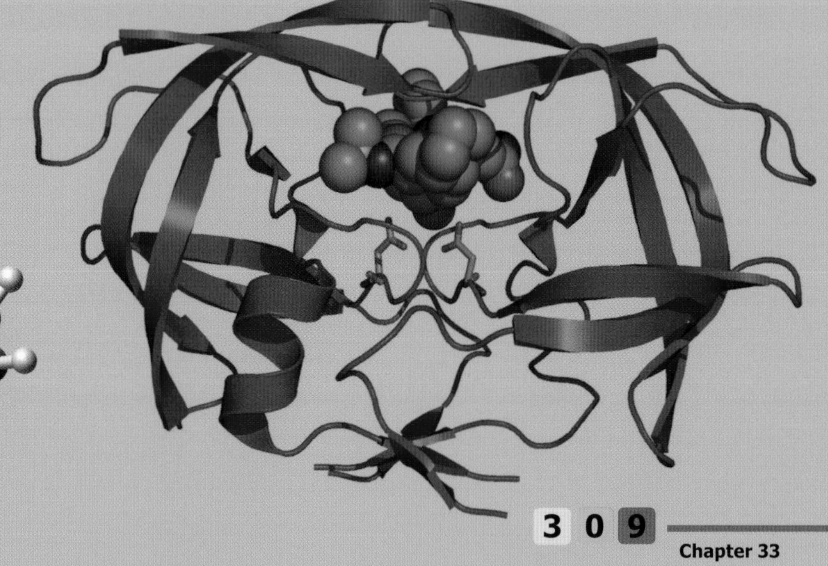

Box 14 Selected other antiviral drugs

Generic name (Trade name)	Structure or description	Mode of action	Marketer	Indication
zanamivir [Relenza®]	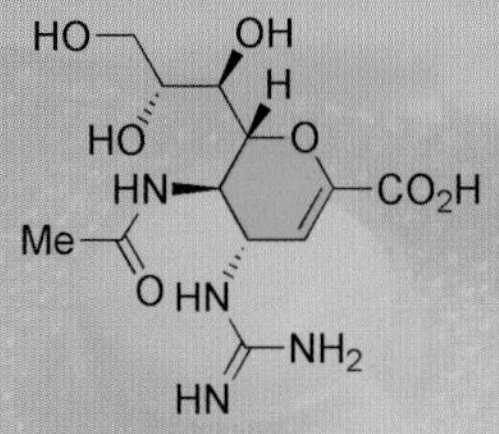	inhibition of influenza virus neuraminidase, alteration of virus particle aggregation and release	**GlaxoSmithKline**	treatment of influenza
oseltamivir phosphate [Tamiflu®]	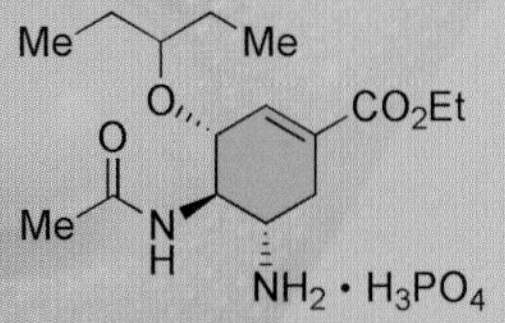	prodrug - after hydrolysis of the ethyl ester to afford the free carboxylic acid, the drug works by inhibition of influenza virus neuraminidase, alteration of virus particle aggregation and release	**Roche**	treatment of influenza
formivirsen sodium [Vitravene®]	antisense drug (designed DNA fragment)	designed DNA fragment that binds tightly to a region of the virus RNA that has been identified as important in viral protein production	**ISIS/ Novartis**	treatment of CMV (cytomegalovirus) eye infections
interferon alpha-2b [PEG-Intron®]	naturally occuring small protein - made through biotechnology	interferons are glycoprotein secreted by cells in response to viral infection, synthetic inducers, or biological inducers. They inhibit protein synthesis and other aspects of viral replication	**Schering-Plough**	treatment of hepatitis C virus - used in combination with ribavirin
ribavirin [Rebetol®]	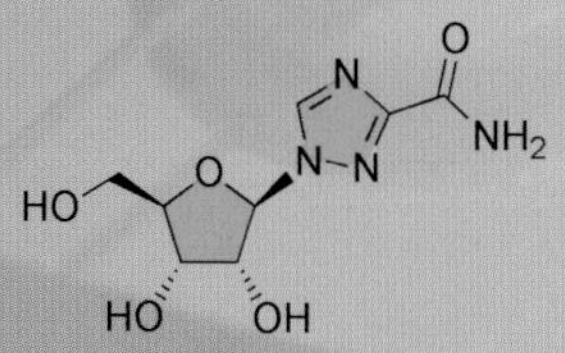	nucleoside analogue - exact mechanism of action unknown	**Schering-Plough**	treatment of hepatitis B and C - used in combination with interferon alpha-2b
pleconaril (drug still at clinical trial phase)	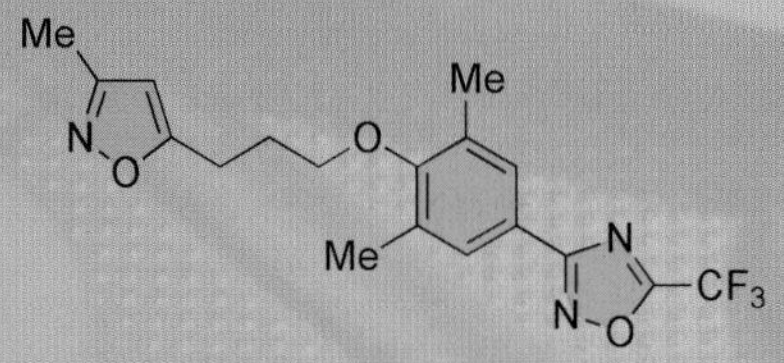	binds to a pocket on rhinovirus thus inhibiting uncoating of the virus	**ViroPharma/ Schering-Plough**	treatment of common cold

to avoid resistance. Among them was David Ho (Aaron Diamond AIDS Research Center, New York City, USA) who described treating HIV sufferers with 'cocktails' of drugs. In this approach, patients were given three or four drugs of different types simultaneously. These patients showed marked reductions in viral load (the number of viral particles in the blood), even to the point that the virus was almost undetectable, and experienced less frequent lapses due to drug resistance. This combination therapy has been adopted widely, and has shown significant success in slowing the progression of the disease and preventing the emergence of resistance, although there is still no cure for HIV. Current combination regimes typically consist of two NRTIs and one NNRTI or a protease inhibitor. Combination therapy has since been applied in the treatment of other conditions, including tuberculosis (a re-emerging epidemic) and certain cancers that are prone to drug resistance. Ho was hailed as the Man of the Year by Time Magazine in 1996 for pioneering this therapeutic approach.

David Ho

Influenza Drugs

While this section has mainly focused on HIV treatments, there have also been advances in the treatment of many other viral infections. A number of these are summarized in Box 14. Some have arisen from the enormous research efforts driven by HIV, while others repre-

Influenza Drugs
•Relenza®
•Tamiflu®

sent new innovative ways to tackle viruses. Zanamivir (Relenza®) and oseltamivir phosphate (Tamiflu®) are both licensed for the treatment of influenza, and are further examples of what can be achieved when rational drug design, using information obtained from x-ray crystallography studies of the corresponding protein target, is employed. Oseltamivir is an orally available prodrug in which the ethyl ester is hydrolyzed to the corresponding carboxylic acid *in vivo*. Due to its poor bioavailability, zanamivir must currently be administered by inhalation. Both drugs function by inhibiting the activity of the neuraminidase enzymes of the influenza virus, blocking the release of the virus from infected cells. All strains of the influenza virus display the glycoprotein enzymes haemagglutinin and neuraminidase on their surface, and these enzymes are critical to the replication strategy of the virus. Haemagglutinin, of which sixteen subtypes are known, is responsible for the initial binding of the virus to the cell that is being infected. After completion of the viral replicative cycle, neuraminidase (nine known subtypes) enables the newly formed virus particles to be released from the infected cell and infect new ones, propagating the spread of the virus throughout the organism. Influenza strains are characterized by the different subtypes of these two enzymes. Oseltamivir was used successfully to help combat the influenza epidemic caused by the H5N1 (haemagglutinin subtype five, neuraminidase subtype one) virus strain ('bird flu') in Southeast Asia in 2005. Fears that mutation of this strain may lead to an influenza pandemic on the scale of the Spanish flu pandemic which swept the world in 1918–1919 (and claimed far more lives than did World War I) has led to the stockpiling of Tamiflu® (oseltamivir) by governments around the globe, and much controversy about the manufacture and supply of this drug. Tamiflu® and Relenza® have spurred the development of further neuraminidase inhibitors, now entering clinical trials.

Drugs for the Gastrointestinal Tract

Many people suffer from gastrointestinal ailments caused by excess gastric acid (mainly hydrochloric acid) coming into contact with tissue that would not normally be exposed to such a harsh chemical environment. Examples include persistent acid reflux disease in which gastric acid enters the esophagus, causing painful heartburn and permanent damage, and also peptic ulcers (where damage to the protective stomach lining allows acid to attack the tissue beneath). Ulcers can eventually perforate, leading to dangerous bleeding. Until the 1980s, stress and lifestyle were blamed for the majority of gastric complaints, however, many have now been found to be caused by bacteria. The presence of *Helicobacter pylori* bacteria in the stomach is common, and is usually asymptomatic, but in 10–15 % of those infected the bacteria cause ulceration. Direct treatment of gastric symptoms is now usually accompanied by a course of antibiotics to tackle the root causes of the condition. The role of these bacteria was discovered by two Australian scientists, Barry J. Marshall and J. Robin Warren, who were awarded the 2005 Nobel Prize in Physiology or Medicine for their work.

H_2-Antagonists. Before the advent of this class of drugs, peptic ulcers could

J. Robin Warren and Barry J. Marshall

Below: X-Ray crystallographic structure of N1 neuraminidase monomer of the H5N1 avian virus in complex with oseltamivir (space filling model), which is bound following hydrolysis of the ethyl ester to the carboxylic acid

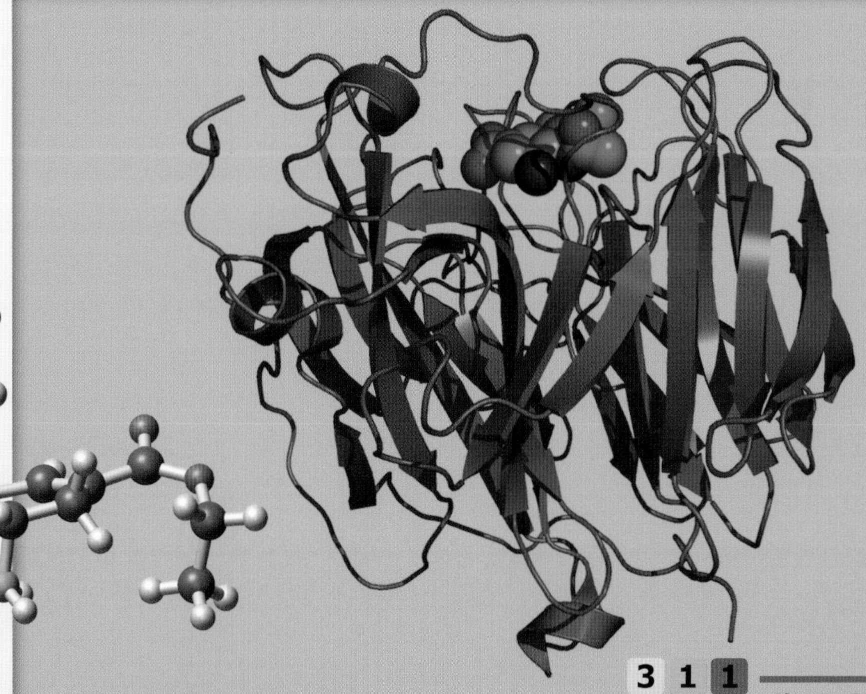

Ball and stick model of the molecule of oseltamivir (Tamiflu®)

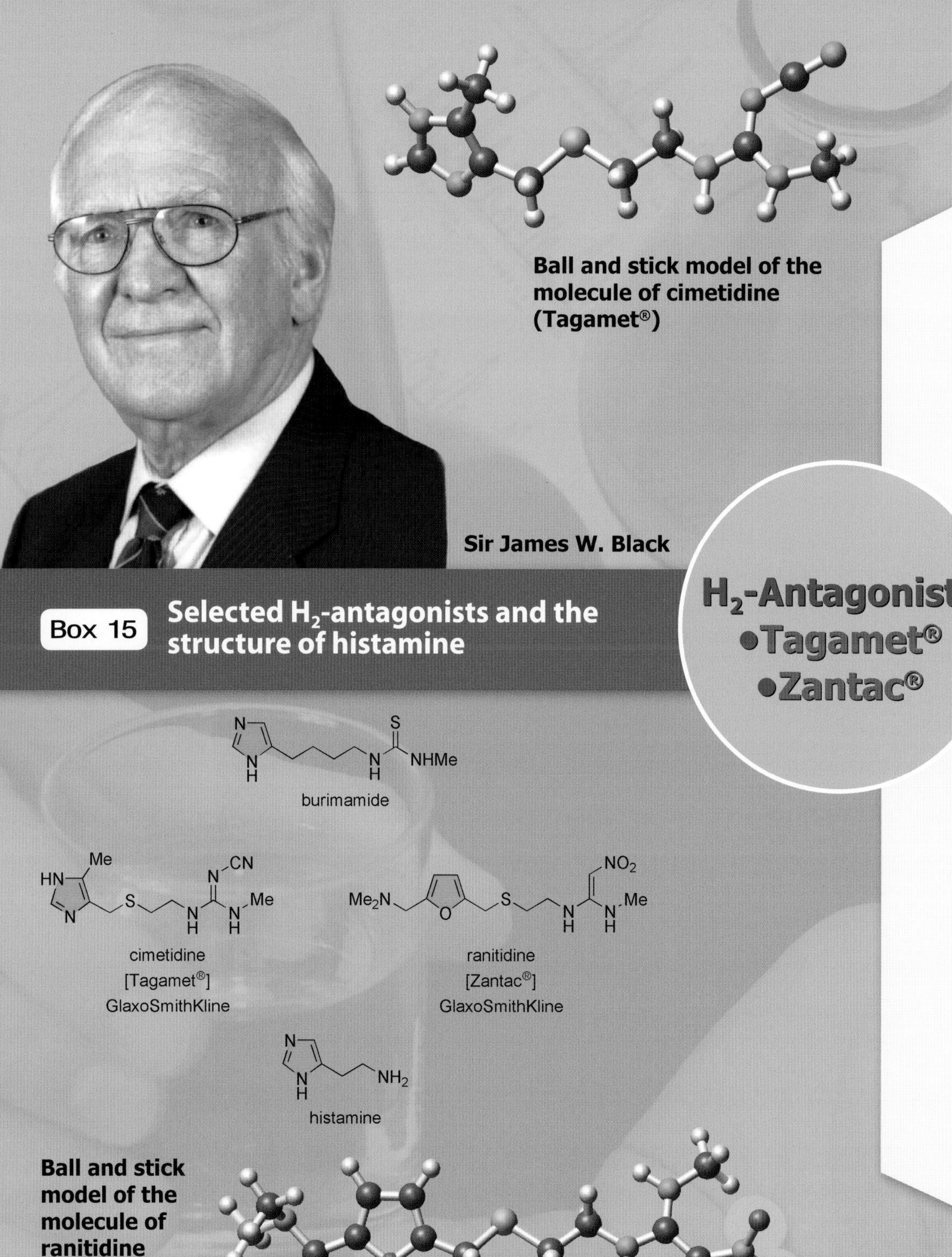

Ball and stick model of the molecule of cimetidine (Tagamet®)

Sir James W. Black

Box 15 Selected H_2-antagonists and the structure of histamine

Ball and stick model of the molecule of ranitidine (Zantac®)

H_2-Antagonists
•Tagamet®
•Zantac®

not be treated effectively. Patients were confined to bed, fed a bland diet, and given antacid tablets. Healing was slow and frequently incomplete, and many would eventually require surgery. The development of the H_2-antagonists began with the work of Sir James W. Black, who determined the link between histamine (Box 15) and increased acid secretion. Histamine is a chemical mediator that plays a key role in many physiological processes, including allergy and inflammation. The wide range of physiological effects of this compound make specificity paramount in any program aimed at disrupting only one of its roles. Four types of histamine receptor (H_1–H_4) are now known, each regulating a different task. Following work by Heinz Schild, Black and his colleagues characterized the H_2 receptor and showed that histamine binding at this site stimulated gastric secretions. Black and his team at the pharmaceutical company of SmithKline and French (UK) began to search for a suitable antagonist (a molecule that inhibits the normal physiological function of a receptor) for the H_2 receptor, with a view to suppressing excess stomach acid. Between 1964 and 1968, the team prepared over two hundred synthetic histamine mimics, but only one compound showed the potential to prevent histamine secretion. After years of further refinement, they prepared burimamide, a thiourea derivative (Box 15). Hailed as a major breakthrough, their results were published in the journal *Nature* in 1972. However, burimamide was not sufficiently potent for oral administration and required injection. In the search for a better drug, a further series of compounds was designed and prepared by a group of chemists from the original team, Graham J. Durant, John C. Emmett, and C. Robin Ganellin. In 1972, they discovered a cyanoguanidine analogue that they called cimetidine (Tagamet®, Box 15). This time the candidate sailed through the approval process and was launched in 1976 as an anti-ulcer drug. By 1979, Tagamet® was the top-selling pharmaceutical in North America. Sir James Black's vision in pioneering the design of molecules to fit known receptors revolutionized the drug discovery process and led to the award of the 1988 Nobel Prize in Physiology or Medicine (shared with Hitchings and Elion).

Meanwhile, the rival firm Glaxo was also searching for H_2-antagonists. Following a quantitative structure-activity relationship (QSAR) strategy, they refined their design, eventually discovering ranitidine (Zantac®, Box 15), which was launched in 1981. Despite being the second drug to make it to the market, Zantac® was also hugely successful, becoming the world's top-selling drug in 1988. The patents on both drugs have now expired and lower doses are available in over-the-counter medications in many countries. Interestingly, the former rivals subsequently joined forces to form the global pharmaceutical giant GlaxoSmithKline (GSK).

Proton Pump Inhibitors. In the late 1980s, the relatively small Swedish pharmaceutical firm Astra launched a new type of drug to compete with the H_2-antagonists. Their drug, called Losec® (omeprazole, Box 16), blocks a gastric H^+/K^+-ATPase, an enzyme that pumps acid into the stomach, and is, therefore, a highly effective antiulcer drug. Losec® was originally marketed as the racemate (an equal

Proton Pump Inhibitors
•Losec®
•Nexium®

mixture of both enantiomeric forms), but was later reformulated as Nexium®, which contains only the active (*S*)-enantiomer. This 'purple pill' has been the subject of one of the biggest pharmaceutical advertising campaigns of recent times in the USA, contributing to its huge success and the emergence of AstraZeneca (as the company became known following a merger with the pharmaceutical arm of Zeneca) as a major player in the pharmaceutical industry. The combined sales of omeprazole (Losec® and Nexium®) in its various guises can now be measured in the tens of billions of dollars.

Drugs for the Heart

There is little need to reiterate the significance of drugs used to treat heart conditions. The story of a number of cholesterol-lowering agents (Chapter 26, Mevacor®, Zaragozic acids, and the CP molecules) highlighted the catastrophic impact of cardiovascular disease on modern society, and outlined some of the measures taken to combat heart problems. Many of the hypocholestemic agents described in Chapter 26 were natural products, or derivatives thereof. In contrast, several other cardiovascular medications comprise rationally designed, small molecule drugs. We shall concentrate here on two classes that have both historical significance and enduring application.

β-Blockers. In the story of Tagamet® above, we encountered the Nobel Laureate Sir James Black. Before Tagamet®, however, this highly successful scientist discovered the β-blockers as treatments for high blood pressure and angina. The story of this discovery begins over two centuries earlier, with the first description of angina in a medical text. In 1768, William Heberden accurately stated, "They who are afflicted with it are seized while they are walking (more especially if it be uphill and soon after eating) with a painful and most disagreeable sensation in the breast, which it seems as if it would extinguish life, if it were to increase or continue." Progress in unraveling the intricate physiology of the heart and the etiology of heart disease was slow during the eighteenth and nineteenth centuries. In 1928, Chester S. Keefer and William H. Resnik published a comprehensive explanation of angina. They confirmed that the cause was oxygen starvation in the coronary tissues, and that this could arise from several underlying conditions, from coronary artery disease and vasospasm, to decreased oxygen saturation in the blood. Heart attacks (myocardial infarctions) were directly linked to coronary artery obstructions and angina, and the risk of sudden death in angina sufferers was highlighted. This work provided the physiologic basis for Black's work in treating these conditions. His great insight was

Medical illustration of a heart

Box 16 Omeprazole, the first proton-pump inhibitor

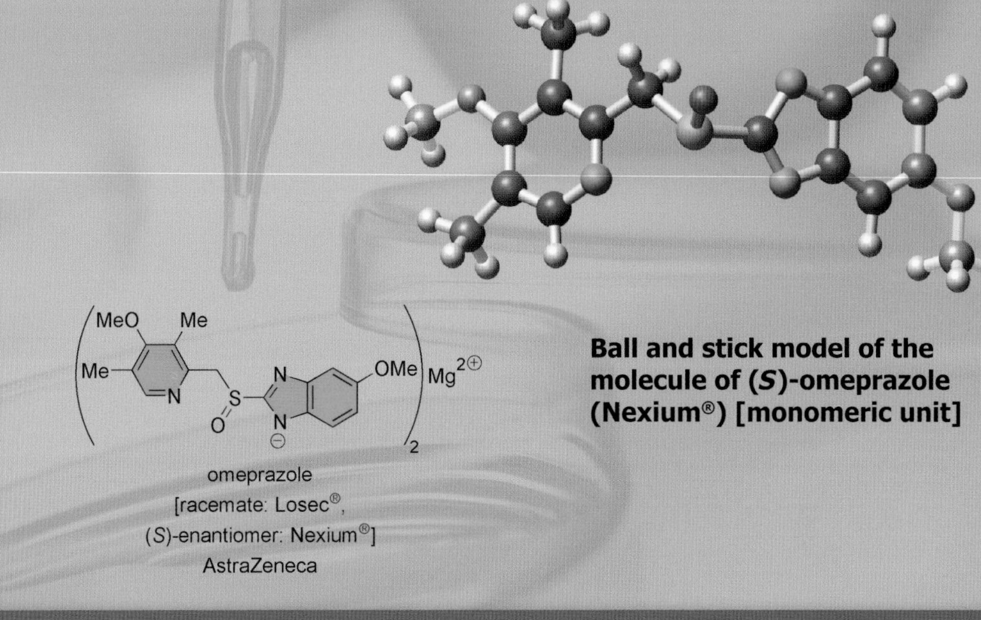

omeprazole
[racemate: Losec®,
(*S*)-enantiomer: Nexium®]
AstraZeneca

Ball and stick model of the molecule of (*S*)-omeprazole (Nexium®) [monomeric unit]

Box 17 Structures of epinephrine (adrenaline), norepinephrine (noradrenaline), and phenoxybenzamine hydrochloride (Dibenzyline®)

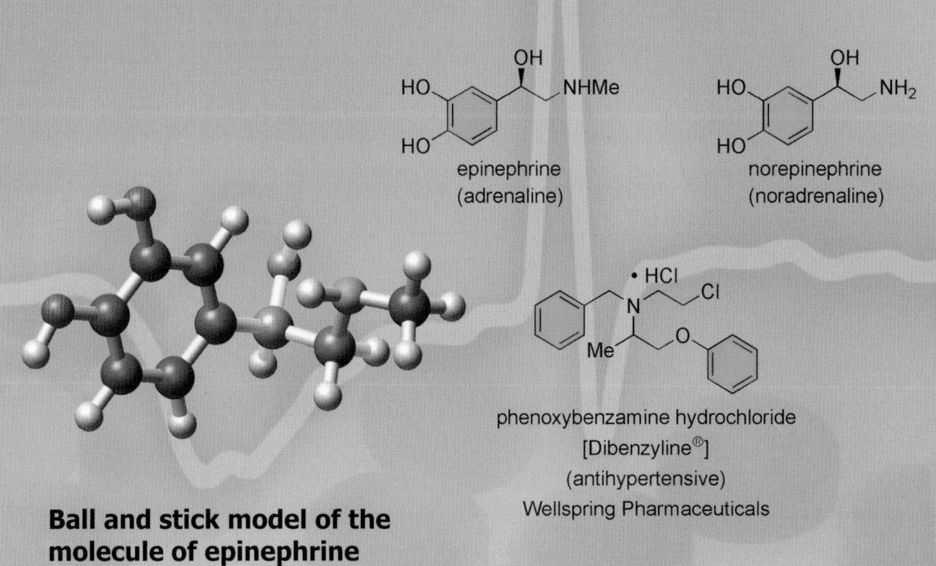

epinephrine
(adrenaline)

norepinephrine
(noradrenaline)

phenoxybenzamine hydrochloride
[Dibenzyline®]
(antihypertensive)
Wellspring Pharmaceuticals

Ball and stick model of the molecule of epinephrine (adrenaline)

Raymond P. Ahlquist

Box 18 Selected β-blockers

non-selective β-blocker

propranolol hydrochloride
[Inderal®]
Wyeth Ayerst

selective β_1-blockers

bisoprolol fumarate
[Zebeta®]
Wyeth Ayerst

atenolol
[Tenormin®]
AstraZeneca

esmolol hydrochloride
[Brevibloc®]
Baxter Healthcare

acebutolol hydrochloride
[Sectral®]
Wyeth Ayerst

Drugs for the Heart
•β-Blockers
•Propranolol
•Adrenaline

to connect the clinical knowledge of heart conditions (he was a clinician himself by training) with the growing body of knowledge of the pharmacology of the autonomic nervous system.

William Bates reported the discovery of adrenaline (or epinephrine, Box 17) in the *New York Medical Journal* in 1886. It was isolated and purified by Polish physiologist Napoleon Cybuski and American pharmacologist John Jacob Abel in 1895 and 1897, respectively. Japanese scientist Jokichi Takamine independently isolated a mixture of adrenaline and noradrenaline (norepinephrine, Box 17) in 1901 and was granted a US patent. He would make a fortune from marketing this mixture under the name Adrenalin®. Animal studies showed that adrenaline causes tachycardia and increased blood pressure. Early anti-adrenaline drugs (e.g. phenoxybenzamine, Box 17) were found to reverse the rise in blood pressure, but to have no effect on heart rate. This dichotomy led Walter B. Cannon and Arturo Rosenblueth to conceive the sympathin hypothesis, invoking different molecular modifiers, in 1939. Unfortunately, this erroneous idea became entrenched, and delayed the discovery of the true nature of the action of adrenaline for many years.

In 1948, Raymond P. Ahlquist proposed that different receptors were responsible for the various effects of adrenaline. He named the receptors alpha (α) and beta (β), names that are still used today. Such was the support for the sympathin hypothesis that Ahlquist found it difficult to publish his results, yet today he is hailed as a visionary scientist. Black was intrigued by the implications of Ahlquist's work, and, in 1958, embarked on a research program to design drugs to reduce the oxygen demand of the heart. This strategy was at odds with standard practice, which focused on vasodilators such as nitroglycerin. Calcium channel blockers, another important class of heart drugs, were originally developed as vasodilators. Black believed that oxygen consumption in the heart was determined by both systemic arterial pressure and heart rate. Disrupting the former is too dangerous because it can cause heart attacks, but reducing heart rate, which is controlled by the autonomic nervous system,

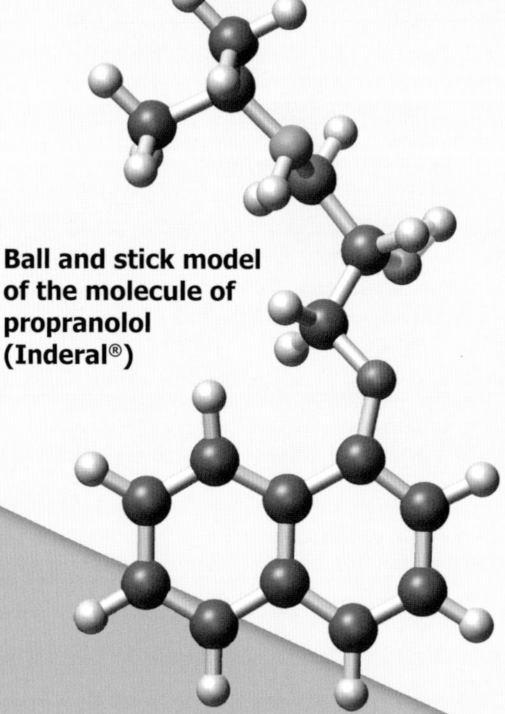

Ball and stick model of the molecule of propranolol (Inderal®)

would have the desired effect. Working at Imperial Chemical Industries (UK) with medicinal chemist John Stephenson, Black began making and testing compounds to block the β-adrenergic receptors. Following a long search, a potent β-antagonist, named nethalide, was discovered, but, while it was effective *in vivo*, it also induced tumors in mice and thus was abandoned. Fortunately, a safer compound, propranolol (Box 18), was found and launched as a drug in 1964 under the brand name Inderal®. Since then, propranolol has changed the face of cardiac medicine. It reduces the incidence of heart attacks in susceptible individuals, treats angina, and lowers dangerously high blood pressure. Curiously, propranolol also dilutes memories, removing the strongest emotions. This occurs because adrenaline is produced in the body in times of excitement or trauma, however, propranolol blocks its action and therefore dulls the memory of the event. Propranolol has been an enduring success, and is still in use today, despite the advent of newer drugs. Many of the newer generation of β-blockers (Box 18) are selective antagonists for the β_1-adrenergic receptors (those found in the heart) and do not affect the β_2-receptors found in smooth muscle. As a result, these drugs have fewer side effects.

Nitroglycerin. Nitroglycerin is closely associated with the name of Nobel. Alfred Nobel was born in Stockholm (Sweden) in October 1833. His father, Immanuel Nobel, was an engineer and inventor who had an interest in the development of explosives. Alfred was highly intelligent and, by age seventeen, was fluent in five languages. His interests spanned science, poetry, and English literature. The young Alfred was sent on an extended tour of scientific institutions, which his father hoped would focus his interests and energy on the physical sciences so he could enter the family business. In Paris, he worked with the famous chemist Théophile-Jules Pelouze. Here he met a young Italian, Ascanio Sobrero, who had prepared an extremely explosive liquid, trinitroglycerin (nitroglycerin), by the reaction of glycerin with a mixture of nitric and sulfuric acids. Sobero believed nitroglycerin to be worthless, as it was too dangerous and unstable to be of practical use. He also noted that exposure to the vapor of this compound resulted in severe headaches, hinting at the medicinal application that would follow. Alfred Nobel was convinced that he could find some way to control the detonation of nitroglycerin, and thus make his fortune.

Meanwhile, Immanuel Nobel had enjoyed great success in St. Petersburg (Russia), where he designed and sold the first naval mines during the Crimean War. However, at the end of the war the market for explosives diminished and Immanuel was bankrupted. Two of his sons remained in St. Petersburg and later became rich in the oil industry. Immanuel and his two other sons, Alfred and Emil, returned to Sweden, where Alfred began investigating a way to tame nitroglycerin. His success did bring fortune, but it came at a high price; Emil Nobel was amongst those killed by an explosion at the factory. Explosions were so frequent that the Stockholm authorities banned Nobel from working within the city limits. He moved to a barge on Lake Mälaren where he found a way to stabilize the explosive by mixing nitroglycerin with

Robert F. Furchgott

Louis J. Ignarro

Ferid Murad

Box 19 Structures of isoamyl nitrite and glycerin trinitrate (trinitroglycerin, nitroglycerin)

isoamyl nitrite

glycerin trinitrate
(trinitroglycerin, nitroglycerin)

Ball and stick model of the molecule of trinitroglycerin (nitroglycerin)

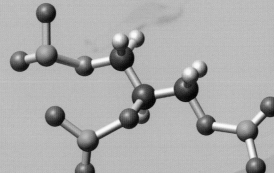

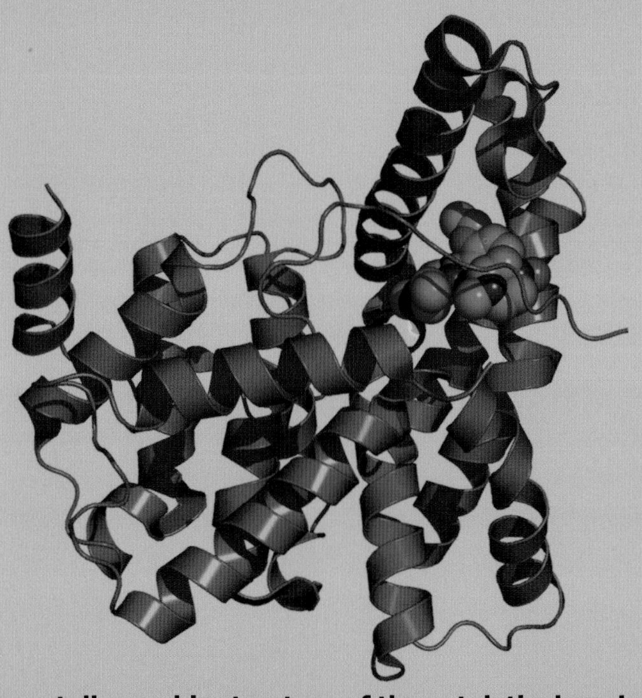

X-Ray crystallographic structure of the catalytic domain of human PDE5A1 in complex with sildenafil (Viagra®) (space filling model)

Drugs for ED
- Viagra®
- Cialis®
- Levitra®

Box 20 Drugs to treat sexual dysfunction

sildenafil citrate [Viagra®] Pfizer

tadalafil [Cialis®] Eli Lilly/Icos

vardenafil hydrochloride [Levitra®] Bayer/GlaxoSmithKline

silica to form a paste that could be molded into charges for use in demolition. The new explosive, called dynamite, was hugely successful in both civil and military applications. Contrary to public perception, Alfred did not initially seem overly disturbed by the potential military application of dynamite and went on to develop other important military explosives. While he was a staunch pacifist who campaigned against war, he believed that peace could be guaranteed only by the invention of a weapon so powerful that it would act as a deterrent.

Nobel endured ill health throughout his life, and in old age he suffered from heart disease and angina. In Paris in 1890, his doctors prescribed nitroglycerin, but he refused to believe it could help and mocked the name, Trinitrin, given to the medicine, believing it was used to avoid alarming patients. The British physician Sir Thomas Lauder Brunton first noted the beneficial effects of organic nitrites in treating angina in 1867. He originally studied amyl nitrite, but nitroglycerin (Box 19) soon became the treatment of choice. In an ironic twist, after over one hundred years of clinical use, the 1998 Nobel Prize in Physiology or Medicine was awarded to Robert F. Furchgott, Louis J. Ignarro and Ferid Murad for determining the mechanism of action of nitroglycerin. They discovered that it stimulates the release of the vasodilator nitric oxide (NO) from endothelial cells of the blood vessels. The NO diffuses into the smooth muscle where it triggers the relaxation of the myofilaments, causing the blood vessels to widen and thus reduce blood pressure. Nitroglycerin is still in use today, although newer and safer drugs have largely supplanted it.

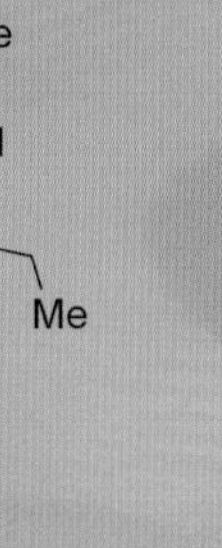

Viagra® pills

Drugs to Sexual Dysfunction

The final drug to be described in this chapter was initially investigated as a treatment for angina, but, while it would fail in this regard, it would open up a new field of medical intervention. In 1992, sildenafil citrate (later branded Viagra®) was being tested as an angina treatment by Pfizer (Sandwich, UK). It was not doing well in early clinical trials; however, an unexpected new property was discovered, saving sildenafil from obscurity. Volunteers in clinical trials are required by law to report every reaction they feel, no matter how insignificant it seems. A number of volunteers overcame their initial embarrassment to report experiencing an increased tendency to get erections. With this result, the drug was reinvestigated as a treatment for impotence, or erectile dysfunction (ED) as it became known. This time the trials progressed successfully, and Viagra® was launched in 1998. Within weeks, more than one million prescriptions had been written for Viagra® in the United States alone. After just one full year on the market, sales of Viagra® had exceeded $1 billion.

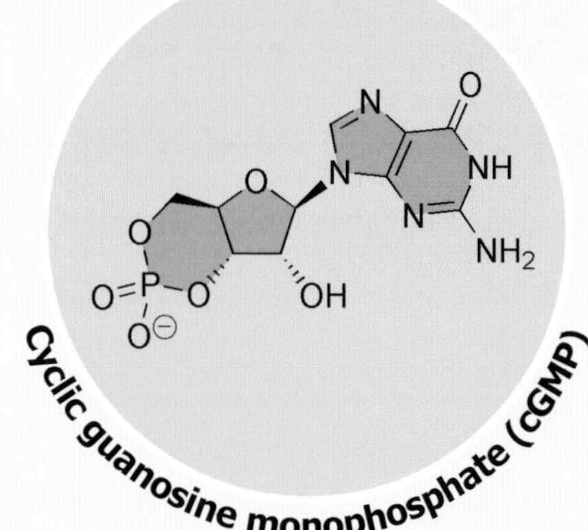

Cyclic guanosine monophosphate (cGMP)

Alfred Nobel and the Nobel Prize

Viagra® (Box 20) is a selective and potent inhibitor of the enzyme phosphodiesterase type 5 (PDE5). PDE5 cleaves and inactivates the chemical messenger cyclic guanosine monophosphate (cGMP). cGMP causes smooth muscle relaxation in the erectile tissue, increasing blood flow. The release of nitric oxide into the blood during sexual arousal triggers the production of cGMP, and it has been suggested that an inability to produce enough nitric oxide causes erectile dysfunction. Inhibiting PDE5 leads to increased levels of cGMP, facilitating erection. The effectiveness of Viagra® may also be enhanced by psychological factors, but such effects are difficult to quantify and are still under investigation. Recent reports suggest that Viagra® can also increase blood flow to the sexual organs in women, making it an effective treatment for female sexual dysfunction. Two other PDE5 inhibitors have subsequently been marketed as treatments for erectile dysfunction, Cialis® (tadalafil, Eli Lilly/Icos) and Levitra® (verdenafil hydrochloride Bayer/GlaxoSmithKline). In 2004, these three drugs boasted combined sales of $2.4 billion.

In June 2005, sildenafil (marketed as Revatio®) was approved as a treatment for

Nobel Prize medal

Alfred Nobel

Box 21 The Nobel Prize

Alfred Nobel bequeathed his substantial fortune for the purposes of awarding annual prizes in the fields of Chemistry, Physics, Physiology or Medicine, Literature, and Peace, reflecting his diverse interests. These five original Nobel Prizes, as the awards became known, were supplemented by the addition of the Nobel Memorial Prize in Economics in 1968. The prizes have been awarded since 1901, and are the most prestigious awards given in each category. Initially, the awards honored the most important work carried out in each field in the previous year, however, the scope has since widened, and the prizes are often given many years after the work has been done. Many of the protagonists in this volume have been awarded the Nobel Prize for their groundbreaking research.

pulmonary arterial hypertension (PAH), a progressive disease characterized by increasing pulmonary vascular resistance, ultimately leading to right ventricular failure of the heart. The connection between erectile dysfunction and PAH may seem tenuous at first, but the same mechanism of action of sildenafil is operative in both cases. PDE5 is primarily distributed in arterial wall smooth muscle in the penis and lungs, thus sildenafil induces vasodilation selectively in these areas of the body. By inhibiting the PDE5-mediated breakdown of cGMP in the smooth muscle of the pulmonary vasculature, sildenafil relaxes the arterial wall, decreasing pulmonary vascular resistance and alleviating the symptoms of PAH.

This chapter has focused on selected important pharmaceuticals, and indicates the central role of organic chemistry in the drug discovery and development process. Of course, the small molecule drugs discussed above are but a few of the many wondrous medications we have available today. The number of treatments and their power to cure will surely increase in the future as science and technology continue to open new opportunities for drug designers to exploit.

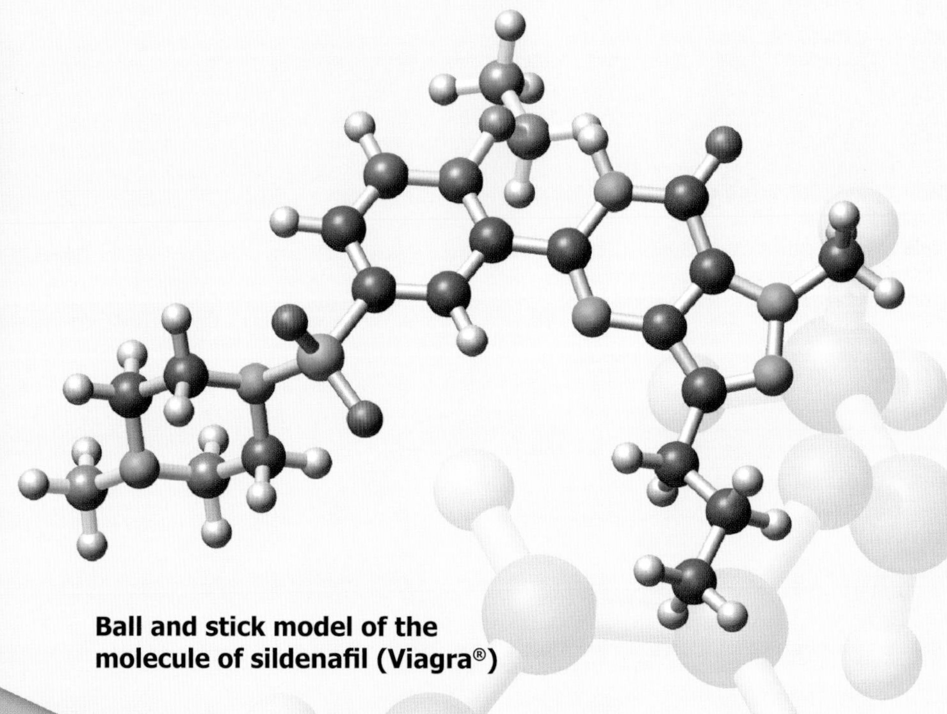

Ball and stick model of the molecule of sildenafil (Viagra®)

Further Reading

M. E. Bowden, A. B. Crow, T. Sullivan, *Pharmaceutical Achievers*, Chemical Heritage Foundation, Philadelphia, **2003**.

Top Pharmaceuticals, *Chem. Eng. News.* **2005**, *83*, 44–136.

T. Bartfai, G. V. Lees, *Drug Discovery, from Bedside to Wallstreet*, Elsevier Academic Press, San Diego, **2006**.

W. Sneader, *Drug Discovery, A History*, John Wiley & Sons, Ltd., Chichester, **2005**.

E. J. Corey, B. Czakó, L. Kürti, *Molecules and Medicine*, John Wiley & Sons, Ltd., Hoboken, **2007**.

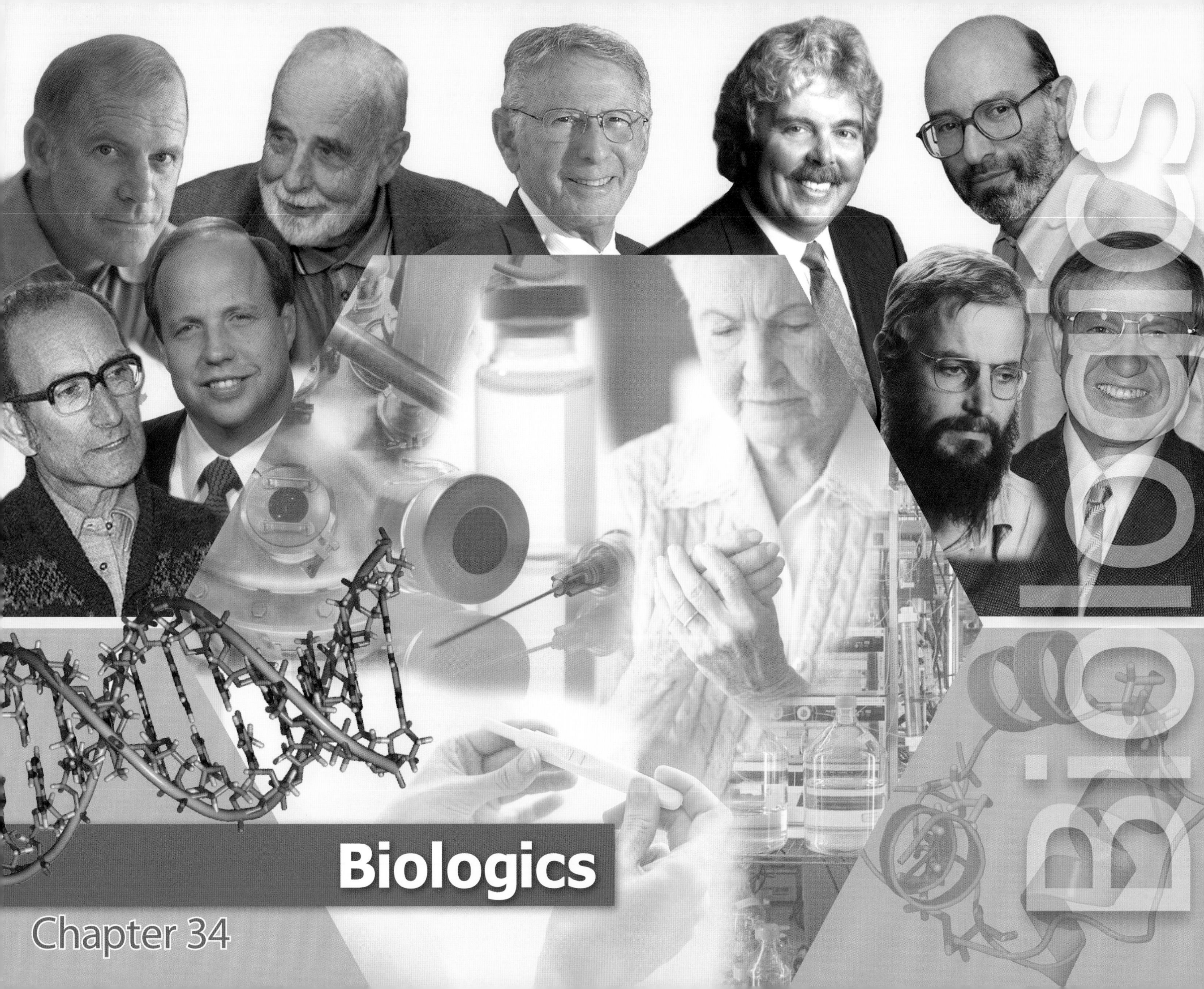
Biologics
Chapter 34
Biologics

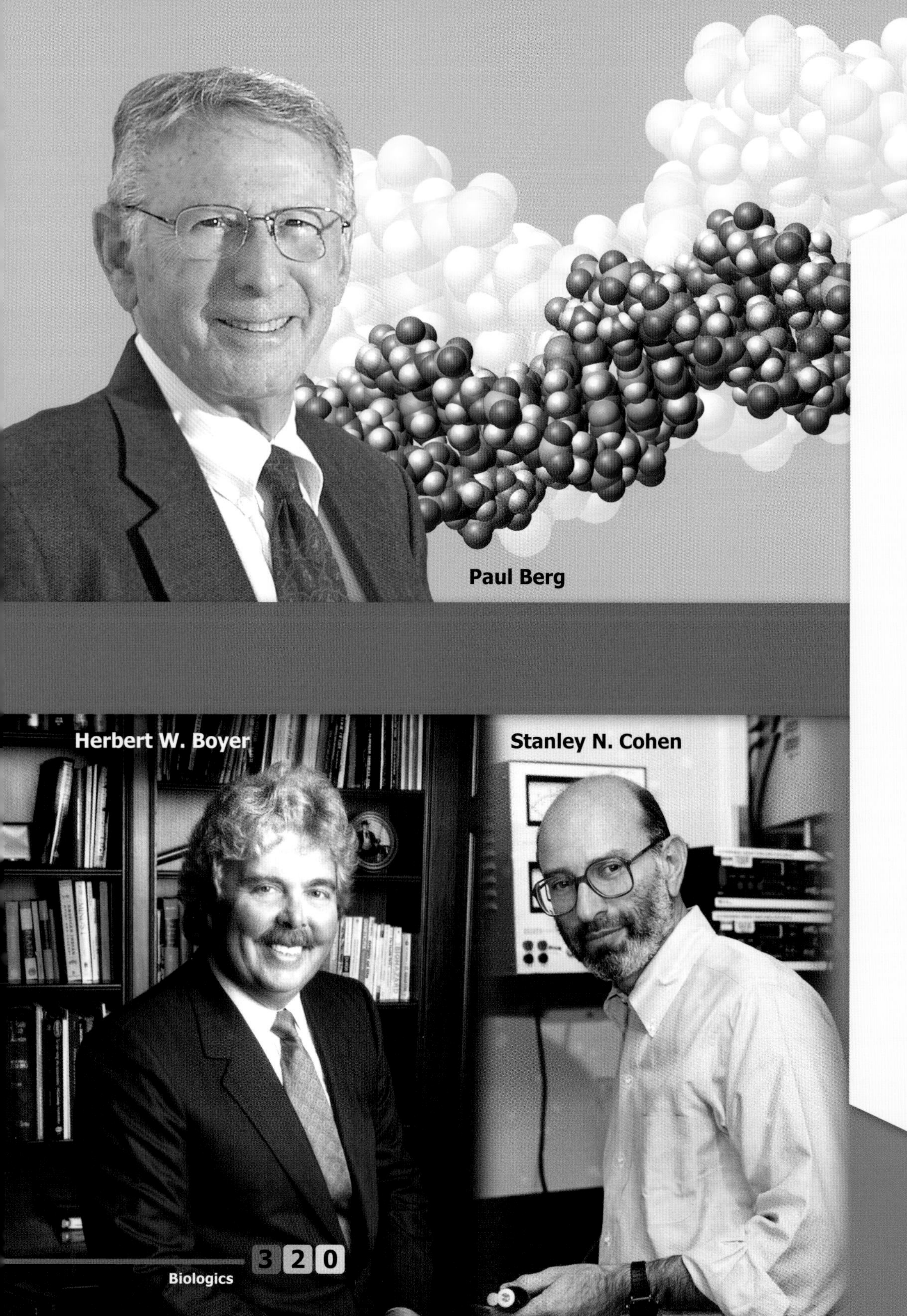

Even the largest natural products discussed in this book are relatively tiny compared to many of the giant biopolymers produced by living organisms. Such polymers include proteins, nucleic acids, and polysaccharides, and it is only recently that we have been able to produce some of them in the laboratory. Until the development of modern biotechnology, these macromolecules had to be harvested from natural sources, a costly and time-consuming process plagued by problems of scale and purification. Transference of disease by products contaminated with, for example, viruses and prions was an additional concern. Relying on natural sources also limited our ability to adapt and manipulate the molecules in question. In this chapter, we will see how the boom of modern biotechnology in the 1970s and 1980s led to a host of new techniques which provided solutions to so many of these issues. We will then discuss a number of macromolecules that have been brought to the market as lifesaving medications, collectively known as biologics.

Escherichia coli

The United States government defines biotechnology (biotech) as any technique that uses living organisms (or parts of organisms) to make or modify products, including foods, medicines, and soap powders, to name but a few items. Under this very broad definition man has been using biotechnology for thousands of years. Activities such as bread, wine, and cheese production all involve the use of colonies of living microorganisms to assist in making the end-product. More recently, antibiotics such as penicillin have been, and continue to be, manufactured using fermentation biotechnologies. However, the triggering of the biotech revolution has only been possible due to the major breakthroughs made in DNA-based technologies over the last few decades, some of which are highlighted below.

The 1970s witnessed the emergence of genetic engineering at the molecular level. Initial research conducted by biochemist Paul Berg (Stanford University, USA) showed that DNA from different species could be integrated with surgical precision to form what would become known as recombinant DNA. Paul Berg's fundamental work in establishing this field led to him being one of the three recipients of the 1980 Nobel Prize in Chemistry. These DNA-manipulating techniques were further developed by a biochemist, Herbert W. Boyer (University of California, San Francisco, USA), and a geneticist, Stanley N. Cohen (Stanford University, USA). In recombinant DNA technologies, sections of DNA that code for an important protein can be identified, spliced from the DNA chain, and inserted into a vector which carries the DNA to a new host where it is replicated. Specially adapted, fast growing bacteria from the *Escherichia coli* (*E. coli*) strain are the most commonly employed hosts. As a consequence of their normal reproductive cycles, the host cells replicate the introduced DNA and begin to synthesize the targeted protein which can be harvested and purified ready for its intended use. In principle, large amounts of the recombinant protein can be produced in factories using vats of genetically engineered organisms.

Escherichia coli

In a great coup in the late 1970s, a chemist turned venture capitalist, Robert A. Swanson, courted Herbert Boyer and persuaded him to set up the first modern biotechnology company, Genentech (South San Francisco). The first meeting between Swanson and Boyer, which took place in 1976, has become legendary. It is said that Swanson had become excited upon reading about Boyer's pioneering recombinant DNA work, so much so that he placed a call to Boyer and requested an immediate meeting. Boyer, absorbed by his science, somewhat grudgingly agreed to give the young venture capitalist a ten minute window in his busy schedule. Swanson's vision and enthusiasm was contagious and the ten minute meeting grew into a three hour discussion. By the time their conversation concluded on that day, Genentech had been conceived. Genentech's initial triumph was to manufacture the first human protein (somatostatin, a growth hormone-releasing inhibitory factor) using a microorganism (*E. coli*). By the end of the 1970s, Genentech scientists had also produced recombinant human insulin and a human growth hormone. Genentech's portfolio allowed it to go public in 1980, and it has gone from strength to strength as a market leader in biologics. In 1982, the Food and Drug Administration (FDA) granted approval to Genentech for the first genetically engineered drug – human insulin produced by bacteria. In addition to marketing its own products, Genentech has also worked closely, through licensing agreements, with many of the biggest pharmaceutical companies in the world to rapidly bring recombinant products

•Recombinant DNA Technologies

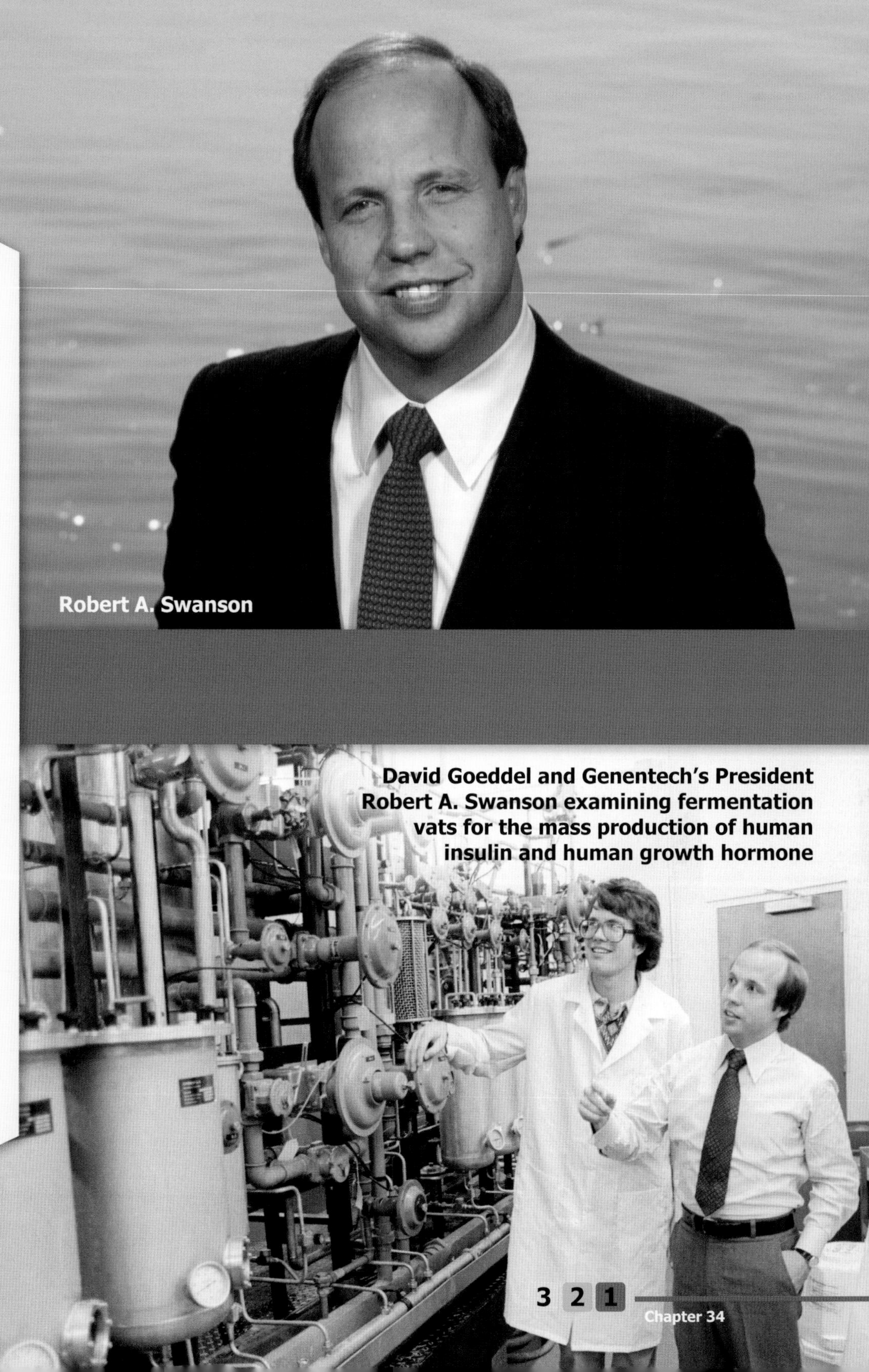

Robert A. Swanson

David Goeddel and Genentech's President Robert A. Swanson examining fermentation vats for the mass production of human insulin and human growth hormone

Modern pregnancy tests use monoclonal antibodies to detect minute traces of the peptide hormone human chorionic gonadotropin (hCG) in the blood or urine to indicate the presence or absence of an implanted embryo.

to the market. The Genentech formula has become the industry standard and many of the biotechnology firms in operation today aspire to emulate the success of Genentech. Indeed, many scientists and venture capitalists strive to repeat this success story as a means to progress their discoveries into marketable drugs.

Besides recombinant proteins, another important class of engineered biomolecules is monoclonal antibodies. Monoclonal antibodies are homogenous populations of immune system proteins designed to bind specifically to a single site on a desired molecular target (antigen). They are produced using special tumor cells (called hybridomas) that have been fused with antibody producing cells. These tumor cells multiply rapidly and indefinitely, generating potentially large amounts of antibodies. Because the tumor cells themselves are all clones of a single parent cell, the antibodies they produce are all identical. The "...discovery of the principle for production of monoclonal antibodies" won Georges J. F. Köhler (Basel Institute for Immunology, Switzerland) and César Milstein (Medical Research Council Laboratory of Molecular Biology, Cambridge, UK) a share of the 1984 Nobel Prize in Physiology or Medicine. They shared this prize with Niels K. Jerne (also at Basel Institute for Immunology), who was recognized "for theories concerning the specificity in development and control of the immune system." Monoclonal antibodies are now widely used as analytical tools in basic research and medical diagnosis. Modern pregnancy tests use monoclonal antibodies to detect minute traces of the peptide hormone human chorionic gonadotropin (hCG) in the blood or urine to indicate the presence or absence of an implanted embryo. The use of monoclonal antibodies is also becoming increasingly important in the treatment of certain diseases, and this chapter will highlight one such application (see also Chapter 23, Calicheamicin γ_1^I).

The discovery of recombinant proteins and monoclonal antibodies, to name but two applications, would not have been possible without concomitant advances in the areas of DNA sequencing, amplification, and chemical synthesis. Paul Berg's fellow recipients of the 1980 Nobel Prize

Niels K. Jerne

- **DNA Structure**
- **DNA Synthesis**
- **DNA Sequencing**
- **PCR**

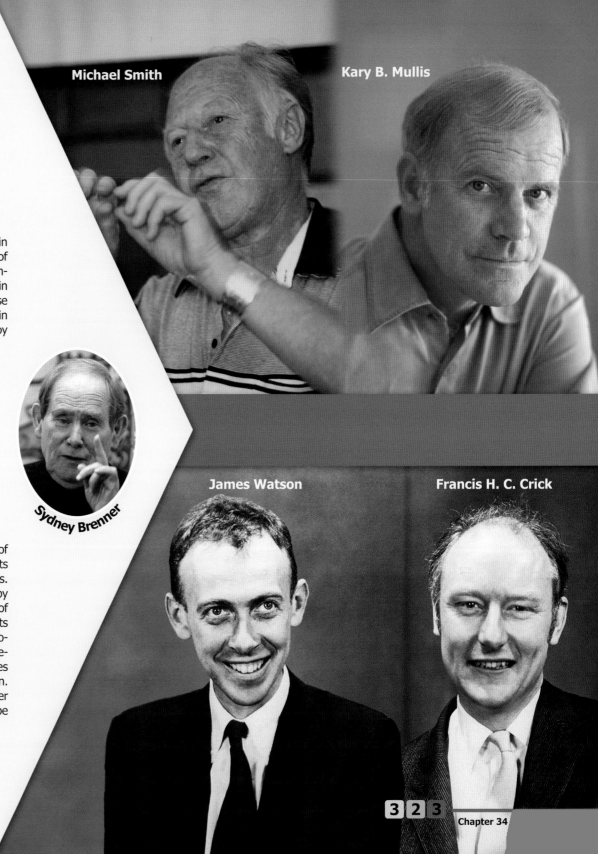

in Chemistry, Walter Gilbert (Harvard University, USA) and Frederick Sanger (Cambridge University, UK), independently developed methods for determining the nucleotide order of DNA fragments, in essence enabling the genetic code to be read in the laboratory. These DNA sequencing techniques are of key importance in the Human Genome Project and related endeavors. In 1983, Kary B. Mullis and co-workers at the biotechnology company Cetus developed a method for enzymatically replicating DNA sequences *in vitro*. Known as the polymerase chain reaction (PCR), this technique enables a small amount of the DNA sequence to be amplified rapidly and exponentially. Today, PCR is used routinely in biological research and other applications; for example, the basis of one type of HIV test involves PCR multiplication of viral RNA, while in another application PCR is used in forensic science to detect and profile minute residual pieces of DNA left at a crime scene. Mullis shared the 1993 Nobel Prize in Chemistry with Michael Smith (University of British Columbia, Canada). Smith pioneered site-directed mutagenesis, the technique by which mutations can be created at defined sites in DNA sequences, and thus specific changes to the amino acid sequence of the protein for which the DNA encodes can be engineered. Also in 1983, Marvin H. Caruthers (University of Colorado, Boulder, USA) developed a chemical method whereby fragments of DNA, in a predetermined sequence of 5–75 base pairs, could be constructed *in vitro*. Within a short time, both Caruthers and Leroy Hood (then at the California Institute of Technology, USA) had invented instruments that automated this DNA synthesis process. This work built upon the foundations laid by H. Gobind Khorana (then at the University of Wisconsin, USA, later at the Massachusetts Institute of Technology, USA), who pioneered methods for the chemical synthesis of oligonucleotides (short sequences of DNA or RNA nucleotides) in solution. Khorana's work, which followed earlier studies by Sydney Brenner (who would be

Frederick Sanger

Walter Gilbert

Marshall W. Nirenberg

Sydney Brenner

Michael Smith

Kary B. Mullis

James Watson

Francis H. C. Crick

Dr. Edward Jenner vaccinating young boy

Box 1 The emergence of vaccines

Date of introduction of first vaccine	Disease
1796	smallpox
1885	rabies
1897	plague
1923	diptheria
1926	pertussis
1927	tetanus
1927	tuberculosis (BCG)
1935	yellow fever
1955	poliomyelitis (injectable)
1962	poliomyelitis (oral)
1964	measles
1967	mumps
1970	rubella
1981	hepatitis B
2006	human papillomavirus

awarded a share of the 2002 Nobel Prize in Physiology or Medicine for later studies on the genetic regulation of organ development), Francis H. C. Crick and others, was also instrumental in elucidating the nature of the genetic information contained in the DNA double helix model proposed by Watson and Crick. Khorana shared the 1968 Nobel Prize in Physiology or Medicine with Robert W. Holley (Cornell University, USA) and Marshall W. Nirenberg (National Institutes of Health, Bethesda, Maryland, USA) "for their interpretation of the genetic code and its function in protein synthesis."

In addition to the selected few scientists discussed above, it should be remembered that a great many other researchers have contributed to the advancement of these fields. Today we can decipher the genetic code of an organism, make DNA strands, produce proteins, and even genetically modify plants and animals, all for the benefit of humanity. To be sure, many more breakthroughs, scientific and technological, are waiting in the wings.

Frederick G. Banting

Vaccines

Vaccines are the oldest and broadest category of biologic drug. Their use has succeeded in eliminating or substantially reducing the devastating impact of many terrible diseases. Today they are used routinely in the developed world, where immunization, especially in childhood, is largely taken for granted. The principle of immunization is much older than many believe. The term 'vaccine' was coined by Edward Jenner in the eighteenth century, but he was not the first to use this technique. Smallpox was probably the first disease to be prevented with vaccines, and it is the only one to have been eradicated in this way. The origins of the practice can be traced to ancient India and China, where physicians used material derived from smallpox sufferers to inoculate healthy people. In China, dried and powdered smallpox pustules were inhaled, while in India smallpox scabs were placed on scratched skin. In 1718, Lady Mary Wortley Montagu, wife of the British ambassador to the Ottoman Empire, reported that the Ottomans inoculated themselves using fluid from the pustules. She endeavored to have this practice adopted in the UK, and even inoculated her own son.

Lady Mary Wortley Montagu

In 1796, Edward Jenner, an English country doctor,

Louis Pasteur

•Vaccines
•Antibodies

noted that several local milkmaids seemed to be immune to smallpox. He attributed this protection to their exposure to the related, but milder, cowpox. If the milkmaids contracted cowpox from their daily contact with cows, they could no longer catch smallpox. In an experiment that would make most people balk today, Jenner took infectious fluid from a milkmaid, Sarah Nelmes, and inoculated a local eight-year-old, James Phipps. Forty-eight days later, when Phipps had recovered from the cowpox, Jenner exposed the child to smallpox. Fortunately, he did not develop any symptoms of the more serious disease. This experiment also gave rise to the term vaccine, named from *vaccina*, the infectious cowpox agent. The great biologist and chemist Louis Pasteur built on Jenner's success, developing vaccines for a number of diseases of viral and bacterial origin, including rabies, cholera and anthrax. Since then, a broad range of vaccines has been developed, increasingly making use of better-characterized formulations (Box 1).

The first public mass vaccinations were carried out as early as 1805, when the Mexican and Guatemalan authorities organized a systematic anti-smallpox inoculation program. This foresight allowed the Americas to be declared smallpox-free long before the rest of the world. A global program did not begin until 1956, when the World Health Organization (WHO) undertook a campaign to eradicate smallpox. They discovered that high coverage and a second line of rapidly targeted vaccinations, to respond to local outbreaks, were needed for success. In 1979, the WHO announced that smallpox had been eradicated. Expanded WHO plans to eradicate other diseases, most notably poliomyelitis, for which cheap vaccines are available, are now underway.

Variola virus that causes smallpox

Vaccines work by preparing our immune systems to fight off the infection efficiently. The vaccine comprises foreign matter that triggers an immune response. Macrophages and lymphocytes attack the foreign material and B-lymphocytes produce antibodies against the invading proteins. When the initial response has passed, memory cells (specialized B-cells) persist in the blood. If the same (or similar) infective agent enters the bloodstream again, the memory cells proliferate rapidly and mount a powerful immune response, preventing the infection from taking hold. Depending on the type of vaccine used, the memory cells may give life-long protection, or they may persist for only a short time, and thus require booster vaccinations. Some details of a number of vaccine types are shown in Box 2.

Vaccines are undoubtedly the most powerful and cost-effective of public health measures, however they are not without controversy. The safety of the vaccines and their formulations must be closely monitored to maintain confidence in vaccination programs. For example, suggested links between some vaccines and autism have led to reduced vaccination uptake in

Box 2 Selected types of vaccines and their properties

Vaccine type	Immunization mode	Strength of immunity	Example
Inactivated/killed	The infecting organism is killed (by soaking in formalin) prior to being injected into the patient	Not strong, booster shots are required	Typhoid vaccine and the Salk poliomyelitis vaccine
Acellular	Part of the organism is injected (e.g. flagella, protein cell wall)	Not strong, booster shots are required	*Haemophilus influenzae* B (HIB-B) vaccine
Live attenuated/ weakened	Weakened organism injected, once inside body multiplies causing a strong immune response. Most dangerous because can mutate back to active form and become infective.	Strong, lifelong protection	Measles, mumps, rubella vaccines
Toxoid	Toxin from organism injected after treatment to render it harmless (e.g. by addition of aluminium salts)	Low level immunity, often administered with an agent that enhances immune response	Tetanus and diphtheria vaccines
Subunit	Genetic engineering used to produce harmless bacteria (or yeast) host cells, which carry part of genetic code of the infective pathogens	Variable, usually good, protection	Hepatitis B vaccine
Similar	Like Jenner's original, a similar but milder pathogen is used to protect the patient	Strong, usually lifelong protection	Bacillus Calmette-Guérin (BCG) vaccine for tuberculosis

Charles H. Best (left) and Frederick G. Banting in the laboratories of John J. R. Macleod

•Diabetes •Humalin® (Insulin)

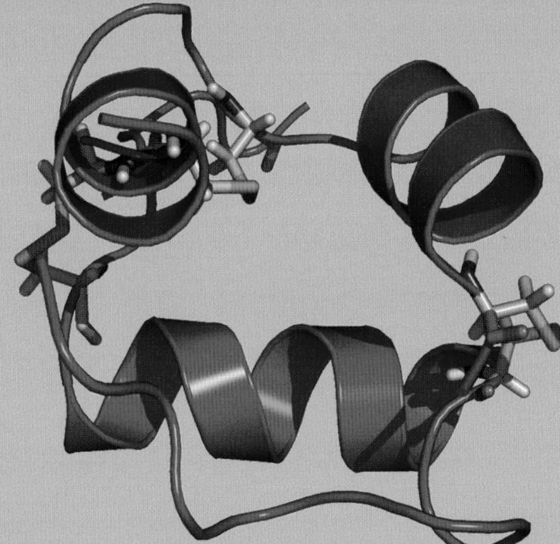

Solution-phase average structure of the highly active monomeric des-[Phe(B25)] human insulin mutant as determined by NMR spectroscopy and molecular dynamics calculations. The one intrastrand and two interstrand disulfide bonds are depicted in stick format.

some countries, leaving populations vulnerable to outbreaks of disease. In addition to proper education, developed nations have a responsibility to increase the availability of vaccines in the developing world, where many children are still dying from preventable diseases.

Humalin® (human insulin)

Insulin provides an excellent gauge of the development of therapies using large biomolecules over the last century. It is a polypeptide hormone produced by β-cells in the pancreas, which regulates blood glucose levels. Diabetes mellitus, the disease that results from a failure to control glucose levels, occurs in two distinct types. Type 1 usually begins in childhood and is characterized by a lack of insulin production. To survive, type 1 diabetics must receive daily injections of insulin. Type 2 diabetes accounts for nearly ninety percent of all cases and, although it is not caused by insulin deficiency, approximately one third of sufferers benefit from insulin therapy. Diabetes is becoming ever more prevalent. The WHO estimates that the number of diabetics worldwide grew from 30 million in 1985 to 177 million in 2000, and that 300 million people will suffer from the disease by the year 2025. It also estimates that nine percent of all deaths are caused by diabetes, with the vast majority of these being considered premature. The cost of treating diabetes is, therefore, enormous; the worldwide cost of insulin alone was $6.75 billion in 2004. This expense cannot be met in many countries, and thus diabetics are left to suffer the consequences such as loss of limbs, blindness, heart disease and kidney failure. Hence, there is the pressing need to discover and develop newer sources of insulin that can satisfy the demand and provide the drug less expensively (to all those patients who need it).

Schack A. S. Krogh

Insulin originally took only twelve months from first isolation to commercialization, due to the dire need for diabetes treatments and the rather less stringent controls placed on drugs in the early twentieth century. Two Canadian scientists, Frederick G. Banting and Charles H. Best, working in the laboratory of John J. R. Macleod at the University of Toronto isolated insulin in 1921. After some initial reluctance they agreed to collaborate with the pharmaceutical industry, and Connaught Laboratories (now part of Sanofi Aventis) and Eli Lilly began producing insulin. The academic scientists also offered royalty-free licenses to a number of European institutions. Danish Nobel Laureate August Krogh, whose wife was a diabetic, exploited this opportunity by founding Nordic Insulin Laboratories (now Novo Nordisk). Through the Medical Research Council, the UK became the first country to approve insulin injections for diabetics, in 1922. Eli Lilly followed suit by winning approval in the USA in 1923 with a much larger scale production. Macleod and Banting received a share of the 1923 Nobel Prize in Physiology or Medicine for their work. The brief gap between the discovery and this award indicates the perception of the importance of their research. At that time, the structure of insulin was unknown, even though it is a relatively small protein, having just fifty-one amino acids (twenty-one in chain A and thirty in chain B, the two chains being joined by two disulfide bonds). The structure was eventually solved in 1969 by Professor Dorothy Crowfoot Hodgkin at the University of Oxford (UK) using x-ray crystallography. She had devoted thirty-five years of work to the problem, beginning in 1934 when she was given a sample of ten milligrams of insulin by another prominent Oxford chemist, Sir Robert Robinson. It took one year to grow suitable crystals, and a further thirty-four years of data analysis in the pre-computer era to determine the full structural details of insulin.

For more than half a century, the demand for insulin was met using supplies extracted from animals. Bovine (cow) and porcine (pig) insulin molecules differ only slightly from the human form and can usually be substituted without problems. However, the growing numbers of diabetics and problems with allergic reactions to the animal products fueled a search for altenative insulin supplies. Insulin is too large for laboratory chemical synthesis to yield practical quantities, so the breakthrough in production had to wait until the advent of genetic engineering (see above). In 1978, the fledgling biotechnology company Genentech collaborated with the City of Hope National Medical Center (California, USA) to produce human insulin using recombinant DNA technology. They synthesized the gene for insulin and introduced it into the genome of *E. coli* bacteria. Fermentation of the bacteria led to the production of large quantities of the hormone, which could be extracted using techniques similar to those developed for penicillin production. At that time, the production of active peptides by genetic engineering was forbidden, so the company prepared the two chains in separate fermenters and joined them chemically after purification. Before such a drug could be launched, the regulatory authorities demanded strict testing to ensure the material was safe. The chemical and biological properties of the recombinant DNA and the active protein had to be determined and the process for joining the two chains understood fully. The folding of the synthetic peptide also had to be examined in detail. Clinical trials began in 1980 and, in 1982, the UK approved the first genetically engineered drug. It was named Humalin® and was produced by Eli Lilly under a licensing agreement from Genentech. This collaboration heralded a new trend in which established pharmaceutical firms use their infrastructure to carry out the development and marketing of discoveries made in smaller, risk-taking biotechnology companies. A new era in drug discovery was on the horizon. Following the launch of Humalin®, analogues of insulin with more favorable pharmacological properties, such as delayed effect and lengthened duration of action, were introduced. More than twenty such drugs are now available. A number of companies are now developing other drug treatments for type 2 diabetes and, in 2005, the FDA approved a new drug to help control blood glucose in patients for whom insulin therapy alone is insufficient. Symlin®, produced

•Anemia
•EPOGEN®
(Erythropoietin)

Dorothy Crowfoot Hodgkin

John J. R. Macleod

William K. Bowes

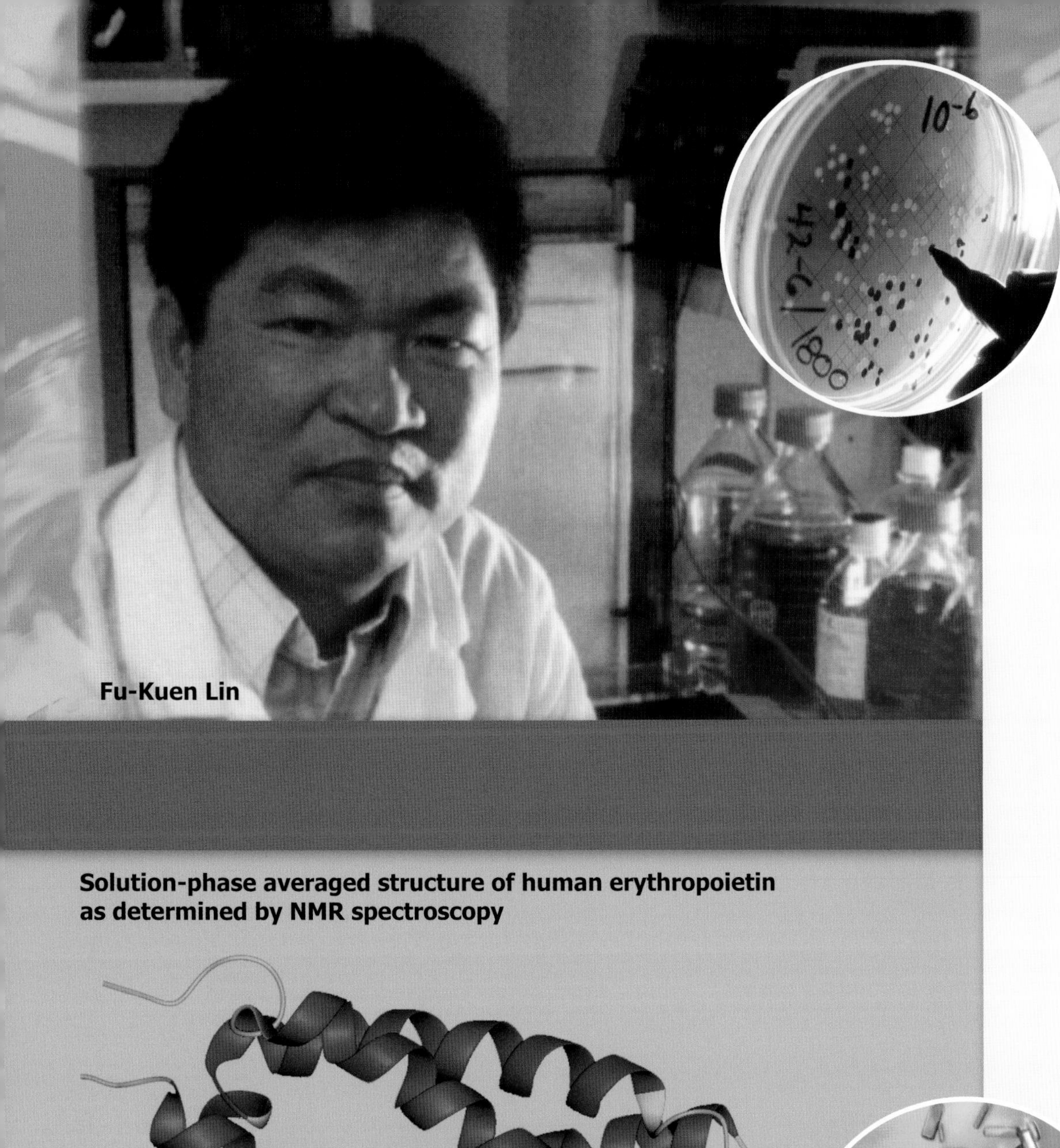
Fu-Kuen Lin

Solution-phase averaged structure of human erythropoietin as determined by NMR spectroscopy

Crystals

by Amylin Pharmaceuticals (USA), is a synthetic analogue of amylin, another pancreatic hormone that contributes to glucose maintenance. And Januvia®, developed by Merck, is the first of a number of emerging small molecule drugs to be approved by the FDA for the treatment of diabetes.

EPOGEN® (erythropoietin)

Erythropoietin (EPO) is a peptide hormone produced in the kidneys (and liver in infants). It stimulates the production of red blood cells from bone marrow and plays pivotal roles in regulating hemoglobin biosynthesis and red blood cell levels. The protein contains 165 amino acids and has four sugar groups attached to it. EPO was first isolated from urine in the 1970s. The gene encoding EPO was identified shortly thereafter, as it was clear that this protein had many potential medical applications, and by using the recombinant DNA techniques described in our introduction to this chapter EPO production was initiated in vitro. The competition in this field was fierce and led to one of the largest patent disputes in history. In 1996, Amgen finally triumphed over the Genetic Institute (now part of Wyeth) to secure full patent protection for its product, named EPOGEN®, seven years after it was first launched.

Amgen (the acronym stands for Applied Molecular Genetics) was founded in 1980 by a small group of scientists and investors led by the venture capitalist William K. Bowes, who recruited the first CEO of this fledgling company, George B. Rathmann. Amgen's first major achievement came in October 1983, when Fu-Kuen Lin successfully cloned the EPO gene. Lin's team then developed a method for producing EPO in a form and quantity that enabled its use in the treatment of anemia. The resulting molecule entered clinical trials in December 1985 and, in 1987, Amgen received its first patent on what would become the company's first medicine. On June 1, 1989, the FDA approved EPOGEN® for anemia management associated with end-stage renal disease. EPOGEN® became biotech's first blockbuster drug (annual sales of at least $1 billion); its commercial success was key in enabling Amgen to flourish from its humble beginnings in the (then) small town of Thousand Oaks, California. Today, Amgen is one of the world's biggest biotechnology companies, rivaling established 'Big Pharma' companies in size and capitalization, and continues to be a leading innovator in the identification, isolation, purification, and use of human proteins as therapeutic targets.

EPO is used to treat severe anemia. Those receiving cancer chemotherapy or HIV combination therapy can suffer serious anemia, which can be overcome using EPO injections. Sufferers of chronic kidney disease can also benefit from EPO. Furthermore, patients preparing for some surgeries can be given EPO prior to the operation in place of blood transfusions during the procedure. Unfortunately, its effects are exploited by some endurance athletes (such as cyclists, distance runners, and skiers) whose performance can be significantly enhanced by the EPO-induced increase in red cell count, and, consequentially, oxygen supply. Abuse of EPO in this way is very dangerous because the associated thickening of the blood can cause heart attacks and strokes, especially when the athlete becomes dehydrated. Indeed, EPO has been implicated in the deaths of a number of high-profile athletes. A legitimate way of gaining this advantage is to train at high altitude (above 6000 feet), where the body adapts to the thinner air by producing more EPO naturally.

In 2004, sales of the α-EPO variant produced by Johnson & Johnson under license from Amgen were $6.2 billion in the USA alone. Sales of other variants outside the USA contributed a further $4.8 billion. This partnership agreement has, however, been far from smooth due to alleged license and patent infringements, an increasing phenomenon between competing pharmaceutical companies. Amgen has recently developed darbepoetin (Aranesp®), an engineered EPO analogue that has two extra sugar units attached to its peptide backbone. These extra groups increase its potency while slowing down its clearance from the body, allowing for marked reductions in dosing.

The story of EPO shows how the lessons learned from insulin have been applied to produce a new blockbuster protein drug, however it also highlights the controversies that can arise from commercializing genetic technology and naturally occurring hormones.

- **Breast Cancer**
- **Herceptin®**

Herceptin® (trastuzumab)

In 1987, Dr. Dennis Slamon and colleagues at UCLA (University of California, Los Angeles, USA) published an article in *Science* in which they linked overexpression of the gene coding for a protein called HER2 (human epidermal growth factor receptor 2) with an aggressive form of breast cancer. Under normal circumstances HER2 transmits signals from outside the cell to the nucleus, directing cell growth. When cancer develops, cellular growth rates change and natural checks and balances controlling the cell cycle fail. The gene that encodes for the protein HER2 is called an oncogene, a gene having the potential to cause a normal cell to become cancerous. While it had been postulated for many years that oncogenes might be associated with growth factors, it was not until the early 1980s that Genentech scientists first proved that this hypothesis was correct when they found and studied an oncogene that was a mutated form of the epidermal growth factor (EGF) cell surface receptor gene. These researchers quickly broadened their investigation and began to search for other oncogenes related to the EGF-receptor gene. They named the first one they found HER2, and, using cloning technology, they were subsequently able to identify the protein for which HER2 coded. It was from a collaboration with the Genentech specialists that Dennis Slamon found high levels of HER2 oncogene expression in tumor samples taken from approximately 25 % of patients with metastatic breast cancer. Slamon found that the HER2 oncogene caused breast cancer cells to produce abnormally high concentrations

George B. Rathmann

Dennis Slamon

Breast cancer cells

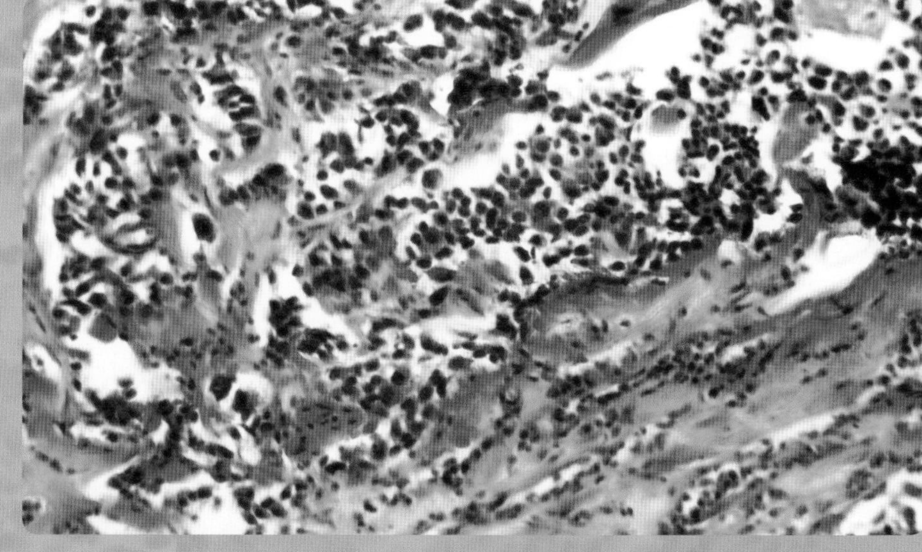

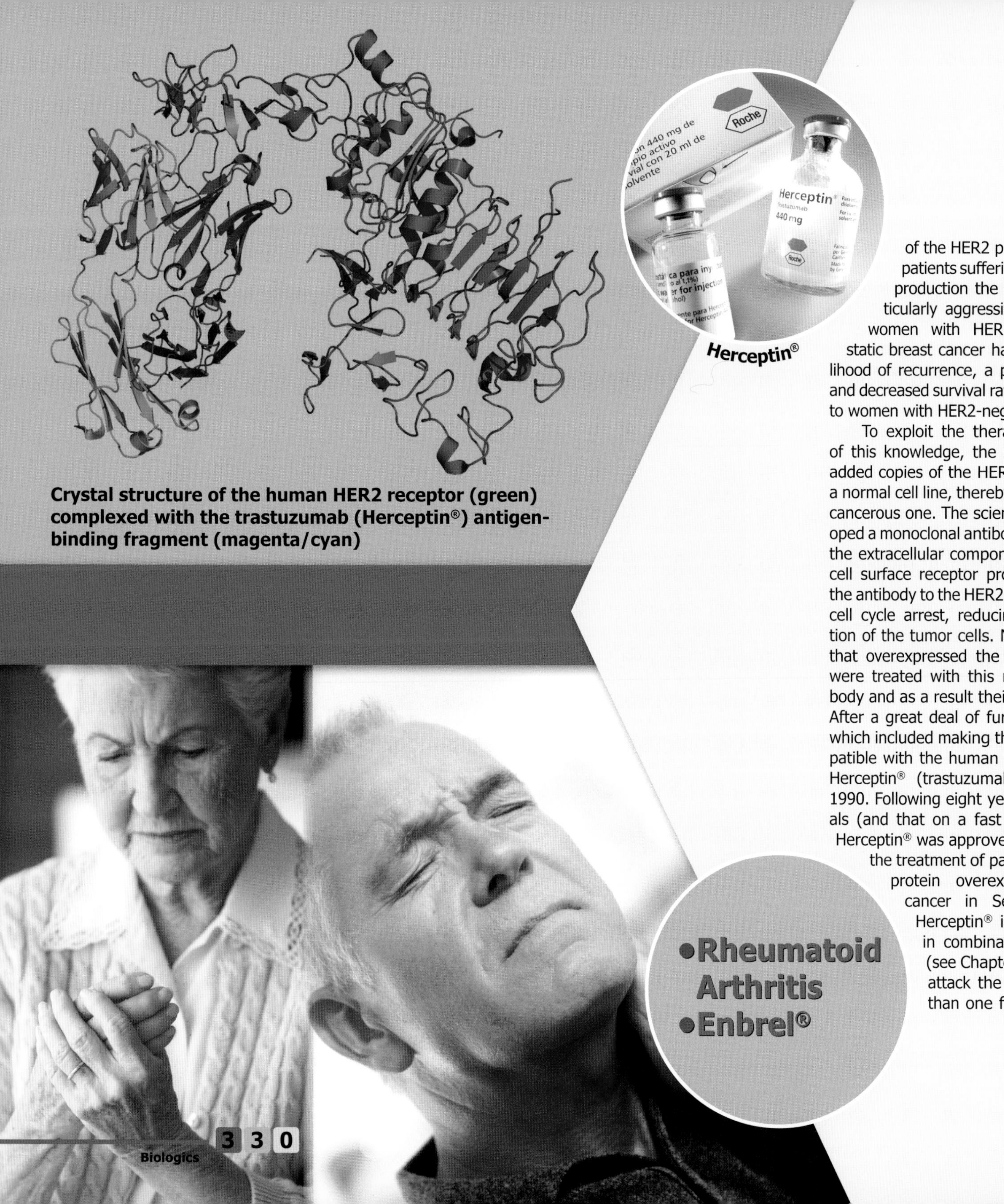

Crystal structure of the human HER2 receptor (green) complexed with the trastuzumab (Herceptin®) antigen-binding fragment (magenta/cyan)

Herceptin®

of the HER2 protein and that in patients suffering from this overproduction the disease was particularly aggressive. Furthermore, women with HER2-positive metastatic breast cancer had a greater likelihood of recurrence, a poorer prognosis, and decreased survival rates in comparison to women with HER2-negative cancer.

To exploit the therapeutic potential of this knowledge, the Genentech team added copies of the HER2 oncogene into a normal cell line, thereby turning it into a cancerous one. The scientists then developed a monoclonal antibody that bound to the extracellular component of the HER2 cell surface receptor protein. Binding of the antibody to the HER2 receptor induces cell cycle arrest, reducing the proliferation of the tumor cells. Mice with tumors that overexpressed the HER2 oncogene were treated with this monoclonal antibody and as a result their tumors shrunk. After a great deal of further refinement, which included making the antibody compatible with the human immune system, Herceptin® (trastuzumab) was born in 1990. Following eight years of clinical trials (and that on a fast track approval), Herceptin® was approved by the FDA for the treatment of patients with HER2 protein overexpressing breast cancer in September 1998. Herceptin® is generally used in combination with Taxol® (see Chapter 25) in order to attack the cancer on more than one front, thus giving patients increased chances of recovery. Herceptin® is thought to posses three separate and distinct types of activity: it blocks tumor cell growth, labels the cancer cells for destruction so the body's immune system becomes involved in fighting the cancer, and works synergistically with chemotherapy to destroy HER2-positive cancer cells. The worldwide distribution and marketing of Herceptin® was aided by a deal signed between Genentech and Hoffmann-La Roche in 1998.

Despite its success in treating breast cancer, it has emerged that the use of Herceptin® can result in the development of cardiac dysfunction in a minority of patients; as a result, those treated with Herceptin® have to be carefully monitored. In some countries, notably the UK, Herceptin® has provoked controversy for an entirely different reason. Herceptin® is considerably more expensive than many traditional chemotherapeutic agents and there is ongoing debate as to whether the drug is an affordable option. In 2006, after an acrimonious legal battle in the highest courts of the country, campaigners won the right for all British women suffering from the early stages of HER2-positive metastic breast cancer to receive Herceptin®. This ruling went against the judgment of the National Institute for Health and Clinical Excellence, the regulatory body charged with appraising which medicines and treatments should be available on the National Health Service (NHS). Critics of this ruling contend that the more liberal use of Herceptin® will inevitably lead to deleterious cutbacks elsewhere in the cancer treatment budget of the NHS. Nevertheless, Herceptin® represents a new direction in targeted therapy using monoclonal antibodies, and, perhaps, one of the first steps on the road to truly personalized medicine.

Enbrel® (etanercept)

Rheumatoid arthritis is a disease in which the body's immune system turns upon itself and attacks apparently healthy tissue, leaving the sufferer physically debilitated and living in constant pain. The exact cause of arthritis is unclear; environment, genetic traits, hormonal factors, and pathogens (viruses and bacteria) have all been implicated. Rheumatoid arthritis results in the joints becoming inflamed and, ultimately, seriously and irrevocably damaged. Damage occurs when the immune system is activated, prompting mediatory cells to migrate to the joints where they initiate an inflammation cascade (see Chapter 15, Prostaglandins & Leukotrienes). The inflammation irritates the locality, thus perpetuating the cycle of trigger followed by exaggerated immune response. Long-term prospects for the patient are grim. Cartilage begins to wear down and swelling of the joint lining (the synovium) means that the joint becomes increasingly damaged and painful. Rheumatoid arthritis affects two million Americans and, contrary to popular belief, can strike at any time in one's life. More women than men suffer from this condition and women are less likely to go into remission with treatment than their male counterparts. Many different approaches exist for the treatment of rheumatoid arthritis. Some drugs target the pain and inflammation directly [e.g. nonsteroidal anti-inflammatory drugs (NSAIDs), topical pain relievers, corticosteroids, and narcotic pain relievers], whilst others attempt to modify the disease progression completely. This latter class of drugs works by disrupting the immune response, and notable recent additions to this category are biologics.

Researchers at the biotechnology company Immunex developed etanercept as a recombinant protein for the treatment of inflammatory diseases. In 1998, the FDA approved etanercept for treatment of the symptoms of rheumatoid arthritis in patients whose disease had failed to respond to prior medication, and the drug was introduced to the market in 1999 under the trade name Enbrel®. Such was the immediate impact and success of Enbrel® that Immunex could not manufacture enough of the drug to keep up with demand. At the same time, the failure of other drug candidates in clinical trials had left the company in a perilous financial position. Seizing this opportunity, Amgen acquired Immunex, and the rights to Enbrel®, in a takeover bid in 2002. Following the merger, Amgen was able to bring additional manufacturing facilities on line, quickly clearing the 40,000-plus waiting list for the drug. Amgen now co-markets Enbrel® in a licensing agreement with Wyeth.

Enbrel® is a so-called dimeric fusion protein in which active domains of two human proteins were fused to generate a new protein that is 934 amino acids long and can be manufactured using recombinant DNA techniques. One protein is the extracellular ligand-binding portion of the receptor for tumor necrosis factor alpha (TNFα). The second protein is a human immunoglobin (IgG1), from which a portion called the Fc domain has been incorporated into Enbrel®. The net effect is an artificially engineered antibody. Enbrel® works by binding to TNFα, thereby blocking its normal interaction with cell surface TNFα

- Humira®
- Remicade®

Phage display

- Kineret®
- Orencia®

Above: X-Ray crystallographic structure of the TNFα trimer, the binding target of etanercept (Enbrel®) and related biologic therapeutics

George P. Smith

Richard A. Lerner

Sir Gregory Winter

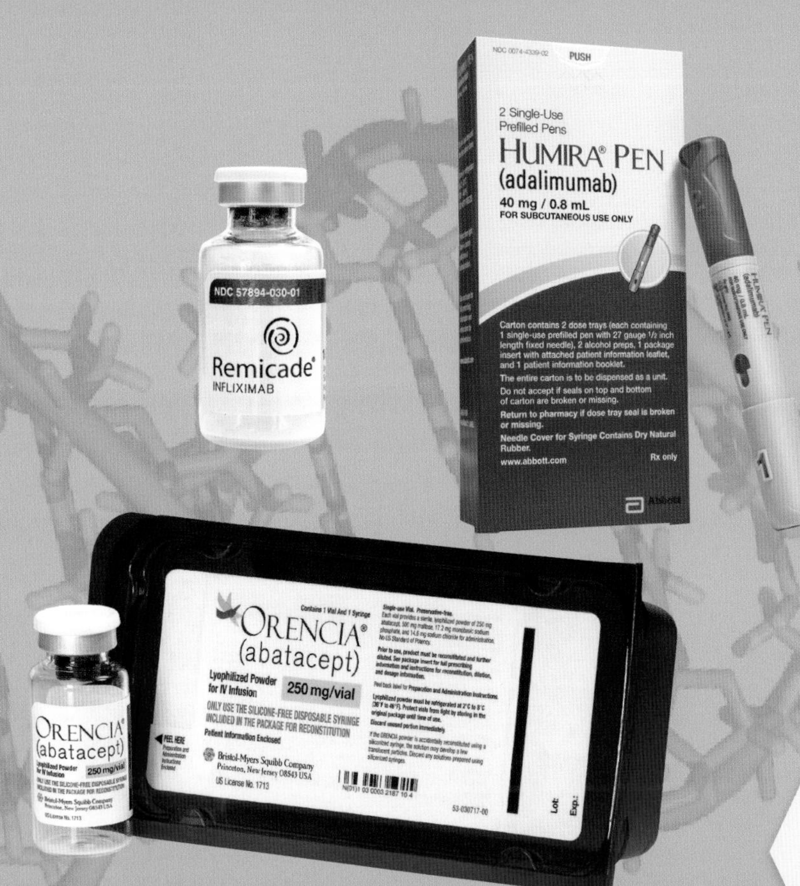

Drugs that relieve pain

receptors. TNFα is a cytokine, a chemical intimately involved in the normal inflammatory and immune responses. Elevated levels of TNFα are found in the tissues of patients suffering from rheumatoid arthritis. By binding to and deactivating TNFα, Enbrel® disrupts the biochemical cascade leading to inflammation, thus alleviating the symptoms of rheumatoid arthritis.

Since its launch, Enbrel® has brought some much needed relief to hundreds of thousands of rheumatoid arthritis patients worldwide. Its indications have been broadened so that it is now regularly used to treat juvenile arthritis, psoriatic arthritis, ankylosing spondylitis, and severe plaque psoriasis. A number of other biologics for the treatment of arthritis are also currently on the market. These products include Humira® (adalimumab, Abbott Laboratories) and Remicade® (infliximab, Centocor Pharmaceuticals), both of which also block TNFα activity, Kineret® (anakinra, Amgen), which blocks the cytokine interleukin-1, and Orencia® (abatacept, Bristol-Myers Squibb), which blocks activation of T-cells. Despite the fact that patients suffering from rheumatoid arthritis speak of these drugs in glowing terms, they are not without their disadvantages. Proteins are not stable to many conditions including stomach acid, therefore these drugs need to be refrigerated and injected subcutaneously. Furthermore, some patients taking these biologics suffer from depression of their immune response, making them more susceptible to infections.

Phage display, a technique for generating large libraries of antibodies, has been pivotal in the discovery of some of these therapeutic antibodies. George P. Smith (University of Missouri, USA), Richard A. Lerner (The Scripps Research Institute, La Jolla, USA), and Sir Gregory Winter (University of Cambridge, UK) contributed to the discovery and development of this powerful method.

We hope that on reading this survey of a few milestones in this area you will be left feeling some of the same excitement and hope that is driving scientists and biotechnology companies to invest ever more time and money in inventing new biologic drugs. Biologics are certain to have a bright future, but it is fair to say that future success in treating diseases will come not from finding one single solution to all problems, but from our ability to innovate and continually build a strong cross-disciplinary approach to research, one in which both chemistry and biology play equally important roles. The human genome project has certainly demonstrated and strengthened the foundation of this indispensible approach to drug discovery.

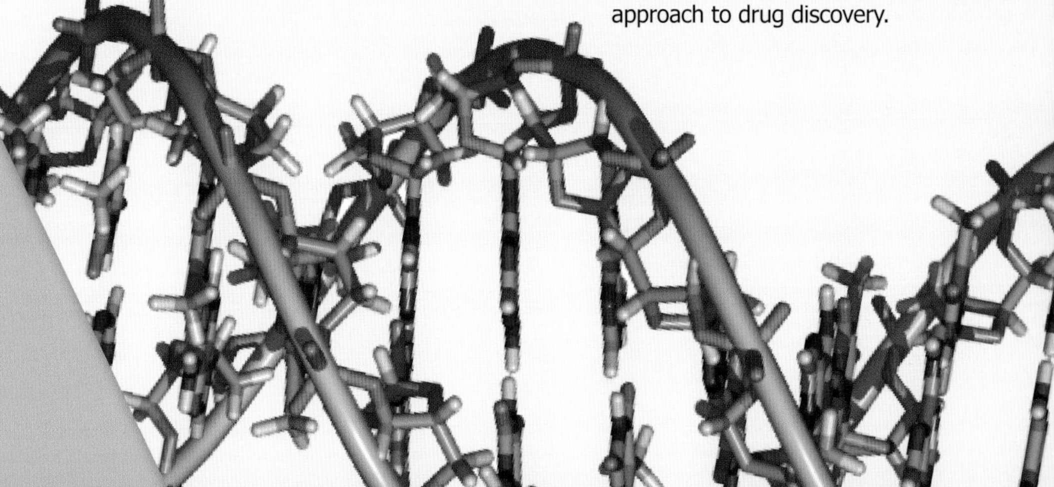

Further Reading

W. Bains, *Biotechnology from A to Z*, Oxford University Press, Oxford, **2004**.

R. J. Y. Ho, M. Gibaldi, *Biotechnology and Biopharmaceuticals: Transforming Proteins and Genes into Drugs*, Wiley-Liss, New Jersey, **2003**.

Top Pharmaceuticals, *Chem. Eng. News* **2005**, *83*, 44–136.

D. E. Duncan, *The Amgen Story*, Tehabi Books, San Diego, **2005**.

R. A. Lerner, Manufacturing Immunity to Disease in a Test Tube: The Magic Bullet Realized, *Angew. Chem. Int. Ed.* **2006**, *45*, 8106–8125.

Epilogue

The stories in this book have been about molecules that changed the world, their chemistry, biology and medicine, and the protagonists that discovered, synthesized and studied them. Because of our background and expertise, this book was written from a perspective amplifying the contributions of chemistry, especially chemical synthesis. We are sure that biologists and medical doctors or even other chemists would have different angles from which to approach the subject. To be sure, however, the fascination and impact of these molecules, and others like them, on society and the excitement of the science behind them remain undisputed.

We hoped that we have conveyed from our perspective the historical development of the art and science of chemical synthesis, its awesome power to create new materials for all imaginable and unimaginable applications for the benefit of humankind, and its importance in shaping the world as we know it today. Due to space and time limitations we focused primarily on medicine and certain other aspects where fine chemicals are playing a role in our lives such as dyes, perfumes, vitamins and agrochemicals. But molecules and chemistry offer so much more. Their central role within the sciences and our world make them enabling, indispensable, and rewarding beyond the boundaries of imagination.

Thus, it was the advent of chemistry and chemical synthesis that led not only to the magic of modern medicine, but also to the refinement of petroleum and the production of gasoline and other energy sources as well as fine chemicals that shaped the world in the twentieth century. Polymers and plastics for all purposes are also products of technologies based on the fundamentals of chemistry and chemical synthesis as are so many other high tech materials from which emerged all kinds of useful tools and devices such as liquid crystal displays, computer chips, engines, instruments of communication, and transportation vehicles. And it is chemistry again that will provide solutions to our future energy and material needs beyond petroleum.

This rich harvest speaks volumes for the future especially if we assume, as it has been said, that the twenty-first century belongs to chemistry and biology. Indeed these sciences stand at a threshold of major developments from which will flow untold, even more revolutionary, benefits to society. Cures for cancer, Alzheimer's disease and viral infections are but a few of such expectations. New energy sources and novel materials for medical devices and other special needs are more good things to come. The appeal of such eventualities and the excitement of being part of such scientific, technological, and medical breakthroughs provides enormous incentives and attractions to young men and women who will surely find the challenges ahead both stimulating and rewarding. We would like to think that, through the pages of this book, we have provided the inspiration for them to choose such careers and the encouragement to pursue them with vigor and enthusiasm fit for a journey full of discoveries and wisdom such as described in the following passage from the Greek poet C. V. Cavafy, with which we leave you.

Ithaca

When you start on your journey to Ithaca,
then pray that the road is long,
full of adventure, full of knowledge,
do not fear the Lestrygonians
and the Cyclopes and the angry Poseidon.
You will never meet such as these on your path,
if your thoughts remain lofty, if a fine
emotion touches your body and your spirit.
You will never meet the Lestrygonians,
the Cyclopes and the fierce Poseidon,
if you do not carry them within your soul,
if your soul does not raise them up before you.

Then pray the road is long.
That the summer mornings are many,
that you will enter ports seen for the first time
with such pleasure, with such joy!
Stop at Phoenician markets,
and purchase fine merchandise,
mother-of-pearl and corals, amber and ebony,
and pleasurable perfumes of all kinds,
buy as many pleasurable perfumes as you can;
visit hosts of Egyptian cities,
to learn and learn from those who have knowledge.

Always keep Ithaca fixed in your mind.
To arrive there is your ultimate goal.
But do not hurry the voyage at all.
It is better to let it last for long years;
and even to anchor at the isle when you are old,
rich withal that you have gained on the way,
not expecting that Ithaca will offer you riches.

Ithaca has given you the beautiful voyage.
Without her you would never have taken the road.
But she has nothing more to give you.

And if you find her poor, Ithaca has not defrauded you.
With the great wisdom you have gained, with so much experience,
you must surely have understood by then what Ithacas mean.

—C. P. Cavafy
(Translation by Rae Dalven)

ΝΙΚΟΛΑΟΥ
K.C. Nicolaou

Image Credits

Numbering refers to page on which image appears in order from top to bottom, then left to right.

Ball and stick models created by Mike Pique, Hao Xu, and William Brenzovich, unless stated otherwise.

Ribbon models created by Alex Perryman, unless stated otherwise.

Chapter 1: Introduction: Atoms, Molecules & Synthesis

1, 3. Demokritos bust. Courtesy of NCSR "Demokritos".
1, 2, 7. Atom. ©Digital Art/Corbis.
1, 2. Robert Boyle. ©Bettmann/Corbis.
1, 2. Antoine Laurent de Lavoisier. ©Stefano Bianchetti/Corbis.
1, 3. John Dalton. Edgar Fahs Smith Collection, University of Pennsylvania Library.
2. Big Bang. NASA/WMAP Science Team.
2. Universe. NASA/ESA/S. Beckwith(STScI) and The HUDF Team.
2. Planet Earth. Image courtesy of Earth Sciences and Image Analysis Laboratory, NASA Johnson Space Center.
4. Dimitri Mendeleev. Edgar Fahs Smith Collection, University of Pennsylvania Library (catalogued Dmitry Mendeleyev).
4. Periodic Table of Elements. Nicolaou group.
6. Acropolis looming over Athens. ©Wolfgang Kaehler/Corbis.
6. Plato. ©Bettmann/Corbis.
6. Aristotle. ©Bettmann/Corbis.
6. Socrates. ©Araldo de Luca/Corbis.
8. Henry Hurd Rusby, Wrestling the Jungle's Secrets. Used with permission from Pfizer Inc. Images by Robert Thom. All rights reserved.
8. *Sorangium cellulosum* cells. Courtesy of Hans Reichenbach.
8. Spiny puffer fish. National Oceanic and Atmospheric Administration/Department of Commerce.
8. Taxol® leaves. Courtesy of Stellios Arseniyadis.
8. Sea squirts – tunicates (*Rhopalaea*). Getty Images/Photographer: Gary Bell.
8. Penicillin and *Staphylococci* in a petri dish. ©Bettmann/Corbis.

Chapter 2: Urea & Acetic Acid

9, 10, 11. Friedrich Wöhler. Edgar Fahs Smith Collection, University of Pennsylvania Library.
9, 10. Urea crystals. Courtesy of Brian Johnston, Microscopy, UK.
9, 11. Vinegar. ©Zeva Oelbaum/Corbis.
9, 11. Hermann Kolbe. Edgar Fahs Smith Collection, University of Pennsylvania Library.
10. Labware. Courtesy of Valer Jeso, Nicolaou group.
10. Friedrich Wöhler. Edgar Fahs Smith Collection, University of Pennsylvania Library.
10. Urine samples in test tubes. ©Bill Varie/Corbis.
10. Justus von Liebig. ©Corbis.
10. Jöns Jakob Berzelius. ©Hulton-Deutsch Collection/Corbis.
12. Robert Boyle. ©Bettmann/Corbis.
12. Carl Wilhelm Scheele. Edgar Fahs Smith Collection, University of Pennsylvania Library.
12. Antoine Laurent de Lavoisier. ©Stefano Bianchetti/Corbis.
12. John Dalton. Edgar Fahs Smith Collection, University of Pennsylvania Library.
13. August Wilhelm von Hofmann. Edgar Fahs Smith Collection, University of Pennsylvania Library.
13. Friedrich August Kekulé. Edgar Fahs Smith Collection, University of Pennsylvania Library.
13. Joseph Priestley. Edgar Fahs Smith Collection, University of Pennsylvania Library.
13. August Wilhelm von Hofmann and students Royal College of Chemistry, London. The Royal Society of Chemistry.
13. Antoine Laurent de Lavoisier in lab in Paris. Used with permission from Pfizer Inc. Images by Robert Thom. All rights reserved.
13. Justus von Liebig in lab in Giessen. ©Bettmann/Corbis.
14. Early 19th century Queen Victoria. ©Stapleton Collection/Corbis.
14. Crowd. ©Rob Lettieri/Corbis.
14. Space shuttle. NASA.
14. Medicine cabinet. ©George B. Diebold/Corbis.
14. Fruit. ©Maximilian Stock Ltd/photocuisine/Corbis.
14. Georgette and Colette Nicolaou. Photo by K. C. Nicolaou.
14. Earth. Image courtesy of Earth Sciences and Image Analysis Laboratory, NASA Johnson Space Center.
14. 1974 Triumph GT6 Coupé. Photo by Adrian Pingstone.
14. Runway models. ©Reuters/Corbis.
14. Vicky Nielsen Armstrong. Courtesy of Vicky Nielsen Armstrong.
14. Perfume bottles. ©Colin Anderson/Corbis.
14. Pills. ©Ed Bohon/Corbis.
14. Sunset. ©Robert Glusic/Corbis.

Chapter 3: Glucose

15, 16. Emil Fischer. Edgar Fahs Smith Collection, University of Pennsylvania Library.
15. Fruit with figs. ©Ryman Cabannes/photocuisine/Corbis.
15. Ice cream scoops. ©Envision/Corbis.
15. Grapes. Photo by K. C. Nicolaou.
15, 20. Emil Fischer in lab. National Library of Medicine (NLM).
16. Looking up through the trees. ©Lester Lefkowitz/Corbis.
16. Spoon of sugar. ©Tom Grill/Corbis.
17. Andreas Sigismund Marggraf. ©Austrian Archives/Corbis.
17. Aleksandr Mikhailovich Butlerov. Edgar Fahs Smith Collection, University of Pennsylvania Library.
17. Emil Fischer. Edgar Fahs Smith Collection, University of Pennsylvania Library.
17. Carafe of milk. ©Adrianna Williams/zefa/Corbis.
17, 18. Beans and various high carbohydrate foods. ©Don Mason/Corbis.

17. Cut sugar cane. ©Jupiterimages/Brand X/Corbis.
17. Red lobsters. ©Richard T. Nowitz/Corbis.
18. Glass of wine. ©Goodshoot/Corbis.
18. Wine bottles, glass and corkscrew. ©Andrew Unangst/Corbis.
18. Wines and fruit. ©Orjan F. Ellingvag/ELLINGVAG/ORJAN/Corbis.
18. Louis Pasteur. Edgar Fahs Smith Collection, University of Pennsylvania Library.
18. Joseph-Achille Le Bel. Edgar Fahs Smith Collection, University of Pennsylvania Library.
18. Jacobus H. van't Hoff. Edgar Fahs Smith Collection, University of Pennsylvania Library.
19. "Lock and key" depiction. Nicolaou group.
19. Wine casks and tanks. ©Charles O'Rear/Corbis.
19. Loaf of Paillasse bread. ©J.Riou/photocuisine/Corbis.
19. Coffee beans. ©Dennis Degnan/Corbis.
19. Pieces of chocolate. ©Christian Schmidt/zefa/Corbis.
19. Protein molecule: triosephosphate isomerase. ©Corbis.
20. Cabernet Sauvignon grapes on the vine. ©Owaki - Kulla/Corbis.
20. Sugarcane field. ©David Muench/Corbis.
20. Woman drinking wine. ©Stefan Schuetz/zefa/Corbis.
20. Ice cream with berries. ©B. Marielle/photocuisine/Corbis.
20. IV bags (glucose bag). ©Randy Faris/Corbis.
20. Insulin injection. ©LWA- JDC/Corbis.

Chapter 4: Aspirin®

21. Early Bayer Pharmaceutical advertisement. ©Bettmann/Corbis.
21, 24. Felix Hoffmann. Bayer Business Services/Corporate History & Archives.
21, 22. Willow tree. ©Richard Hamilton Smith/Corbis.
21, 25, 26, 27, 28. Aspirin® tablets. ©Michael A. Keller/Corbis.
22. The House of Life. Used with permission from Pfizer Inc. Images by Robert Thom. All rights reserved.
21. Hippocrates illustration. Courtesy of Bayer HealthCare AG.
22. Hippocrates bust. ©Corbis.
22. Ebers papyrus. Courtesy of Bayer HealthCare AG.
23, 24. Crystals of acetylsalicylic acid. Courtesy of Bayer HealthCare AG.
23. Brain with Aspirin® pill. Courtesy of Bayer HealthCare AG.
23. Hermann Kolbe. Edgar Fahs Smith Collection, University of Pennsylvania Library.
23. *100 Years Aspirin®* book cover. Courtesy of Bayer HealthCare AG.
23. Meadowsweet. ©iStockphoto.com/hmproudlove.
24, 28. Bayer boxes. Courtesy of Bayer HealthCare AG.
24. Arthur Eichengrün. Bayer Business Services/Corporate History & Archives.
24. Heinrich Dreser. Bayer Business Services/Corporate History & Archives.
24. Heart and Aspirin® pill. Courtesy of Bayer HealthCare AG.
24. Ulf Svante von Euler-Chelpin. ©Hulton-Deutsch Collection/Corbis.
25. Sune K. Bergstrøm. National Library of Medicine (NLM).
25. Bengt I. Samuelsson. National Library of Medicine (NLM).
25. Sir John R. Vane. National Library of Medicine (NLM).
25. White tablets. ©iStockphoto.com/Canoneer.
26. Aspirin® poster (woman in blue). Courtesy of Bayer HealthCare AG.
26. Aspirin® poster (woman in orange). Courtesy of Bayer HealthCare AG.
26. Bottle of medicine. ©iStockphoto.com/mdilsiz.
26. Blood platelets and thrombus formation. Courtesy of Bayer HealthCare AG.
27. E. J. Corey. K. C. Nicolaou archive.
27. Aspirin® bottle. Courtesy of Bayer HealthCare AG.
27. Aspirin® definition. ©iStockphoto.com/LDF.
28. Eicosanoids in action. Courtesy of Bayer HealthCare AG.
28. Tree illustration. Courtesy of Bayer HealthCare AG.

Chapter 5: Camphor

29. Gustaf Komppa. Teknillisen Korkeakoulun Ylioppilaskunta/The Student Union of Helsinki University of Technology.
29, 30. Camphor tree. K. C. Nicolaou.
29, 30. Brushes in black ink. ©Jack Hollingsworth/Corbis.
30. Gustaf Komppa in lab. Teknillisen Korkeakoulun Ylioppilaskunta/The Student Union of Helsinki University of Technology.
30. Varied test tubes. ©Tom Grill/Corbis.
30. K. Julius Bredt. Courtesy of Achim Lenzen, RWTH Aachen University.
30. Marco Polo. ©Hulton-Deutsch Collection/Corbis.
31, 32. *Cinnamomum camphora*. Köhler's Medizinal-Pflanzen.
31, 32. *Cinnamomum camphora*. Kenpei.
31. William Henry Perkin, Jr. Edgar Fahs Smith Collection, University of Pennsylvania Library.
31. Measuring liquid in beaker. ©Tom Grill/Corbis.
32. Gustaf Komppa in classroom. Teknillisen Korkeakoulun Ylioppilaskunta/The Student Union of Helsinki University of Technology.

Chapter 6: Terpineol

33, 37. William Henry Perkin, Jr. Science Museum/Science and Society Picture Library.
33, 34. Knobs on the fir tree branch. ©iStockphoto.com/Pingwin.
33. Shalimar. Courtesy of Guerlain, Inc.
33, 36. Roses. Photo by K. C. Nicolaou.
34. Fir tree, St. Andrews, Canada. ©Erica Shires/zefa/Corbis.
34. Ceramic pot collecting resin for turpentine. ©Ric Ergenbright/Corbis.
34. Geranium bud. ©Clive Druett; Papilio/Corbis.
34. Egyptian art. The Metropolitan Museum of Art, Rogers Fund, 1930. (30.4.114) Photograph ©1979 The Metropolitan Museum of Art.

35. Ancient Greek cosmetic box (*pyxis*). The Metropolitan Museum of Art, Rogers Fund, 1907. (07.286.36) Photograph ©2001 The Metropolitan Museum of Art.
35. Perfume bottles from the Middle Ages. The Metropolitan Museum of Art, Gift of Henry G. Marquand, 1883; Edward C. Moore Collection, Bequest of Edward C. Moore, 1891; and Gift of J. Pierpont Morgan, 1917.
35. Marc Antony and Cleopatra as portrayed by Richard Burton and Elizabeth Taylor. ©Bettmann/Corbis.
35. Perfume bottles. ©Colin Anderson/Corbis.
36. Jean-Paul Guerlain tests new range of smells in his Paris lab. ©Reuters/Corbis.
36. Chanel No 5. Courtesy of Chanel N5.
36. Machines distilling perfume. ©Gail Mooney/Corbis.
37. William Henry Perkin, Sr. Edgar Fahs Smith Collection, University of Pennsylvania Library.
37. Lilac. ©Mark Bolton/Corbis.
37. White rose. Photo by K. C. Nicolaou.
38. Victor Grignard. ©Bettmann/Corbis.
38. Duilio Arigoni. Courtesy of Duilio Arigoni.
38. Perfume bottles. Photo by K. C. Nicolaou.
38. Yellow rose. Photo by K. C. Nicolaou.
39. Spindle bottle. Antiquities Department, Cyprus.
39. Perfume flask from mid-17th century India. The Metropolitan Museum of Art, Purchase, Mrs. Charles Wrightsman Gift, 1993. (1993.18) Photograph ©1993 The Metropolitan Museum of Art.
39. 18th Century pocket-sized French perfume bottle. The Metropolitan Museum of Art, Bequest of Catherine D. Wentworth, 1948. (48.187.482) Photograph ©1999 The Metropolitan Museum of Art.
39. Lancôme ad from "Femina", March-April 1952. The Metropolitan Museum of Art, Irene Lewisohn Costume Reference Library. Photograph ©1998 The Metropolitan Museum of Art.
40. Pink plumeria. ©Terry W. Eggers/Corbis.
40. Pink roses. Photo by K. C. Nicolaou.
40. Florida Water poster. The Metropolitan Museum of Art, The Jefferson R. Burdick Collection, Gift of Jefferson R. Burdick. (Album 34) Photograph ©1998 The Metropolitan Museum of Art.
40. Cleopatra. ©Bettmann/Corbis.

Chapter 7: Tropinone

41, 46. Carl Wilhelm Scheele. Used with permission from Pfizer Inc. Images by Robert Thom. All rights reserved.
41, 46. Friedrich Wilhelm Adam Sertürner. Used with permission from Pfizer Inc. Images by Robert Thom. All rights reserved.
41. University of Oxford. ©iStockphoto.com/AlamarPhotography.
41, 42, 45. *Atropa belladonna* (deadly nightshade). ©Naturfoto Honal/Corbis.
41, 48. Sir Robert Robinson (painting). The Royal Society.
41, 45. Richard Willstätter. ©Bettmann/Corbis
42. *Datura stramonium* plant. ©Stapleton Collection/Corbis.
42, 43. Sir Robert Robinson (photo). Dr. John Jones, University of Oxford.
42. Brazilwood sawdust. ©2002 Natural Pigments LLC www.naturalpigments.com.
43. Gilbert N. Lewis. Edgar Fahs Smith Collection, University of Pennsylvania Library.
43. Irving Langmuir. Edgar Fahs Smith Collection, University of Pennsylvania Library.
43. Sir Christopher Ingold. Photograph courtesy of John D. Roberts.
43. Test tube. ©Tom Grill/Corbis.
43. Friedrich August Kekulé. Edgar Fahs Smith Collection, University of Pennsylvania Library.
44. Labware background. ©Mario Beauregard/Corbis.
44. Atropine. ©Reuters/Corbis.
44. Atropine emergency medicine. ©Mediscan/Corbis.
44. Cocaine. ©Reuters/Corbis.
44. Linus Pauling. ©Bettmann/Corbis.
44. Erich Hückel. Hückel, E. *Ein Gelehrtenleben*. Ernst und Satire, Verlag Chemie, Weinheim, 1975.
44. Robert S. Mulliken. ©Bettmann/Corbis.
45. *Datura stramonium* (jimsonweed). ©Hal Horwitz/Corbis.
45. Lab beakers. ©Mario Beauregard/Corbis.
46, 47. Datura Metel. Artist: Eudoxia Woodward.
47. Robert B. Woodward. ©Bettmann/Corbis.
48. University of Oxford spires. Courtesy of Nasir Hamid.

Chapter 8: Haemin

49, 50. Hans Fischer. Edgar Fahs Smith Collection, University of Pennsylvania Library.
49. Colette Nicolaou. Photo by K. C. Nicolaou.
49. Red blood cells. ©Micro Discovery/Corbis.
49, 52. Johann Friedrich Wilhelm Adolf von Baeyer. Edgar Fahs Smith Collection, University of Pennsylvania Library.
50. Red blood cells. ©iStockphoto.com/jgroup.
51. Greek painting. ©Gianni Dagli Orti/Corbis.
51. Murex shell. ©Lawson Wood/Corbis.
51, 52. Silk fabric. ©Macduff Everton/Corbis.
52. Indigo fabric. ©Jacqui Hurst/Corbis.
52. Alexander, Georgette and Christopher Nicolaou. Photo by K. C. Nicolaou.
52. Cochineal insects feeding on Nopal cacti. ©Macduff Everton/Corbis.
53. Buckingham Palace guards. ©iStockphoto.com/texasmary.
53. August Wilhelm von Hofmann. Edgar Fahs Smith Collection, University of Pennsylvania Library.
53. William Henry Perkin, Sr. Edgar Fahs Smith Collection, University of Pennsylvania Library.
53. Perkin's original mauve shawl, 1856. Science Museum/Science and Society Picture Library.
53. Perkin's original mauve dye, 1856. Science Museum/Science and Society Picture Library.
54. Michael Faraday lecturing. ©Corbis.
54. Justus von Liebig. Edgar Fahs Smith Collection, University of Pennsylvania Library.
54. Friedrich Wöhler. Edgar Fahs Smith Collection, University of Pennsylvania Library.
54. Jöns Jakob Berzelius. Edgar Fahs Smith Collection, University of Pennsylvania Library.

54. August Wilhelm von Hofmann. Edgar Fahs Smith Collection, University of Pennsylvania Library.
55. Thomas J. Katz. K. C. Nicolaou archive.
55. Friedrich August Kekulé. Edgar Fahs Smith Collection, University of Pennsylvania Library.
55. Josef Loschmidt. Masarykova Univerzita V Brně.
55. Erich Hückel. Hückel, E. *Ein Gelehrtenleben.* Ernst und Satire, Verlag Chemie, Weinheim, 1975.
55. Franz Sondheimer. K. C. Nicolaou archive.
55. Peter J. Garratt. K. C. Nicolaou archive.
55. Emmanuel Vogel. Courtesy of Emmanuel Vogel.
56. Red madder plant (*Rubia tinctorum*). Köhler's Medizinal-Pflanzen.
56. Empress Eugenie. Portrait by Franz Winterhalter.
56. Queen Victoria. ©By kind permission of the Trustees of the Wallace Collection, London.

Chapter 9: Quinine

57, 62. William von Eggers Doering and Robert B. Woodward. Courtesy of Time Life Syndication.
57, 64. Cinchona tree. United States Geological Survey, United States Department of Interior.
57, 58. Ruins in Machu Picchu. ©Frans Lanting/Corbis.
57, 63. Gilbert Stork. K. C. Nicolaou archive.
57, 61. Paul Rabe. L. Harmajakivi: Leben und Werk dreier Hamburger Chemiker, Paul Rabe, Heinrich Remy, Heinrich Schlubach. Studienarbeit, Hamburg 1987.
58. Alexander the Great. ©Araldo de Luca/Corbis.
58, 64. *Cinchona officinalis L.* Photo: Joaquina Albán Castillo, Dpto. Etnobotanica y Botánica económica Museo de Historia Natural. UNMSM.
58. Hippocrates. ©Bettmann/Corbis.
59. National coat of arms of Peru. Guillermo Romero.
59. 19th Century Print, A Soldier Grasps Atahalpa, King of the Incas. ©Stapleton Collection/Corbis.
59. Countess of Chinchon. ©Arte & Immagini srl/Corbis.
59. Mosquito. ©iStockphoto.com/altmarkfoto.
59. Malaria sign. ©iStockphoto.com/LanceB.
59. Transmission of malaria by mosquitoes illustration. Nicolaou group.
60. August Wilhelm von Hofmann. Edgar Fahs Smith Collection, University of Pennsylvania Library.
60. William Henry Perkin, Sr. Edgar Fahs Smith Collection, University of Pennsylvania Library.
60. Adolph Strecker. Edgar Fahs Smith Collection, University of Pennsylvania Library.
60. Ira Remsen. ©Bettmann/Corbis.
61. Oliver Cromwell in battle. ©Bettmann/Corbis.
61. Louis Pasteur. ©Archivo Iconografico, S.A./Corbis.
62. Vladimir Prelog. Michigan State University, Chemistry Portrait Library.
62. Joseph Bienaimé Caventou and Pierre Joseph Pelletier. Used with permission from Pfizer Inc. Images by Robert Thom. All rights reserved.
63. Milan Uskoković. Courtesy of Milan Uskoković.
63. White pills. ©Kelly Redinger/Design Pics/Corbis.
63. *Cinchona officinalis.* Köhler's Medizinal-Pflanzen.
64. Eric N. Jacobsen. Courtesy of Eric N. Jacobsen.
65. *Cinchona officinalis L.* Courtesy of Raintree Nutrition, Inc.
65. The habitat of *Cinchona officinalis L.* Photo: Joaquina Albán Castillo, Dpto. Etnobotanica y Botánica económica Museo de Historia Natural. UNMSM.
65. *Cinchona officinalis L.* Photo: Joaquina Albán Castillo, Dpto. Etnobotanica y Botánica económica Museo de Historia Natural. UNMSM.
65. Bill and Melinda Gates. Courtesy of Bill and Melinda Gates Foundation.
66. Gilbert Stork. K. C. Nicolaou archive.
66. Drying cinchona bark. ©Hulton-Deutsch Collection/Corbis.

Chapter 10: Morphine

67, 76. Sir Robert Robinson. The Royal Society.
67, 71. Friedrich Wilhelm Adam Sertürner. Julius Giere.
67. Opium poppies in field. ©Guenter Rossenbach/zefa/Corbis.
67, 68, 72. Two opium poppy heads. ©Ashley Cooper/Corbis.
67, 76. Marshall Gates. Photo courtesy of the Department of Rare Books and Special Collections, University of Rochester Library and the University Public Relations Department, University of Rochester.
67, 77. Barry M. Trost. Courtesy of Barry M. Trost.
68. Sumarian city. ©Nik Wheeler/Corbis.
68. Japanese Buddhist statue. ©Jed & Kaoru Share/Corbis.
68. Dioscorides. ©Stapleton Collection/Corbis.
68. Paracelsus. ©Stapleton Collection/Corbis.
69. George Washington. ©Brooklyn Museum/Corbis.
69. Florence Nightengale. ©Bettmann/Corbis.
69. Dioscorides collecting herbs. Used with permission from Pfizer Inc. Images by Robert Thom. All rights reserved.
69. Heroin field in Poland. ©Ed Kashi/Corbis.
70. Opium poppies. ©Josh Westrich/zefa/Corbis.
70. Opium in France. ©Stefano Bianchetti/Corbis.
70, 77. Opium poppy field. ©Jeffrey L. Rotman/Corbis.
70. Greek god Hypnos. ©Bettmann/Corbis.
70. Samuel Taylor Coleridge. ©Michael Nicholson/Corbis.
71. Opium poppy flower. ©Chris Stewart/The Chronicle/Corbis.
71. Felix Hoffmann. Bayer Business Services/ Corporate History & Archives.
71. Beakers. ©Mario Beauregard/Corbis.
72. Heroin background. ©iStockphoto.com/yoepro.
73. Hydrocodone tablets. ©Marianna Day Massey/Corbis.
73. Choh Hao Li. Reprinted with permission by University of California, San Francisco.
73. Solomon H. Snyder. Courtesy of Albert and Mary Lasker Foundation.
73. Avram Goldstein. ©Jeff Albertson/Corbis.
73. Hans Kosterlitz. Courtesy of Albert and Mary Lasker Foundation.
73. Poppies in field as background and inset. ©iStockphoto.com/AtWaG.
73. Illustration of body. Nicolaou group.

74. Close-up of chocolate. ©Image100/Corbis.
74. Ice cold soda. ©Lew Robertson/Corbis.
74. Three Coke Bottles by Andy Warhol. ©Andy Warhol Foundation/Corbis.
74. Espresso glass and coffee beans. ©Volker Möhrke/Corbis.
74. Cup of tea with sugar cubes. ©Steve Lupton/Corbis.
74, 75. Cannabis plant. ©Mika/zefa/Corbis.
74. Unharvested coffee beans. ©Pablo Corral V/Corbis.
74. Coca plant. ©Jorge Uzon/Corbis.
74. Cola nuts. ©Marc Garanger/Corbis.
75. Various tablets of the drug ecstasy. ©Scott Houston/Sygma/Corbis.
75. Mexican peyote cactus. ©iStockphoto.com/bcphotobiz.
75. Ecstasy dancing at a club. ©S. Hammid/zefa/Corbis.
76. Old morphine tube. ©Tim Wright/Corbis.
78. Otto Diels. Otto Diels-Institute for Organic Chemistry, Christian-Albrechts-Universität.
78. Kurt Alder. ©Bettmann/Corbis.
78. Kenichi Fukui. Michigan State University, Chemistry Portrait Library.
78. *Angewandte Chemie* journal cover, Diels and Alder. Wiley-VCH Publisher, Weinheim.
78. Poppies. ©iStockphoto.com/Chuew.

Chapter 11: Steroids & the Pill

79, 83. Sir John W. Cornforth. ©Bettmann/Corbis.
79, 82. John D. Bernal. Getty Images/Photographer: Nat Farbman/Collection: Time & Life Pictures.
79, 82. Edward A. Doisy. National Library of Medicine (NLM).
79, 83. Sir Robert Robinson. John Jones, University of Oxford.
79, 83. Tadeus Reichstein. ©Bettmann/Corbis.
79, 83. Philip S. Hench. ©Bettmann/Corbis.
79, 82. Heinrich O. Wieland. Edgar Fahs Smith Collection, University of Pennsylvania Library.
79, 83. Young woman using an inhaler. ©iStockphoto/danielle71.
79. Line drawing of head. ©Images.com/Corbis.
79, 83. Edward C. Kendall. National Library of Medicine (NLM).
79, 82. Adolf F. Butenandt. akg-images, London.
79, 83. Birth control pills. ©Michael A. Keller/Corbis.
79. Sun through trees. ©iStockphoto/AVTG.
79. Model of DNA Strand. ©Tom Grill/Corbis.
79, 83, 87. K. Peter C. Vollhardt. Courtesy of K. P. C. Vollhardt.
79, 83, 86. William S. Johnson. Courtesy of Stanford, Department of Chemistry.
79, 83. Willard Allen. Christopher Hoolihan, Edward G. Miner Library, University of Rochester Medical Center.
79, 83. Werner E. Bachmann. Courtesy of Werner E. Bachmann.
79, 85. Carl Djerassi. Courtesy of Carl Dejerassi.
79, 83, 84. Robert B. Woodward. ©Bettmann/Corbis.
79, 82. Adolf O. R. Windaus. Georg-August-Universität Göttingen.
79, 85. Percy L. Julian. ©Bettmann/Corbis.
80. Silhouette of man. ©A. Sneider/zefa/Corbis.
80. Mexican farmer holding Mexican yam. Courtesy of Carl Dejerassi.
81, 82, 85. Mexican yam (*Dioscorea macrostachya*). Mark W. Skinner @ USDA PLANTS.
81. Mexican farmer. Courtesy of Carl Dejerassi.
82. Woman applying gel to back of hand. ©T & L/Image Point FR/Corbis.
82. Pharmacist with medication. ©Gaetano/Corbis.
83, 90. Woman choosing to use birth control. ©Michael A. Keller/Corbis.
84. Test tubes in rack. ©Tom Grill/Corbis.
84. Petri dishes. ©Tom Grill/Corbis.
84. Margaret Sanger. ©Corbis.
84. Gregory Pincus. ©Bettmann/Corbis.
85. Russell E. Marker with Mexican yams. Courtesy of Carl Dejerassi.
85. Birth control pills in hand. ©Tom & Dee Ann McCarthy/Corbis.
86. Erlenmeyer flasks. ©Tom Grill/Corbis
86. Dropper and test tubes. ©Tom Grill/Corbis.
86. Enovid® bottle. Library of Congress, Prints & Photographs Division, NYWT&S Collection [LC-USZ62-128827].
87. Beaker and dropper. Nicolaou group.
88. George Olah. Courtesy of George Olah.
88. Saul Winstein. Courtesy of UCLA, Department of Chemistry and Biochemistry, Organic Chemistry Division.
88. Odd Hassel. Chemistry Library, Department of Chemistry, University of Oslo.
88. Sir Derek H. R. Barton. K. C. Nicolaou archive.
89. Sir Derek H. R. Barton. K. C. Nicolaou archive.
89. Gilbert Stork. Courtesy of Gilbert Stork.
89. Gerald Pattenden. Courtesy of Gerald Pattenden.
90. Treasure of the Jungle. Courtesy of Carl Dejerassi; by Mexican muralist David Siqueiros.

Chapter 12: Strychnine

91, 92. Pierre Joseph Pelletier and Joseph Bienaimé Caventou. Used with permission from Pfizer Inc. Images by Robert Thom. All rights reserved.
91. Sir Robert Robinson. The Royal Society.
91, 96. *Strychnos ignatii*. Köhler Medizinal Pflanzen.
91. Robert B. Woodward. ©Bettmann/Corbis.
91, 92. Strychnine bottle. Courtesy of Museum of Medical History, Sierra Sacramento Valley Medical Society.
91, 95. Larry E. Overman. Courtesy of Larry E. Overman.
92. Strychnine alkaloid bottle. Köhler Medizinal Pflanzen.
92. *Strychnos nux vomica*. Köhler Medizinal Pflanzen.
93. Sir Robert Robinson. John Jones, University of Oxford.
93. Death of Socrates painting. ©Francis G. Mayer/Corbis.
93. Cigarettes. ©iStockphoto.com/macho.
94. Robert B. Woodward. Estate of Francis Bello/Photo Researchers, Inc.
94. Flasks. ©iStockphoto.com/DNY59.
94. Emil Fischer. ©Corbis.
95. Masakatsu Shibasaki. K. C. Nicolaou archive.

95. Test tubes. ©iStockphoto.com/chrisboy2004.
95. Microscope. ©iStockphoto.com/dra_schwartz.
96. Apothecary jar. Courtesy of Hawaii Medical Library, Mamiya Medical Heritage Center.
96. *Strychnos descussata*. ©H.G. Robertson, Iziko; South Africa Museum.

Chapter 13: Penicillin

97, 101. Sir Alexander Fleming. Courtesy of Pfizer Inc.
97, 100. Colonies of *Penicillium notatus*. ©Bettmann/Corbis.
97, 104. John C. Sheehan. Courtesy of the MIT Museum.
98. Sir Alexander Fleming in his lab. Courtesy of Alexander Fleming Laboratory Museum, St. Mary's NHS Trust.
98. Louis Pasteur. ©Nom de l'auteur/Institut Pasteur.
98. Microscope. Image Courtesy of the Blocker History of Medicine Collections, Moody Medical Library, The University of Texas Medical Branch, Galveston, Texas, USA.
98, 101. Fleming's famous penicillin and *staphylococci* in a petri dish. ©Bettmann/Corbis.
99. *Vibrio cholerae*. ©Mediscan/Corbis.
99. Joseph Lister. ©Nom de l'auteur/Institut Pasteur.
99. Robert Koch. ©Nom de l'auteur/Institut Pasteur.
99. Joseph Lister, pioneer of antisepsis. Used with permission from Pfizer Inc. Images by Robert Thom. All rights reserved.
100. Paul Ehrlich. Photo Courtesy of Paul Ehrlich Institute.
100. Gerhard Domagk. Wiley-VCH Publisher, Weinheim.
100. Labware. Courtesy of Valer Jeso, Nicolaou group.
101. Petri dish background. ©Tom Grill/Corbis.
101. *Leptospira interrogans spirochete* bacteria. ©Visuals Unlimited/Corbis.
101. *Staphylococcus aureus* bacteria. ©Visuals Unlimited/Corbis.
101. *Proteus mirabilis* bacteria. ©Visuals Unlimited/Corbis.
102. Alexander Fleming and Oxford team. Used with permission from Pfizer Inc. Images by Robert Thom. All rights reserved.
102. Microscope. ©Tom Grill/Corbis.
102. *Penicillium chysogenum*. Courtesy of Satoshi Ômura.
103. Cantaloupe on vine. ©Ed Young/Corbis.
103. Crystals. Nicolaou group.
103. Dorothy Hodgkin and husband. Courtesy of University of Oxford.
103. General Dwight D. Eisenhower conferring with General Bernard Montgomery. ©Bettmann/Corbis.
104. Test tubes. Courtesy of Valer Jeso, Nicolaou group.
105. Sir Jack Baldwin. Courtesy of Sir Jack Baldwin.
106. Labware. Nicolaou group.
106. University of Oxford. Courtesy of Simon Y. W. Ho.

Chapter 14: Longifolene

107, 108. E. J. Corey holding Logic book. Courtesy of E. J. Corey.
107, 109. *Pinus roxburghii*. Christopher J. Earle, www.conifers.org.
107. Calcomp 565 drum plotter. Courtesy of E. J. Corey.
108, 110. *Chamaecyparis obtusa* (*cripsii*). ©WJ Hayden, University of Richmond.
109. *Chamaecyparis obtusa* (evergreen). ©WJ Hayden, University of Richmond.
109. *Pinus ponderosa*. ©iStockphoto.com/katsgraphicslv.
109. Scientific labware. ©George B. Diebold/Corbis.
110. Row of test tubes. ©Mario Beauregard/Corbis.
111. E. J. Corey at computer. Courtesy of E. J. Corey.
111. Chemistry lab containers. ©iStockphoto.com/KeeTron.
112. E. J. Corey at blackboard. Courtesy of E. J. Corey.
112. *Pinus roxburghii* (branches). Christopher J. Earle, www.conifers.org.

Chapter 15: Prostaglandins & Leukotrienes

113, 116. Sune K. Bergstrøm. National Library of Medicine (NLM).
113, 116. Bengt I. Samuelsson. National Library of Medicine (NLM).
113, 116. Sir John R. Vane. National Library of Medicine (NLM).
113, 114. Heart and lungs in a human male. ©Mediscan/Corbis.
113. Respiratory inhaler. ©Southern Stock/Brand X/Corbis.
113, 118. E. J. Corey in front of picture. K. C. Nicolaou archive.
114. Doctor examining boy. ©Michael A. Keller/zefa/Corbis.
114. X-Ray of lungs. ©Steve Allen/Brand X/Corbis.
115. Mast cell. ©Visuals Unlimited/Corbis.
115. Singulair®. ©Visuals Unlimited/Corbis.
115. Woman using a tissue. ©Schultheiss Productions/zefa/Corbis.
116. Graduated glasses. ©iStockphoto.com/ElementalImaging.
116. Ulf Svante von Euler-Chelpin. ©Hulton-Deutsch Collection/Corbis.
117. Herd of sheep. ©George B. Diebold/Corbis.
117. Ibuprofen bottle and tablets. ©Krista Kennell/ZUMA/Corbis.
117. Bottle of Tylenol® capsules. ©Roger Ressmeyer/Corbis.
117. Aleve®. ©Ruaridh Stewart/ZUMA/Corbis
117, 119. Aspirin® spilling out of bottle. ©Envision/Corbis.
118. Erlenmeyer lab flask. ©iStockphoto.com/PixHook.
118. Blood clot in the brain. ©Howard Sochurek/Corbis.
119. E. J. Corey. K. C. Nicolaou archive.
119. Lab glassware. ©iStockphoto.com/Kalashnikov_O.
120. *Angewante Chemie* journal cover, Diels and Alder. Wiley-VCH Publisher, Weinheim.
120. Labware. Nicolaou group.
121. *Plexaura homomalla*. ©James Beveridge/Visuals Unlimited.

121. Endothelium, blood platelets, and a red blood cell. Courtesy of Bayer HealthCare AG.
122. Harvard University. ©Kevin Fleming/Corbis.
122. Karolinska Institute. Oscar Franzén.

Chapter 16: Vitamin B_{12}

123, 128. Dorothy Crowfoot Hodgkin. National Media Museum/Science and Society Picture Library.
123, 133. *Propionibacterium shermanii*. Courtesy of A. Ian Scott.
123, 127, 128. Vitamin B_{12} crystals. ©Beat Ernst, Basel.
123, 129. Albert Eschenmoser. Courtesy of Albert Eschenmoser.
123, 129. Robert B. Woodward. Courtesy of Ian Fleming, University of Cambridge.
124. James Lind. Used with permission from Pfizer Inc. Images by Robert Thom. All rights reserved.
124. Vasco da Gamma. Library of Congress, Prints & Photographs Division, [LC-USZC4-2069].
125, 127. Oranges. ©iStockphoto.com/Tschon.
125. Christiaan Eijkman. ©Museum Boerhaave.
125. Pills on finger tip. ©John Lund/Tiffany Schoepp/Blend Images/Corbis.
126. Dorothy Crowfoot Hodgkin in lab. Courtesy of University of Oxford.
126. Albert Szent-Györgyi. Photo by J.W. McGuire of the National Institutes of Health.
126. Walter N. Haworth. ©Hulton-Deutsch Collection/Corbis.
126. Sir Frederick Gowland Hopkins. Edgar Fahs Smith Collection, University of Pennsylvania Library.
126. Tadeus Reichstein. ©Bettmann/Corbis.
127. Paul Karrer. ©Bettmann/Corbis.
127. Richard Kuhn. ©Hulton-Deutsch Collection/Corbis.
128. George R. Minot. ©Bettmann/Corbis.
128. William P. Murphy. National Library of Medicine (NLM).
128. George H. Whipple. ©Bettmann/Corbis.
128. Test tubes and dropper. ©Tom Grill/Corbis.
128. Lord Alexander R. Todd. ©Bettmann/Corbis.
128. Vladimir Prelog. Photograph courtesy of John D. Roberts.
128. Leopold Ružička. Photograph courtesy of John D. Roberts.
129. Volumetric flasks. ©Tom Grill/Corbis.
129. Petri dish and dropper. ©Tom Grill/Corbis.
130. Flasks and cylinders. ©Mario Beauregard/Corbis.
130. Jack D. Dunitz. Courtesy of Jack D. Dunitz.
131. Robert B. Woodward and Albert Eschenmoser. Reprint from *Nachrichten aus Chemie*, Technik und Laboratorium, Permissions from Wiley-VCH, 1972.
131. Woman's hands holding tablets. ©iStockphoto.com/molka.
132. Robert B. Woodward upon receipt of the Nobel Prize. RBW Nobel photos courtesy of Eudoxia, Crystal and Eric Woodward.
133. Sir Alan R. Battersby. University of Cambridge, Photographer: Doug Young.
133. A. Ian Scott. Courtesy of A. Ian Scott
134. Robert B. Woodward illustration. Wiley-VCH Publisher, Weinheim.
134. Roald Hoffmann. Gary Hodges, Jon Reis Photography.
135. Kenichi Fukui. Michigan State University, Chemistry Portrait Library.
135. E. J. Corey. K. C. Nicolaou archive.
135. Emmanuel Vogel. Courtesy of Emmanuel Vogel.
135. Colorful medications. ©William Whitehurst/Corbis.
136. Composition of fruit. ©Maximilian Stock Ltd/photocuisine/Corbis.

Chapter 17: Erythronolide B & Erythromycin A

137, 142. E. J. Corey. Courtesy of E. J. Corey.
137, 139. *Streptomyces griseus*. Courtesy of Satoshi Ômura.
137, 138, 141, 144. Petri dishes *Saccharopolyspora erythraea*. Courtesy of John Ward, University College London.
137. Hands holding dirt with sprout. ©Robert Llewellyn/Corbis.
137, 143. Robert B. Woodward. ©Bettmann/Corbis.
138. Henry David Thoreau. ©Bettmann/Corbis.
139. Selman A. Waksman. ©Bettmann/Corbis.
139. Beakers, flasks, and data results. ©Don Carstens/Brand X/Corbis.
140. Stained tuberculosis bacteria under microscope. ©CDC/PHIL/Corbis.
140. Egyptian mummy of Amonred. ©North Carolina Museum of Art/Corbis.
141. Assortment of lab beakers. ©Mario Beauregard/Corbis.
141. E. J. Corey and K. C. Nicolaou. K. C. Nicolaou archive.
142. Petri dish. ©iStockphoto.com/Jarrod1.
143. Petri dishes. ©iStockphoto.com/Jarrod1.
144. Harvard seal. ©Corbis.

Chapter 18: Monensin

145, 146. Donald J. Cram. Photo by University of California, Los Angeles (UCLA) Photography.
145, 146. Jean-Marie Lehn. ©Fernando Villar/epa/Corbis.
145, 146. Charles J. Pedersen. Courtesy of DuPont Public Relations.
145, 150. Cattle. ©Jim Richardson/Corbis.
145, 148. Rumensin®. Courtesy of Ellanco, Aukland.
145, 149. Yoshito Kishi. Courtesy of Yoshito Kishi.
146. Crop dusting. ©Owaki - Kulla/Corbis.
146. Sprinkler system on farmland. ©Photo 24/Brand X/Corbis.
146. Labware. Courtesy of Valer Jeso, Nicolaou group.
147. Labware test tubes. ©Mario Beauregard/Corbis.
147. Peter Agre. ©Joe Giza/Reuters/Corbis.
147. Roderick MacKinnon. ©Stefan Lindblom/Corbis.
148, 152. Cattle in a pasture. ©Royalty-Free/Corbis.
150. Two test tubes. ©Tom Grill/Corbis.
150, 152. Curious goat. ©Tom Stewart/Corbis.
151. Herbert C. Brown. Courtesy of Department of Chemistry, Purdue University.
151. Nobel Prize medal. ®The Nobel Foundation.

151. HCB - Boranes Rendering. Courtesy of Department of Chemistry, Purdue University.
151. Barley field. ©Jim Craigmyle/Corbis.
152. Close up of measurements on a vial. ©Image100/Corbis.
152. W. Clark Still. Courtesy of W. Clark Still.
152. White rooster. ©Little Blue Wolf Productions/Corbis.

Chapter 19: Avermectin

153, 154. Satoshi Ômura. Courtesy of Satoshi Ômura.
153. Barley field. ©Jim Craigmyle/Corbis.
153. Desert locust on stalk. ©Anthony Bannister; Gallo Images/Corbis.
153, 154. *Streptomyces avermitilis* MA-4680. Courtesy of Satoshi Ômura.
153, 156. Stephen Hanessian. Courtesy of Stephen Hanessian.
153, 158. Samuel J. Danishefsky. Photo credit: Sarah May.
154. Crops growing in a field. ©Royalty-Free/Corbis.
154. Golf ball in the rough. ©Peter Adams/zefa/Corbis.
154. Kitasato Medal. Photo by K. C. Nicolaou.
155. Greenhouse worker. ©Holger Winkler/zefa/Corbis.
155. Flasks and cylinders. ©Mario Beauregard/Corbis.
156. *Streptomyces avermitilis*. ©2006 Merck & Co., Inc., Whitehouse Station, New Jersey, USA. All rights reserved.
157. Close up of a vial. ©Image100/Corbis.
158. Labware. ©iStockphoto.com/duncan1890.
158. Refugees in Goma camps. ©Jon Jones/ Sygma/Corbis.
158. *Onchocerca volvulus*. CDC/Ladene Newton (PHIL #4637), 1975.
159. P. Roy Vagelos. 2007©Donald Danforth Plant Science Center. All rights reserved.
160. African continent on globe. ©Digital Art/Corbis.
160. Refugee from Rwanda. ©Jon Jones/Sygma/ Corbis.
160. Two flies. ©Ben Welsh/zefa/Corbis.

Chapter 20: Amphotericin B

161, 164. Collection of various mushrooms. ©Ashley Cooper/Corbis.
161, 164. *Candida albicans* fungi. ©CDC/PHIL/Corbis.
161. Confluence of rivers in Venezuela. ©Yann Arthus-Bertrand/Corbis.
161. K. C. Nicolaou. K. C. Nicolaou archive.
162. Rye and wild flower. ©Markus Botzek/zefa/Corbis.
162. Engraving of Pliny the Elder. ©Bettmann/Corbis.
162. Lithograph by George H. Walker of *The Witch Number 3* by J. E. Baker. ©Bettmann/Corbis.
162. Camembert cheese. ©iStockphoto.com/lucgillet.
162. Roquefort cheese. ©Ryman Cabannes/photocuisine/Corbis.
163. Hippocrates. ©Bettmann/Corbis.
163. Hippocrates with boy. Used with permission from Pfizer Inc. Images by Robert Thom. All rights reserved.
163. Pedicure. ©Tom Grill/Corbis.
164, 166. *Streptomyces griseus*. ©Scimat/Photo Researchers/Corbis.
164. Two toadstools in forest. ©Herbert Zettl/zefa/Corbis.
165. Yellow morel. ©Robert Marien/Corbis.
165. K. C. Nicolaou holding flask. K. C. Nicolaou archive.
166. K. Barry Sharpless. K. C. Nicolaou archive.
166. Tsutomu Katsuki. Courtesy of T. Katsuki.
166. Glenn J. McGarvey. Glenn J. McGarvey.
166. Satoru Masamune. Massachusetts Institute of Technology (MIT).
167. Georg Wittig. ©Bettmann/Corbis.
167. Nobel Prize medal. ®The Nobel Foundation.
167. Leopold Horner. Courtesy of Horst Kunz.
168. Sculpture of Benjamin Franklin at the University of Pennsylvania. Courtesy of Bruce Andersen.

Chapter 21: Ginkgolide B

169. Leaves of the ginkgo in front of red background. ©Josh Westrich/zefa/Corbis.
169, 172. Koji Nakanishi. Courtesy of Koji Nakanishi.
169, 172, 176. Ginkgo tree leaf. ©Wolfgang Kaehler/Corbis.
169, 170, 176. *Ginkgo biloba* branch. ©Guy Cali/Corbis.
169, 174. E. J. Corey. ©Rick Friedman/Corbis.
170. Buddhist monks in monastery. ©Angelo Cavalli/zefa/Corbis.
170. Geisha. ©Bob Krist/Corbis.
170. Goethe original "*Ginkgo biloba*" poem. Johann Wolfgang von Goethe, public domain.
170. Goethe and Schiller Monument in Weimar, Germany. ©Dietrich Rose/zefa/Corbis.
170. Charles Darwin. ©Bettmann/Corbis.
171. Ginkgo tree (Hiroshima, 1945). Courtesy of Koji Nakanishi.
171. Ginkgo fossils. Courtesy of Koji Nakanishi.
171. Ginkgo tree on the campus of Tohoku University, Sendai. Courtesy of Koji Nakanishi.
172. Ginkgo tree. Courtesy of Koji Nakanishi.
172. Ginkgo bonsai tree. Courtesy of Koji Nakanishi.
172. Ginkgo female leaves and nuts. Courtesy of Koji Nakanishi.
172. Ginkgo fossil. ©DK Limited/Corbis.
173. *Ginkgo biloba* in front of Riverside Church, Manhattan, New York City. Courtesy of Koji Nakanishi.
173. Gingkolide crystal (left). Courtesy of Koji Nakanishi.
173. Koji Nakanishi holding structure. Courtesy of Koji Nakanishi.
174. Inverted phase control microscope. ©Tom Stewart/Corbis.
174. Test tubes in tray. ©Tom Grill/Corbis.
174. Michael T. Crimmins. Courtesy of Michael T. Crimmins.
175. Ginkgo leaf. ©iStockphoto.com/aloha_17.
175. Nobel Prize medal. ®The Nobel Foundation.
175. E. J. Corey receiving the Nobel Prize. Courtesy of E. J. Corey.
176. E. J. Corey receiving the National Medal of

Science. Courtesy of E. J. Corey.
176. *Ginkgo biloba* bottle and capsules. ©Visuals Unlimited/Corbis.

Chapter 22: Cyclosporin, FK506 & Rapamycin

177, 182. John A. Findlay. Courtesy of John A. Findlay.
177. Suren Sehgal. Courtesy of Magid Abou-Gharbia, Wyeth Research.
177, 180. Blood cells. ©iStockphoto.com/sgame.
177. Statues on Easter Island. ©Goodshoot/Corbis.
177, 184. Stuart L. Schreiber. Courtesy of Stuart L. Schreiber.
177, 187. K. C. Nicolaou. K. C. Nicolaou archive.
178. Alexis Carrell. ©Corbis.
178. Labware. Courtesy of Valer Jeso, Nicolaou group.
178. Scientific research flasks. ©iStockphoto.com/DNY59.
178. Immune system illustration. Nicolaou group.
179. Richard and Ronald Herrick. ©Bettmann/Corbis.
179. Sir Peter B. Medawar. ©Hulton-Deutsch Collection/Corbis.
179. Joseph E. Murray. Brigham and Women's Hospital, Public Affairs Office.
179. Erlenmeyer flasks. ©Tom Grill/Corbis.
179. George H. Hitchings. ©Bettmann/Corbis.
179. Gertrude B. Elion. ©Bettmann/Corbis.
179, 184. Labware. Courtesy of Valer Jeso, Nicolaou group.
180. Jean-François Borel. Novartis Pharma AG, Switzerland/DT-Natural Product Unit and Novartis Archive.
180. Hartmann F. Stähelin. Novartis Pharma AG, Switzerland/DT-Natural Product Unit and Novartis Archive.
180. Flasks and cylinders. ©Mario Beauregard/Corbis.
180. E. Donnall Thomas. ©1999 Susie Fitzhugh.
181. Selman A. Waksman. Library of Congress, Prints & Photographs Division, NYWT&S Collection, LC-USZ62-119821.
181. Hands holding soil. ©L. Clarke/Corbis.
182. Row of statues on Easter Island. ©Mark A. Johnson/Corbis.
182. David A. Evans. Courtesy of David A. Evans.
182, 183. Labware. Courtesy of Valer Jeso, Nicolaou group.
182. *Tolypocladium inflatum*. Novartis Pharma AG, Switzerland/DT-Natural Product Unit and Novartis Archive.
182. *Streptomyces hygroscopicus*. Courtesy of Satoshi Omura.
183. Thomas E. Starzl. Courtesy of Thomas E. Starzl.
183. Suren Sehgal and Joe Camardo. The photo is used with the permission of Wyeth. All rights reserved to Wyeth.
183. Doctors perform surgery. ©Minnesota Historical Society/Corbis.
184. Illustration of B-cells signaling T-cells to attack cancer. ©Mediscan/Corbis.
184. Prograf® capsules. ©Astellas Pharma, US.
185. Petri dish and dropper. ©Tom Grill/Corbis.
185. Labware. Courtesy of Valer Jeso, Nicolaou group.
186. Nicolaou rapamycin team. K. C. Nicolaou archive.
186. Petri dishes. ©Tom Grill/Corbis.
186. Amos B. Smith, III. Courtesy of Amos B. Smith, III.
186. Steven V. Ley. Courtesy of Steven V. Ley.
186. Samuel J. Danishefsky. Photo credit: Sarah May.
187. Irving Weissman. Courtesy of Irving Weissman.
188. Labware. ©iStockphoto.com/duncan1890.
189. Labware. ©iStockphoto.com/JanPietruszka.
189. Barry M. Trost. Courtesy of Barry M. Trost.
189. Akira Suzuki. Courtesy of Akira Suzuki.
189. Ei-ichi Negishi. K. C. Nicolaou archive.
189. Richard F. Heck. Courtesy of Douglass Taber.
189. John K. Stille. Courtesy of Robert Williams.
189. David Milstein. Courtesy of David Milstein.
189. John F. Hartwig. Courtesy of John F. Hartwig.
189. Stephen L. Buchwald. Courtesy of Stephen L. Buchwald.
189. J. Tsuji. Courtesy of Takashi Takahashi.
190. Labware. Courtesy of Valer Jeso, Nicolaou group.
190. Labware. Courtesy of Valer Jeso, Nicolaou group.
190. Rapamune® bottle. Courtesy of Magid Abou-Gharbia, Wyeth Research.

Chapter 23: Calicheamicin γ_1^I

191, 197. George Ellestad. Courtesy of George Ellestad.
191, 197. Janis Upeslacis. Courtesy of Janis Upeslacis.
191, 197. Philip Hamann. Courtesy of Philip Hamann.
191, 192. Calichea stone. Courtesy of Professor John Belew, 1994.
191, 195. K. C. Nicolaou. K. C. Nicolaou archive.
192. Robert Bergman. Courtesy of Robert Bergman.
192, 195. Petri dish with dropper. ©Tom Grill/Corbis.
192. *Micromonospora echinospora* NRRL 18149. Courtesy of Michael Greenstein.
193. *Micromonospora echinospora* ATCC 27299. Courtesy of Michael Greenstein.
193. Franz Sondheimer. K. C. Nicolaou archive.
193. Satoru Masamune. Massachusetts Institute of Technology (MIT).
193. Labware. Courtesy of Valer Jeso, Nicolaou group.
194. Scientific labware. ©George B. Diebold/Corbis.
194. Double strand cuts. Nicolaou group.
194. Calicheamicin γ_1^I bound to DNA. Nicolaou group.
195. Samuel J. Danishefsky. Photo credit: Sarah May.
195. Andrew G. Myers. Courtesy of Andrew G. Myers.
195. Masahiro Hirama. Courtesy of Wiley-VCH Publisher, Weinheim.
195. Double Helix Model. ©Tom Grill/Corbis.
196. Nicolaou calicheamicin γ_1^I team with inset. K. C. Nicolaou archive.
196. César Milstein. ©Bettmann/Corbis.
196. Georges J. F. Köhler and Niels K. Jerne. ©Bettmann/Corbis.
197. Herceptin®. Courtesy of F. Hoffmann-La Roche Ltd.
197. Leukemia cell. ©Mediscan/Corbis.
197. Mylotarg® bottle and box. Courtesy of Magid Abou-Gharbia, Wyeth.
197. Humira® box. John Kaprielian/Photo Researchers, Inc.
198. Labware. Courtesy of Valer Jeso, Nicolaou group.

198. Total synthesis of calicheamicin γ_1^I. Nicolaou group.

Chapter 24: Palytoxin

199, 201. Paul J. Scheuer. Courtesy of University of Hawaii, Hamilton Library.
199, 202. Yoshimasa Hirata. Courtesy of Yoshito Kishi.
199, 201. Richard E. Moore. Cancer Research Center of Hawaii, University of Hawaii.
199. Maui. Courtesy of Tom Patterson, US National Park Service.
199, 200. *Limu* (seaweed). Courtesy of Ian Lind, www.ilind.net.
199, 201. Colonial anemones (*Palythoa*). ©Robert Yin/Corbis.
199, 203. Yoshito Kishi. Courtesy of Yoshito Kishi.
200, 206. Waipi'o Valley lookout and Hamakua coast in Hawaii. ©Stuart Westmorland/Corbis.
200. Koko Head. ©Danny Lehman/Corbis.
201. *Palythoa* coral. ©Yuko Stender, MarinelifePhotography.com.
201, 205. Puffer fish (fugu). ©Jeffrey L. Rotman/Corbis.
201. Puffer fish (*Tetrodontoidea*). ©Jeffrey L. Rotman/Corbis.
201. Shellfish (*Pinna muricata*). ©2007 Guido & Philippe Poppe – www.conchology.be.
201. Colombian poison arrow frog (*Dendrobates histrionicus*). Art Wolfe/Photo Researcher, Inc.
201. Colombian poison arrow frog (*Phyllobates bicolor*). Dennis Nilsson, www.dartfrog.tk.
201. Alaska butter clams (*Saxidomus giganteus*). Photo by K. C. Nicolaou.
201. *Palythoa* coral. Courtesy of Terry Siegel, *Advanced Aquarist*.
202. Oahu. ©Douglas Peebles/Corbis.
203. *Palythoa* coral (closed). ©Yuko Stender, MarinelifePhotography.com.
203. Hitosi Nozaki. Courtesy of Hitosi Nozaki.
203. Tamejiro Hiyama. Courtesy of Tamejiro Hiyama.
203. Sparse *palythoa*. ©Keoki Stender, MarinelifePhotography.com.
204. Giant kelp. ©Ralph A. Clevenger/Corbis.
205. *Palythoa lesueuri*. ©Robert Yin/Corbis.
205. Chef preparing fugu fish. ©James Marshall/Corbis.
206. Spiny puffer fish spouting water. ©Jonathan Blair/Corbis.
206. Maui coast. Ryan Oelke.

Chapter 25: Taxol®

207, 210. Mansukh E. Wani and Monroe C. Wall. RTI International/Jimmy W. Crawford, Photographer.
207, 210. Susan B. Horwitz. Courtesy of Hongliang Zhou.
207. Fruit on a sprig of common yew. ©Eric and David Hosking/Corbis.
207, 209. Pacific yew bark. ©Charles Mauzy/Corbis.
207, 211. K. C. Nicolaou. K. C. Nicolaou archive.
207, 213. Robert A. Holton. Courtesy of Taxolog, Inc.
208. Moss-Hung yew in Olympic rainforest. ©Kevin Schafer/Corbis.
208. Evergreen berries. ©iStockphoto.com/YawningDog.
208. *Nature* journal cover, Taxol®. Reprinted by permission from Nature, ©1994, Macmillan Publishers, Inc.
208. Cancer cell. Credit: hybrid medical animation/Photo Researchers, Inc.
209. Needles of *Taxus baccata*. Courtesy of Stellios Arseniyades.
209. *Taxus brevifolia* branch. Courtesy of Stellios Arseniyades.
209. Julius Caesar. ©iStockphoto.com/goldhafen.
209. Tubulin polymerization and microtubules. Nicolaou group.
210. The cell cycle and Taxol®'s mechanism of action. Nicolaou group.
211, 213. Semisynthetic Taxol® bottle. ©Bristol-Myers Squibb.
211. The Nicolaou Taxol® team. K. C. Nicolaou archive.
211. *Taxus baccata* berries. ©iStockphoto.com/Whiteway.
213. *Taxus baccata* berry. ©iStockphoto.com/YawningDog.
213. The Holton Taxol® team. Courtesy of Robert A. Holton.
214. Pierre Potier, F. Guéritte-Voegelein and Daniel Guénard. Courtesy of Simeon Arseniyades.
214. Dropper and test tubes. ©Tom Grill/Corbis.
214. *Angewandte Chemie* journal cover, Taxol®. Wiley-VCH Publisher, Weinheim.
215, 216. Capsules in medication cups. ©Visuals Unlimited/Corbis.
217. *Taxus baccata*, 'Standishii', flank the entrance to a house. ©Eric Crichton/Corbis.
217. Lab bench. Nicolaou group.
217. Teruaki Mukaiyama. Courtesy of Teruaki Mukaiyama.
217. Iwao Ojima. Courtesy of Iwao Ojima.
217. Samuel J. Danishefsky. Courtesy of Samuel J. Danishefsky.
217. Paul A. Wender. Courtesy of Paul A. Wender.
217. Isao Kuwajima. Courtesy of Isao Kuwajima.
218. K. C. Nicolaou conferring with students. K. C. Nicolaou archive.
218. *Classics in Total Synthesis*. Wiley-VCH Publisher, Weinheim.
218. *Classics in Total Synthesis II*. Wiley-VCH Publisher, Weinheim.
218. *Taxus baccata* branches. ©iStockphoto.com/anouchka.
218. Taxol® box and bottle. Courtesy of Bristol-Myers Squibb.

Chapter 26: Mevacor®, Zaragozic Acids & CP Molecules

219, 223. Joseph L. Goldstein. ©Jacques Langevin/Corbis Sygma.
219, 227. Akira Endo. Courtesy of Akira Endo.
219, 223. Michael S. Brown. ©Jacques Langevin/Corbis Sygma.
219. Beakers and test tubes. ©Mario

Beauregard/Corbis.
219. Jalon River waterfall. Courtesy of Ignacio Ferrando Margelí. ©Ábaco Digital.
219, 230. Matthew D. Shair. Courtesy of Matthew D. Shair.
219, 228. David A. Evans. Courtesy of David A. Evans.
219, 228. Erick M. Carreira. Courtesy of Georg Wuitschik.
219, 228. K. C. Nicolaou. K. C. Nicolaou archive.
219, 230. Tohru Fukuyama. K. C. Nicolaou archive.
219, 230. Samuel J. Danishefsky. Courtesy of Samuel J. Danishefsky.
220. Cardiology health care. ©Steve Allen/Brand X/Corbis.
220. Varied test tubes. ©Tom Grill/Corbis.
220. Adolf O. R. Windaus. Göttingen Museum of Chemistry.
220. John D. Bernal. Birkbeck Photographic Unit.
221. Monocytes. ©Lester V. Bergman/Corbis.
221. The development of atherosclerosis. Nicolaou group.
222, 223. Heartrate and pills. ©Image Source/Corbis.
224. Giant sunflower. ©Richard Klune/Corbis.
224. Butter and knife. ©Bagros/photocuisine/Corbis.
224. Steak. ©Olivier Pojzman/Olivier Pojzman/ZUMA/Corbis.
224. Half coconut and flower on bamboo mat. ©Jamie Grill/Corbis.
224. Trout with white butter sauce. ©P.Desgrieux/photocuisine/Corbis.
224. Fruit. ©SIE Productions/zefa/Corbis.
225. Three eggplants. ©iStockphoto.com/WinterWitch.
225. Lipitor®. ©Tannen Maury/epa/Corbis.
225. Zocor® bottle. Courtesy of Merck.
225. Zetia® bottle and tablets. Courtesy of Schering-Plough Corporation.
225. Pravachol® tablet. Leonard Lessin/Photo Researcher, Inc.
225. Zocor® single tablet and bottles. ©Jeff Zelevansky/Reuters/Corbis.
226. Zocor® bottles. ©Jeff Zelevansky/Reuters/Corbis.
226. Konrad E. Bloch. ©Bettmann/Corbis.
226. Feodor Lynen. ©Bettmann/Corbis.
226. *Penicillium citrinum*. ©2006 Merck & Co., Inc., Whitehouse Station, New Jersey, USA. All rights reserved.
227. Masahiro Hirama. Courtesy of Masahiro Hirama.
227. Charles J. Sih. Courtesy of Pamela French.
227. Paul A. Grieco. Courtesy of Paul A. Grieco.
227. Fruits and vegetables. ©Fukuhara, Inc./Corbis.
227. Mevacor® bottle. Courtesy of Merck.
227. *Aspergillus terreus*. ©2006 Merck & Co., Inc., Whitehouse Station, New Jersey, USA. All rights reserved.
228. Petri dishes. ©Tom Grill/Corbis.
229. Takushi Kaneko. Courtesy of Takushi Kaneko.
229. Nicolaou CP molecule team. K. C. Nicolaou archive.
229. Microscope. ©iStockphoto.com/dra_schwartz.
230. Beakers, flasks, and data results. ©Don Carstens/Brand X/Corbis.
230. K. C. Nicolaou. K. C. Nicolaou archive.
231. R. Criegee. Courtesy of Stefan Bräse.
232. K. Barry Sharpless. Courtesy of K. Barry Sharpless.
232. Tsutomu Katsuki. Courtesy of Tsutomu Katsuki.

Chapter 27: Brevetoxin B

233, 236. Koji Nakanishi. Courtesy of Koji Nakanishi.
233, 236. Jon C. Clardy. Courtesy of Jon C. Clardy.
233. Red tide with fisheries. Kunie Suzuki, Suisan Aviation Co., Ltd.
233, 234. *Karenia brevis*. Courtesy of Florida Marine Research Institute.
233. K. C. Nicolaou. K. C. Nicolaou archive.
234, 238. Red tide. Courtesy of Peter J. S. Franks, Scripps Institution of Oceanography, University of California, San Diego (UCSD).
234, 237. *Classics in Total Synthesis*. Wiley-VCH Publisher, Weinheim.
235. Puffer fish. ©Jeffrey L. Rotman/Corbis.
235. *Classics in Total Synthesis II*. Wiley-VCH Publisher, Weinheim.
235. K. C. Nicolaou. K. C. Nicolaou archive.
235, 240. Red tide. Kunie Suzuki, Suisan Aviation Co., Ltd.
236. *Eleutherobia grayi* coral. ©2004 Joseph Dougherty, Ecology Photographic, www.ecology.org.
237. K. C. Nicolaou. K. C. Nicolaou archive.
237. Crystals of synthetic brevetoxin B. K. C. Nicolaou archive.
237. Nicolaou brevetoxin B team. K. C. Nicolaou archive.
238. Paul J. Scheuer. Courtesy of Paul Scheuer.
238. Yuzuru Shimizu. Courtesy of Yuzuru Shimizu.
238. Masahiro Hirama. Courtesy of Masahiro Hirama.
238. Fish. ©iStockphoto.com/PhotoPhly.
238. Bottlenosed dolphins breaching. ©Stuart Westmorland/Corbis.
238. Stingray on sea floor. ©Carson Ganci/Design Pics/Corbis.
239. The Nicolaou azaspiracid team. K. C. Nicolaou archive.
239. Brevetoxin diagram. Nicolaou group.
239. Azaspiracid-1. Nicolaou group.
240. Puffer fish. Courtesy of National Oceanic & Atmospheric Administration (NOAA).
240. *Chemistry – A European Journal* cover, Brevetoxin A. Wiley-VCH Publisher, Weinheim.

Chapter 28: Ecteinascidin 743

241, 243. Kennith L. Rinehart. Courtesy of Kennith L. Rinehart.
241, 244. Tunicates (*Rhopalaea crassa*). Getty Images/Photographer: Gary Bell.
241, 242, 250. *Ecteinascidia turbinata*, orange. John Easley Photography, www.johneasley.com.
241, 244. E. J. Corey. ©Rick Friedman/Corbis.
242, 248. Caribbean sea. ©Atlantide Phototravel/Corbis.
242. Test tubes. ©Tom Grill/Corbis.
243. Leukemia cells. ©Howard Sochurek/Corbis.
244. Alfred Nobel bust. Courtesy of E. J. Corey.
244. Test tubes in tray. ©Tom Grill/Corbis.

245. Andrew G. Myers. Courtesy of Andrew G. Myers.
245. Tohru Fukuyama. Courtesy of Tohru Fukuyama.
245. Inverted phase control microscope. ©Tom Stewart/Corbis.
245. Petri dishes. ©Tom Grill/Corbis.
246. *Salinospora tropica* CNB-392. Courtesy of William Fenical, Scripps Institution of Oceanography, University of California, San Diego (UCSD).
246. *Hyrtios erecta*. Courtesy of G. Robert Pettit.
246. William H. Fenical. Courtesy of William H. Fenical, Scripps Institution of Oceanography, University of California, San Diego (UCSD).
246. *Ecteinascidia turbinata*. Courtesy of PharmaMar.
247. *Trididemnum solidum*. Courtesy of Madeleine M. Joullié.
247. D. John Faulkner. Courtesy D. John Faulkner, Scripps Institution of Oceanography, University of California, San Diego (UCSD).
247. Ian Paterson. Courtesy of Ian Paterson.
247. David A. Evans. Courtesy of David A. Evans.
247. Michael T. Crimmins. Courtesy of Michael T. Crimmins.
247. Amos B. Smith, III. Courtesy of Amos B. Smith, III.
247. Steven V. Ley. Courtesy of Steven V. Ley.
247. Clayton H. Heathcock. Courtesy of Clayton H. Heathcock.
247. Madeleine M. Joullié. Courtesy of Madeleine M. Joullié.
247. Yoshito Kishi. Courtesy of Yoshito Kishi.
248. Acropolis looming over Athens. ©Wolfgang Kaehler/Corbis.
248. Bust of Greek philosopher Plato. ©Bettmann/Corbis.
248. Bust of Greek philosopher Aristotle. ©Bettmann/Corbis.
248. Oceanous Earth. Courtesy of NASA Jet Propulsion Laboratory.
248. *Discodermia dissoluta*. Courtesy of Amos B. Smith, III.
249. Paul Sabatier. Courtesy of Dr. A. Latte Archive, ©YAN-Toulouse, Rue Erasme.
249. Geoffrey Wilkinson. Courtesy of Michigan State University, Chemistry Portrait Library.
249. William S. Knowles. ©Greenblatt Bill/Corbis Sygma.
249. Henri B. Kagan. Courtesy of Henri B. Kagan.
249. Ryoji Noyori. K. C. Nicolaou archive.
249. S. Akabori. Courtesy of Osaka University.
250. Goldfish reef. ©DiMaggio/Kalish/Corbis.

Chapter 29: Epothilones

251, 252. Hans Reichenbach and Gerhard Höfle. K. C. Nicolaou archive.
251. Rainbow over Victoria Falls. ©Brian A. Vikander/Corbis.
251, 252. *Sorangium cellulosum* cells. Courtesy of Hans Reichenbach.
251, 255. K. C. Nicolaou. K. C. Nicolaou archive.
251, 254. Deiter Schinzer. Courtesy of Deiter Schinzer.
251, 254. Samuel J. Danishefsky. Courtesy of Samuel J. Danishefsky.
252. Zambezi River. ©Peter Johnson/Corbis.
253. Cancer cell. Credit: hybrid medical animation/ Photo Researchers, Inc.
253. Tubulin polymerization. Nicolaou group.
253, 260. PtK2 cells. Courtesy of Florenz Sasse and Hans Reichenbach.
254. Petri dishes. ©Tom Grill/Corbis.
254. Cancer cells. ©Digital Art/Corbis.
255. The Nicolaou epothilone team. K. C. Nicolaou archive.
255. Test tube. ©Tom Grill/Corbis.
256. R. Bruce Merrifield. ©Bettmann/Corbis.
256. Microkan® reactors. Courtesy of NEXUS Biosystems.
257. Steve V. Ley. Courtesy of Steve V. Ley.
257. Epothilones diagram. Wiley-VCH Publisher, Weinheim.
257. Erlenmeyer flasks. ©Tom Grill/Corbis.
257. Illustration of epothilones. Nicolaou group.
257. Microscope. ©Thom Lang/Corbis.
258. Assortment of lab beakers. ©Mario Beauregard/Corbis.
258. Samuel J. Danishefsky. K. C. Nicolaou archive.
258. K. C. Nicolaou. K. C. Nicolaou archive.
258. Thomas J. Katz. K. C. Nicolaou archive.
258. Nissim Calderon. Courtesy of The Goodyear Tire & Rubber Company.
258. Yves Chauvin. ©Christophe Gaye/Reuters/Corbis.
258. Richard R. Schrock. ©Rick Friedman/Corbis.
258. Robert H. Grubbs. California Institute of Technology.
260. Microscope. ©Tom Grill/Corbis.
260. Epothilone illustration. Nicolaou group.

Chapter 30: Resiniferatoxin

261, 268. Peter M. Blumberg. Courtesy of Peter M. Blumberg.
261. Red pepper. Wiley-VCH Publisher, Weinheim.
261. Spices. ©J.Garcia/photocuisine/Corbis.
261, 271. Paul A. Wender. Courtesy of Paul A. Wender.
262. Wooden globe featuring Asia. ©Peter Dazeley/zefa/Corbis.
262. Different sorts of fruit. ©ImageShop/Corbis.
262. Red chili pepper. ©Darren Greenwood/Design Pics/Corbis.
263. Ba Cho Mee (hot food). ©Justin Guariglia/Corbis.
263. Richard Axel. ©Jennifer Altman/epa/Corbis.
263. Linda B. Buck. ©Dan Lamont/Corbis.
264, 268. Engraving of Pliny the Elder. ©Bettmann/Corbis.
264. Statue of Herodotus. ©Corbis.
264. Cinnamon sticks and powder. ©J.Garcia/photocuisine/Corbis.
264. *Piper nigrum*. ©Cynthia Hart Designer/Corbis.
265. Cloves and powdered cloves. ©Maximilian Stock Ltd/photocuisine/Corbis.
265. *Myristica fragrans*. ©Susanne Borges/A.B./zefa/Corbis.
265. Ferdinand Magellan. ©Bettmann/Corbis.
265. *Myristica fragrans*. ©Stapleton Collection/Corbis.
265. Map of the Americas with portraits of European explorers by Theodor de Bry. ©Corbis.

265. Statue of Christopher Columbus. ©Joseph Sohm/Visions of America/Corbis.
266. Red chili peppers. ©Trizeps Photography/photocuisine/Corbis.
266. Green, white and black pepper. Rainer Zenz.
266. Pepper grinder. ©Lawrence Manning/Corbis.
266. Pyramid of Kukulcan. ©Roger Ressmeyer/Corbis.
267. Piles of lentils, beans and spices. ©Spencer Jones/PictureArts/Corbis.
267. Red chili peppers. ©Martin Jepp/zefa/Corbis.
267. Woman wincing. ©Harry Vorsteher/Corbis.
268. Woman showing Dioscorides a mandrake plant. ©Stapleton Collection/Corbis.
268. *Euphorbia resinifera*. Valérie75; Valérie Chansigaud.
268. David Julius. Courtesy of David Julius.
269. Coin of King Juba II. Photo: PHGCOM.
270. Stanford University. ©iStockphoto.com/DanCardiff.
270. *Euphorbia resinifera*. Köhlers Medizinal-Pflanzen.
271. *Euphorbia resinifera*. BS Thurner Hof.
272. Richard Axel and Linda B. Buck. Courtesy of Richard Axel.

Chapter 31: Vancomycin

273, 277. Dudley H. Williams. University of Cambridge/Photographer: John Holman.
273. *Escherichia coli*. Rocky Mountain Laboratories, NIAID, NIH.
273. Vancocin® (vancomycin) capsules. Courtesy of ViroPharma Incorporated.
273. K. C. Nicolaou. K. C. Nicolaou archive.
273, 278, 283. David A. Evans. Courtesy of David A. Evans.
273, 279. Dale L. Boger. Courtesy of Dale L. Boger.
274. Sir Alexander Fleming. ©Corbis.
274. Penicillium mold. ©Lester V. Bergman/Corbis.
275. Vancocin® (vancomycin) box. Courtesy of ViroPharma Incorporated.
275. *Amycolatopsis orientalis*. Courtesy of Yasuhiro Gyobu, Meiji Seika Kaisha, LTD.
275. Drug-resistance illustration. Nicolaou group.
276, 280. *Staphylococcus aureus* bacteria. ©CDC/PHIL/Corbis.
276. Selman A. Waksman. New York World-Telegram and the Sun staff photographer: Higgins, Roger, photographer.
276. Drug-resistant *enterococcus faecium* bacteria. ©Visuals Unlimited/Corbis.
276. Bacterial cell wall illustration. Nicolaou group.
276. *Streptomyces griseus*. Courtesy of Satoshi Ômura.
277, 279. *Staphylococcus aureus*. ©CDC/PHIL/Corbis.
277, 284. *Lactobacillus fermentum*. ©Mediscan/Corbis.
277. Patrice Courvalin. Courtesy of Patrice Courvalin.
277. Christopher T. Walsh. Courtesy of Christopher T. Walsh.
277. Microscope. ©Thom Lang/Corbis.
278. *Enterococcus faecalis*. CDC/Pete Wardell.
280. K. C. Nicolaou. K. C. Nicolaou archive.
280. Pipette and petri dish. ©Andrew Brookes/Corbis.
281. The Nicolaou vancomycin team. K. C. Nicolaou archive.
281. Test tubes in tray. ©Tom Grill/Corbis.
283. Teruaki Mukaiyama. Courtesy of Teruaki Mukaiyama.
283. Satoru Masamune. Massachusetts Institute of Technology (MIT).
283. Ian Paterson. Courtesy of Ian Paterson.
283. Clayton H. Heathcock. Courtesy of Clayton H. Heathcock.

Chapter 32: Thiostrepton

285, 287. George W. Kenner. ©1978, Nature Publishing Group.
285, 287. Miklos Bodanszky. Courtesy of Novartis.
285, 288. Dorothy Crowfoot Hodgkin. ©Bettmann/Corbis.
285, 286. Veterinarian examining dog. ©Tom Grill/Corbis.
285, 286. Puppies. Photo: ©www.stuewer-tierfoto.de.
285, 293. K. C. Nicolaou. K. C. Nicolaou archive.
286. Labware. Courtesy of Valer Jeso, Nicolaou group.
286. Panalog® ointment. Novartis Animal Health, Inc.
287. Performing surgery on a dog. ©Tom Stewart/Corbis.
287. Test tubes. ©Tom Grill/Corbis.
287. Labware. Courtesy of Valer Jeso, Nicolaou group.
288. Lab beakers. ©Mario Beauregard/Corbis.
288. Bernard Weisblum. Courtesy of Bernard Weisblum.
289. Roger D. Kornberg. Courtesy of Roger D. Kornberg.
289. Flasks and cylinders. ©Mario Beauregard/Corbis.
289. Protein synthesis illustration. Nicolaou group.
290. Cats. Photo credit: www.ronkimballstock.com.
290. Petri dishes. ©Tom Grill/Corbis.
290. Methicillin-resistant *Staphylococcus aureus*. CDC/Janice Carr/Jeff Hageman, M.H.S.
290. *Enterococcus faecium*. Courtesy of Satoshi Ômura.
291. Christopher J. Moody. Courtesy of Christopher J. Moody.
291, 293. Beakers, flasks, and data results. ©Don Carstens/Brand X/Corbis.
292. Glass tubes and pipette. ©Andrew Brookes/Corbis.
292. Labware. Courtesy of Valer Jeso, Nicolaou group.
292. Barrie W. Bycroft. Courtesy of Barrie W. Bycroft.
292. Heinz G. Floss. Courtesy of Heinz G. Floss.
293. K. C. Nicolaou. K. C. Nicolaou archive.
293. Nicolaou thiostrepton team. K. C. Nicolaou archive.
293. Thiostrepton crystals. Courtesy of Charlie Bond.
294. Blue-footed boobies. Courtesy of Leopold Flohé.
294. Panda. Courtesy of Valery V. Fokin.
294. Tiger. ©iStockphoto.com/stephenmeese.
294. Horse. Photo ©Mark J. Barrett, www.markjbarrett.com.

Chapter 33: Small Molecule Drugs

295, 306. Gertrude B. Elion. ©Bettmann/Corbis
295, 306. George H. Hitchings. ©Bettmann/Corbis
295. Lab chemist mixing pink and blue chemicals.

©JLP/Deimos/zefa/Corbis.
295, 308, 312. Beakers, flasks and data. ©Don Carstens/Brand X/Corbis.
295, 312. Sir James W. Black. Courtesy of the James Black Foundation.
295. Petri dishes. ©Tom Grill/Corbis.
296, 303. Volumetric flasks. ©Royalty-Free/Corbis.
296. Lab mouse. Janet Stephens (photographer).
296. Research in hospital laboratory. ©Michael Prince/Corbis.
296. Mother taking temperature of her son. ©Brigitte Sporrer/zefa/Corbis.
296. FDA logo. Food and Drug Administration (FDA).
296. Wooden gavel. ©iStockphoto.com/spxChrome.
296. White and red pills. ©iStockphoto.com/clintscholz.
296. American money. ©iStockphoto.com/LUke1138.
296. Green, blue and pink pills. Chaos.
297. Caduceus symbol. ©Royalty-Free/Corbis.
297, 313. Pipette petri dish. ©Andrew Brookes/Corbis.
297. Chemical laboratory depicted in Robert Thom painting. Used with permission from Pfizer Inc. Images by Robert Thom. All rights reserved.
298. Doctor and interns talking to patient. ©Artiga Photo/Corbis.
298. Elli Kile, cancer patient. Courtesy of Rita Kile.
298. Paul Jason Nicolaou. Photo by K. C. Nicolaou.
299. Aspirin® box. Permission by Bayer.
299. Bayer logo. ©Gerten Martin/dpa/Corbis.
299. Microscope. ©Thom Lang/Corbis.
299. Johann Friedrich Wilhelm Adolf von Baeyer. Edgar Fahs Smith Collection, University of Pennsylvania Library.
299. White pills. ©Kelly Redinger/Design Pics/Corbis.
300. Marilyn Monroe. ©Bettmann/Corbis.
300. Jimi Hendrix. ©Henry Diltz/Corbis.
301. Leo H. Sternbach. ©F. Hoffmann-La Roche Ltd, Corporate Communications.
301. Erlenmeyer flasks. ©Tom Grill/Corbis.
302. Prozac®. ©Najlah Feanny/Corbis.
302. Holding bottle of Prozac®. ©David Butow/Corbis Saba.
303. Bryan B. Molloy. Photo courtesy of Eli Lilly and Company Archives.
303. Klaus K. Schmiegel. Photo courtesy of Eli Lilly and Company Archives.
304, 305. Ritalin® pills. Matze6587.
304. Sir George F. Still. National Library of Medicine (NLM).
304. Charles Bradley. Courtesy of Bradley Hospital.
305. Dropper and test tubes. ©Tom Grill/Corbis.
305. Kids playing. ©Randy Faris/Corbis.
305. Microscope. ©Tom Stewart/Corbis.
306. Helix structure. ©Matthias Kulka/zefa/Corbis.
306. Petri dish with dropper. ©Tom Grill/Corbis.
306. Epstein-Barr virus. ©Visuals Unlimited/Corbis.
306. Zovirax® box. ©GlaxoSmithKline. Used with permission.
307. Luc Montagnier. ©Andanson James/Corbis Sygma.
307. James Watson and Francis Crick. ©Bettmann/Corbis.
307. Retrovir®. ©GlaxoSmithKline. Used with permission.
308. Flasks and cylinders. ©Mario Beauregard/Corbis.
310. Ampules, syringe and tablets. ©iStockphoto.com/shulz.
310. David Ho. Courtesy of David Ho.
311. J. Robin Warren and Barry J. Marshall. ©Oliver Berg/epa/Corbis.
312. Two Aspirin® and water. ©iStockphoto.com/sdominick.
313. Medical illustration of a heart. ©iStockphoto.com/mstroz.
313. Heartrate and pills. ©Image Source/Corbis
314. Cardiology health care. ©Steve Allen/Brand X/Corbis.
314. Raymond P. Ahlquist. National Library of Medicine (NLM).
315. Nobel Prize medal. ©Ted Spiegel/Corbis.
315. Robert F. Furchgott. ©Maiman Rick/Corbis Sygma.
315. Louis J. Ignaro. ©Pizzoli Alberto/Corbis Sygma.
315. Ferid Murad. ©F. Carter Smith/Sygma/Corbis.
315. Petri dishes. ©Tom Grill/Corbis.
316. Viagra® pills. ©Jerzy Dabrowski/dpa/Corbis.
317. Alfred Nobel. ©Bettmann/Corbis.
317. Nobel ceremony. Courtesy of Ryoji Noyori.
317. Nobel Prize medal. ®The Nobel Foundation.
318. Nature versus nurture helix. ©Mike Agliolo/Corbis.

Chapter 34: Biologics

319, 323. Kary B. Mullis. Courtesy of Kary B. Mullis.
319, 322. César Milstein. ©Bettmann/Corbis.
319, 321. Robert A. Swanson. ©Roger Ressmeyer/Corbis.
319, 329. George B. Rathmann. Courtesy of Amgen.
319. Fermentation equipment. Courtesy of Amgen.
319, 320. Paul Berg. Courtesy of Paul Berg.
319, 320. Herbert W. Boyer. University of California, San Francisco (UCSF) Public Affairs.
319, 324, 325. Vials and hypodermic syringe.©Kelly Redinger/Design Pics/Corbis.
319, 322. Woman checking pregnancy test kit. ©Mike Watson Images/Corbis.
319, 320. Stanley N. Cohen. Courtesy of Stanley N. Cohen.
319, 330. Arthritis pain (woman's hands). ©Ariel Skelley/Corbis.
319. Amgen lab. ©Ann Johansson/Corbis.
319, 322. Georges J. F. Köhler. Max-Planck-Society/Filser.
319, 322. Marvin H. Caruthers. Courtesy of Marvin H. Caruthers.
320. *Escherichia coli*. Photo by Eric Erbe, digital colorization by Christopher Pooley. Agricultural Research Service (ARS), USDA.
321. *Escherichia coli*. Rocky Mountain Laboratories, NIAID, NIH.
321. David Goeddel and Robert A. Swanson at Genentech. ©Roger Ressmeyer/Corbis.
322. Row of test tubes. ©Mario Beauregard/Corbis.
322. Niels K. Jerne. ©Bettmann/Corbis.

322. Har Gobind Khorana. ©Bettmann/Corbis.
322. Robert W. Holley. ©Bettmann/Corbis.
323. Frederick Sanger. ©Bettmann/Corbis.
323. Walter Gilbert. National Library of Medicine (NLM).
323. Marshall W. Nirenberg. ©Bettmann/Corbis.
323. Sydney Brenner. ©Reuters/Corbis.
323. Michael Smith. ©Relke Christopher/Corbis Sygma.
323. James Watson and Francis H. C. Crick. ©Bettmann/Corbis.
324. Edward Jenner. Used with permission from Pfizer Inc. Images by Robert Thom. All rights reserved.
324. Frederick G. Banting. ©Bettmann/Corbis.
324. Lady Mary Wortley Montagu. ©Hulton-Deutsch Collection/Corbis.
324. Louis Pasteur. Félix Nadar.
325. *Variola* virus. ©Murphy-Whitfield/Image Point FR/Corbis.
326. Charles H. Best & F. G. Banting. Used with permission from Pfizer Inc. Images by Robert Thom. All rights reserved.
326. Schack A. S. Krogh. National Library of Medicine (NLM).
327. John J. R. Macleod. National Library of Medicine (NLM).
327. Dorothy Crowfoot Hodgkin at work. ©Hulton-Deutsch Collection/Corbis.
327. William K. Bowes statue. Courtesy of Amgen.
328, 329. Chemistry lab containers. ©MedioImages/Corbis.
328. Fu-Kuen Lin. Courtesy of Amgen.
328. Petri dish. Courtesy of Amgen.
328. Erythropoietin ribbon model. Courtesy of Amgen.
328. Crystals. Courtesy of Amgen.
329. Dennis Slamon. Courtesy of University of California, Los Angeles (UCLA).
329. Breast cancer cells. ©Visuals Unlimited/Corbis.
330. Senior man in pain. ©Norbert Schaefer/Corbis.
330. Herceptin®. ©F. Hoffmann-La Roche Ltd., Corporate Communications.
331. George P. Smith. Courtesy of George P. Smith.
331. Richard A. Lerner. Courtesy of Richard A. Lerner.
331. Sir Gregory Winter. Courtesy of Sir Gregory Winter.
332. Orencia® bottle and box. Courtesy of ©Bristol-Myers Squibb.
332. Remicade® bottle. Courtesy of Centocor Pharmaceuticals.
332. Humira® box and pen. Courtesy of Abbott Laboratories.

Register of Persons

Subject Index

B

C

D

E

O

P

Q

R

W

X

Y

Z